Methods in Enzymology

Volume 302

GREEN FLUORESCENT PROTEIN

METHODS IN ENZYMOLOGY

Methods in Enzymology

Volume 302

Green Fluorescent Protein

EDITED BY

P. Michael Conn

OREGON HEALTH SCIENCES UNIVERSITY, PORTLAND, OREGON, AND
OREGON REGIONAL PRIMATE RESEARCH CENTER
BEAVERTON, OREGON

ACADEMIC PRESS
San Diego London Boston New York Sydney Tokyo Toronto

This book is printed on acid-free paper. 

Academic Press
a division of Harcourt Brace & Company
525 B Street, Suite 1900, San Diego, California 92101-4495, USA
http://www.academicpress.com

Academic Press Limited
24-28 Oval Road, London NW1 7DX, UK
http://www.hbuk.co.uk/ap/

International Standard Book Number: 0-12-182203-6

PRINTED IN THE UNITED STATES OF AMERICA
99 00 01 02 03 04 MM 9 8 7 6 5 4 3 2 1

Table of Contents

Section I. Monitoring of Physiological Processes

Section II. Localization of Molecules

Section III. Special Uses

Section IV. Mutants and Variants of Green Fluorescent Protein

Contributors to Volume 302

Article numbers are in parentheses following the names of contributors.
Affiliations listed are current.

MICHAEL T. ANDERSON (26), *Department of Pathology, University of Alabama at Birmingham, Birmingham, Alabama 35294*

TAKASHI AOKI (23), *Department of Biochemistry, Faculty of Pharmaceutical Sciences, Health Sciences University of Hokkaido, Ishikari-Tobetsu, Hokkaido 061-0293, Japan*

HOWARD J. ATKINSON (27), *Centre for Plant Sciences, University of Leeds, Leeds LS2 9JT, United Kingdom*

JOSEPH C. AYOOB (15), *Department of Cell and Developmental Biology, University of Pennsylvania School of Medicine, Philadelphia, Pennsylvania 19104-6058*

GEORGE BANTING (1), *Department of Biochemistry, University of Bristol School of Medical Sciences, Bristol BS8 1TD, United Kingdom*

LARRY S. BARAK (14), *Howard Hughes Medical Institute and Department of Cell Biology, Duke University, Durham, North Carolina 27710*

K. K. BENCE (19), *Brown University, Providence, Rhode Island 02912*

LUIZ E. BERMUDEZ (25), *Kuzell Institute for Arthritis and Infectious Diseases, California Pacific Medical Center, San Francisco, California 94115*

L. A. C. BLAIR (19), *Brown University, Providence, Rhode Island 02912*

JENNIFER K. BLUMSOM (27), *Centre for Plant Sciences, University of Leeds, Leeds LS2 9JT, United Kingdom*

MARC G. CARON (14), *Howard Hughes Medical Institute and Department of Cell Biology, Duke University, Durham, North Carolina 27710*

WAYNE E. CASCIO (29), *Department of Medicine, School of Medicine, University of North Carolina—Chapel Hill, Chapel Hill, North Carolina 27599*

ROBERT CHERVENAK (16, 17), *Department of Microbiology and Immunology, The Center for Excellence in Cancer Research, and The Center for Excellence in Arthritis and Rheumatology, Louisiana State University Medical Center, Shreveport, Louisiana 71130*

STEPHEN R. CRONIN (8), *Department of Biology, University of California, San Diego, La Jolla, California 92093-0347*

GUISSOU A. DABIRI (15), *Department of Cell and Developmental Biology, University of Pennsylvania School of Medicine, Philadelphia, Pennsylvania 19104-6058*

TRISHA N. DAVIS (10), *Department of Biochemistry, University of Washington, Seattle, Washington 98195*

TOMMY DUONG (4), *CLONTECH Laboratories, Inc., Palo Alto, California 94303-4230*

YU FANG (18), *CLONTECH Laboratories, Inc., Palo Alto, California 94303-4230*

STEPHEN S. G. FERGUSON (14), *John P. Robarts Research Institute, London, Ontario, Canada N6A 5K8*

MARK R. FLORY (10), *Molecular and Cellular Biology Program, University of Washington, Seattle, Washington 98195*

TEIICHI FURUICHI (20), *Department of Molecular Neurobiology, The Institute of Medical Science, The University of Tokyo, Shirokanedai, Minato-ku, Tokyo 108-8639, Japan*

DAVID W. GALBRAITH (26), *Department of Plant Sciences, University of Arizona, Tucson, Arizona 85721*

VIRGINIE GEORGET (12), *Montpellier School of Medicine, INSERM U-439, 34090 Montpellier, France*

HANS-HERMANN GERDES (2), *Department of Neurobiology, University of Heidelberg, D-69120 Heidelberg, Germany*

GÜNTHER GERISCH (7), *Max-Planck-Institut für Biochemie, D-82152 Martinsried, Germany*

GORDON L. HAGER (9), *Laboratory of Receptor Biology and Gene Expression, National Cancer Institute, National Institutes of Health, Bethesda, Maryland 20892-5055*

RANDOLPH Y. HAMPTON (8), *Department of Biology, University of California, San Diego, La Jolla, California 92093-0347*

FRANK HANAKAM (7), *Micromet GmbH, D-82152 Martinsried, Germany*

ROGER HEIM (34), *Aurora Biosciences Corporation, San Diego, California 92121*

JOHN C. HERR (24), *Department of Cell Biology, University of Virginia, Charlottesville, Virginia 22908*

LEONARD A. HERZENBERG (26), *Department of Genetics, Stanford University Medical School, Stanford, California 94305-5318*

ROBERT M. HOFFMAN (3), *AntiCancer, Inc., San Diego, California 92111*

CHIAO-CHIAN HUANG (4, 18, 36), *CLONTECH Laboratories, Inc., Palo Alto, California 94303-4230*

SATOSHI INOUYE (37), *Scripps Institution of Oceanography, University of California, San Diego, La Jolla, California 92093-0202*

XIN JIANG (36), *CLONTECH Laboratories, Inc., Palo Alto, California 94303-4230*

CHRISTOPH KAETHER (2), *Department of Neurobiology, University of Heidelberg, D-69120 Heidelberg, Germany*

STEVEN R. KAIN (4, 5, 18, 28, 30, 36), *CLONTECH Laboratories, Inc., Palo Alto, California 94303-4230*

MARKO KALLIO (24), *Department of Cell Biology, University of Virginia, Charlottesville, Virginia 22908*

MOTOYA KATSUKI (21), *Department of DNA Biology and Embryo Engineering, Research Center of Animal Models for Human Diseases, The Institute of Medical Science, The University of Tokyo, Shirokanedai, Minato-ku, Tokyo 108-8639, Japan*

YUKIO KIMATA (31), *Research and Education Center for Genetic Information, Nara Institute of Science and Technology, Ikoma, Nara 630-01, Japan*

LINDA A. KING (33), *School of Biological and Molecular Sciences, Oxford Brookes University, Oxford OX3 0BP, United Kingdom*

KATHERINE S. KOCH (23), *Department of Pharmacology, School of Medicine, University of California, San Diego, La Jolla, California 92093-0636*

KENJI KOHNO (31), *Research and Education Center for Genetic Information, Nara Institute of Science and Technology, Ikoma, Nara 630-01, Japan*

SABINE KUPZIG (1), *Department of Biochemistry, University of Bristol School of Medical Sciences, Bristol BS8 1TD, United Kingdom*

STEPHANE A. LAPORTE (14), *Howard Hughes Medical Institute and Department of Cell Biology, Duke University, Durham, North Carolina 27710*

SAN SAN LEE (1), *Department of Biochemistry, University of Bristol School of Medical Sciences, Bristol BS8 1TD, United Kingdom*

HYAM L. LEFFERT (23), *Department of Pharmacology and Center for Molecular Genetics, School of Medicine, University of California, San Diego, La Jolla, California 92093-0636*

JOHN J. LEMASTERS (29), *Department of Cell Biology and Anatomy, School of Medicine, University of North Carolina—Chapel Hill, Chapel Hill, North Carolina 27599*

JOHN P. LEVY (28, 30, 35), *Molecular Oncology Laboratory, Human Gene Therapy Research Institute, Iowa Health System, Des Moines, Iowa 50309-3202*

XIANQIANG LI (4, 18, 36), *CLONTECH Laboratories, Inc., Palo Alto, California 94303-4230*

CHUN REN LIM (31), *Research and Education Center for Genetic Information, Nara Institute of Science and Technology, Ikoma, Nara 630-01, Japan*

CHARLES J. LINK, JR. (28, 30, 35), *Molecular Oncology Laboratory, Human Gene Therapy Research Institute, Iowa Health System, Des Moines, Iowa 50309-3202*

LONNIE LYBARGER (16, 17), *Department of Microbiology and Immunology, Louisiana State University Medical Center, Shreveport, Louisiana 71130*

JING-TYAN MA (5), *CLONTECH Laboratories, Inc., Palo Alto, California 94303-4230*

MARKUS MANIAK (6), *MRC Laboratory for Molecular Cell Biology, University College London, London WC1E 6BT, England*

J. MARSHALL (19), *Brown University, Providence, Rhode Island 02912*

JULIE M. MATHESON (20), *Department of Molecular Neurobiology, The Institute of Medical Science, The University of Tokyo, Shirokanedai, Minato-ku, Tokyo 108-8639, Japan*

ILYA A. MAZO (28, 30), *CLONTECH Laboratories, Inc., Palo Alto, California 94303-4230*

TAKAYUKI MICHIKAWA (20), *Department of Molecular Neurobiology, The Institute of Medical Science, The University of Tokyo, Shirokanedai, Minato-ku, Tokyo 108-8639, Japan*

KATSUHIKO MIKOSHIBA (20), *Calciosignal Net Project, Exploratory Research for Advanced Technology, Tokyo 153-0064, Japan*

ATSUSHI MIYAWAKI (20), *Department of Molecular Neurobiology, The Institute of Medical Science, The University of Tokyo, Shirokanedai, Minato-ku, Tokyo 108-8639, Japan*

SIMON G. MØLLER (27), *Centre for Plant Sciences, University of Leeds, Leeds LS2 9JT, United Kingdom*

REBECCA R. MULDOON (28, 30, 35), *Molecular Oncology Laboratory, Human Gene Therapy Research Institute, Iowa Health System, Des Moines, Iowa 50309-3202*

AKIRA MUTO (20), *Calciosignal Net Project, Exploratory Research for Advanced Technology, Tokyo 153-0064, Japan*

KENJI NAKAMURA (21), *Department of DNA Biology and Embryo Engineering, Research Center of Animal Models for Human Diseases, The Institute of Medical Science, The University of Tokyo, Shirokanedai, Minato-ku, Tokyo 108-8639, Japan*

KAZUKI NAKAO (21), *Department of DNA Biology and Embryo Engineering, Research Center of Animal Models for Human Diseases, The Institute of Medical Science, The University of Tokyo, Shirokanedai, Minato-ku, Tokyo 108-8639, Japan*

JEAN-CLAUDE NICOLAS (12), *Montpellier School of Medicine, INSERM U-439, 34090 Montpellier, France*

HISAYUKI OHATA (29), *Showa University, Tokyo 142, Japan*

J. B. OLMSTED (11), *Department of Biology, University of Rochester, Rochester, New York 14627*

KEITH R. OLSON (11), *Cellomics, Inc., Pittsburgh, Pennsylvania 15238*

GOTTFRIED J. PALM (32), *Department of Structural Biology and Crystallography, Institute of Molecular Biotechnology, D-07745 Jena, Germany*

AMY PARKER (25), *Kuzell Institute for Arthritis and Infectious Diseases, California Pacific Medical Center, San Francisco, California 94115*

ROBERT D. POSSEE (33), *NERC Institute of Virology and Environmental Microbiology, Oxford OX1 3SR, United Kingdom*

TING QIAN (29), *Department of Cell Biology and Anatomy, School of Medicine, University of North Carolina—Chapel Hill, Chapel Hill, North Carolina 27599*

P. PRABHAKARA REDDI (24), *Department of Cell Biology, University of Virginia, Charlottesville, Virginia 22908*

ORNA RESNEKOV (13), *Laboratory of Molecular Genetics, National Institute of Child Health and Human Development, National Institutes of Health, Bethesda, Maryland 20892-2785*

FELIX J. SANGARI (25), *Kuzell Institute for Arthritis and Infectious Diseases, California Pacific Medical Center, San Francisco, California 94115*

JEAN M. SANGER (15), *Department of Cell and Developmental Biology, University of Pennsylvania School of Medicine, Philadelphia, Pennsylvania 19104-6058*

JOSEPH W. SANGER (15), *Department of Cell and Developmental Biology, University of Pennsylvania School of Medicine, Philadelphia, Pennsylvania 19104-6058*

LEE G. SAYERS (20), *Department of Molecular Neurobiology, The Institute of Medical Science, The University of Tokyo, Shirokanedai, Minato-ku, Tokyo 108-8639, Japan*

TATIANA SEREGINA (35), *Molecular Oncology Laboratory, Human Gene Therapy Research Institute, Iowa Health System, Des Moines, Iowa 50309-3202*

SUMIO SUGANO (21), *Department of Virology, The Institute of Medical Science, The University of Tokyo, Shirokanedai, Minato-ku, Tokyo 108-8639, Japan*

CHARLES SULTAN (12), *Montpellier School of Medicine, INSERM U-439, 34090 Montpellier, France*

TATSUYUKI TAKADA (21), *Department of Experimental Radiology, Shiga University of Medical Science, Ohtsu, Shiga 520-2192, Japan*

BÉATRICE TEROUANNE (12), *Montpellier School of Medicine, INSERM U-439, 34090 Montpellier, France*

CAROLE J. THOMAS (33), *School of Biological and Molecular Sciences, Oxford Brookes University, Oxford OX3 0BP, United Kingdom*

DONNA R. TROLLINGER (29), *Department of Cell Biology and Anatomy, School of Medicine, University of North Carolina—Chapel Hill, Chapel Hill, North Carolina 27599*

FREDERICK I. TSUJI (37), *Scripps Institution of Oceanography, University of California, San Diego, La Jolla, California 92093-0202*

GOZOH TSUJIMOTO (21), *Department of Molecular, Cell Pharmacology, National Children's Medical Research Center, Taishido, Setagaya-ku, Tokyo 154-8509, Japan*

KENAN K. TURNACIOGLU (15), *Department of Cell and Developmental Biology, University of Pennsylvania School of Medicine, Philadelphia, Pennsylvania 19104-6058*

KAZUHIKO UMESONO (37), *Institute for Virus Research, Kyoto University, Shogoin, Sakyo-ku, Kyoto 606-8507, Japan*

PETER E. URWIN (27), *Centre for Plant Sciences, University of Leeds, Leeds LS2 9JT, United Kingdom*

ALAN S. VERKMAN (22), *Cardiovascular Research Institute, University of California, San Francisco, San Francisco, California 94143-0521*

SUMING WANG (35), *Molecular Oncology Laboratory, Human Gene Therapy Research Institute, Iowa Health System, Des Moines, Iowa 50309-3202*

HIROYUKI WATABE (23), *Department of Biochemistry, Faculty of Pharmaceutical Sciences, Health Sciences University of Hokkaido, Ishikari-Tobetsu, Hokkaido 061-0293, Japan*

CHRIS D. WEBB (13), *Information Technology Office, Defense Advanced Research Projects Agency, Arlington, Virginia 22203-1714*

NICOLA WILKINSON (33), *School of Biological and Molecular Sciences, Oxford Brookes University, Oxford OX3 0BP, United Kingdom*

ALEXANDER WLODAWER (32), *Macromolecular Structure Laboratory, ABL–Basic Research Program, National Cancer Institute–Frederick Cancer Research and Development Center, Frederick, Maryland 21702*

KENICHI YOSHIDA (21), *Department of Virology, The Institute of Medical Science, The University of Tokyo, Shirokanedai, Minato-ku, Tokyo 108-8639, Japan*

JIE ZHANG (14), *Howard Hughes Medical Institute and Department of Cell Biology, Duke University, Durham, North Carolina 27710*

XIAONING ZHAO (4, 36), *CLONTECH Laboratories, Inc., Palo Alto, California 94303-4230*

Preface

This *Methods in Enzymology* volume deals with the utility of green fluorescent protein (GFP). The OVID database (including MEDLINE, Current Contents, and other sources) lists nine references to GFP for the ten-year period 1985–1994. In contrast, in less than four years thereafter, more than 500 are listed, a testament to the rapid growth of interest in this probe.

This volume documents many diverse uses for this interesting molecule in disciplines that broadly span biology. The methods presented include shortcuts and conveniences not included in the sources from which they were taken. The techniques are described in a context that allows comparisons to other related methodologies. The authors were encouraged to do this in the belief that such comparisons are valuable to readers who must adapt extant procedures to new systems. Also, so far as possible, methodologies are presented in a manner that stresses their general applicability and potential limitations. Although for various reasons some topics are not covered, the volume provides a substantial and current overview of the extant methodology in the field and a view of its rapid development.

Particular thanks go to the authors for their attention to meeting deadlines and for maintaining high standards of quality, to the series editors for their encouragement, and to the staff of Academic Press for their help and timely publication of the volume.

P. Michael Conn

METHODS IN ENZYMOLOGY

VOLUME XVII. Metabolism of Amino Acids and Amines (Parts A and B)
Edited by HERBERT TABOR AND CELIA WHITE TABOR

VOLUME XVIII. Vitamins and Coenzymes (Parts A, B, and C)
Edited by DONALD B. MCCORMICK AND LEMUEL D. WRIGHT

VOLUME XIX. Proteolytic Enzymes
Edited by GERTRUDE E. PERLMANN AND LASZLO LORAND

VOLUME XX. Nucleic Acids and Protein Synthesis (Part C)
Edited by KIVIE MOLDAVE AND LAWRENCE GROSSMAN

VOLUME XXI. Nucleic Acids (Part D)
Edited by LAWRENCE GROSSMAN AND KIVIE MOLDAVE

VOLUME XXII. Enzyme Purification and Related Techniques
Edited by WILLIAM B. JAKOBY

VOLUME XXIII. Photosynthesis (Part A)
Edited by ANTHONY SAN PIETRO

VOLUME XXIV. Photosynthesis and Nitrogen Fixation (Part B)
Edited by ANTHONY SAN PIETRO

VOLUME XXV. Enzyme Structure (Part B)
Edited by C. H. W. HIRS AND SERGE N. TIMASHEFF

VOLUME XXVI. Enzyme Structure (Part C)
Edited by C. H. W. HIRS AND SERGE N. TIMASHEFF

VOLUME XXVII. Enzyme Structure (Part D)
Edited by C. H. W. HIRS AND SERGE N. TIMASHEFF

VOLUME XXVIII. Complex Carbohydrates (Part B)
Edited by VICTOR GINSBURG

VOLUME XXIX. Nucleic Acids and Protein Synthesis (Part E)
Edited by LAWRENCE GROSSMAN AND KIVIE MOLDAVE

VOLUME XXX. Nucleic Acids and Protein Synthesis (Part F)
Edited by KIVIE MOLDAVE AND LAWRENCE GROSSMAN

VOLUME XXXI. Biomembranes (Part A)
Edited by SIDNEY FLEISCHER AND LESTER PACKER

VOLUME XXXII. Biomembranes (Part B)
Edited by SIDNEY FLEISCHER AND LESTER PACKER

VOLUME XXXIII. Cumulative Subject Index Volumes I–XXX
Edited by MARTHA G. DENNIS AND EDWARD A. DENNIS

VOLUME XXXIV. Affinity Techniques (Enzyme Purification: Part B)
Edited by WILLIAM B. JAKOBY AND MEIR WILCHEK

VOLUME XXXV. Lipids (Part B)
Edited by JOHN M. LOWENSTEIN

VOLUME XXXVI. Hormone Action (Part A: Steroid Hormones)
Edited by BERT W. O'MALLEY AND JOEL G. HARDMAN

VOLUME XXXVII. Hormone Action (Part B: Peptide Hormones)
Edited by BERT W. O'MALLEY AND JOEL G. HARDMAN

VOLUME XXXVIII. Hormone Action (Part C: Cyclic Nucleotides)
Edited by JOEL G. HARDMAN AND BERT W. O'MALLEY

VOLUME XXXIX. Hormone Action (Part D: Isolated Cells, Tissues, and Organ Systems)
Edited by JOEL G. HARDMAN AND BERT W. O'MALLEY

VOLUME XL. Hormone Action (Part E: Nuclear Structure and Function)
Edited by BERT W. O'MALLEY AND JOEL G. HARDMAN

VOLUME XLI. Carbohydrate Metabolism (Part B)
Edited by W. A. WOOD

VOLUME XLII. Carbohydrate Metabolism (Part C)
Edited by W. A. WOOD

VOLUME XLIII. Antibiotics
Edited by JOHN H. HASH

VOLUME XLIV. Immobilized Enzymes
Edited by KLAUS MOSBACH

VOLUME XLV. Proteolytic Enzymes (Part B)
Edited by LASZLO LORAND

VOLUME XLVI. Affinity Labeling
Edited by WILLIAM B. JAKOBY AND MEIR WILCHEK

VOLUME XLVII. Enzyme Structure (Part E)
Edited by C. H. W. HIRS AND SERGE N. TIMASHEFF

VOLUME XLVIII. Enzyme Structure (Part F)
Edited by C. H. W. HIRS AND SERGE N. TIMASHEFF

VOLUME XLIX. Enzyme Structure (Part G)
Edited by C. H. W. HIRS AND SERGE N. TIMASHEFF

VOLUME L. Complex Carbohydrates (Part C)
Edited by VICTOR GINSBURG

VOLUME LI. Purine and Pyrimidine Nucleotide Metabolism
Edited by PATRICIA A. HOFFEE AND MARY ELLEN JONES

VOLUME LII. Biomembranes (Part C: Biological Oxidations)
Edited by SIDNEY FLEISCHER AND LESTER PACKER

VOLUME LIII. Biomembranes (Part D: Biological Oxidations)
Edited by SIDNEY FLEISCHER AND LESTER PACKER

VOLUME LIV. Biomembranes (Part E: Biological Oxidations)
Edited by SIDNEY FLEISCHER AND LESTER PACKER

Volume LV. Biomembranes (Part F: Bioenergetics)
Edited by Sidney Fleischer and Lester Packer

Volume LVI. Biomembranes (Part G: Bioenergetics)
Edited by Sidney Fleischer and Lester Packer

Volume LVII. Bioluminescence and Chemiluminescence
Edited by Marlene A. DeLuca

Volume LVIII. Cell Culture
Edited by William B. Jakoby and Ira Pastan

Volume LIX. Nucleic Acids and Protein Synthesis (Part G)
Edited by Kivie Moldave and Lawrence Grossman

Volume LX. Nucleic Acids and Protein Synthesis (Part H)
Edited by Kivie Moldave and Lawrence Grossman

Volume 61. Enzyme Structure (Part H)
Edited by C. H. W. Hirs and Serge N. Timasheff

Volume 62. Vitamins and Coenzymes (Part D)
Edited by Donald B. McCormick and Lemuel D. Wright

Volume 63. Enzyme Kinetics and Mechanism (Part A: Initial Rate and Inhibitor Methods)
Edited by Daniel L. Purich

Volume 64. Enzyme Kinetics and Mechanism (Part B: Isotopic Probes and Complex Enzyme Systems)
Edited by Daniel L. Purich

Volume 65. Nucleic Acids (Part I)
Edited by Lawrence Grossman and Kivie Moldave

Volume 66. Vitamins and Coenzymes (Part E)
Edited by Donald B. McCormick and Lemuel D. Wright

Volume 67. Vitamins and Coenzymes (Part F)
Edited by Donald B. McCormick and Lemuel D. Wright

Volume 68. Recombinant DNA
Edited by Ray Wu

Volume 69. Photosynthesis and Nitrogen Fixation (Part C)
Edited by Anthony San Pietro

Volume 70. Immunochemical Techniques (Part A)
Edited by Helen Van Vunakis and John J. Langone

Volume 71. Lipids (Part C)
Edited by John M. Lowenstein

Volume 72. Lipids (Part D)
Edited by John M. Lowenstein

VOLUME 73. Immunochemical Techniques (Part B)
Edited by JOHN J. LANGONE AND HELEN VAN VUNAKIS

VOLUME 74. Immunochemical Techniques (Part C)
Edited by JOHN J. LANGONE AND HELEN VAN VUNAKIS

VOLUME 75. Cumulative Subject Index Volumes XXXI, XXXII, XXXIV–LX
Edited by EDWARD A. DENNIS AND MARTHA G. DENNIS

VOLUME 76. Hemoglobins
Edited by ERALDO ANTONINI, LUIGI ROSSI-BERNARDI, AND EMILIA CHIANCONE

VOLUME 77. Detoxication and Drug Metabolism
Edited by WILLIAM B. JAKOBY

VOLUME 78. Interferons (Part A)
Edited by SIDNEY PESTKA

VOLUME 79. Interferons (Part B)
Edited by SIDNEY PESTKA

VOLUME 80. Proteolytic Enzymes (Part C)
Edited by LASZLO LORAND

VOLUME 81. Biomembranes (Part H: Visual Pigments and Purple Membranes, I)
Edited by LESTER PACKER

VOLUME 82. Structural and Contractile Proteins (Part A: Extracellular Matrix)
Edited by LEON W. CUNNINGHAM AND DIXIE W. FREDERIKSEN

VOLUME 83. Complex Carbohydrates (Part D)
Edited by VICTOR GINSBURG

VOLUME 84. Immunochemical Techniques (Part D: Selected Immunoassays)
Edited by JOHN J. LANGONE AND HELEN VAN VUNAKIS

VOLUME 85. Structural and Contractile Proteins (Part B: The Contractile Apparatus and the Cytoskeleton)
Edited by DIXIE W. FREDERIKSEN AND LEON W. CUNNINGHAM

VOLUME 86. Prostaglandins and Arachidonate Metabolites
Edited by WILLIAM E. M. LANDS AND WILLIAM L. SMITH

VOLUME 87. Enzyme Kinetics and Mechanism (Part C: Intermediates, Stereochemistry, and Rate Studies)
Edited by DANIEL L. PURICH

VOLUME 88. Biomembranes (Part I: Visual Pigments and Purple Membranes, II)
Edited by LESTER PACKER

VOLUME 89. Carbohydrate Metabolism (Part D)
Edited by WILLIS A. WOOD

VOLUME 90. Carbohydrate Metabolism (Part E)
Edited by WILLIS A. WOOD

VOLUME 91. Enzyme Structure (Part I)
Edited by C. H. W. HIRS AND SERGE N. TIMASHEFF

VOLUME 92. Immunochemical Techniques (Part E: Monoclonal Antibodies and General Immunoassay Methods)
Edited by JOHN J. LANGONE AND HELEN VAN VUNAKIS

VOLUME 93. Immunochemical Techniques (Part F: Conventional Antibodies, Fc Receptors, and Cytotoxicity)
Edited by JOHN J. LANGONE AND HELEN VAN VUNAKIS

VOLUME 94. Polyamines
Edited by HERBERT TABOR AND CELIA WHITE TABOR

VOLUME 95. Cumulative Subject Index Volumes 61–74, 76–80
Edited by EDWARD A. DENNIS AND MARTHA G. DENNIS

VOLUME 96. Biomembranes [Part J: Membrane Biogenesis: Assembly and Targeting (General Methods; Eukaryotes)]
Edited by SIDNEY FLEISCHER AND BECCA FLEISCHER

VOLUME 97. Biomembranes [Part K: Membrane Biogenesis: Assembly and Targeting (Prokaryotes, Mitochondria, and Chloroplasts)]
Edited by SIDNEY FLEISCHER AND BECCA FLEISCHER

VOLUME 98. Biomembranes (Part L: Membrane Biogenesis: Processing and Recycling)
Edited by SIDNEY FLEISCHER AND BECCA FLEISCHER

VOLUME 99. Hormone Action (Part F: Protein Kinases)
Edited by JACKIE D. CORBIN AND JOEL G. HARDMAN

VOLUME 100. Recombinant DNA (Part B)
Edited by RAY WU, LAWRENCE GROSSMAN, AND KIVIE MOLDAVE

VOLUME 101. Recombinant DNA (Part C)
Edited by RAY WU, LAWRENCE GROSSMAN, AND KIVIE MOLDAVE

VOLUME 102. Hormone Action (Part G: Calmodulin and Calcium-Binding Proteins)
Edited by ANTHONY R. MEANS AND BERT W. O'MALLEY

VOLUME 103. Hormone Action (Part H: Neuroendocrine Peptides)
Edited by P. MICHAEL CONN

VOLUME 104. Enzyme Purification and Related Techniques (Part C)
Edited by WILLIAM B. JAKOBY

VOLUME 105. Oxygen Radicals in Biological Systems
Edited by LESTER PACKER

VOLUME 106. Posttranslational Modifications (Part A)
Edited by FINN WOLD AND KIVIE MOLDAVE

VOLUME 107. Posttranslational Modifications (Part B)
Edited by FINN WOLD AND KIVIE MOLDAVE

Volume 108. Immunochemical Techniques (Part G: Separation and Characterization of Lymphoid Cells)
Edited by Giovanni Di Sabato, John J. Langone, and Helen Van Vunakis

Volume 109. Hormone Action (Part I: Peptide Hormones)
Edited by Lutz Birnbaumer and Bert W. O'Malley

Volume 110. Steroids and Isoprenoids (Part A)
Edited by John H. Law and Hans C. Rilling

Volume 111. Steroids and Isoprenoids (Part B)
Edited by John H. Law and Hans C. Rilling

Volume 112. Drug and Enzyme Targeting (Part A)
Edited by Kenneth J. Widder and Ralph Green

Volume 113. Glutamate, Glutamine, Glutathione, and Related Compounds
Edited by Alton Meister

Volume 114. Diffraction Methods for Biological Macromolecules (Part A)
Edited by Harold W. Wyckoff, C. H. W. Hirs, and Serge N. Timasheff

Volume 115. Diffraction Methods for Biological Macromolecules (Part B)
Edited by Harold W. Wyckoff, C. H. W. Hirs, and Serge N. Timasheff

Volume 116. Immunochemical Techniques (Part H: Effectors and Mediators of Lymphoid Cell Functions)
Edited by Giovanni Di Sabato, John J. Langone, and Helen Van Vunakis

Volume 117. Enzyme Structure (Part J)
Edited by C. H. W. Hirs and Serge N. Timasheff

Volume 118. Plant Molecular Biology
Edited by Arthur Weissbach and Herbert Weissbach

Volume 119. Interferons (Part C)
Edited by Sidney Pestka

Volume 120. Cumulative Subject Index Volumes 81–94, 96–101

Volume 121. Immunochemical Techniques (Part I: Hybridoma Technology and Monoclonal Antibodies)
Edited by John J. Langone and Helen Van Vunakis

Volume 122. Vitamins and Coenzymes (Part G)
Edited by Frank Chytil and Donald B. McCormick

Volume 123. Vitamins and Coenzymes (Part H)
Edited by Frank Chytil and Donald B. McCormick

Volume 124. Hormone Action (Part J: Neuroendocrine Peptides)
Edited by P. Michael Conn

VOLUME 125. Biomembranes (Part M: Transport in Bacteria, Mitochondria, and Chloroplasts: General Approaches and Transport Systems)
Edited by SIDNEY FLEISCHER AND BECCA FLEISCHER

VOLUME 126. Biomembranes (Part N: Transport in Bacteria, Mitochondria, and Chloroplasts: Protonmotive Force)
Edited by SIDNEY FLEISCHER AND BECCA FLEISCHER

VOLUME 127. Biomembranes (Part O: Protons and Water: Structure and Translocation)
Edited by LESTER PACKER

VOLUME 128. Plasma Lipoproteins (Part A: Preparation, Structure, and Molecular Biology)
Edited by JERE P. SEGREST AND JOHN J. ALBERS

VOLUME 129. Plasma Lipoproteins (Part B: Characterization, Cell Biology, and Metabolism)
Edited by JOHN J. ALBERS AND JERE P. SEGREST

VOLUME 130. Enzyme Structure (Part K)
Edited by C. H. W. HIRS AND SERGE N. TIMASHEFF

VOLUME 131. Enzyme Structure (Part L)
Edited by C. H. W. HIRS AND SERGE N. TIMASHEFF

VOLUME 132. Immunochemical Techniques (Part J: Phagocytosis and Cell-Mediated Cytotoxicity)
Edited by GIOVANNI DI SABATO AND JOHANNES EVERSE

VOLUME 133. Bioluminescence and Chemiluminescence (Part B)
Edited by MARLENE DELUCA AND WILLIAM D. MCELROY

VOLUME 134. Structural and Contractile Proteins (Part C: The Contractile Apparatus and the Cytoskeleton)
Edited by RICHARD B. VALLEE

VOLUME 135. Immobilized Enzymes and Cells (Part B)
Edited by KLAUS MOSBACH

VOLUME 136. Immobilized Enzymes and Cells (Part C)
Edited by KLAUS MOSBACH

VOLUME 137. Immobilized Enzymes and Cells (Part D)
Edited by KLAUS MOSBACH

VOLUME 138. Complex Carbohydrates (Part E)
Edited by VICTOR GINSBURG

VOLUME 139. Cellular Regulators (Part A: Calcium- and Calmodulin-Binding Proteins)
Edited by ANTHONY R. MEANS AND P. MICHAEL CONN

VOLUME 140. Cumulative Subject Index Volumes 102–119, 121–134

Volume 141. Cellular Regulators (Part B: Calcium and Lipids)
Edited by P. Michael Conn and Anthony R. Means

Volume 142. Metabolism of Aromatic Amino Acids and Amines
Edited by Seymour Kaufman

Volume 143. Sulfur and Sulfur Amino Acids
Edited by William B. Jakoby and Owen Griffith

Volume 144. Structural and Contractile Proteins (Part D: Extracellular Matrix)
Edited by Leon W. Cunningham

Volume 145. Structural and Contractile Proteins (Part E: Extracellular Matrix)
Edited by Leon W. Cunningham

Volume 146. Peptide Growth Factors (Part A)
Edited by David Barnes and David A. Sirbasku

Volume 147. Peptide Growth Factors (Part B)
Edited by David Barnes and David A. Sirbasku

Volume 148. Plant Cell Membranes
Edited by Lester Packer and Roland Douce

Volume 149. Drug and Enzyme Targeting (Part B)
Edited by Ralph Green and Kenneth J. Widder

Volume 150. Immunochemical Techniques (Part K: *In Vitro* Models of B and T Cell Functions and Lymphoid Cell Receptors)
Edited by Giovanni Di Sabato

Volume 151. Molecular Genetics of Mammalian Cells
Edited by Michael M. Gottesman

Volume 152. Guide to Molecular Cloning Techniques
Edited by Shelby L. Berger and Alan R. Kimmel

Volume 153. Recombinant DNA (Part D)
Edited by Ray Wu and Lawrence Grossman

Volume 154. Recombinant DNA (Part E)
Edited by Ray Wu and Lawrence Grossman

Volume 155. Recombinant DNA (Part F)
Edited by Ray Wu

Volume 156. Biomembranes (Part P: ATP-Driven Pumps and Related Transport: The Na,K-Pump)
Edited by Sidney Fleischer and Becca Fleischer

Volume 157. Biomembranes (Part Q: ATP-Driven Pumps and Related Transport: Calcium, Proton, and Potassium Pumps)
Edited by Sidney Fleischer and Becca Fleischer

Volume 158. Metalloproteins (Part A)
Edited by James F. Riordan and Bert L. Vallee

VOLUME 176. Nuclear Magnetic Resonance (Part A: Spectral Techniques and Dynamics)
Edited by NORMAN J. OPPENHEIMER AND THOMAS L. JAMES

VOLUME 177. Nuclear Magnetic Resonance (Part B: Structure and Mechanism)
Edited by NORMAN J. OPPENHEIMER AND THOMAS L. JAMES

VOLUME 178. Antibodies, Antigens, and Molecular Mimicry
Edited by JOHN J. LANGONE

VOLUME 179. Complex Carbohydrates (Part F)
Edited by VICTOR GINSBURG

VOLUME 180. RNA Processing (Part A: General Methods)
Edited by JAMES E. DAHLBERG AND JOHN N. ABELSON

VOLUME 181. RNA Processing (Part B: Specific Methods)
Edited by JAMES E. DAHLBERG AND JOHN N. ABELSON

VOLUME 182. Guide to Protein Purification
Edited by MURRAY P. DEUTSCHER

VOLUME 183. Molecular Evolution: Computer Analysis of Protein and Nucleic Acid Sequences
Edited by RUSSELL F. DOOLITTLE

VOLUME 184. Avidin–Biotin Technology
Edited by MEIR WILCHEK AND EDWARD A. BAYER

VOLUME 185. Gene Expression Technology
Edited by DAVID V. GOEDDEL

VOLUME 186. Oxygen Radicals in Biological Systems (Part B: Oxygen Radicals and Antioxidants)
Edited by LESTER PACKER AND ALEXANDER N. GLAZER

VOLUME 187. Arachidonate Related Lipid Mediators
Edited by ROBERT C. MURPHY AND FRANK A. FITZPATRICK

VOLUME 188. Hydrocarbons and Methylotrophy
Edited by MARY E. LIDSTROM

VOLUME 189. Retinoids (Part A: Molecular and Metabolic Aspects)
Edited by LESTER PACKER

VOLUME 190. Retinoids (Part B: Cell Differentiation and Clinical Applications)
Edited by LESTER PACKER

VOLUME 191. Biomembranes (Part V: Cellular and Subcellular Transport: Epithelial Cells)
Edited by SIDNEY FLEISCHER AND BECCA FLEISCHER

VOLUME 192. Biomembranes (Part W: Cellular and Subcellular Transport: Epithelial Cells)
Edited by SIDNEY FLEISCHER AND BECCA FLEISCHER

VOLUME 193. Mass Spectrometry
Edited by JAMES A. MCCLOSKEY

VOLUME 194. Guide to Yeast Genetics and Molecular Biology
Edited by CHRISTINE GUTHRIE AND GERALD R. FINK

VOLUME 195. Adenylyl Cyclase, G Proteins, and Guanylyl Cyclase
Edited by ROGER A. JOHNSON AND JACKIE D. CORBIN

VOLUME 196. Molecular Motors and the Cytoskeleton
Edited by RICHARD B. VALLEE

VOLUME 197. Phospholipases
Edited by EDWARD A. DENNIS

VOLUME 198. Peptide Growth Factors (Part C)
Edited by DAVID BARNES, J. P. MATHER, AND GORDON H. SATO

VOLUME 199. Cumulative Subject Index Volumes 168–174, 176–194

VOLUME 200. Protein Phosphorylation (Part A: Protein Kinases: Assays, Purification, Antibodies, Functional Analysis, Cloning, and Expression)
Edited by TONY HUNTER AND BARTHOLOMEW M. SEFTON

VOLUME 201. Protein Phosphorylation (Part B: Analysis of Protein Phosphorylation, Protein Kinase Inhibitors, and Protein Phosphatases)
Edited by TONY HUNTER AND BARTHOLOMEW M. SEFTON

VOLUME 202. Molecular Design and Modeling: Concepts and Applications (Part A: Proteins, Peptides, and Enzymes)
Edited by JOHN J. LANGONE

VOLUME 203. Molecular Design and Modeling: Concepts and Applications (Part B: Antibodies and Antigens, Nucleic Acids, Polysaccharides, and Drugs)
Edited by JOHN J. LANGONE

VOLUME 204. Bacterial Genetic Systems
Edited by JEFFREY H. MILLER

VOLUME 205. Metallobiochemistry (Part B: Metallothionein and Related Molecules)
Edited by JAMES F. RIORDAN AND BERT L. VALLEE

VOLUME 206. Cytochrome P450
Edited by MICHAEL R. WATERMAN AND ERIC F. JOHNSON

VOLUME 207. Ion Channels
Edited by BERNARDO RUDY AND LINDA E. IVERSON

VOLUME 208. Protein–DNA Interactions
Edited by ROBERT T. SAUER

VOLUME 209. Phospholipid Biosynthesis
Edited by EDWARD A. DENNIS AND DENNIS E. VANCE

VOLUME 210. Numerical Computer Methods
Edited by LUDWIG BRAND AND MICHAEL L. JOHNSON

VOLUME 211. DNA Structures (Part A: Synthesis and Physical Analysis of DNA)
Edited by DAVID M. J. LILLEY AND JAMES E. DAHLBERG

VOLUME 212. DNA Structures (Part B: Chemical and Electrophoretic Analysis of DNA)
Edited by DAVID M. J. LILLEY AND JAMES E. DAHLBERG

VOLUME 213. Carotenoids (Part A: Chemistry, Separation, Quantitation, and Antioxidation)
Edited by LESTER PACKER

VOLUME 214. Carotenoids (Part B: Metabolism, Genetics, and Biosynthesis)
Edited by LESTER PACKER

VOLUME 215. Platelets: Receptors, Adhesion, Secretion (Part B)
Edited by JACEK J. HAWIGER

VOLUME 216. Recombinant DNA (Part G)
Edited by RAY WU

VOLUME 217. Recombinant DNA (Part H)
Edited by RAY WU

VOLUME 218. Recombinant DNA (Part I)
Edited by RAY WU

VOLUME 219. Reconstitution of Intracellular Transport
Edited by JAMES E. ROTHMAN

VOLUME 220. Membrane Fusion Techniques (Part A)
Edited by NEJAT DÜZGÜNEŞ

VOLUME 221. Membrane Fusion Techniques (Part B)
Edited by NEJAT DÜZGÜNEŞ

VOLUME 222. Proteolytic Enzymes in Coagulation, Fibrinolysis, and Complement Activation (Part A: Mammalian Blood Coagulation Factors and Inhibitors)
Edited by LASZLO LORAND AND KENNETH G. MANN

VOLUME 223. Proteolytic Enzymes in Coagulation, Fibrinolysis, and Complement Activation (Part B: Complement Activation, Fibrinolysis, and Nonmammalian Blood Coagulation Factors)
Edited by LASZLO LORAND AND KENNETH G. MANN

VOLUME 224. Molecular Evolution: Producing the Biochemical Data
Edited by ELIZABETH ANNE ZIMMER, THOMAS J. WHITE, REBECCA L. CANN, AND ALLAN C. WILSON

VOLUME 225. Guide to Techniques in Mouse Development
Edited by PAUL M. WASSARMAN AND MELVIN L. DEPAMPHILIS

VOLUME 226. Metallobiochemistry (Part C: Spectroscopic and Physical Methods for Probing Metal Ion Environments in Metalloenzymes and Metalloproteins)
Edited by JAMES F. RIORDAN AND BERT L. VALLEE

VOLUME 263. Plasma Lipoproteins (Part C: Quantitation)
Edited by WILLIAM A. BRADLEY, SANDRA H. GIANTURCO, AND JERE P. SEGREST

VOLUME 264. Mitochondrial Biogenesis and Genetics (Part B)
Edited by GIUSEPPE M. ATTARDI AND ANNE CHOMYN

VOLUME 265. Cumulative Subject Index Volumes 228, 230–262

VOLUME 266. Computer Methods for Macromolecular Sequence Analysis
Edited by RUSSELL F. DOOLITTLE

VOLUME 267. Combinatorial Chemistry
Edited by JOHN N. ABELSON

VOLUME 268. Nitric Oxide (Part A: Sources and Detection of NO; NO Synthase)
Edited by LESTER PACKER

VOLUME 269. Nitric Oxide (Part B: Physiological and Pathological Processes)
Edited by LESTER PACKER

VOLUME 270. High Resolution Separation and Analysis of Biological Macromolecules (Part A: Fundamentals)
Edited by BARRY L. KARGER AND WILLIAM S. HANCOCK

VOLUME 271. High Resolution Separation and Analysis of Biological Macromolecules (Part B: Applications)
Edited by BARRY L. KARGER AND WILLIAM S. HANCOCK

VOLUME 272. Cytochrome P450 (Part B)
Edited by ERIC F. JOHNSON AND MICHAEL R. WATERMAN

VOLUME 273. RNA Polymerase and Associated Factors (Part A)
Edited by SANKAR ADHYA

VOLUME 274. RNA Polymerase and Associated Factors (Part B)
Edited by SANKAR ADHYA

VOLUME 275. Viral Polymerases and Related Proteins
Edited by LAWRENCE C. KUO, DAVID B. OLSEN, AND STEVEN S. CARROLL

VOLUME 276. Macromolecular Crystallography (Part A)
Edited by CHARLES W. CARTER, JR., AND ROBERT M. SWEET

VOLUME 277. Macromolecular Crystallography (Part B)
Edited by CHARLES W. CARTER, JR., AND ROBERT M. SWEET

VOLUME 278. Fluorescence Spectroscopy
Edited by LUDWIG BRAND AND MICHAEL L. JOHNSON

VOLUME 279. Vitamins and Coenzymes (Part I)
Edited by DONALD B. MCCORMICK, JOHN W. SUTTIE, AND CONRAD WAGNER

VOLUME 280. Vitamins and Coenzymes (Part J)
Edited by DONALD B. MCCORMICK, JOHN W. SUTTIE, AND CONRAD WAGNER

VOLUME 281. Vitamins and Coenzymes (Part K)
Edited by DONALD B. MCCORMICK, JOHN W. SUTTIE, AND CONRAD WAGNER

VOLUME 302. Green Fluorescent Protein
Edited by P. MICHAEL CONN

VOLUME 303. cDNA Preparation and Display (in preparation)
Edited by SHERMAN M. WEISSMAN

VOLUME 304. Chromatin (in preparation)
Edited by PAUL M. WASSERMAN AND ALAN P. WOLFFE

VOLUME 305. Bioluminescence and Chemiluminescence (Part C) (in preparation)
Edited by THOMAS O. BALDWIN AND MIRIAM M. ZIEGLER

VOLUME 306. Expression of Recombinant Genes in Eukaryotic Cells (in preparation)
Edited by JOSEPH C. GLORIOSO AND MARTIN C. SCHMIDT

VOLUME 307. Confocal Microscopy (in preparation)
Edited by P. MICHAEL CONN

VOLUME 308. Enzyme Kinetics and Mechanism (Part E) (in preparation)
Edited by VERN L. SCHRAMM AND DANIEL L. PURICH

VOLUME 309. Amyloids, Prions, and Other Protein Aggregates (in preparation)
Edited by RONALD WETZEL

VOLUME 310. Biofilms (in preparation)
Edited by RON J. DOYLE

Section I

Monitoring of Physiological Processes

[1] Membrane Trafficking

By SABINE KUPZIG, SAN SAN LEE, and GEORGE BANTING

Introduction

Immunofluorescence microscopy has played a central role in providing information concerning the localization of specific integral membrane proteins. However, immunofluorescence microscopy simply generates a "snapshot" of what is going on and fails to provide direct information concerning the movement of integral membrane proteins in living cells. Time course experiments, for example, designed to study the effects of specific hormones and drugs on intracellular morphology and membrane trafficking, are also particularly laborious to perform if based on the use of immunofluorescence microscopy. Such studies, and many others on the dynamic nature of membrane traffic within eukaryotic cells, would be much easier to perform and would not be subject to possible fixation artifacts if specific integral membrane proteins could be visualized in living cells. The expression of recombinant, green fluorescent protein (GFP)-tagged integral membrane proteins has provided us with precisely the right tools with which to perform such experiments. This chapter describes the use of a Leica [Leica Microsystems GmbH (Heidelberg, Germany); http://www.llt.de/index1.html] DM IRBE inverted epifluorescence microscope attached to a Leica TCS-NT confocal laser scanning system to monitor the localization and movement of GFP-tagged integral membrane proteins in living cells; however, the general principles are applicable to a wide range of microscope systems.

Methods

Construct Design

Conventional molecular biological techniques should be used to generate constructs encoding GFP-tagged integral membrane proteins. We have found that GFP does not affect the intracellular distribution of the tagged membrane protein, and that GFP remains fluorescent when placed on either side of the membrane (i.e., cytosolic or within the lumen of the organelle to which the tagged membrane protein is localized).[1] In the constructs we have made, we have introduced a short (four- to six-amino acid), flexible (glycine-rich) linker region between the GFP sequence and that of the

[1] M. Girotti and G. Banting, *J. Cell Sci.* **109,** 2915 (1996).

0076-6879/99 $30.00

tagged protein. This may or may not be necessary, but is done to increase the chances of the two components of the hybrid protein folding independently and correctly.

Construct Expression

The hybrid DNA construct should be placed in an appropriate eukaryotic expression vector. Our experience with GFP-tagged ratTGN38 [a type I integral membrane protein that is predominantly localized to the trans-Golgi network (TGN) at steady state[2]] is that the protein is not correctly localized when expressed in transiently transfected cells.[1] However, when the same hybrid protein is expressed in stably transfected cells it is found to be correctly localized,[1] as has been described for another GFP-tagged, Golgi-localized integral membrane protein (*N*-acetylglucosamine transferase 1) expressed in stably transfected cells.[3] Others have expressed GFP-tagged Golgi membrane proteins in transiently transfected cells and observed appropriate localization (e.g., see Refs. 4 and 5). It is unclear why some constructs should be correctly localized only when expressed in stably transfected cells, but this may be due to the level of expression of the recombinant protein, because expression levels in stable transfectants are generally lower than those in transiently transfected cells. Thus, if a GFP-tagged integral membrane protein is not appropriately localized in transiently transfected cells it may yet be in stable transfectants.

Imaging

Stably transfected cells expressing the GFP-tagged membrane protein of interest are cultured on round (22-mm diameter) sterile glass coverslips until approximately 60–70% confluent. Immediately prior to image acquisition these coverslips are transferred to an appropriate holder for use on the laser scanning confocal microscope. We generally use a homemade aluminum holder as illustrated in Fig. 1. Approximately 0.2 ml of CO_2-independent medium (GIBCO-BRL, Gaithersburg, MD), prewarmed to 37°, is then added to bathe the cells on the upper surface of the coverslip. The slide holder is then placed into the center of a heated jacket (Medical Systems Corp., Greenvale, NY; http://www.medicalsystems.com/msc.htm)

[2] J. P. Luzio, B. Brake, G. Banting, K. E. Howell, P. Braghetta, and K. K. Stanley, *Biochem. J.* **270,** 97 (1990).

[3] D. T. Shima, K. Haldar, R. Pepperkok, R. Watson, and G. Warren, *J. Cell. Biol.* **137,** 1211 (1997).

[4] N. B. Cole, C. L. Smith, N. Sciaky, M. Terasaki, M. Edidin, and J. Lippincott-Schwartz, *Science* **273,** 797 (1996).

[5] N. Cole, M. Terasaki, N. Sciaky, and J. Lippincott-Schwartz, *Mol. Biol. Cell* **6,** 8 (1995).

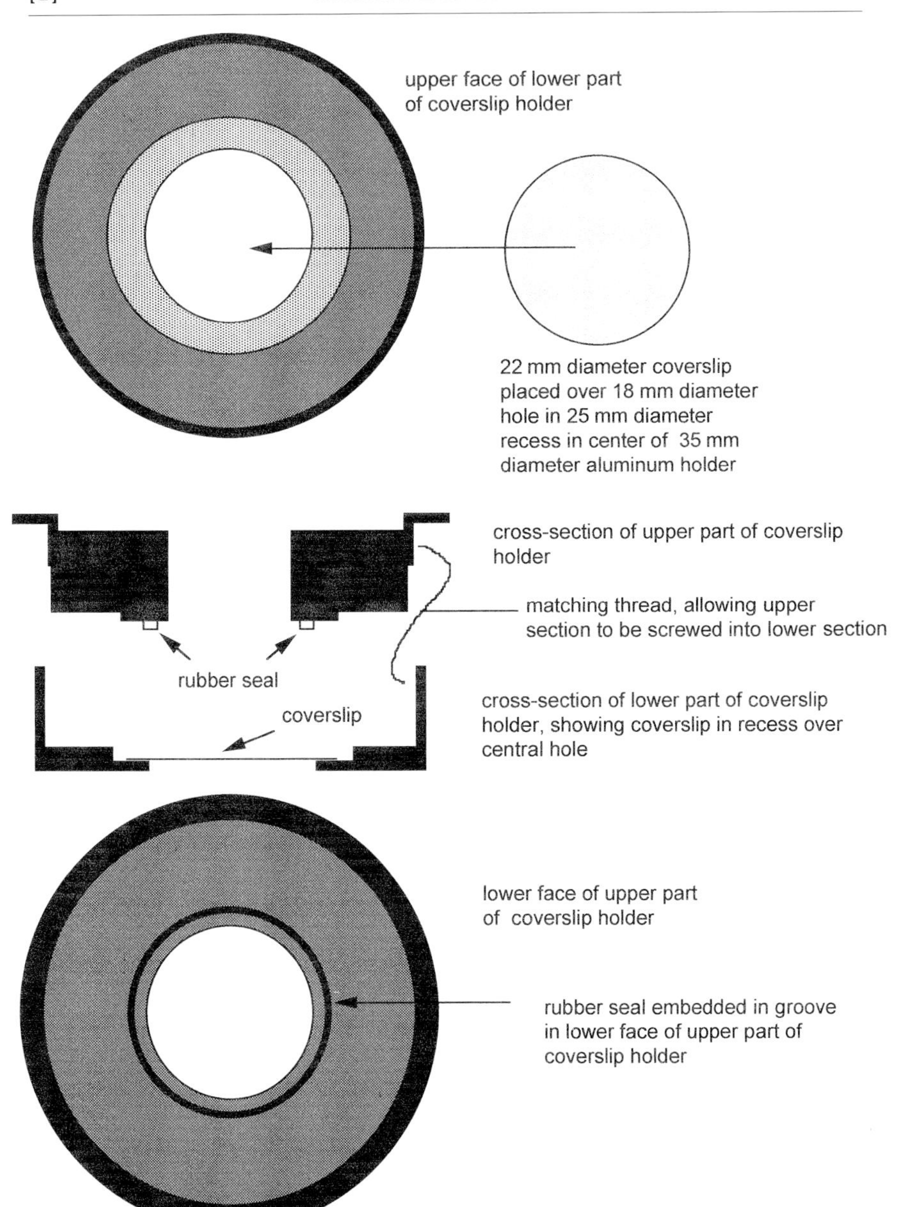

FIG. 1. Diagrammatic representation of coverslip holder for use in live cell imaging.

attached to a thermal regulator unit (Medical Systems Corp.) that has been precalibrated to maintain the temperature of the medium bathing the cells on the coverslip at 37°. This unit is then placed on the stage of a Leica DM IRBE inverted epifluorescence microscope attached to a Leica TCS-NT confocal laser scanning head equipped with a krypton–argon laser (488-, 568-, and 647-nm lines). The cells are initially viewed, under illumination from a halogen lamp, through a ×63 (1.4 NA) oil immersion, objective lens using a fluorescein filter set. Once a suitable cell has been brought into focus and appropriately aligned on the stage, the microscope is switched to the confocal laser scanning format. The slide is then scanned at 488 nm and images acquired using the Leica TCS-NT software. Initial scanning should be performed at medium scan speed and with laser intensity set below 50% in order to minimize photobleaching of the sample prior to image acquisition. The relevant photomultiplier tube (PMT) can be used to adjust the intensity of the signal obtained; increasing the PMT value will give a brighter signal, but resolution will be lost. There is thus always a trade-off between low laser intensity plus high PMT value (minimal photobleaching, detectable signal, poor resolution) and high laser intensity plus low PMT value (increased photobleaching, detectable signal, increased resolution). Acceptable image resolution is generally obtained with the PMT value set below 600 (preferably closer to 300). The system can be converted to time-lapse image acquisition once a suitable image has been obtained. Note that the zoom facility in the Leica TCS-NT software allows the operator to increase the displayed image size up to 32 times that of the original scanned and displayed image. Before time-lapse image acquisition the scan speed should be changed to "slow" in order to obtain brighter images and the software should be configured (according to manufacturer instructions) to scan and record images at defined intervals (generally every 1–20 sec). The Leica TCS-NT software allows the operator to set up a simple time-lapse procedure; for example, scanning a defined number of times (to generate an average image) at defined intervals for a specified number of frames. In reality, it is impractical to perform multiple scans at a particular point in time, since there is movement within the sample during the time period in which the multiple scans, which are used to generate the average image, are performed; thus single scans are performed at defined intervals for a specified number of frames in order to generate a time-lapse movie. The Leica TCS-NT software also allows the operator to set up a more complex time-lapse procedure in which the frequency of frame acquisition may be varied. Once acquired, the movie can be played back using the TCS-NT software and can be saved in either scanner file format (for further manipulation using the TCS-NT software) or in export file format (for transfer as a folder of individual frames to other systems—

Macintosh, PC, Silicon Graphics, etc.—for video compilation and/or image analysis).

The preceding describes the basic procedures involved in imaging live cells expressing GFP-tagged proteins. Clearly, the effects of various factors (changes in temperature, different drugs, growth factors, hormones, etc.) on the morphology of the organelle in which a specific GFP-tagged integral membrane protein is primarily located can be assayed using this system [e.g., the effects of the fungal metabolite brefeldin A (BFA) on the morphology of the TGN can be demonstrated using GFP-tagged ratTGN38 as a reporter; Fig. 2], as can the movement of populations of GFP-tagged integral membrane proteins over time [e.g., GFP-tagged ratTGN38 in stably transfected normal rat kidney (NRK) cells; Fig. 3].

Fluorescence Recovery after Photobleaching

The basic procedure described in the preceding section can be modified to permit the study of the diffusional mobility in the plane of the lipid bilayer of GFP-tagged integral membrane proteins. Such studies involve fluorescence recovery after photobleaching (FRAP) (also known as fluorescence photobleaching recovery, FPR) analysis. The principle underlying FRAP is that transiently increasing the intensity of the light illuminating a particular region of a fluorescently labeled cell will lead to photobleaching

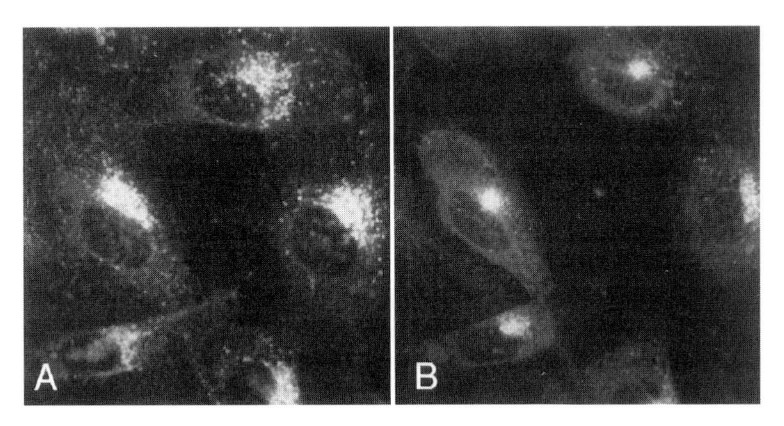

FIG. 2. Appearance of cells before and after BFA is administered. Live, stably transfected NRK cells expressing GFP-tagged ratTGN38 were imaged at 37° on a Leica TCS-NT confocal laser scanning microscope (as described in text) before (A) and after (B) incubation in brefeldin A (BFA, 5 μg/ml) for 30 min. These images show the "before and after" effects of BFA. In fact, the cells were imaged regularly during the incubation in BFA. This provided a series of images that have been used to generate a time-lapse movie showing what happens to the morphology of the TGN during incubation of cells in BFA.

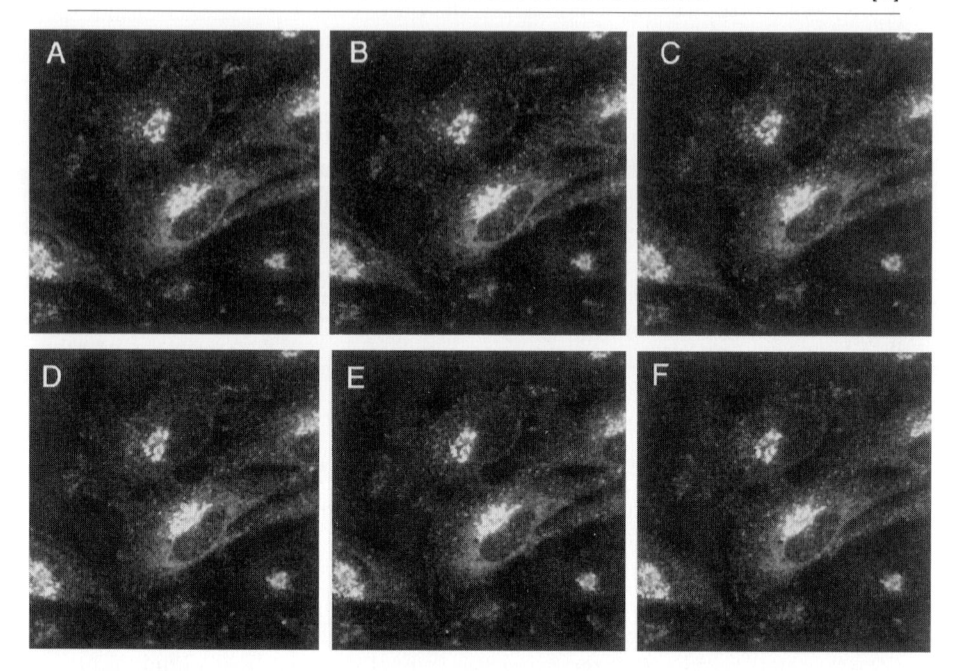

FIG. 3. Membrane movement in living cells. Live, stably transfected NRK cells expressing GFP-tagged ratTGN38 were imaged at 37° on a Leica TCS-NT confocal laser scanning microscope (as described in text). Images were recorded every 5 sec over a 5-min period in order to visualize membrane movement within the cells. (A–F) The first six images in this series. Multiple vesicle-like profiles can be seen to have moved during this time period and the morphology of the TGN has also changed slightly. The directional "vesicle" movement is dependent on the presence of intact microtubules, because incubation of cells in the presence of the microtubule-depolymerizing drug nocodazole (20 μg/ml) leads to cessation of this movement.

of that region of the cell that has been subjected to increased illumination, that is, a "black hole" will be generated in the region of fluorescence. Once the photobleached area has been generated, further time-lapse imaging (under conditions of reduced light intensity) will allow analysis of how rapidly the photobleached area is repopulated with fluorescent molecules from the surrounding area, that is, how rapidly fluorescence recovers after photobleaching. This procedure has been used to monitor the lateral diffusional mobility of integral membrane proteins and lipids in the plane of the lipid bilayer at the cell surface (e.g., see Refs. 6–11), and more recently

[6] E. Yeichel and M. Edidin, *J. Cell Biol.* **105,** 755 (1987).

[7] M. P. Sheetz, M. Schindler, and D. E. Koppel, *Nature* (*London*) **285,** 510 (1980).

[8] M. Shields, P. La Celle, R. E. Waugh, M. Scholz, R. Peters, and H. Passow, *Biochim. Biophys. Acta* **905,** 181 (1987).

to study the lateral diffusional mobility of integral membrane proteins in the plane of the lipid bilayer in intracellular membranes.[4] The earlier work on lateral diffusional mobility of integral membrane proteins and lipids in the plane of the lipid bilayer at the cell surface used fluorescently labeled lipid, fluorescently labeled integral membrane proteins, fluorescently labeled lectins, or fluorescently labeled antibodies attached to target membrane proteins as reporter molecules; the more recent studies on the lateral diffusional mobility of integral membrane proteins in the plane of the lipid bilayer in intracellular membranes have used GFP-tagged integral membrane proteins as reporters. To perform FRAP experiments, cells are prepared and image acquisition proceeds initially as described in the previous section. However, once the first image has been obtained the zoom function should be used to focus in on a specific, fluorescent region of the cell under study. This region should be positioned in the center of the screen prior to zooming in, since the zoom function will automatically zoom in on the center of the screen. The operator should zoom in either 16 or 32 times (depending on the area it is intended to photobleach). The scan should continue (at slow scan speed) and the laser power should be increased to maximum until the fluorescent signal from GFP is lost. It is advisable in these experiments to use a readily photobleachable version of GFP that is not expressed at particularly high levels. The reason for this is that the aim is to obtain maximal photobleaching while minimizing the potential for photodamage to the membranes; by using the S65T/Q80R/I167T triple mutant of GFP we have found that two or three slow scans at maximum laser power is generally sufficient to photobleach the desired region. The laser power should be returned to that previously used for image acquisition during scanning as soon as photobleaching is complete. Time-lapse image acquisition should then proceed as already described in order to record fluorescence recovery of the photobleached area. An example of FRAP is given in Fig. 4.

Combining Green Fluorescent Protein Imaging with Immunofluorescence

We have observed that the GFP fluorescence remains after both methanol and paraformaldehyde fixation of cells.[1] It is therefore possible to combine conventional immunofluorescence microscopy, using a tetramethylrhodamine B isothiocyanate (TRITC-) or rhodamine-conjugated second-

[9] D. E. Golan and W. Veatch, *Proc. Natl. Acad. Sci. U.S.A.* **77,** 2537 (1980).
[10] M. Edidin and I. Stroynowski, *J. Cell Biol.* **112,** 1143 (1990).
[11] M. Edidin, M. C. Zuniga, and M. P. Sheetz, *Proc. Natl. Acad. Sci. U.S.A.* **91,** 3378 (1994).

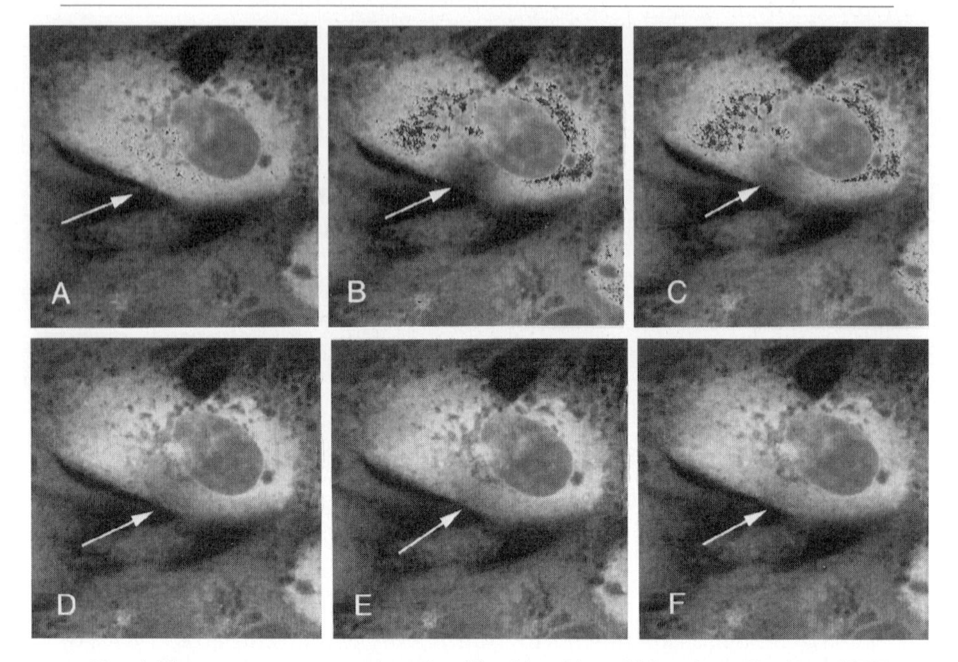

FIG. 4. Fluorescence recovery after photobleaching. Live, chick embryo fibroblast (CEF) cells expressing GFP-tagged sialyltransferase (ST) were incubated in the presence of BFA (5 μg/ml) for 10 min in order to redistribute the ST into the endoplasmic reticulum (ER), and were then subjected to FRAP analysis (as described in text). (A) A cell immediately prior to photobleaching of an area of the ER (the area to be photobleached is indicated by an arrow). (B) The same cell immediately after photobleaching of an area of the ER (the area that has been photobleached is indicated by an arrow). (C–F) Images (taken at 2-sec intervals) of the same cell as it recovers from photobleaching.

ary antibody, with detection of GFP-tagged membrane proteins for colocalization studies.[1]

Future Developments

The ever-expanding range of spectral variants of GFP will provide the opportunity to tag various integral membrane proteins with GFP molecules with various excitation and emission spectra. Thus, given a confocal microscope with a spectrophotometer scanner head that can be tuned to appropriate wavelengths and bandwidths, it will be possible to colocalize different integral membrane proteins in the same living cell. It should also be possible to perform fluorescence resonance energy transfer (FRET) experiments in living cells in order to address the proximity of different integral membrane proteins in living cells.

Acknowledgments

The authors thank David Stephens for critical reading of the manuscript, the Medical Research Council for providing an Infrastructure Award (G4500006) to establish the School of Medical Sciences Cell Imaging Facility, and Dr. Mark Jepson for assistance with confocal image processing. The authors also thank Prof. Colin Hopkins for providing the cells used for the experiment presented in Fig. 4.

[2] Monitoring of Protein Secretion with Green Fluorescent Protein

By CHRISTOPH KAETHER and HANS-HERMANN GERDES

Introduction

In eukaryotic cells, the transport of secretory proteins from the endoplasmic reticulum (ER) to the plasma membrane (PM) is orchestrated by a number of vesicular transport steps. For each of these steps carrier vesicles bud from topologically separate compartments of the secretory pathway and specifically fuse with the membrane of the subsequent compartment.[1] In addition, the steady state size of the various compartments is maintained by a membrane flux opposite to the exocytic route. Despite the complexity of this traffic, the delivery of proteins to the PM occurs with remarkable speed, emphasizing the dynamics as a prevalent feature of the secretory pathway.

Biochemical approaches largely based on *in vitro* reconstitution assays led to the discovery of molecules of involved in the budding and fusion reactions along this pathway.[2] Furthermore, morphological studies on fixed cells provided snapshots of the underlying membranous structures. Despite this progress, the spatial and temporal resolution of secretory traffic as it occurs in live cells remained largely elusive, because appropriate tools to address this issue were absent.

The availability of the green fluorescent protein (GFP) opened the door for real-time studies and made it possible to monitor distinct vesicular transport steps of the secretory pathway in real time (see also [1] in this volume).[2a] Transport from the ER to the Golgi was studied using the

[1] G. E. Palade, *Science* **189,** 347 (1975).
[2] J. E. Rothman, *Nature* (*London*) **372,** 55 (1994).
[2a] S. Kupzig, S. S. Lee, and G. Banting, *Methods Enzymol.* **302,** [1] 1999 (this volume).

0076-6879/99 $30.00

vesicular stomatitis virus (VSV) G protein tagged with GFP.[3,4] Our group is focusing on the transport of proteins from the trans-Golgi network (TGN) to the PM.[5] This transport is either constitutive, if TGN-derived secretory vesicles fuse continuously with the PM, or regulated, if the vesicles fuse only in response to a stimulus. The regulated pathway is a special feature of certain cell types such as exocrine, endocrine, and neuronal cells, whereas the constitutive pathway is common to all eukaryotic cells. To monitor these two pathways we have tagged a secretory protein, human chromogranin B (hCgB), with GFP. hCgB is a soluble protein of neuroendocrine secretory granules (SGs),[6] the storage organelles of the regulated pathway. Ectopic expression of this protein in cells with a regulated pathway results in its sorting to SGs.[7] Expression in cells lacking a regulated pathway of protein secretion, however, leads to its constitutive secretion.[8] Thus, depending on the cell type in which hCgB is expressed, it serves as a marker for SGs or constitutive secretory vesicles (CVs). In this chapter we describe methods for monitoring constitutive and regulated protein secretion in living cells, using hCgB tagged with GFP.

Constitutive Secretion

Principle of the Trans-Golgi Network-to-Plasma Membrane Secretion Assay

To study constitutive secretion from the TGN to the PM we use hCgB[9] that is C-terminally tagged with GFP(S65T).[10] The fusion protein, hCgB-GFP(S65T),[8] is expressed under the control of a cytomegalovirus (CMV) promoter in a CDM8 vector (Invitrogen, Leek, The Netherlands). We express hCgB–GFP(S65T) transiently in Vero or HeLa cells (Fig. 1), or take advantage of a stable Vero cell line, V7, expressing hCgB–GFP(S65T).[5] Vero and HeLa cells have only a constitutive pathway of protein secretion.

[3] J. F. Presley, N. B. Cole, T. A. Schroer, C. Hirschberg, K. J. M. Zaal, and J. Lippincott-Schwartz, *Nature (London)* **389,** 81 (1997).

[4] S. J. Scales, R. Pepperkok, and T. E. Kreis, *Cell* **90,** 1137 (1997).

[5] I. Wacker, C. Kaether, A. Krömer, A. Migala, W. Almers, and H.-H. Gerdes, *J. Cell Sci.* **110,** 1453 (1997).

[6] P. Rosa and H.-H. Gerdes, *J. Endocrinol. Invest.* **17,** 207 (1994).

[7] A. Krömer, M. Glombik, W. B. Huttner and H.-H. Gerdes, *J. Cell Biol.* **140,** 1331 (1998).

[8] C. Kaether and H.-H. Gerdes, *FEBS Lett.* **369,** 267 (1995).

[9] U. M. Benedum, A. Lamouroux, D. S. Konecki, P. Rosa, A. Hille, P. A. Baeuerle, R. Frank, F. Lottspeich, J. Mallet, and W. B. Huttner, *EMBO J.* **6,** 1203 (1987).

[10] R. Heim, A. B. Cubitt, and R. Y. Tsien, *Nature (London)* **373,** 663 (1995).

Under normal culture conditions at 37° no fluorescent hCgB–GFP(S65T) is detected (Fig. 1, top row). However, when biosynthetic protein transport is arrested at the trans-Golgi network by a 20° secretion block, fluorescent hCgB–GFP(S65T) is observed in the TGN (Fig. 1, middle row). Subsequent reversal of the secretion block results in the formation of fluorescent constitutive secretory vesicles (Fig. 1, bottom row), which can be analyzed by fluorescence microscopy. Thirty to 60 min after the release of the secretion block no cellular GFP fluorescence is detected, reflecting the complete secretion of fluorescent hCgB–GFP(S65T).

Comment. By using hCgB–GFP(S65T) in combination with a secretion block and a subsequent release we have detected green fluorescent CVs in all cell lines that have only a constitutive pathway of protein secretion.

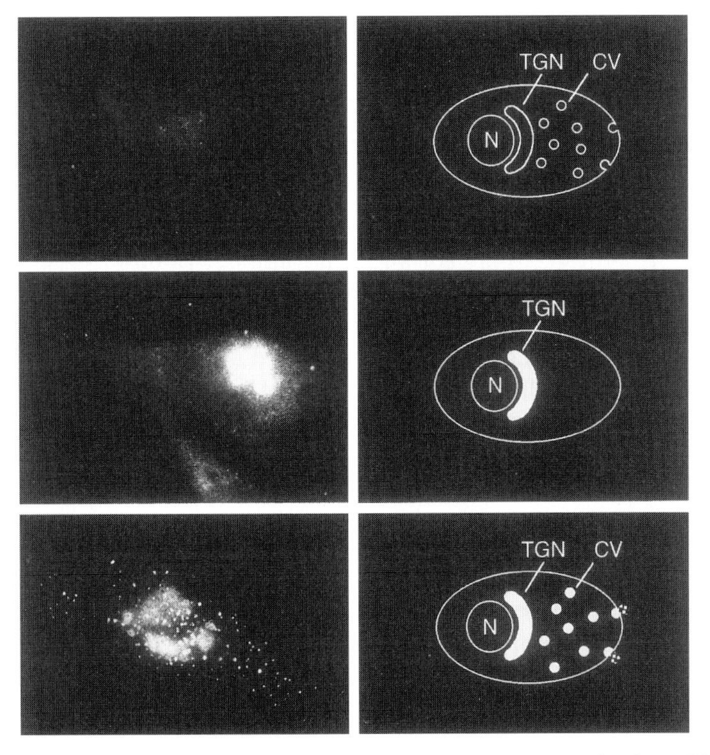

FIG. 1. Trans-Golgi network-to-plasma membrane secretion assay. *Left column:* Photomicrographs of HeLa cells transiently transfected with hCgB–GFP(S65T). *Right column:* Schematics symbolizing the principle of the assay; structures filled with white indicate the presence of GFP fluorescence. *Top row:* Cells incubated at 37°. *Middle row:* Cells incubated at 20° for 2 hr. *Bottom row:* Cells incubated for 2 hr at 20° and then for 30 min at 37°. N, Nucleus; TGN, trans-Golgi network; CV, constitutive secretory vesicle.

We have tested baby hamster kidney (BHK), HeLa, Madin–Darby canine kidney (MDCK), Vero, NIH 3T3, RK13, and human thymidine kinase-deficient ($HuTK^-$) cells.

Experimental Procedure

Cell Culture and Transfection. Vero cells (African green monkey kidney cells, ATCC CCL81) are grown at 37° and 5% CO_2 in Eagle's minimal essential medium (EMEM; GIBCO-BRL, Eggenstein, Germany) supplemented with 10% (v/v) fetal calf serum and 2 m*M* glutamine. Transfection is performed either by standard calcium phosphate precipitation after plating the cells on coverslips in 3.5-mm dishes, or by electroporation using a Bio-Rad Gene Pulser (Bio-Rad Laboratories, Richmond CA). For electroporation, confluent cells on a 15-cm dish are trypsinized, pelleted, and resuspended in 0.8 ml of phosphate-buffered saline (PBS). The cell suspension is transferred to a 0.8-ml cuvette and 50 μg of DNA purified on a Nucleobond AX500 column (Macherey–Nagel GmbH, Düren, Germany) is added. Thereafter the suspension is kept for 5 min at room temperature, with occasional mixing. Cells are then electroporated at 300 V and 960 μF, diluted in culture medium, and plated immediately.

Stable Cell Lines. For generation of a stable cell line expressing hCgB–GFP(S65T), Vero cells are cotransfected with hCgB–GFP(S65T)/pCDM8 linearized with *Sfi*I, and with pMC1-Neo (Stratagene, La Jolla, CA) linearized with *Sca*I, and then incubated in cell culture medium. Two days later the medium is replaced with fresh cell culture medium supplemented with G418 (0.5 mg/ml; GIBCO, Grand Island, NY). After 3–6 weeks of selection single G418-resistant colonies are picked (by drawing them into a 200-μl pipette tip), triturated, and replated in 24-well dishes. Clones are analyzed for expression of GFP fusion proteins after treatment with 2 m*M* butyrate for 21 hr (see the next section) by Western blotting, immunofluorescence staining, and detection of GFP fluorescence. For Western blotting a peptide antibody directed against the C terminus of GFP is used.[5] Immunofluorescence microscopy is performed with a polyclonal anti-GFP antibody (Clontech Laboratories, Palo Alto, CA). For detection of GFP fluorescence cells are analyzed after application of a 20° secretion block for 2 hr (see Secretion Block and Release, below).

Sodium Butyrate Treatment. To enhance expression of the GFP fusion proteins we often use sodium buyrate, which induces expression of CMV promoter-driven genes. We have found that treatment with sodium butyrate leads to morphological changes of the cells (e.g., Vero and HeLa cells become spindle-shaped). For sufficient induction and minimal morphological changes in Vero cells the treatment has been optimized to a 21-hr incubation period in 2 m*M* sodium butyrate. To prepare induction medium we use cell culture medium and a stock solution of 500 m*M* sodium butyrate in H_2O stored at 4°.

Culture Chambers for Microscopy. To record living cells, disposable culture chambers that have a coverslip as bottom can be used (LabTek chambered coverglass; Nalge Nunc International, Naperville, IL). Alternatively, a culture chamber equipped with a 42-mm coverslip for holding cells (POC chamber; Bachofer, Reutlingen, Germany) may be assembled. The Bachofer chamber is fitted with an adaptor (available from Leica) for the stage of the microscope. For both setups the temperature is controlled by a custom-made, water-heated objective jacket.

Secretion Block and Release. For a 20° secretion block a heating bath is placed in a 4° cold room and adjusted to 20°. The block is started by replacing the culture medium with precooled medium supplemented with 10 m*M* HEPES, pH 7.4. The dishes are then placed for 2 hr on a glass plate, which is located just below the water surface of the heating bath.

If cells are grown in LabTek chambers, the chambers are placed directly after the 20° incubation onto the microscope stage. The medium is subsequently replaced with prewarmed medium containing 10 m*M* HEPES, pH 7.4, or with prewarmed PBS. If cells are grown on 42-mm coverslips, the Bachofer chamber is assembled, perfused with prewarmed medium or PBS, and transferred to the microscope stage. Biogenesis of constitutive secretory veiscles is observed 10–15 min after release of the block and their movement is recorded.

COMMENT. In principle, the secretion block and its subsequent release can be accomplished on the microscope stage. However, during the warming from 20° to 37° thermal extensions of the objective and the microscope stage cause a continuous shift of the focal plane, disabling cell recordings during the warming-up period. Therefore the secretion block is performed separately in a water bath, and thereafter the cells are transferred to the microscope, the objective of which has been prewarmed for at least 30 min.

Confocal Microscopy. The confocal microscope is used when optical sectioning or simultaneous detection of two different fluorescent probes is required. In this way the dynamics of the TGN can be monitored, and exo- and endocytic transport can be visualized simultaneously.

We use a Leica TCS-4D confocal system equipped with an argon–krypton laser (Leica Microsystems GmbH, Heidelberg, Germany) and a ×63/1.4 NA PlanApo objective. GFP fluorescence is detected with 488-nm excitation and a Leica filter set optimized for GFP detection (TK 500 and BP 525/50).

Images are recorded with 512 × 512 pixel resolution in the fast scan mode and four-times line averaging. These settings allow a time resolution of 1.7 sec. Series of images are stored as TIFF files on a magneto-optical disk. We usually record series of 20 to 40 images per cell. During these recordings minor photobleaching and no effects of phototoxicity are observed.

Two-Color Imaging. To record exo- and endocytic traffic simultaneously, Vero cells expressing hCgB–GFP(S65T) are fed with Texas Red-labeled transferrin (Molecular Probes, Eugene, OR) during the 20° secretion block as described.[5] The confocal settings for two-channel scanning are as follows: excitation 488/568 nm, double dichroic (DD) 488/568, BP 525/50 for GFP detection, LP 590 for Texas Red detection, fast scan mode, four-times line averaging, 512 × 512 pixel resolution. The images of the two channels can then be combined using the multicolor analysis option of the confocal software.

COMMENT. A two-color video can be viewed on the World Wide Web (http://www.nbio.uni-heidelberg.de/Gerdes.html). An example of two-color imaging of two different GFP mutants, using a Leica confocal system, is described by Zimmerman and Siegert.[11]

Fluorescence Video Microscopy. For better time resolution we employ a cooled charge-coupled device (CCD) camera. A Quantix-G2 camera (Photometrics GmbH, Munich, Germany) is mounted onto a Zeiss Axiophot (Carl Zeiss Jena GmbH, Jena, Germany) equipped with a 100-W HBO lamp and a Zeiss Plan Apochromat ×63/1.4 objective. GFP fluorescence is detected with a special filter set from Zeiss (BP 470/20, FT 493, BP 505–530). For the Axiophot a Bachofer chamber with a temperature-regulated holder frame (Bachofer) is used. The camera is controlled by IP Lab Spectrum software (Signal Analytics Corp., Vienna, VA) that runs on a MacIntosh 9500/150 Power PC. With this setup moving vesicles can be recorded with an exposure time down to 0.2 sec, allowing sufficient resolution of rapid movements in the range of 2 μm/sec.

COMMENT. When Vero cells are imaged with a CCD camera the secretion of hCgB–GFP(S65T) into the medium leads to strong background fluorescence. This can be reduced by frequent changes of the medium or by perfusion of the Bachofer chamber. Confocal imaging precludes background problems caused by secreted hCgB–GFP(S65T), owing to the confocal sectioning.

Image Processing. Images obtained by confocal or CCD video microscopy are already digitized and can be transferred to various image-processing programs. For quantitation of fluorescence intensities or animation of a series of images we use NIH-Image, a shareware program from the National Institutes of Health (available at http://rsb.info.nih.gov/nih-image/).

[11] T. Zimmerman and F. Siegert, *BioTechniques* **24,** 458 (1998).

Adobe Photoshop (Adobe Systems, Mountain View, CA) is used to lay out images, and Avid Videoshop (Avid Technology, Tewksbury, MA) is used to create video clips.

Tracking. To analyze series of images we track vesicles using a macro written for NIH-Image.[5] From these tracks parameters such as velocities, travel distances, and direction of movement are calculated. In this way the effect of manipulations such as drug treatment or injection of antibodies on vesicle motility can be quantitated. By following this procedure we have analyzed the effect of nocodazole on vesicle motility and were able to demonstrate that CV movement (from the TGN to the PM) is microtubule dependent.[5]

Fluorometry. Secretion of fluorescent hCgB–GFP(S65T) can be measured by fluorometry. For this the stable clone V7 and Vero wild-type cells (as a control) are grown for 1 day to subconfluency and then treated for 21 hr with 2 m*M* sodium butyrate. To accumulate GFP fluorescence in the TGN, cells are incubated at 20° for 2 hr in external solution (ES), consisting of 140 m*M* NaCl, 5 m*M* KCl, 2 m*M* $CaCl_2$, 1 m*M* $MgCl_2$, 1 m*M* glucose, and 10 m*M* HEPES (pH 7.4). Thereafter ES (20°) is replaced by ES (37°) and cells are incubated for 2 hr at 37°. The ES is collected every 30 min during the 37° incubation. Emission spectra are recorded with excitation at 470 nm (Aminco Bowman fluorimeter) from the ES collected after various incubation times at 37° and from cells after 2 hr of incubation at 37°. To obtain spectra of cellular hCgB–GFP(S65T) fluorescence cells are lysed by three cycles of freezing and thawing in hypotonic buffer [10 m*M* HEPES (pH 7.4), 1 m*M* EDTA, 1 m*M* magnesium acetate, 1.25 m*M* phenylmethylsulfonyl fluoride (PMSF), leupeptin (10 μg/ml)] followed by preparation of high-speed (100,000*g*) supernatants. Emission spectra are normalized to total cellular protein. Difference spectra are calculated by subtracting the spectra of wild-type cells from those of V7 cells (Fig. 2). For calculation of secretion kinetics the peak values at 510 nm can be used.[5]

Regulated Secretion

Rationale

A crucial point for the *in vivo* analysis of SGs is the specificity of labeling. In the TGN of neuroendocrine cells regulated secretory proteins are sorted from constitutively secreted proteins and packaged into SGs.[12] Inefficient sorting of a regulated protein tagged with GFP would result in labeling of both SGs and CVs, thus disabling the unequivocal identification

[12] R. B. Kelly, *Science* **230,** 25 (1985).

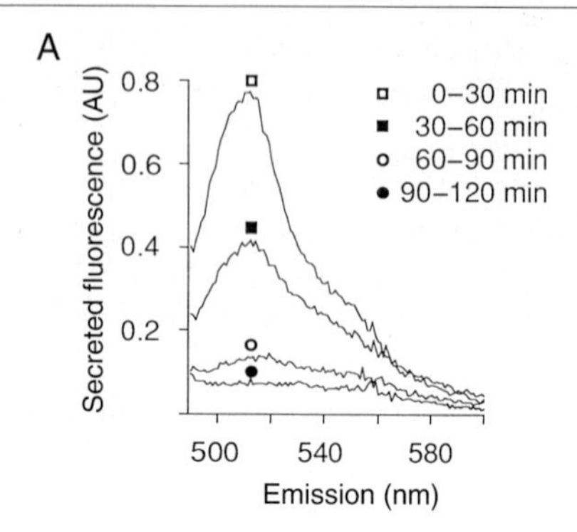

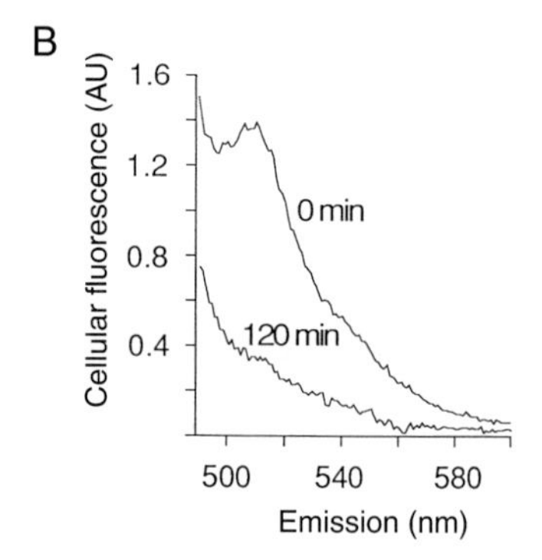

FIG. 2. Fluorometric analysis of hCgB–GFP(S65T) secretion. (A) Difference spectra of external solutions collected during release at 37° as indicated. (B) Difference spectra of cell lysates after a 2-hr secretion block (0 min) or after 2 hr of release at 37° (120 min). AU, Arbitrary unit.

of SGs by fluorescence microscopy. Therefore, one must first determine the sorting efficiency of the GFP fusion protein at the level of the TGN. In our opinion, the most appropriate way to address this is by pulse–chase labeling in combination with gradient centrifugation. This method permits the separation of CVs from newly formed secretory granules on a gradient. A detailed protocol for this type of analysis has been described.[13] By using this protocol we have been able to show that hCgB–EGFP (another GFP mutant; see the next section) exits from the TGN specifically into SGs, thus representing a highly specific fluorescent probe for SGs.[14] With this probe movement of SGs, their docking and subsequent exocytosis can be

[13] S. A. Tooze and W. B. Huttner, *Methods Enzymol.* **219,** 81 (1992).

[14] C. Kaether, T. Salm, M. Glombik, W. Almers, and H.-H. Gerdes, *Eur. J. Cell Biol.* **74,** 133 (1997).

addressed at the single-cell level. The fact that GFP elicits bright fluorescence in SGs demonstrates its applicability as an *in vivo* reporter under extreme conditions, that is, when it is aggregated and exposed to a pH 5.0 to 5.5.

Experimental Procedure

To fluorescently label SGs we use another GFP mutant, EGFP (Clontech Laboratories), fused to the C terminus of hCgB (hCgB–EGFP).[14] Because of the more efficient and temperature-insensitive folding of EGFP a detectable amount of green fluorescence is generated without a 20° secretion block. We now also use hCgB–GFP(S65T) in cells with a regulated pathway of secretion in combination with a 20° secretion block.[14a]

As a regulated cell line we use PC12 cells, a rat pheochromocytoma cell line (clone 251)[15]; these cells are grown at 37° and 10% CO_2 in Dulbecco's modified Eagle's medium (DMEM) supplemented with 10% (v/v) horse serum and 5% (v/v) fetal calf serum. PC12 cells are prepared for electroporation in the same manner as Vero cells, except that electroporation is performed at a density of 2–3 $\times$ 10^7 cells/ml and at 250 V and 960 μF. Stable cell lines are generated as described for Vero cells. Expression of hCgB–EGFP can be enhanced by incubation in 10 m*M* sodium butyrate for 17 hr.

Living cell recordings are done by confocal microscopy as described for constitutive secretion. To examine the motility of SGs near the PM we have taken confocal sections at the bottom of the cells, where the PM is attached to the coverslip.[14]

Comment. An elegant way of visualizing GFP-labeled SGs in the immediate vicinity of the PM has been demonstrated by Lang *et al.*, using evanescent-wave microscopy.[16]

Acknowledgments

We thank D. Corbeil, J. Garwood, and R. Rudolf for critical comments on the manuscript. C. Kaether was supported by the Graduiertenkolleg Molekulare und Zelluläre Neurobiologie. H.-H. Gerdes is a recipient of a grant from the Deutsche Forschungsgemeinschaft (SFB 317/C7).

[14a] T. Salm, R. Rudolf, and H.-H. Gerdes, manuscript in preparation (1998).

[15] R. Heumann, V. Kachel, and H. Thoenen, *Exp. Cell Res.* **145,** 179 (1983).

[16] T. Lang, I. Wacker, J. Steyer, C. Kaether, I. Wunderlich, T. Soldati, H.-H. Gerdes, and W. Almers, *Neuron* **18,** 857 (1997).

[3] Green Fluorescent Protein to Visualize Cancer Progression and Metastasis

By ROBERT M. HOFFMAN

Chinese hamster ovary cells and the human lung adenocarcinoma cell lines ANIP 973 and H-460 were transfected with the dicistronic expression vector containing the humanized green fluorescent protein (GFP) cDNA. Stable GFP-expressing clones were selected in 1.5 μM methotrexate *in vitro* and injected subcutaneously in nude mice. Stable, high-level expression of GFP was maintained in the subcutaneously growing tumors. To utilize GFP expression for metastasis studies, fragments of subcutaneously growing tumor, which were composed of GFP-expressing cells, were implanted by surgical orthotopic implantation (SOI) in the ovary and lung, respectively, of nude mice. Subsequent micrometastases were visualized in systemic organs by GFP fluorescence in the lung, liver, brain, skeleton, and other organs down to the single-cell level. With this fluorescence tool, we detected and visualized for the first time tumor cells at the microscopic level in fresh viable tissue in their normal host organ. The results with the GFP-transfected tumor cells, combined with the use of SOI, demonstrate a fundamental advance in the visualization and study of lung cancer metastasis in process. Lung tissue seeded with GFP-expressing ANIP 973 human lung carcinoma cells was incubated in three-dimensional sponge-gel matrix-supported histoculture. Tumor progression was continuously visualized by GFP fluorescence in the same individual cultures over a 52-day period, during which time the tumors spread throughout the histocultured lung. Histoculture tumor colonization was selective for the growth of lung cancer cells on lung tissue, as no growth occurred on histocultured mouse liver tissue, as also observed *in vivo*. The ability to support selective organ colonization in histoculture and visualize tumor progression by GFP fluorescence allows the *in vitro* study of tumor progression *in situ*.

Introduction

Our understanding of the cancer metastatic process has advanced considerably. However, analysis of the early stages of tumor progression and micrometastasis formation has been difficult and has been hampered by the inability to identify small numbers of tumor cells against a background of many host cells. The visualization of tumor cell emboli and micrometas-

0076-6879/99 $30.00

tases, and their progression over real time during the course of the disease, have been difficult to study in current models of metastasis. Previous studies used transfection of tumor cells with the *Escherichia coli* β-galactosidase (*lacZ*) gene to detect micrometastases.[1,2] However, detection of LacZ protein requires extensive histological preparation, and therefore it is impossible to detect and visualize tumor cells in viable fresh tissue or the live animal at the microscopic level. The visualization of tumor invasion and micrometastasis formation in viable fresh tissue or the live animal is necessary for a critical understanding of tumor progression and its control.

To enhance the resolution of the visualization of micrometastases in fresh tissue, we have utilized the green fluorescent protein (GFP) gene, cloned from the bioluminescent jellyfish *Aequorea victoria*.[3] GFP has demonstrated its potential for use as a marker of gene expression in a variety of cell types.[4,5] The GFP cDNA encodes a 283-amino acid polypeptide with a molecular weight of 27,000.[6,7] The monomeric GFP requires no other *Aequorea* proteins, substrates, or cofactors to fluoresce.[8] GFP gene gain-of-function mutants have been generated by various techniques.[9–12] For example, the GFP-S65T clone has the Ser-65 codon substituted with a threonine codon, which results in a single excitation peak at 490 nm.[9] Moreover, to develop higher expression in human and other mammalian cells, a humanized hGFP-S65T clone was isolated.[13] The much brighter fluorescence in the mutant clones allows for easy detection of GFP expression in transfected cells. We have demonstrated the isolation of stable

[1] W. C. Lin, T. P. Pretlow, T. G. Pretlow, and L. A. Culp, *Cancer Res.* **50,** 2808 (1990).
[2] W. C. Lin and L. A. Culp, *Invasion Metastasis* **12,** 197 (1992).
[3] J. Morin and J. Hastings, *J. Cell Physiol.* **77,** 313 (1972).
[4] M. Chalfie, Y. Tu, G. Euskirchen, W. W. Ward, and D. C. Prasher, *Science* **263,** 802 (1994).
[5] L. Cheng, J. Fu, A. Tsukamoto, and R. G. Hawley, *Nature Biotechnol.* **14,** 606 (1996).
[6] D. C. Prasher, V. K. Eckenrode, W. W. Ward, F. G. Prendergast, and M. J. Cormier, *Gene* **111,** 229 (1992).
[7] F. Yang, L. G. Miss, and G. N. Phillips, Jr., *Nature Biotechnol.* **14,** 1252 (1996).
[8] C. W. Cody, D. C. Prasher, V. M. Welstler, F. G. Prendergast, and W. W. Ward, *Biochemistry* **32,** 1212 (1993).
[9] R. Heim, A. B. Cubitt, and R. Y. Tsien, *Nature (London)* **373,** 663 (1995).
[10] S. Delagrave, R. E. Hawtin, C. M. Silva, M. M. Yang, and D. C. Youvan, *Bio/Technology* **13,** 151 (1995).
[11] B. Cormack, R. Valdivia, and S. Falkow, *Gene* **173,** 33 (1996).
[12] A. Cramer, E. A. Whitehorn, E. Tate, and W. P. C. Stemmer, *Nature Biotechnol.* **14,** 315 (1996).
[13] S. Zolotukhin, M. Potter, W. W. Hauswirth, J. Guy, and N. Muzycka, *J. Virol.* **70,** 4646 (1996).

transfectants of human lung adenocarcinoma cells (ANIP 973 and H-460) and Chinese hamster ovary (CHO) cells that express high-level GFP fluorescence *in vitro*.[14–17] The transfectants are highly fluorescent *in vivo* in tumors formed from the cells. Using these fluorescent transfectants, orthotopic transplant animal models[18–20] were utilized for visualizing the metastatic processes in fresh tissue down to the single-cell level, an achievement heretofore not possible.

We also developed a rapid, nondestructive means for examining tumor colonization by GFP-expressing ANIP 973 lung cancer cells of histocultured[21–27] normal lung tissue, such that continuous measurements could be made over a long time period. Metastatic host organ colonization was followed by GFP visualization of the tumor cells over a 52-day period in histoculture.[28]

These developments with GFP-expressing tumor cells provide an invaluable new tool for understanding the most important steps in tumor host–organ interaction, tumor progression, and metastasis.

[14] T. Chishima, Y. Miyagi, X. Wang, H. Yamaoka, H. Shimada, A. R. Moossa, and R. M. Hoffman, *Cancer Res.* **57,** 2042 (1997).

[15] T. Chishima, Y. Miyagi, X. Wang, M. Yang, Y. Tan, H. Shimada, A. R. Moossa, and R. M. Hoffman, *Clin. Exp. Metastasis* **15,** 547 (1997).

[16] T. Chishima, Y. Miyagi, X. Wang, Y. Tan, H. Shimada, A. R. Moossa, and R. M. Hoffman, *Anticancer Res.* **17,** 2377 (1997).

[17] T. Chishima, Y. Miyagi, L. Li, Y. Tan, E. Baranov, M. Yang, H. Shimada, A. R. Moossa, and R. M. Hoffman, *In Vitro Cell. Dev. Biol.* **33,** 745 (1997).

[18] M. Yang, S. Hasegawa, P. Jiang, X. Wang, Y. Tan, T. Chishima, H. Shimada, A. R. Moossa, and R. M. Hoffman, *Cancer Res.* **58,** 4217 (1998).

[19] P. Astoul, H. G. Colt, X. Wang, and R. M. Hoffman, *Anticancer Res.* **14,** 85 (1994).

[20] X. Wang, X. Fu, and R. M. Hoffman, *Int. J. Cancer* **51,** 992 (1992).

[21] A. Freeman and R. M. Hoffman, *Proc. Natl. Acad. Sci. U.S.A.* **83,** 2694 (1986).

[22] R. A. Vescio, C. H. Redfern, T. J. Nelson, S. Ugoretz, P. H. Stern, and R. M. Hoffman, *Proc. Natl. Acad. Sci. U.S.A.* **84,** 5029 (1987).

[23] R. M. Hoffman, A. Z. Monosov, K. M. Connors, H. Herrera, and J. H. Price, *Proc. Natl. Acad. Sci. U.S.A.* **86,** 2013 (1989).

[24] R. M. Hoffman, *Stem Cells* **11,** 105 (1993).

[25] R. M. Hoffman, *Crit. Rev. Oncol./Hematol.* **15,** 99 (1993).

[26] T. Furukawa, T. Kubota, and R. M. Hoffman, *Clin. Cancer Res.* **1,** 305 (1995).

[27] T. Kubota, N. Sasano, O. Abe, I. Nakao, E. Kawamura, T. Saito, M. Endo, K. Kimura, H. Demura, H. Sasano, H. Nagura, N. Ogawa, and R. M. Hoffman, *Clin. Cancer Res.* **1,** 1537 (1995).

[28] T. Chishima, M. Yang, Y. Miyagi, L. Li, Y. Tan, E. Baranov, H. Shimada, A. R. Moossa, S. Penman, and R. M. Hoffman, *Proc. Natl. Acad. Sci. U.S.A.* **94,** 11573 (1997).

Materials and Methods

DNA Manipulations and Expression Vector Constructions

The dicistronic expression vector (pED-mtx^r) is obtained from the Genetics Institute (Cambridge, MA).[29] The expression vectors containing the codon-optimized hGFP-S65T gene is purchased from Clontech Laboratories (Palo Alto, CA). To construct the hGFP-S65T-containing expression vector, phGFP-S65T is digested with *Hind*III, blunted at the end. The entire hGFP-coding region is then excised with *Xba*I. The pED-mtx^r vector is digested with *Pst*I, blunted at the end, and further digested with *Xba*I. The hGFP-S65T cDNA fragment is then unidirectionally subcloned into pED-mtx^r.

Cell Culture, Transfection, Selection

CHO-K1 cells are cultured in Dulbecco's modified Eagle's medium (DMEM; GIBCO, Grand Island, NY) containing 10% fetal calf serum (FCS) (Gemini Bio-products, Calabasas, CA), 2 m*M* L-gulutamine, and 100 μM nonessential amino acids (Irvine Scientific, Santa Ana, CA). ANIP 973 and H-460 cells are cultured in RPMI 1640 (GIBCO) containing 10% FCS (Gemini Bio-products), 2 m*M* L-glutamine, and 100 μM nonessential amino acid (Irvine Scientific).

For transfection, near-confluent CHO-K1 or ANIP cells are incubated with a precipitated mixture of LipofectAMINE reagent (GIBCO), and saturating amounts of plasmids, for 6 hr before being replenished with fresh medium. H-460 is transduced with GFP-containing vector pLEIN (Clontech Laboratories Inc., Palo Alto, CA) which also contains the neomycin resistance gene.[18]

CHO-K1 and ANIP cells are harvested by trypsin–EDTA 48 hr post-transfection, and subcultured at a ratio of 1 : 15 into selective medium that contains 1.5 μM methotrexate (MTX). Cells with stably integrated plasmids are selected by growing transiently transfected cells in the MTX-containing medium. H-460 is selected in 800 μg/ml G418 (Life Technologies, Grand Island, NY). Clones are isolated with cloning cylinders (Bel-Art Products, Pequannock, NJ) by trypsin–EDTA. They are amplified and transferred by conventional culture methods. CHO Clone-38, or ANIP Clone-26 and H-460-GFP were chosen because of their high-intensity GFP fluorescence and stability.

Doubling Time of Stable Green Fluorescent Protein Clones

CHO-K1 parental cells and Clone-38 cells are seeded at 2.0×10^5 in 60-mm culture dishes. The cells are harvested and counted every 24 hr,

[29] R. J. Kaufman, M. V. Davies, L. C. Wasley, and D. Michnick, *Nucleic Acids Res.* **19,** 4485 (1991).

using a hemocytometer (Reichert Scientific Instruments, Buffalo, NY). The doubling time is calculated from the cell growth curve over 6 days.

Subcutaneous Tumor Growth

Three 6-week-old BALB/c *nu/nu* female mice are injected subcutaneously with a single dose of 10^7 CHO Clone-38 cells, ANIP 973 Clone-26 cells or H-460 cells. Cells are first harvested by trypsinization and washed three times with cold serum-containing medium, and then kept on ice. Cells are injected in a total volume of 0.4 ml within 40 min of harvesting. The nude mice are sacrificed to harvest the tumor fragments 3 weeks after tumor cell injection.

Surgical Orthotopic Implantation of CHO-K1 Clone-38 in Nude Mice

Tumor fragments (1 mm^3) derived from the nude mouse subcutaneous CHO-K1 Clone-38 tumor are implanted by surgical orthotopic implantation (SOI) on the ovarian serosa in six nude mice[30] as follows. The mice are anesthetized by isofluran inhalation. An incision is made through the left lower abdominal pararectal line and peritoneum. The left ovary is exposed and part of the serosal membrane is scraped with a forceps. Four 1-mm^3 tumor pieces are fixed on the scraped site of the serosal surface with an 8-0 nylon suture (Look, Norwell, MA). The ovary is then returned into the peritoneal cavity, and the abdominal wall and the skin are closed with 6-0 silk sutures. Four weeks later, the mice are sacrificed and the lungs and the other organs are removed.

Surgical Orthotopic Implantation of ANIP 973 and H-460 Human Lung Cancer in Nude Mice

Tumor pieces (1 mm^3) derived from the nude mouse subcutaneous tumor formed from ANIP Clone-26 or H-460 are implanted by SOI onto the left visceral pleura in four nude mice[18–20] as follows. The mice are anesthetized by isofluran inhalation. A small 1-cm transverse incision is made on the left-lateral chest of the nude mice via the fourth intercostal space. A small incision provides access to the pleural space, and results in total lung collapse. Five tumor pieces are sewn together with an 8-0 nylon (Look) surgical suture and fixed by making one knot. The lung is taken up by forceps and the tumor sewn into the lower part of the lung with one suture. The lung tissue is then returned into the chest cavity, and the chest muscles and skin are closed with a single layer of 6-0 silk sutures. The lung is reinflated by withdrawing air from the chest cavity with a 23-gauge needle.

[30] X. Fu and R. M. Hoffman, *Anticancer Res.* **13,** 283 (1993).

Analysis of Metastases

Mice are sacrificed 4 weeks after implantation and the systemic organs are removed. The fresh specimens are sliced with a blade at approximately 1-mm thickness and observed directly by fluorescence microscopy. Frozen sections of samples are also examined histologically for fluorescence. The slides are then rinsed with phosphate-buffered saline (PBS) and fixed for 10 min at 4° in 2% formaldehyde plus 0.2% glutaraldehyde in PBS. The slides are washed with PBS and stained with hematoxylin and eosin, using standard techniques.

Stability of Green Fluorescent Protein Expression

The subcutaneous CHO-K1 Clone-38 tumors from the nude mice are minced for *in vitro* culture. Cells are subcloned in cell culture medium in the absence of MTX. Parental Clone-38 cells (10^7) maintained in 1.5 μM MTX, and Clone-38 cells from a nude mouse tumor maintained in the absence of MTX, are harvested. Cell extracts are prepared by lysis in 0.1% IGEPAL CA-630 (Sigma, St. Louis, MO) with 1 m*M* EDTA in PBS. The cell extracts are diluted 1:10 with PBS. GFP fluorescence is measured with a fluorescence photometer [Hitachi (Tokyo, Japan) F-2000; excitation, 490 nm; emission, 515 nm].

Microscopy

Light and fluorescence microscopy are carried out using a Nikon (Tokyo, Japan) microscope equipped with a xenon lamp power supply and a GFP filter set (Chromatechnology, Brattleboro, VT). The confocal microscopy system is an MRC-600 confocal imaging system (Bio-Rad, Hercules, CA) mounted on a Nikon microscope with an argon laser.

Intravenous Injection

Eight nude mice are injected in the tail vein with a single dose of 1×10^7 ANIP 973 Clone-26 cells. Cells are first harvested by trypsinization and washed three times with cold serum-containing medium, and then kept on ice. Cells are injected in a total volume of 0.8 ml of serum-free medium within 40 min of harvesting. CHO-K1 Clone-38 cells (5×10^6) are injected into a nude mouse through the tail vein. After 2 min, the mouse is sacrificed and fresh visceral organs are analyzed by fluorescence microscopy.

Tumor Progression in Histoculture

Whole lung tissues seeded with ANIP Clone-26 cells by intravenous injection are aseptically removed from the nude mice. The lung tissues are divided into pieces ~2–3 mm in diameter, which are then placed on prehydrated collagen sponge-gels (Upjohn, Kalamazoo, MI). The gels are floated in 24-well plates at the air–water interface in RPMI 1640 containing 20% FCS (Gemini Bio-products), 2 m*M* L-glutamine, and penicillin. The histocultures are incubated at 37° in a humidified atmosphere containing 95% air and 5% CO_2. The lung tumor colony growth in the histocultured host lung tissue is repeatedly observed in the same cultures with fluorescence photomicroscopy of the GFP expression on days 6, 14, 24, and 52 of histoculture.

Results

Isolation of Stable High Level-Expression Green Fluorescent Protein Transfectants

The expression vector-transfected CHO-K1 and ANIP cells were able to grow in levels of MTX up to 1.5 and 50 n*M*, respectively. The transduced H-460 cells were able to grow in 800 μg/ml. The selected MTX-resistant CHO and ANIP cells had a striking increase in GFP fluorescence compared with the transiently transfected cells. Subclone CHO-K1 GFP 38 (Clone-38) proved to be stable in 1.5 μ*M* MTX, possibly owing to stable chromosomal integration of the amplified GFP genes.[14] There was no difference in the cell proliferation rates of parental cells and selected transfectants, determined by comparing their doubling times. A subclone of ANIP 973 that expressed the strongest GFP was isolated (Fig. 1, see color insert) and called ANIP 973-hGFP-S65T-Clone-26 (Clone 26).[15]

Stable High-Level Expression of Green Fluorescent Protein in Tumors in Nude Mice

Three weeks after injection of CHO-K1 Clone-38 cells, the mice were sacrificed. All mice had a subcutaneous tumor that ranged in diameter from 13.0 to 18.5 mm (mean diameter, 15.2 ± 2.9 mm). The tumor tissue was strongly fluorescent, thereby demonstrating stable high-level GFP expression *in vivo* during tumor growth. GFP was extracted from Clone-38 cells and cells subcloned from the tumor formed from Clone-38 cells. The extraction experiments showed that GFP expression of the transfectants did not decrease *in vivo* even in the absence of MTX, as determined by fluorescence spectrometry.[14] A >1-cm tumor was formed 5 weeks after inoculation of 1×10^7 ANIP Clone-26 or H-460-GFP cells on the flank of a nude mouse. This tumor fluoresced brightly *in vivo*.[15]

Green Fluorescent Protein-Expressing Macro- and Micrometastases in Nude Mice

Six nude mice were implanted with 1-mm^3 cubes of CHO-K1 Clone-38 tumor into the ovary and were sacrificed at 4 weeks.[14] All mice had tumors in the ovaries. The tumor had also seeded throughout the peritoneal cavity, including the colon, cecum, small intestine, spleen, and peritoneal wall. The primary tumor and peritoneal metastases were strongly fluorescent. Numerous micrometastases were detected by fluorescence on the lungs of all mice. Multiple micrometastases were also detected by fluorescence on the liver, kidney, contralateral ovary, adrenal gland, paraaortic lymph node, and pleural membrane at the single-cell level. Single-cell micrometastases could not be detected by standard histological techniques. Even these multiple-cell small colonies were difficult to detect by hematoxylin and eosin staining, but they could be detected and visualized clearly by GFP fluorescence. Some colonies were observed by confocal microscopy. As these colonies developed, the density of tumor cells was markedly decreased in the center of the colonies.

Patterns of Lung Tumor Metastases after Surgical Orthotopic Implantation Visualized by Green Fluorescent Protein Expression

Primary tumors grew in the operated left lungs of all mice after SOI of GFP-transfected ANIP Clone-26 or H-460-GFP. GFP expression allowed visualization of the advancing margin of the tumor spreading in the ipsilateral lung. All animals explored had evidence of chest wall invasion and local and regional spread. Metastatic contralateral tumors involved the mediastinum, contralateral pleural cavity, and the contralateral visceral pleura, occurring in all four mice. Whereas the ipsilateral tumor has a continuous and advancing margin, the contralateral tumor seems to have been formed by multiple seeding events (Fig. 2, see color insert). These observations were made possible by GFP fluorescence of the fresh tumor tissue.[15,16,18] When non-GFP-transfected ANIP cells were compared with GFP-transformed ANIP cells for metastatic capability, similar results were seen.[15] Contralateral hilar lymph nodes were also involved, as well as cervical lymph nodes, shown by GFP expression. A cervical lymph node metastasis was brightly visualized by GFP in fresh tissue.[15,16] H-460-GFP also metastasized widely in the skeleton.

Green Fluorescent Protein-Expressing Metastases after Intravenous Injection in Nude Mice

CHO-K1 Clone-38 transfectants injected via the tail vein were detected and visualized in the peritoneal wall vessels to the single-cell level (Fig. 3, see color insert).[14] These cells formed emboli in the capillaries of the lung, liver, kidney, spleen, ovary, adrenal gland, thyroid gland, and brain.

ANIP Clone-26 cells (1.0×10^7) were injected into the tail vein of nude mice, and the mice were sacrificed at 4 and 8 weeks. In both groups, numerous micrometastatic colonies were detected in the whole lung tissue by GFP expression.[16] Even 8 weeks after injection, most of the colonies were not obviously further developed compared with those of mice sacrificed at 4 weeks.[16] Numerous small colonies, which ranged in number down to less than 10 cells, were detected at the lung surface in both groups. Brain metastases ranging in diameter from 7.8 to 22.5 mm were visualized in mice after 4 weeks; after 8 weeks the metastases ranged from 9.7 to 585.5 mm in diameter.[16] After 8 weeks, a mouse had systemic metastases in the brain (Fig. 4, see color insert), the submandibular gland, the whole lung, the pancreas, the bilateral adrenal glands, the peritoneum, and the pulmonary hilum lymph nodes.[16] All metastases were detected by GFP expression in fresh tissue.

Host Lung Colony Growth by Green Fluorescent Protein-Transfected Lung Tumor Cells in Histoculture

ANIP Clone-26-seeded mouse lungs were removed from the mice and then histocultured on collagen sponge-gels. Tumor colonies grew and spread rapidly in the lung tissue over time in histoculture. The progressive colonization of normal lung tissue by the lung tumor cells in individual cultures was visualized at multiple time points.[17] After 6 days in histoculture, the tumor colonies were still classifiable as microcolonies. However, by day 14, extensive growth of the colonies had occurred, with three different areas of GFP-labeled malignant cells visible. By day 24 of histoculture, the tumor colonies had grown significantly, reaching sizes of 750 μm in diameter and involving approximately one-half of the histocultured mouse lung (Fig. 5, see color insert). By 52 days of histoculture, tumor cells had involved the lung even more extensively and appeared to form multiple layers and histologically suggestive structures on the histocultured lung. Also by day 52, GFP-expressing satellite tumor colonies formed in the sponge-gel distant from the primary colonies in the lung tissue.[17]

Specificity of Host Tissue for Successful Colonization

To determine the specificity of tumor host tissue colonization, GFP-expressing ANIP 973 tissue fragments were placed in small wells made in mouse lung tissue and in liver tissue that had already been placed in histoculture. The results were striking in their contrast. When the lung

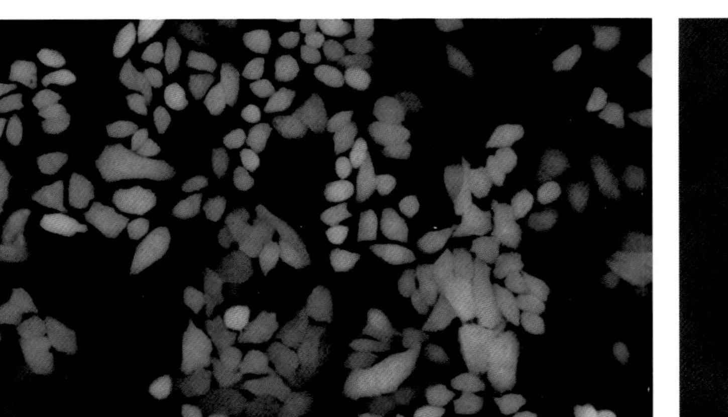

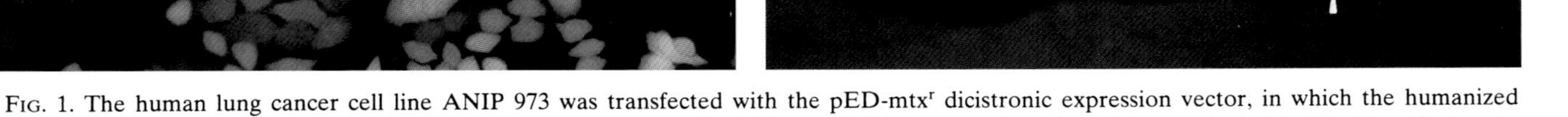

FIG. 1. The human lung cancer cell line ANIP 973 was transfected with the pED-mtx^r dicistronic expression vector, in which the humanized green fluorescent protein (GFP)-S65T and DHFR genes were transcribed. The stable high-expression Clone-26 was selected in 50 n*M* methotrexate (MTX) *in vitro*. (Original magnification: ×200.)

FIG. 2. Metastasis of GFP-expressing human lung cancer cell line ANIP 973 to the contralateral right lung in surgically orthotopically implanted nude mice. GFP-expressing ANIP 973 tumor tissue was surgically orthotopically implanted in the left lung of nude mice. At sacrifice, both large metastatic colonies and small metastatic seeds were observed in the contralateral lung by GFP expression (arrowheads).

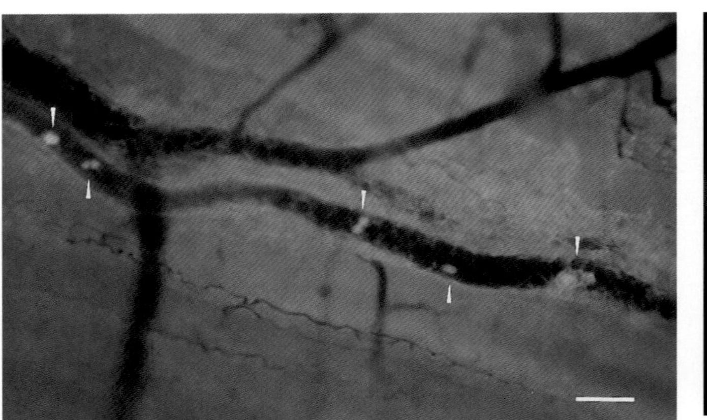

FIG. 3. GFP-expressing CHO cells in veins and capillaries. To study the limit of detection of GFP transfectants *in vivo*, a nude mouse was sacrificed 2 min after tail vein injection of CHO Clone-38 cells. The fresh organ tissues removed from the mouse were observed directly by fluorescence microscopy with no treatment. Arrowheads mark GFP-expressing Clone-38 cells in a peritoneal vessel. Bar: 100 μm.

FIG. 4. Brain metastases in a mouse injected in the tail vein with GFP-expressing ANIP human lung cancer cells. Colonies ranged from 48.5 to 818.2 μm. (Original magnification: $\times 8$.)

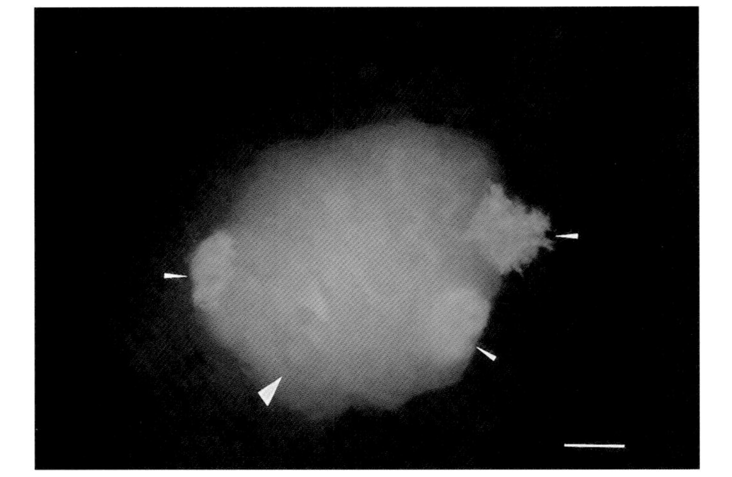

FIG. 5. Mouse lung colonization by GFP human lung tumor cells in histoculture. A parallel, 24-day histocultured lung had brightly fluorescing tumor colonies that invaded the supporting sponge-gel matrix as well as the lung itself.

tumor fragments were put on the mouse lung, extensive growth and GFP fluorescence were seen on day 29. However, when the lung tumor fragments were put on the liver tissue, no tumor could be detected. Even by day 29, only a dark hole was apparent in the well of the liver where the tumor had been implanted.[17]

Discussion

We have demonstrated the effectiveness and sensitivity of the GFP gene as a marker with which to visualize micrometastases in live tissue. To use GFP as a marker for *in vivo* experiments, it was necessary to establish stable transfectants that can express GFP long term under nonselective conditions. Previous studies have shown that retroviral transfer on the GFP gene can result in stable transfectants of human cancer cells *in vitro*.[31] For the studies described in this article, a dicistronic GFP plasmid was transfected into CHO-K1 and ANIP 973 cells by liposome-mediated transfection (lipofection). The pED-mtx^r vector[29] utilizes a putative internal ribosomal entry site (IRES) derived from the encephalomyocarditis virus (ECMV). Insertion of the GFP gene upstream of the IRES to produce a dicistronic mRNA did not reduce the translation of the dihydrofolate reductase (DHFR) gene, which was inserted downstream of it. The transfectants had to amplify not only DHFR but also GFP during MTX selection because of the dicistronic structure. With this plasmid, we could isolate stable, high-level GFP-expressing CHO and ANIP 973 subclones.[14–17] Furthermore, the stable GFP transfectants CHO-K1 Clone-38 and ANIP Clone-26 were also stable *in vivo*, even in the absence of MTX. The pLEIN vector with a dicistronic GFP and neo gene was found effective for lung cancer H-460.[18]

To study tumor progression and the metastatic process at the earliest stages, it was important to use a metastatic model that could allow spontaneous metastasis in the animal. It has become clear that orthotopic sites of implantation are critical for the metastatic expression of transplanted tumors in nude mice. We have found that the use of tissue fragments rather than dispersed cells allows full expression of the metastatic capability of surgically orthotopically implanted (SOI) tumors.[32] GFP gene-transfected CHO-K1 Clone-38 cells were successfully used to visualize extensive peritoneal seeding, as well as distant macro- and micrometastases (including the

[31] J. P. Levy, R. R. Muldoon, S. Zolotukhin, and C. J. Link, Jr., *Nature Biotechnol.* **14,** 610 (1996).

[32] R. M. Hoffman, *J. Cell. Biochem.* **56,** 1 (1994).

lung), in nude mice following the SOI in the ovary of subcutaneous tumor fragments. We utilized SOI model and GFP-expressing ANIP Clone-26 and H-460-GFP tumor to visualize human lung tumor growth and metastases to the mediastinum and contralateral lung, as well as to lymph nodes and for H-460-GFP to the skeleton.[18] This is an important clinical event and can now be visualized much more realistically by GFP in fresh tissue. These models provided the opportunity to demonstrate the capability of GFP for detection and visualization of metastases. Standard fluorescence microscopy allows high-resolution visualization of the GFP-expressing tumor.

In previous studies, the *lacZ* gene was transfected into mammalian cells to detect micrometastasis.[1,2] However, detection of the LacZ protein requires extensive histologic preparation and results in a high background due to endogenous β-galactosidase activity in certain cells.[33] In contrast, GFP fluorescence does not need any preparation and can be seen in fresh tissues without interference from endogenous GFP.

Our findings show that screening for micrometastases can be done easily and quickly in all systemic organs. We have observed in these studies that as a metastatic lesion progresses and grows, the packing density of the colonies decreases. This finding suggests that development of the interstitial tissue, including angiogenesis, is an important trigger for the expansion of metastatic colonies.[34] These are among the major advantages of using GFP rather than other reporter genes, such as *lacZ*, to monitor metastasis.

For an *in vivo* study of tumor progression at the single-cell level, the GFP transfectants were injected into a nude mouse intravenously. GFP expression allowed visualization at the single-cell level of tumor emboli or seeding by fluorescence in the fresh tissues and blood vessels.[14]

Chambers *et al.*[35,36] utilized intravital videomicroscopy with fluorescence-tagged cells and directly observed the fate of tumor cells after intravenous injection. They demonstrated that more than 80% of injected cells survive and extravasated by 24 hr. Their results suggested that the most important process of metastasis was the migration and growth of colonies at a distant site after extravasation. However, tumor cells tagged with dyes decrease their fluorescence after a few generations and are, therefore, not useful for the long periods of time needed to study metastasis.

A previous study by Margolis and co-workers also described the seeding

[33] L. Zdenek, *Histochemie* **23,** 289 (1970).

[34] L. Holmgren, M. S. O'Reilly, and J. Folkman, *Nature Med.* **1,** 149 (1995).

[35] A. F. Chambers, I. C. MacDonald, E. E. Schmidt, S. Koop, V. L. Morris, R. Khokha, and A. C. Groom, *Cancer Metastasis Rev.* **14,** 279 (1995).

[36] S. Koop, I. C. MacDonald, K. Luzzi, E. E. Schmidt, V. L. Morris, M. Grattan, R. Khokha, A. Chambers, and A. C. Groom, *Cancer Res.* **55,** 2520 (1995).

and migration of mouse tumor cells externally labeled with the fluorescent dye DiI C18 in host mouse lung tissue in three-dimensional histoculture.[37] However, tumor cells tagged with dyes lose their fluorescence after only a few generations and are, therefore, not useful for the long periods of time needed to study metastasis. In contrast, GFP transfectants can be followed in the primary and target organs, since the fluorescence gene has been integrated and is passed on to subsequent generations.

In the present studies, we visualized actively colonizing as well as dormant tumor cells in the lung. Many tumor cells in the lung have remained as small, but live, colonies more than 8 weeks after intravenous injection.[16] Dormant micrometastasis is one of the most important steps to understand in tumor progression.[34] The mechanism of this important phenomenon has been studied with regard to angiogenesis and other chemical regulators of tumor colonization.[34] However, these experimental models did not allow direct observation of the dormant colonies in fresh live tissue as it occurs over time, as do the present studies.

We have also demonstrated the replication and visualization of selective organ colonization and colony growth *in vitro* using histoculture of host organs and a GFP-transfected human lung adenocarcinoma.[17,28] It has previously proved difficult to study the mechanisms of metastasis and tumor progression, because micrometastases and colonization usually develop in internal organs *in vivo* and can be visualized by current imaging procedures only at relatively late stages. Current *in vivo* imaging procedures do not permit continual direct observation of the ongoing process of tumor colonization and progression over the time course of the disease. This histoculture system[38,39] can be visualized continuously over very long time periods of 1 to 2 months or more by GFP expression of the tumor cells colonizing a host organ, which may be the governing step of cancer metastasis.[40] The tumor host-organ chimeric histoculture system we have developed with GFP fluorescing tumor cells can significantly advance our ability to understand and treat human metastatic cancer.

Using the methods described in this chapter, GFP fluorescence will facilitate our understanding of tumor growth and progression, including seeding and target-organ colonization, which should provide new insights into metastatic mechanisms and treatment of metastatic disease.

[37] L. B. Margolis, S. E. Glushakova, B. A. Baibakov, C. Collin, and J. Zimmerberg, *In Vitro Cell. Dev. Biol.* **31,** 2211 (1995).

[38] J. Leighton, *Cancer Res.* **17,** 929 (1957).

[39] R. M. Hoffman, *Cancer Cells* **3,** 86 (1991).

[40] T. Kuo, T. Kubota, M. Watanbe, T. Furukawa, T. Teramoto, K. Ishibiki, M. Kitajima, A. R. Moossa, S. Penman, and R. M. Hoffman, *Proc. Natl. Acad. Sci. U.S.A.* **92,** 12085 (1995).

[4] Comparison of Enhanced Green Fluorescent Protein and Its Destabilized Form as Transcription Reporters

By XIAONING ZHAO, TOMMY DUONG, CHIAO-CHIAN HUANG, STEVEN R. KAIN, and XIANQIANG LI

Introduction

Studies of signal transduction and gene expression, as well as of the regulation of these processes, often rely on an *in vivo* reporter system. The key element in such a system is the reporter gene that is used to measure the change in these biological processes. Commonly, this reporter gene is linked to the regulatory sequence of interest, such as an inducible promoter or enhancer. Assay of the reporter gene product provides a quantitative measure of the level of gene expression, and thus cis-acting sequences and trans-acting factors can be characterized.

Currently, several reporter genes are commonly used, such as β-galactosidase,[1] luciferase,[2] and secreted alkaline phosphatase (SEAP).[3] The activities of such reporters are detected by measuring the light absorption or emission of the enzymatic products. These reporter enzymes have several advantages including high sensitivity, accuracy, and consistency. However, their assays require the preloading of substrate or the preparation of cell lysate. Therefore, they are not ideal for applications that require handling of a large number of samples, such as high-throughput screening. Also, for some reporters, the cost of substrates and other assay reagents can be quite high. Alternative reporter genes, ideally, would combine the sensitivity of the enzymatic analysis with a convenient assay involving minimal handling and, in particular, no preparation of cell lysate.

In this chapter, we describe the utility of green fluorescent protein (GFP) as a reporter gene in the study of gene expression. GFP emits fluorescent light on excitation, without the addition of substrate or cofactor, both *in vivo* and *in vitro*.[4,5] The GFP fluorescence activity can be detected using a fluorescence microscope, fluorometer, fluorescence-activated cell sorting (FACS) machine, or imaging microplate reader. Enhanced GFP (EGFP) is a GFP mutant with brighter fluorescence that makes the detec-

[1] G. An, K. Hidaka, and L. Siminovitch, *Mol. Cell. Biol.* **2,** 1628 (1982).
[2] S. J. Gould and S. Subramani, *Anal. Biochem.* **7,** 5 (1988).
[3] B. R. Cullen and M. H. Malim, *Methods Enzymol.* **216,** 362 (1992).
[4] M. Chalfie, *Photochem. Photobiol.* **60,** 651 (1995).
[5] J. Marshall, R. Molloy, G. Moss, J. R. Howe, and T. E. Hughes, *Neuron* **14,** 211 (1995).

0076-6879/99 $30.00

tion much more sensitive.[6] Use of GFP as a reporter gene offers a number of advantages. These include real-time analysis, minimal sample handing, the possibility of large-quantity analysis, and high sensitivity. However, GFP is a stable protein, so that it is easily accumulated when expressed in cells. The accumulation makes the inducible expression of the reporter insensitive to any change in induction and thus it would be difficult to use in kinetics studies. In [36] of this volume[6a] we describe the generation of a destabilized EGFP (dEGFP) by fusing the degradation domain of mouse ornithine decarboxylase to EGFP. The fusion protein dEGFP, without any significant change in its fluorescence properties, has a short half-life of 2 hr. In this chapter, we test the utility of EGFP and dEGFP as transcription reporters by fusing them with NF-κB-binding sequence and thymicline kinase (TK) promoter, and comparing the difference in expression between EGFP and dEGFP. We demonstrate that both EGFP and dEGFP can be used as reporters in transcription studies. We also show that dEGFP is more sensitive in response to changes in tumor necrosis factor (TNF) treatment owing to its faster turnover rate.

Principle of Methods

In this chapter, we describe the utility of EGFP and dEGFP as reporters to monitor TNF-mediated NF-κB activation. NF-κB is an inducible trans-activator of a large array of genes encoding cytokines, chemokines, other transcription factors, and receptors essential to the immune response.[7,8] NF-κB exists in an inactive form in the cytoplasm of most mammalian cells, and is activated in response to a variety of inducers, such as TNF treatment. On activation, NG-κB is rapidly released from its inhibitor, IκB, translocates to the nucleus, binds to specific recognition sequences in DNA, and induces transcription of the genes downstream of the NF-κB response element.[9] We put the reporter gene EGFP or dEGFP under the regulatory control of four copies of the NF-κB response element and TK promoter. After transfection into cells, the reporter gene expression was induced by TNF-α treatment. The expression level of EGFP or dEGFP was quantitated by flow cytometry.

[6] T. T. Yang, L. Cheng, and S. R. Kain, *Nucleic Acids Res.* **24,** 4592 (1996).

[6a] X. Zhao, X. Jiang, C.-C. Huang, S. R. Kain, and X. Li, *Methods Enzymol.* **302,** [36], 1999 (this volume).

[7] A. S. Baldwin, *Annu. Rev. Immunol.* **14,** 649 (1996).

[8] P. A. Baeuerle and D. Baltimore, *Cell* **87,** 13 (1996).

[9] G. Peltz, *Curr. Opin. Biotechnol.* **8,** 467 (1997).

Materials

TNF-α (Clontech, Palo Alto, CA): Dissolve in phosphate-buffered saline (PBS) to make a stock of (concentration, 0.2 mg/ml); keep at 4°
CalPhos Maximizer transfection kit (Clontech): For use in DNA transfection
Luciferase assay system (Promega, Madison, WI)

Procedure

Time Course of Induction

1. 293 cells are seeded sparsely in six-well plates 1 day before transfection.
2. The culture medium is replaced with fresh medium 1 hr before transfection.
3. The CalPhos transfection kit from Clontech is used for transfection. A master transfection mixture is made according to the manufacturer manual. One microgram of purified plasmid DNA is used for each well of the transfection. Two hundred microliters of the master transfection mixture is added to each well.
4. After 4 hr of incubation at 37°, the transfection mixture is removed and replaced with fresh medium after rinsing once with PBS.
5. TNF-α is added to a final concentration of 0.1 μg/ml in 2-hr intervals.
6. Before collection, cells are washed once with PBS. For the luciferase assay, cell lysates are prepared according to the manual of the provider (Promega). For flow cytometry, cells are treated with 1 ml of 2 m*M* EDTA in PBS, centrifuged at 10,000 rpm for 1 min, and resuspended in 0.5 ml of PBS.

Four copies of NF-κB-binding sequences are inserted in front of the simian virus 40 (SV40) early promoter of the plasmid pSEAP2-Promoter vector (Clontech) at the *Nhe*I and *Bgl*II sites. The TK minimal promoter is used to replace the SV40 early promoter. The SEAP gene of the plasmid is then substituted by cDNA encoding EGFP, dEGFP, or luciferase, resulting in pNF-κB-EGFP, pNF-κB-dEGFP, and pNF-κB-luciferase, respectively (Fig. 1). The reporter systems are used to monitor NF-κB activation induced by TNF-α.

On addition of TNF-α, the expression of the reporter genes is induced. EGFP- or dEGFP-transfected cells are analyzed by flow cytometry. As shown in Fig. 2, increased fluorescence intensity is observed in TNF-α-treated cells as a function of induction time. In both EGFP- and dEGFP-transfected cells, the total mean fluorescence of cells increases following

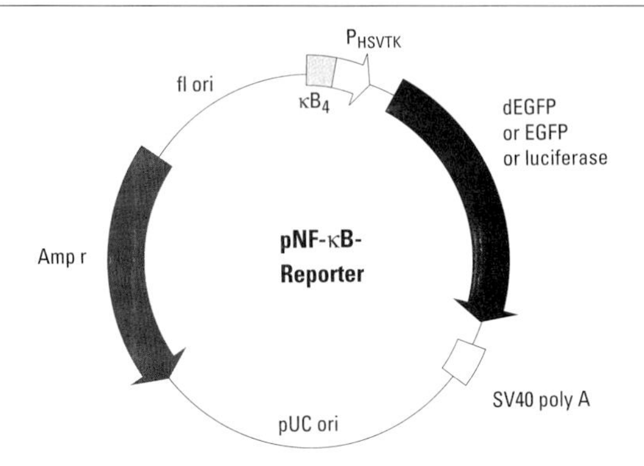

FIG. 1. Schematic map of luciferase, EGFP, and dEGPF reporter plasmid.

the addition of TNF-α. However, they show different kinetic responses to TNF-α induction (Fig. 2A). In EGFP-transfected cells, a constant increase in EGFP expression is observed throughout the 8-hr period. In contrast, dEGFP-transfected cells show a rapid increase in fluorescence activity for the first several hours and a plateau after 6 hr of induction. After 8 hr, a slight decrease in dEGFP expression occurs. This kinetic response is more comparable to that of the conventional reporter gene luciferase (Fig. 2B), although it reaches maximal induction 2 hr later than luciferase. The delayed response could be due to the posttranslational folding of dEGFP, which requires more time and is necessary for its active fluorescence. These results demonstrate that dEGFP, with a faster degradation rate and less accumulation, responds to TNF-α more quickly than does the stable protein EGFP. Therefore, it could be a better reporter in kinetics studies of gene expression and regulation.

Dose Response to Tumor Necrosis Factor α

1. 293 cells are seeded into a six-well plate 1 day before transfection.
2. One microgram of each construct is transfected to each well of the six-well plate.
3. Twenty-four hours after transfection, TNF-α is added to each well at a series of concentrations.
4. Cells are treated for an additional 6 hr and collected for FACS analysis or luciferase assay.

As the concentration of TNF-α is elevated in the culture medium, the fluorescence activity of the dEGFP-transfected cells increases. However, no such dose-dependent response is observed when EGFP is used as the receptor (Fig. 3A). The stability of EGFP results in its accumulation even

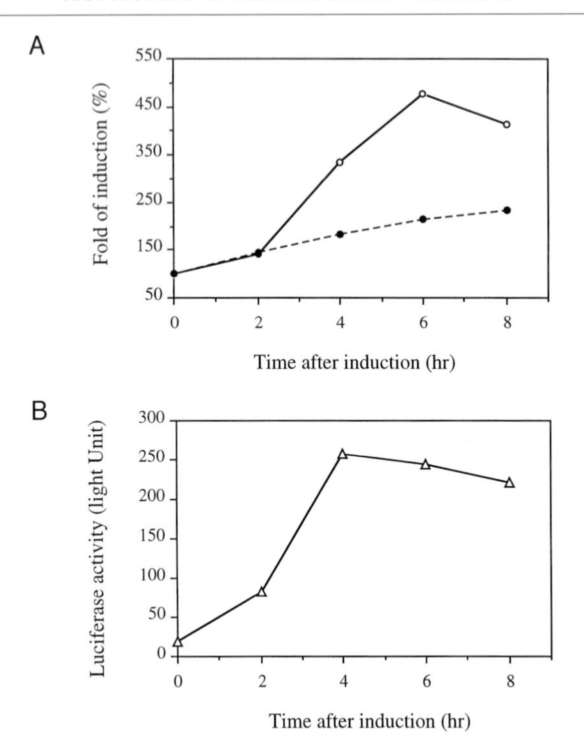

FIG. 2. Time course study of TNF-α induction. 293 cells were transiently transfected with pNF-κB-luc, pNF-κB-EGFP, or pNF-κB-dEGFP. Twenty-four hours after transfection, cells were treated with TNF-α (0.1 μg/ml, final concentration) for various times, and then collected for FACS analysis or luciferase assay. (A) EGFP and dEGFP as reporters in time course study. After TNF-α induction, cells were subjected to FACS analysis, and mean fluorescence intensity was measured. Data are represented by normalizing the uninduced cells to 100%. (○) dEGFP as reporter; (●) EGFP as reporter. (B) Luciferase as reporter in time course study.

under the influence of the TK minimal promoter before TNF-α induction. The higher basal level of the reporter, therefore, makes it insensitive to TNF-α-mediated induction. The dose-dependent response of dEGFP is more comparable to that of the conventional reporter gene luciferase (Fig. 3B), which shows maximum induction at a 0.1-μg/ml concentration of TNF-α.

Conclusion and Remarks

In this chapter, we have compared the utility of EGFP and dEGFP as reporters in a kinetics study of gene expression. By coupling the reporters to TNF-α-mediated NF-κB activation, we are able to monitor the activation

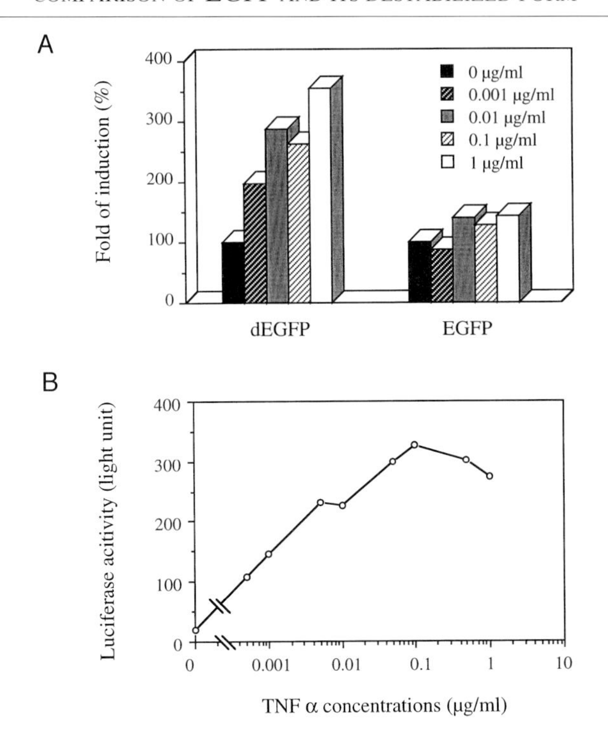

FIG 3. Dose-dependent induction of reporters by TNF-α. 293 cells were transiently transfected with pNF-κB-luc, pNF-κB-EGFP, or pNF-κB-dEGFP. Twenty-four hours after transfection, various concentrations of TNF-α were added to the culture. Six hours after addition of TNF-α, cells were collected for FACS analysis or luciferase assay. (A) Dose-dependent induction of EGFP and dEGFP. For EGFP and dEGFP, data are represented by normalizing the uninduced cells to 100%. (B) Dose-dependent induction of luciferase.

of NF-κB by measuring the changes in EGFP or dEGFP expression by flow cytometry. Furthermore, we have demonstrated that dEGFP is a better reporter than EGFP when used in a kinetics study. In any inducible expression system, expression of reporter genes includes both basal and regulated expression. If the reporter is a stable protein, such as EGFP, it will be easily accumulated even at the basal expression level. This accumulation leads to a narrowing of the range of induction. Hence, the induction becomes insensitive to the treatment of the inducer. In contrast, the accumulation of dEGFP is low, owing to its rapid turnover. Therefore, the induction range of dEGFP is greater than that of the stable protein and the kinetics of induction is faster. This rate determines the sensitivity of the reporter in the process of monitoring changes in treatment, as demonstrated in this study. In conclusion, this reporter system with dEGFP or EGFP can be used in both basic study and pharmaceutical screening. Because the detection is easy, of low cost, and reasonably sensitive, the system will be especially useful in high-throughput drug screening.

[5] Early Detection of Apoptosis with Annexin V–Enhanced Green Fluorescent Protein

By STEVEN R. KAIN and JING-TYAN MA

Introduction

Apoptosis is an important and well-regulated form of cell death observed under a variety of physiological and pathological conditions. Inappropriate apoptosis has been implicated in many disease processes such as Alzheimer's disease, autoimmune disorders, leukemia, lymphomas, and several other malignancies.[1,2] Mammalian cell apoptosis can be triggered by many stimuli, and proceeds by a regulated series of events beginning with activation of caspase proteases, and leading eventually to DNA fragmentation, rupture of the plasma membrane, and death.[3,4] A relatively early event in the apoptotic cascade is the acquisition of cell surface changes by apoptotic cells that result in the recognition and uptake of these cells by phagocytes.[5–7] This recognition is due to the externalization of a plasma membrane phospholipid called phosphatidylserine (PS), an event that occurs regardless of the stimulus used to trigger apoptosis.[7] Most of the PS in healthy cells is localized to the cytoplasmic leaflet of the plasma membrane. Within 1–2 hr after the induction of apoptosis, the PS redistributes to the outer leaflet of the plasma membrane. On the basis of the kinetics of this process, and the requirement for active transport of other membrane phospholipids, it is thought the PS is translocated by an energy-dependent "flipase."[7] By assaying for PS externalization one can detect apoptosis as it begins, providing the means to analyze the early events that regulate programmed cell death.

[1] E. Duvall and A. H. Wyllie, *Immunol. Today* **7,** 115 (1986).
[2] D. W. Nicholson, *Nature Biotechnol.* **14,** 297 (1996).
[3] J. J. Cohen, R. C. Duke, V. A. Fadok, *et al., Annu. Rev. Immunol.* **10,** 267 (1992).
[4] S. J. Martin and D. R. Green, *Crit. Rev. Oncol./Hematol.* **18,** 137 (1995).
[5] E. Duvall, A. H. Wyllie and R. G. Morris, *Immunology* **56,** 351 (1985).
[6] G. Koopman, C. P. M. Reutelingsperger, G. A. M. Kuijten, *et al., Blood* **84,** 1415 (1994).
[7] S. J. Martin, C. P. M. Reutelingsperger, A. J. McGahon, *et al., J. Exp. Med.* **192,** 1545 (1995).

METHODS IN ENZYMOLOGY, VOL. 302
0076-6879/99 $30.00

Enhanced Green Fluorescent Protein Variant of Green Fluorescent Protein

Several investigators have described a number of different "red-shifted" variants of wild-type green fluorescent protein (wt GFP), most of which contain one or more amino acid substitutions in the chromophore region of the protein. The red-shifted terminology refers to the position of the major fluorescence excitation peak, which is shifted for each of these variants toward the red, from 395 nm in wt GFP, to 488–505 nm. The emission spectra for such variants are largely unaffected, and these mutants still produce green light with a wavelength maximum of approximately 507–511 nm. The major excitation peak of the red-shifted variants encompasses the excitation wavelength of commonly used fluorescence filter sets, and thus the resulting signal is much brighter relative to wt GFP. Similarly, the argon ion laser used in most flow cytometers and confocal scanning laser microscopes emits at 488 nm, and thus excitation of such variants is more efficient than excitation of wt GFP. In practical terms, this means the detection limits are considerably lower with the red-shifted variants.

The two most commonly used red-shifted GFP variants are S65T[8–10] which contains a serine-to-threonine substitution at position 65 in the chromophore, and GFPmut1 or EGFP (enhanced GFP),[11,12] having the same S65T change plus a phenylalanine-to-leucine mutation at position 64. GFPmut1 and EGFP have identical amino acid sequences, but the EGFP coding sequence has been further modified with 190 silent base changes to contain codons preferentially found in highly expressed human proteins.[13] The "humanized" backbone used in EGFP contributes to efficient expression of this variant in mammalian cells and subsequently very bright fluorescent signals. On the basis of spectral analysis of equal amounts of soluble protein, EGFP fluoresces approximately 35-fold more intensely than wt GFP when excited at 488 nm,[11] owing to an increase in its extinction coefficient (Em). The Em for EGFP is ~55,000 cm^{-1} M^{-1} for 488 nm excitation,[14] compared with 7000 cm^{-1} M^{-1} for wt GFP measured under similar conditions. In addition to improved sensitivity, other advantages of the EGFP and S65T variants over wt GFP include (1) improved solubility,

[8] K. Brejc, T. K. Sixma, P. A. Kitts, *et al., Proc. Natl. Acad. Sci. U.S.A.* **94,** 2306 (1997).
[9] R. Heim, A. B. Cubitt, and R. Y. Tsien, *Nature (London)* **373,** 663 (1995).
[10] R. Heim and R. Y. Tsien, *Curr. Biol.* **6,** 178 (1996).
[11] B. P. Cormack, R. Valdivia, and S. Falkow, *Gene* **173,** 33 (1996).
[12] T. T. Yang, L. Cheng, and S. R. Kain, *Nucleic Acids Res.* **24,** 4592 (1996).
[13] J. Haas, E.-C. Park, and B. Seed, *Curr. Biol.* **6,** 315 (1996).
[14] G. H. Patterson, S. M. Knobel, W. D. Sharif, *et al., Biophys. J.* **73,** 2782 (1997).

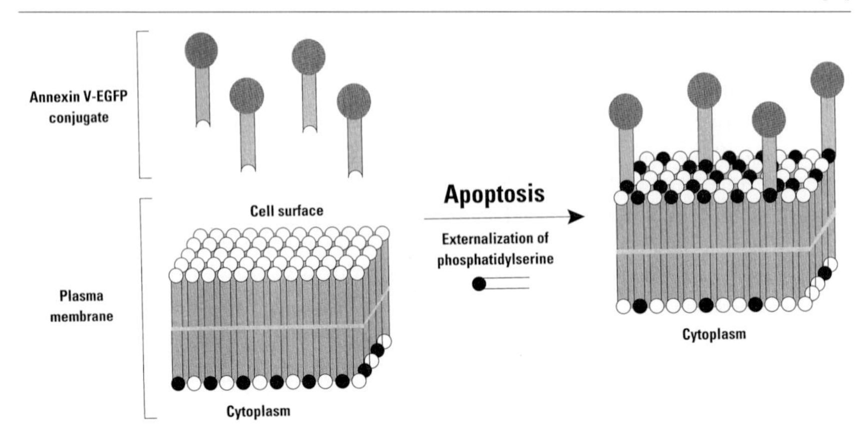

FIG. 1. The principle of annexin V–EGFP detection of apoptosis. In healthy cells, PS is predominantly localized to the inner leaflet of the plasma membrane. Following induction of the apoptotic cascade, PS is translocated to the outer leaflet. Annexin V–EGFP binds to PS exposed on the cell surface in a Ca^{2+}-dependent fashion, thereby making the cell fluorescent.

(2) more efficient protein folding, (3) faster chromophore oxidation to form the fluorescent form of protein, and (4) reduced rates of photobleaching.

Annexin V–Enhanced Green Fluorescent Protein

Annexin V is a calcium-dependent phospholipid-binding protein that preferentially binds to PS.[7,15,16] Therefore, by conjugation of this protein to chemical fluorophores such as fluorescein isothiocyanate (FITC), one can generate a useful probe for apoptotic cells that is compatible with both fluorescence microscopy and flow cytometry.[6,7,17] The principle of the annexin V assay is shown in Fig. 1. The assay is a simple one-step procedure that takes less than 10 min to perform. It is important to emphasize that the assay uses living cells, thereby allowing for recovery of the cells for analysis of later stages in apoptosis.

To improve sensitivity, and reduce complications for quantitative studies owing to photobleaching and loss of signal, which is characteristic of chemical fluorophores, we constructed a genetic fusion between annexin V and

[15] H. A. M. Andree, C. P. M. Reutelingsperger, R. Hauptmann, *et al., J. Biol. Chem.* **265,** 4923 (1990).

[16] P. Thiagarajan and J. F. Tait, *J. Biol. Chem.* **265,** 17420 (1990).

[17] G. Zhang, V. Gurtu, S. R. Kain, *et al., BioTechniques* **23,** 525 (1997).

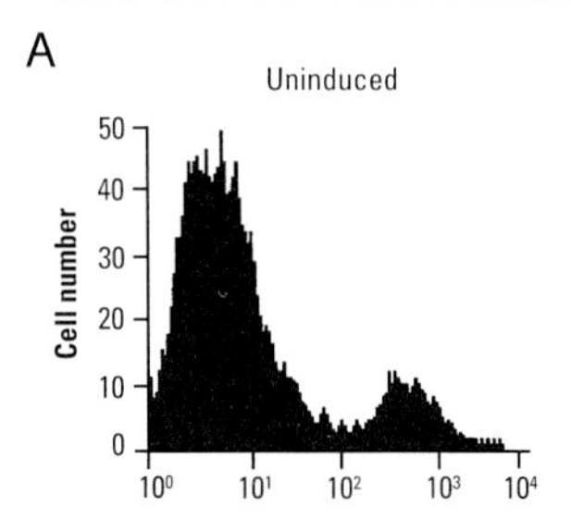

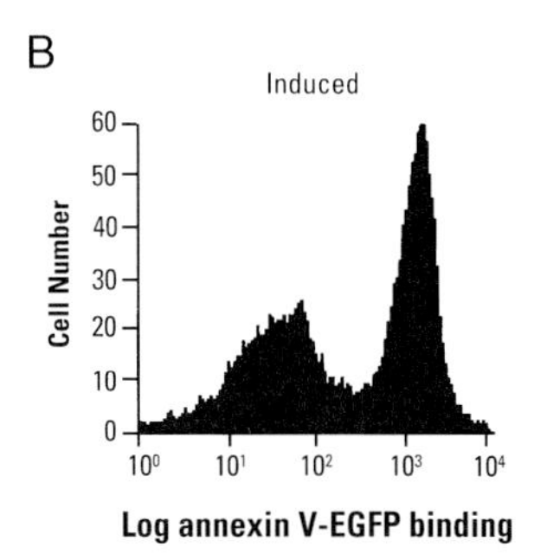

FIG 2. Detection of annexin V–EGFP binding by flow cytometry. Apoptosis was induced in Jurkat cells by incubation with a 200-ng/ml concentration of anti-Fas monoclonal antibody (clone CH-11) for 6 hr (B). Cells were collected by centrifugation, and incubated with annexin V–EGFP (0.5 μg/ml) for 10 min in the dark. (A) Uninduced control cells; (B) cells induced with anti-Fas antibody.

EGFP. As shown in Fig. 2, we observed that 65% of Jurkat cells treated with anti-Fas monoclonal antibody underwent PS externalization as reflected by binding of annexin V–EGFP (compare Fig. 2A and B) measured by using a flow cytometer. The sensitivity provided by annexin V–EGFP in flow cytometric analysis of apoptosis was approximately 20-fold greater than that provided by a similar fluorescent conjugate of annexin V using FITC (data not shown). These results suggest that annexin V–EGFP may provide adequate sensitivity for analysis of apoptosis in multiwell plates designed for high-throughput screening and apoptosis-related drug discovery.

Materials

Annexin V–EGFP (20 μg/ml in Tris-NaCl buffer; Clontech Laboratories, Palo Alto, CA)
Binding buffer, 1X (Clontech)

Propidium iodide (50 μg/ml in 1× binding buffer). *Note:* Propidium iodide is highly toxic and should be handled with extreme caution

Phosphate-buffered saline (PBS; 58 m*M* Na_2HPO_4, 17 m*M* NaH_2PO_4, 68 m*M* NaCl, pH 7.4

Formaldehyde solution, 2% (Sigma Chemical Co., St. Louis, MO)

Apoptosis Detection Protocol

Because apoptosis is a rapid and dynamic process, we recommend that the analysis be performed immediately after staining with annexin V–EGFP.

Incubation of Cells with Annexin V–Enhanced Green Fluorescent Protein

1. Induce apoptosis by the desired method.
2. Wash 1×10^5–10^6 cells with PBS by resuspension and centrifugation at 200 *g* for 10 min.
3. Resuspend the cell pellet in 200 μl of 1× binding buffer.
4. Add 5.0 μl of annexin V–EGFP to a final concentration of 0.5 μg/ml plus 5 μl of propidium iodide solution (50 μg/ml) (optional).
5. Incubate the cells at room temperature for 5–15 min in the dark.
6. Detect apoptotic cells by either flow cytometry or fluorescence microscopy. Apoptotic cells will be annexin V–EGFP positive (green), but propidium iodide negative (red).

Detection by Flow Cytometry

1. Analyze the cells by flow cytometry, using a single laser excitation at 488 nm. The laser conditions, as well as filter sets, are generally the same as those used to detect fluorescein. The signal generated by propidium iodide can be monitored by the detector reserved for phycoerythrin emission.
2. It is advisable to being by running mock cells not incubated with annexin V–EGFP to establish baselines and autofluorescence pattern. Place the negative baseline fluorescence within the first decade of the log scale.
3. Use forward angle light scatter versus propidium iodide to gate on the viable cells. Exclude debris, aggregates, and all propidium iodide-positive cells (dead or dying cells) from the rest of the analysis.
4. Plot gated viable cell fluorescence versus scatter.

Cytometers

There are a variety of commercial instruments that one can use for detection of apoptotic cells with bound annexin V–EGFP, or GFP expression in general. Published flow cytometry data have been reported for the following instruments from Becton Dickinson (Mountain View, CA): FACScan, FACStar^{Plus}, FACS Calibur, and FACS Vantage. There have also been reports on using the EPICS Elite-ESP cytometer from Coulter Corporation (Hialeah, FL). Most instruments used for detection of GFP and the red-shifted GFP variants are equipped with argon ion lasers tuned to 488 nm excitation.

Detection by Fluorescence Microscopy

1. Place the cell suspension from step 5 (see above, Incubation of Cells with Annexin V–EGFP) on a glass slide. Cover with a glass coverslip. For adherent cells, grow the cells directly on the coverslip. Following step 5, invert the coverslip onto a glass slide and visualize the cells. The cells can also be washed and fixed in 2% formaldehyde prior to visualization.
2. The cells must be incubated with annexin V–EGFP before fixation since cell membrane disruption may cause nonspecific binding of the annexin reagent to PS on the internal side of the plasma membrane.
3. Observe the cells using a fluorescence microscope equipped with an FITC filter set for annexin V–EGFP, and rhodamine filters for propidium iodide.
4. Apoptotic cells that have bound annexin V–EGFP will have a "green halo" by virtue of binding to the plasma membrane. Cells that have lost an intact plasma membrane (late-stage apoptosis or necrosis) will also show red staining throughout the cytoplasm owing to the uptake of propidium iodide.

[6] Green Fluorescent Protein in the Visualization of Particle Uptake and Fluid-Phase Endocytosis

By MARKUS MANIAK

Introduction

The green fluorescent protein (GFP) has been used to analyze a variety of steps in endocytosis, in particular to tag ligands, receptors, and components of the endocytic machinery involved in the formation of vesicles.

One obvious application is to replace chemically labeled synthetic particles used in uptake studies with GFP-labeled natural particles. An example is the expression of GFP in bacteria that are able to infect and survive

0076-6879/99 $30.00

within host cells.[1,2] This approach has been taken one step further by expressing GFP from bacterial promoters, which are regulated in response to the environment in the phagosome.[3] The strategy of these infectious bacteria is to prevent the transit from early to later endocytic compartments, where they would become degraded.[4] A number of endocytosed toxin molecules of bacterial or plant origin are resistant to the conditions encountered in endosomes and are then transported in retrograde direction through the Golgi apparatus into the endoplasmic reticulum.[5] These processes can now be studied in living cells using GFP-tagged ricin[6] as a marker for endocytosis.

Traffic through the endocytic pathway can also be followed by GFP fused to a variety of plasma membrane receptors. Once a receptor interacts with a ligand, it becomes internalized and translocated to endosomal compartments, where it can encounter two fates. The receptor becomes either degraded or recycled to the plasma membrane to undergo a new round of endocytosis. A receptor that is subject to degradation in lysosomes is the β_2-adrenergic receptor, whose trafficking has been studied by means of GFP fused to its C terminus.[7] On the other hand, GFP-tagged cholecystokinin receptor type A has been observed to unload its ligand in an endosomal compartment, from where the receptor recycles to the cell surface, while only its ligand is sorted to the lysosome for degradation.[8] In spite of the success in monitoring the localization and dynamics of GFP fusion proteins, one must take into account that the GFP moiety might interfere with a function residing in the terminus of the target protein that is modified by the fusion. One example is the glucose transporter GLUT4. A fusion of GFP to the C terminus of the protein reveals the expected behavior: insulin addition causes the insertion of GLUT4–GFP-containing vesicles into the plasma membrane; withdrawal of insulin results in retrieval of the transporter. A chimera in which GFP is fused to the N terminus of GLUT4 is correctly inserted into the plasma membrane but cannot be endocytosed

[1] L. Kremer, A. Baulard, J. Estaquier, O. Poulain-Godefroy, and C. Locht, *Mol. Microbiol.* **17,** 913 (1995).

[2] S. Dhandayuthapani, L. E. Via, C. A. Thomas, P. M. Horowitz, D. Deretic, and V. Deretic, *Mol. Microbiol.* **17,** 901 (1995).

[3] R. H. Valdivia and S. Falkow, *Mol. Microbiol* **22,** 367 (1996).

[4] L. E. Via, D. Deretic, R. J. Ulmer, N. S. Hibler, L. A. Huber, and V. Deretic, *J. Biol. Chem.* **272,** 13326 (1997).

[5] K. Sandvig and B. Van Deurs, *Physiol. Rev.* **76,** 949 (1996).

[6] E. Tagge, B. Harris, C. Burbage, P. Hall, J. Vesely, M. Willingham, and A Frankel, *Bioconjugate Chem.* **8,** 743 (1997).

[7] L. Kallal, A. W. Gagnon, R. B. Penn, and J. L. Benovic, *J. Biol. Chem.* **273,** 322 (1998).

[8] N. I. Tarasova, R. H. Stauber, J. K. Choi, E. A. Hudson, G. Czerwinski, J. L. Miller, G. N. Pavlakis, C. J. Michejda, and S. A. Wank, *J. Biol. Chem.* **272,** 14817 (1997).

in the absence of insulin, suggesting that the N terminus of GLUT4 is important for retrieval.[9]

Ligand binding to a cell surface receptor allows association of the receptor with components of the endocytic machinery. In the case of receptors internalized by clathrin-dependent endocytosis, the direct binding partners are adapter molecules. In human immunodeficiency virus (HIV)-infected cells, a GFP-tagged variant of the viral Nef protein is thought to connect adapter complexes to sites at the plasma membrane where CD4 is internalized.[10] To ultimately cut off a vesicle from the plasma membrane, the activity of the GTPase dynamin is required. Some dynamin isoforms have been tagged with GFP, and one isoform indeed localizes to the plasma membrane and to vesicles, but also to tubules of the endoplasmic reticulum.[11] The combined action of clathrin and dynamin, however, is not the only mechanism to produce endocytic vesicles at the plasma membrane. The main driving force to shape large, fluid-filled or particle-containing vesicles is the actin cytoskeleton. Many fusions of GFP to actin-binding proteins have been made in the yeast system. Of these, the fimbrin homolog[12] and verprolin fusions[13] are capable of complementing endocytosis defects in the corresponding mutants. The fusion proteins localize to so-called cortical patches, which on the ultrastructural level coincide with membrane invaginations,[14] and are therefore believed to represent sites of endocytosis. The small size of these structures and the fact that the yeast cell is surrounded with a thick cell wall have precluded the direct visualization of solute uptake at these sites *in vivo*.

Choice of Experimental System

A model system that combines several of the advantages of yeast with the structural features of a mammalian cell is *Dictyostelium discoideum*. Well known for its use in the study of cell type differentiation, it was among the first organisms in which GFP was functionally expressed to label individual cell types.[15,16] The other main area of *Dictyostelium* research

[9] S. P. Dobson, C. Livingstone, G. W. Gould, and J. M. Tavare, *FEBS Lett.* **393,** 179 (1996).

[10] M. E. Greenberg, S. Bronson, M. Lock, M. Neumann, G. N. Pavlakis, and J. Skowronski, *EMBO J.* **16,** 6964 (1997).

[11] Y. Yoon, K. R. Pitts, S. Dahan, and M. A. McNiven, *J. Cell Biol.* **140,** 779 (1998).

[12] T. Doyle and D. Botstein, *Proc. Natl. Acad. Sci. U.S.A.* **93,** 3886 (1996).

[13] G. Vaduva, N. C. Martin, and A. K. Hopper, *J. Cell Biol.* **139,** 1821 (1997).

[14] J. Mulholland, D. Preuss, A. Moon, A. Wong, D. Drubin, and D. Botstein, *J. Cell Biol.* **125,** 381 (1994).

[15] P. Fey, K. Compton, and E. C. Cox, *Gene* **165,** 127 (1995).

[16] S. Hodgkinson, *Trends Genet.* **11,** 327 (1995).

deals with cell motility, which is best studied in the aggregation stage of development. Nutrient-deprived cells that migrate toward a source of chemoattractant are elongated and polarized into a leading lamella and a retracting end. The rear end is prominently stained by GFP–myosin II[17] and dominated by contractile activity, whereas the leading edge is enriched in GFP–actin[18] and expanded by the polymerization of monomeric actin into filaments. One of the proteins associated with filamentous actin is coronin. Coronin tagged at its C terminus accumulates at sites beneath the cell surface where pseudopodia are induced after chemotactic stimulation.[19] These early studies have been repeated using a GFP-tagged actin-binding domain of ABP120,[20] and confirm our previous findings.[19] In the vegetative phase, *Dictyostelium* cells feed on soil bacteria by phagocytosis. Laboratory strains, in addition, are able to grow on liquid media by macropinocytosis. Thus, both particles and fluid can be used to study initial steps of uptake.[21,22] Changes in vesicular pH value allow the identification of early or late endosomal stages.[23–26]

Choice of Imaging System

Dictyostelium cells adhere to untreated glass surfaces. On top of 5 × 5 cm coverslips, a plastic ring of nearly the same diameter and 4 mm in height, previously dropped into molten paraffin, is positioned. When the wax hardens, a water-tight seal is formed. This chamber for use on inverted microscopes holds 1 ml of medium containing 2–3 × 10^6 cells. The chamber also allows the rapid change of medium or access with a microcapillary during microscopic observation. For prolonged recording, it is covered with another 5 × 5 cm coverslip to reduce evaporation. To avoid light-induced damage to the cells, the illumination with a mercury lamp on a conventional fluorescence microscope (Axiovert 135; Zeiss, Oberkochen, Germany) requires that the light intensity be reduced with gray filters to around 1% of the output of a 50-W lamp. The quality of digital images acquired with a

[17] S. L. Moores, J. H. Sabry, and J. A. Spudich, *Proc. Natl. Acad. Sci. U.S.A.* **93,** 443 (1996).
[18] M. Westphal, A. Jungbluth, M. Heidecker, B. Mühlbauer, C. Heizer, J. M. Schwartz, G. Marriott, and G. Gerisch, *Curr. Biol.* **7,** 176 (1997).
[19] G. Gerisch, R. Albrecht, C. Heizer, S. Hodgkinson, and M. Maniak, *Curr. Biol.* **5,** 1280 (1995).
[20] K. M. Pang, E. Lee, and D. A. Knecht, *Curr. Biol.* **8,** 405 (1998).
[21] M. Maniak, R. Rauchenberger, R. Albrecht, J. Murphy, and G. Gerisch, *Cell* **83,** 915 (1995).
[22] U. Hacker, R. Albrecht, and M. Maniak, *J. Cell Sci.* **110,** 105 (1997).
[23] L. Aubry, G. Klein, J. L. Martiel, and M. Satre, *J. Cell Sci.* **105,** 861 (1993).
[24] H. Padh, J. Ha, M. Lavasa, and T. L. Steck, *J. Biol. Chem.* **268,** 6742 (1993).
[25] R. Rauchenberger, U. Hacker, J. Murphy, J. Niewöhner, and M. Maniak, *Curr. Biol.* **7,** 215 (1997).
[26] N. Jenne, R. Rauchenberger, U. Hacker, T. Kast, and M. Maniak, *J. Cell Sci.* **111,** 61 (1998).

silicon intensified tube (SIT) camera (C2400-08; Hamamatsu, Herrsching, Germany) under these conditions[19] is, however, inferior to the images obtained with a confocal microscope (Zeiss LSM 410; Fig. 1, see color insert).[21,22] Confocal microscopes offer a number of advantages. First, the laser power can be conveniently attenuated. Second, time-control options of the scanning software allow its use as a shutter. Together, these two features reduce the exposition of the cells to damaging light. Third, it is possible to record images in one or more fluorescence channels simultaneously with a transmission image. Fourth, the confocal mode allows the imaging of a translucent cell in a bath of fluorescent medium (see below).

Choice of Endocytic Markers

Particles

Dictyostelium cells are capable of phagocytosing particles varying in size from bacteria to erythrocytes.[27] The yeast particle (*Saccharomyces cerevisiae* type II; Sigma, Deisenhofen, Germany) as a substrate for phagocytosis has several advantages: it is inexpensive, homogeneous in size, large enough to allow the distinction of several phases of phagocytosis over time, visible in phase contrast, and efficiently labeled with isothiocyanate derivatives of fluorochromes.[28] The obvious dye to be used in combination with GFP is tetramethylrhodamine isothiocyanate (TRITC). It is even possible to achieve a labeling intensity high enough to excite the yeast particle at a suboptimal wavelength with the argon ion laser (488 nm) used for GFP illumination[21] (Fig. 1A). Alternatively, the yeast cell wall can be noncovalently stained with Calcofluor (American Cyanamid, Wayne, NJ), an ultraviolet (UV)-excitable compound that is easily detected with 4,6-diamidino-2-phenyl-indol (DAPI) filter sets. Another useful property of the yeast cell wall is its spongelike binding of quenchers such as trypan blue, a vital dye that allows one to distinguish adherent from ingested particles.[28] This feature is the basis for a quantitative phagocytosis assay based on TRITC-labeled yeast particles,[21] which allows a comparison of qualitative, microscopic results with quantitative, spectrofluorometric data.

Fluid-Phase Tracers

For measuring pinocytosis, dextran of 70-kDa molecular mass has been the most commonly used fluid-phase marker. As discussed for the measure-

[27] G. Vogel, L. Thilo, H. Schwarz, and R. Steinhart, *J. Cell Biol.* **86,** 456 (1980).
[28] J. Hed, *Methods Enzymol.* **132,** 198 (1986).

ment of phagocytosis, one advantage is that a dextran-based quantitative assay has been established, which allows the correlation with microscopic data.[22] During passage through the endocytic pathway dextran is not degraded by *Dictyostelium*.[29] Various types of fluorescently labeled dextrans are commerically available. Care must be taken when fixable dextrans (Molecular Probes, Eugene, OR) are used, which are covalently derivatized with lysine in order to provide amino groups for aldehyde fixation. Possibly owing to the increased positive charge, these tracers tend to adsorb to the cell surface and therefore do not represent true fluid-phase markers.[30]

Again, for use with GFP fusion proteins, the most obvious fluorescence tag for dextran is TRITC (Sigma). Concentrations of 1 mg/ml or less require excitation with a 543-nm helium–neon laser (Fig. 1B), while TRITC–dextran well above 2 mg/ml is also excited with the suboptimal 488-nm argon ion laser (data not shown). Whereas TRITC–dextran serves as a tracer for the fluid phase in vesicles, fluorescein isothiocyanate (FITC)–dextran (Sigma) provides information about the pH value of endocytic compartments, because its fluorescence decreases in acidic environments. This property can be exploited by combining the pH-insensitive dye TRITC with the pH indicator FITC, by mixing the fluorescent dextrans in a 10-to-1 ratio. A color switch then allows the monitoring of endosomal acidification *in vivo*[26] (Fig. 1C). To distinguish the FITC fluorescence from the GFP fluorescence (both are excited by the 488-nm laser and emit at similar wavelengths) that is the subject of investigation, the TRITC fluorescence, that colocalizes with the FITC signal, is used as a contrasting color (Fig. 1D). After image analysis, this method allows extraction of the individual contribution of GFP from the image (Fig. 1E).

In principle, pH changes in endosomes can also be visualized with fluorescent dyes such as acridine orange[31] or the red-emitting dye neutral red.[25] The mode of their accumulation is different from that of labeled dextrans. These dyes can diffuse through membranes in their uncharged form and accumulate specifically in acidic compartments on protonation. Because the accumulation is diffusion limited, and because fluorescence detection requires a high signal-to-noise ratio, these dyes cannot be used for observing rapid pH changes within individual vesicles as in Fig. 1C and D. They may be used, however, for identifying the steady state pH value of cellular compartments decorated with a GFP fusion protein.[25,26]

[29] G. Klein and M. Satre, *Biochem. Biophys. Res. Commun.* **138,** 1146 (1986).

[30] M. Maniak, unpublished observation (1996).

[31] G. Buczynski, J. Bush, L. Zhang, J. Rodriguez-Paris, and J. Cardelli, *Mol. Biol. Cell* **8,** 1343 (1997).

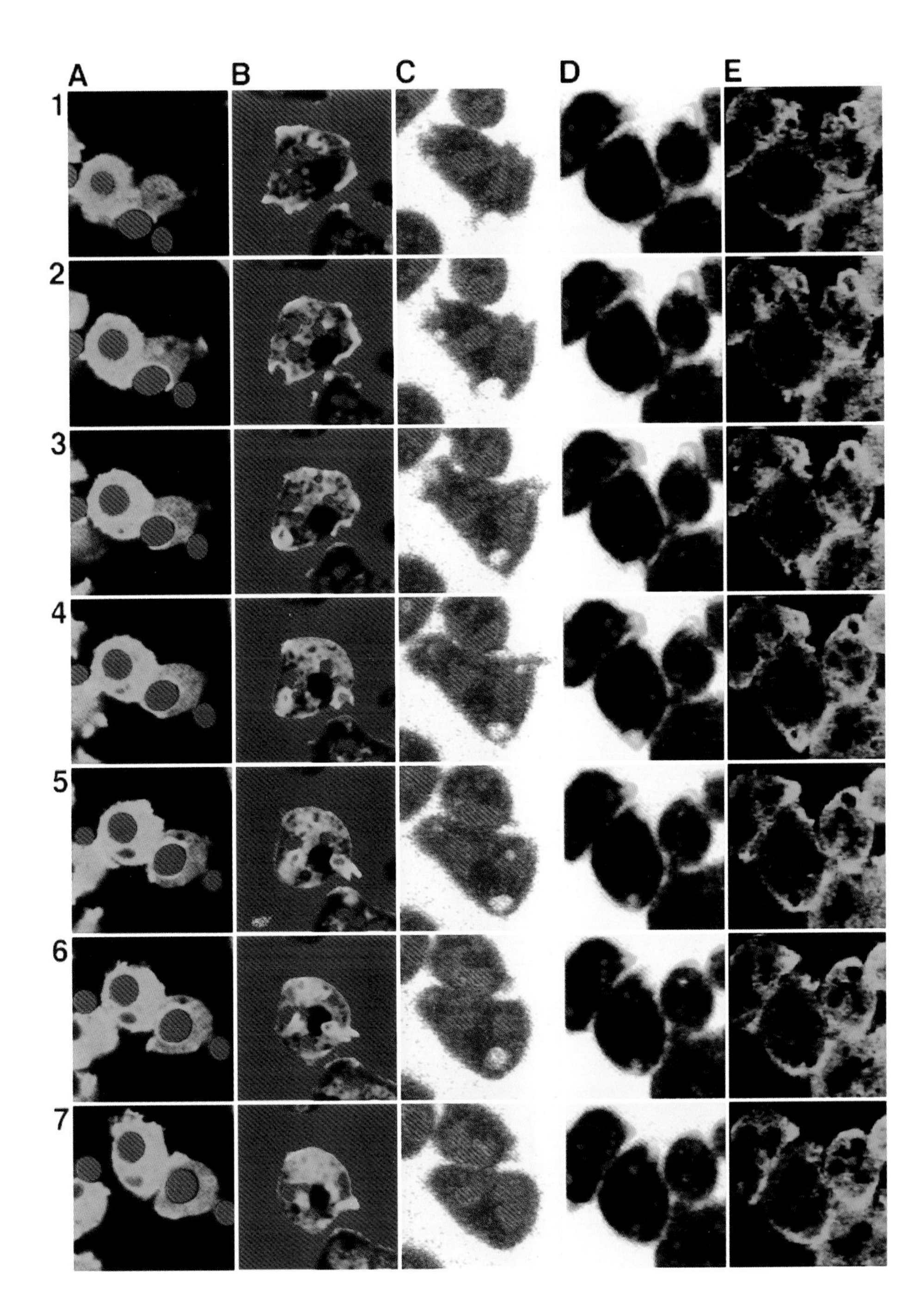

A
B
C
D
E
1
2
3
4
5
6
7

FIG. 1. Endocytic events shown as a time series of confocal sections through coronin–GFP-expressing cells at intervals of 15 sec (A, B, D, and E), or a wild-type cell every 10 sec (C). One frame is 20 μm wide. In (A) two cells compete for a TRITC-labeled yeast particle (red). Frame 1 shows the particle attached to the lower edge of the cells and coronin–GFP (green) accumulates at the site of contact. In frames 2 and 3 the right cell expands the contact zone around the particle, until in frame 4 the engulfment is complete. The particle remains within a coat of coronin–GFP, seen as a ring (frames 5 and 6), which dissociates from the phagosome and reintegrates into the cell cortex, where it initiates two new protrusions (frame 7). The left cell fails to maintain contact to the particle (frame 4), is hit by a cell with a bright green leading edge coming in from the left (frame 5), and begins to leave the scene toward the top of the image in frame 7. In (B) a cell shows two sequential events of macropinocytic uptake in nutrient medium containing TRITC–dextran (red). Initially coronin–GFP (green) is enriched at three sites of the cell cortex. In frame 2 the lower left edge of the cell develops a C-shaped protrusion, which closes in frame 3 to engulf an aliquot of the medium. The vesicle surrounded by the coronin–GFP coat detaches from the cell cortex (frame 4) and migrates toward the center of the cell (frame 5). In frame 6 the vesicle is released from its coat. The coat seems to adhere to the lower half of the vesicle, possibly explaining its almost triangular shape. Finally free of a cytoskeletal coat, the vesicle rounds up in frame 7. When the first act of macropinocytosis was completed, the same cell is seen to extend pseudopods at its lower right side (frames 3 to 4), which result in the formation of another coronin-coated macropinosome (frame 5), underscoring that macropinocytosis occurs at a high frequency in *Dictyostelium*. (C) illustrates the pH changes that accompany endocytic events in nutrient medium containing FITC–dextran (green) and TRITC–dextran (red). The regions of overlap appear as yellow. The cell protrudes pseudopods at its lower edge (frames 1 and 2), which fuse at their distal tips and release a fluid-filled vesicle into the cytoplasm (frame 3). The vesicle is easily identified by its size and yellow color, indicating that the lumenal pH value corresponds to the outside medium (frames 4 to 6). In frame 7 the shape of the vesicle becomes irregular again, suggesting that it has lost its cytoskeletal coat. Most importantly, the color change from yellow to red indicates that its lumen was rapidly (within less than 10 sec) acidified. The color change is brought about by a decrease in FITC–dextran fluorescence (green) in an acidic environment, where the fluorescence intensity of TRITC–dextran (red) is unaffected. (D) and (E) show an attempt to visualize the change in vesicular pH value [yellow and red in (D)] together with the behavior of the coronin–GFP fusion [green in (D) and (E)]. To obtain the GFP image in (E), the red channel containing the signal from TRITC–dextran was subtracted from the green channel containing the signals from both FITC–dextran and coronin–GFP. The two cells in the center of each frame display events of internalization and acidification [compare (D) and (E)]: the cell to the left protrudes pseudopods pointing downward in frames 1 to 3 and closes off the macropinosome in frame 4. The pH change occurs in frames 6 and 7. The right cell forms a macropinosome at its upward facing edge in frames 1 and 2. A ring of coronin–GFP persists around the vesicle in frames 3 to 4. Concomitant with the dissociation of coronin–GFP from the vesicle and integration of the cytoskeleton into the cortex (frames 5 and 6), the lumen of the vesicle becomes acidified. Further improvement in optical and temporal resolution will allow analysis in detail of the causal relationship between pH change and vesicle uncoating.

Choice of Green Fluorescent Protein Variant

The original GFP cDNA from *Aequorea victoria*[32] has been extensively mutated for two purposes: (1) to introduce silent mutations in order to optimize the codon bias for use in mammalian cells without changing the amino acid sequence, and (2) to change the amino acid sequence and select for spectral variants.

The wild-type gene corresponds well to the codon usage of *Dictyostelium*, in that position 3 of a codon is most frequently A or T. Adaptation of the codon usage to mammalian preferences by introducing mostly G or C in the wobble position of 88 codons[33] is predicted to decrease the suitability for *Dictyostelium* expression in 84 positions. Overall, the changes result in a drop of the preference index from about 51 to 12%. This is a considerable change in view of the fact that unbiased codon usage would result in a value of 30% (18 amino acids encoded by 59 redundant codons).

Mutants selected on the basis of spectral characteristics of GFP can be grouped into two classes. The first class includes mutants that are brighter because their folding at higher temperatures is improved.[34,35] These thermotolerant mutations are useful for mammalian cells, but may not provide a significant advantage for *Dictyostelium*, which grows at room temperature and is killed by a 34° heat shock.[36] The second class of mutations affects the GFP chromophore and therefore changes the spectral properties. In the classic mutation S65T[37,38] the UV-excitable peak of the wild-type molecule is shifted toward blue. This mutation improves the efficiency of excitation of GFP with the 488-nm argon ion laser. One adverse effect, however, needs to be mentioned. *Dictyostelium* cells cultivated in nutrient medium show a marked greenish fluorescence of internal vesicles if excited with blue light. In cells expressing a GFP fusion protein at low levels, this signal cannot be distinguished from autofluorescence. In cells excited with UV light the autofluorescence appears blue. If these cells express a wild-type GFP (as opposed to S65T), it will also be excited by UV light but emit in green, allowing a clear distinction between the respective signals. Therefore,

[32] D. C. Prasher, V. K. Eckenrode, W. W. Ward, F. G. Prendergast, and M. J. Cormier, *Gene* **111,** 229 (1992).

[33] S. Zolotukhin, M. Potter, W. W. Hauswirth, J. Guy, and N. Muzyczka, *J. Virol.* **70,** 4646 (1996).

[34] A. Crameri, E. A. Whitehorn, E. Tate, and W. P. C. Stemmer, *Nature Biotech.* **14,** 315 (1996).

[35] K. R. Siemering, R. Golbik, R. Sever, and J. Haseloff, *Curr. Biol.* **6,** 1653 (1996).

[36] W. F. Loomis and S. A. Wheeler, *Dev. Biol.* **79,** 399 (1980).

[37] R. Heim, D. C. Prasher, and R. Y. Tsien, *Proc. Natl. Acad. Sci. U.S.A.* **91,** 12501 (1994).

[38] S. Delagrave, R. E. Hawtin, C. M. Silva, M. M. Yang, and D. C. Youvan, *Bio/Technology* **13,** 151 (1995).

especially for the study of endocytosis, where GFP fusions are made to label vesicular compartments, using the wild-type chromophore may be advantageous.[30]

Outlook

Future experiments will make increased use of the simultaneous detection of spectral GFP variants in *Dictyostelium.* This method has been used for labeling different *Dictyostelium* cell types in different colors,[39] and will be valuable for tagging two proteins in a single cell. With this possibility in mind, there are a number of questions to be addressed that relate to the topic discussed here. (1) Is it possible to resolve the spatiotemporal sequence of the accumulation of different actin-binding proteins in a leading edge or endocytic structure? Indirect comparison suggests that coronin–GFP is already present when the lamella is protruded by actin polymerization,[19,21,22] while a protein that links polymerized actin into stable filament bundles, GFP–cofilin, is lagging a few seconds behind.[40] (2) Are the same pseudopods that accumulate coronin–GFP during protrusion rich in GFP–myosin II during retraction?[17] (3) Is endosome maturation accompanied by the switching of protein markers that decorate its cytoplasmic surface? Such a relationship has been inferred from the timing of endocytic trafficking using a fluid-phase tracer, which appears earlier in coronin–GFP-decorated vacuoles than in GFP–vacuolin-labeled compartments.[25]

Altogether, live cell imaging of GFP fusion proteins will enhance our understanding of the dynamics of the molecules involved in the intricate process of endocytosis.

Acknowledgments

I thank Ulrike Hacker, Nicole Jenne, and Robert Rauchenberger for contributions to the experimental work. Richard Albrecht, Jean-Marc Schwartz, and John Murphy are acknowledged for help with confocal microscopy and maintenance of the computer network. I am grateful to Günther Gerisch for reading the manuscript.

[39] T. Zimmermann and F. Siegert, *Biotechniques* **24,** 458 (1998).
[40] H. Aizawa, Y. Fukui, and I. Yahara, *J. Cell Sci.* **110,** 2333 (1997).

[7] Monitoring Intracellular Shuttling of Histidine-Rich pH Sensor Proteins Tagged with Green Fluorescent Protein

By FRANK HANAKAM and GÜNTHER GERISCH

Introduction

Control of cytoplasmic pH is necessary for a cell to optimize and regulate enzyme activity and protein interactions. Na^+/H^+ antiporters, anion exchangers, and proton pumps control the pH by regulating the exchange of protons between the cytoplasm and the external medium or specialized intracellular compartments. Modest changes in pH can arise from external stimulation and may act as a secondary message in signal transduction pathways.[1,2] A prerequisite for pH homeostasis in the cytoplasm, and for a role of pH changes in signal transduction, is the presence of pH sensors.

As switching devices in pH sensor proteins histidine residues play a unique role, because histidine is the only amino acid whose protonation is strongly altered within the physiological pH range of the cytoplasm.[3,4] In the context of a polypeptide chain the p*K* of histidine residues is on the order of 5.6 to 7.0.[5] A role for a single histidine residue has been demonstrated in the Na^+/H^+ antiporter A of *Escherichia coli*, where a shift in pH sensitivity is caused by converting His-226 into arginine.[2] In the AE2 anion exchanger of nucleated (nonerythroid) cells, a cluster of four histidine residues at the cytoplasmic domain plays a key role in pH sensitivity.[3]

This chapter deals with a group of small histidine-rich proteins, hisactophilins, whose pH-sensing activity is manifested in their translocation within the cells. Hisactophilins I and II associate under slightly acidic conditions with the plasma membrane and the nucleus. Under neutral or slightly alkaline conditions, hisactophilins dissociate from the membrane or from their nuclear-binding sites and accumulate in the cytoplasm. Using hisactophilin–green fluorescent protein (GFP) fusions and fluorescence scanning techniques with appropriate temporal and spatial resolution, pH changes can be monitored in real time within living cells. The description in this

[1] W. H. Moolenar, *Annu. Rev. Physiol.* **48,** 363 (1986).

[2] E. Padan and S. Schuldiner, *J. Bioenerg. Biomembr.* **25,** 647 (1993).

[3] I. Sekler, S. Kobayashi, and R. R. Kopito, *Cell* **86,** 929 (1996).

[4] E. Padan and S. Schuldiner, *Biochim. Biophys. Acta* **1187,** 206 (1994).

[5] H. B. Bull, "An Introduction to Physical Biochemistry," 2nd Ed., p. 94. Davis, Philadelphia 1971.

0076-6879/99 $30.00

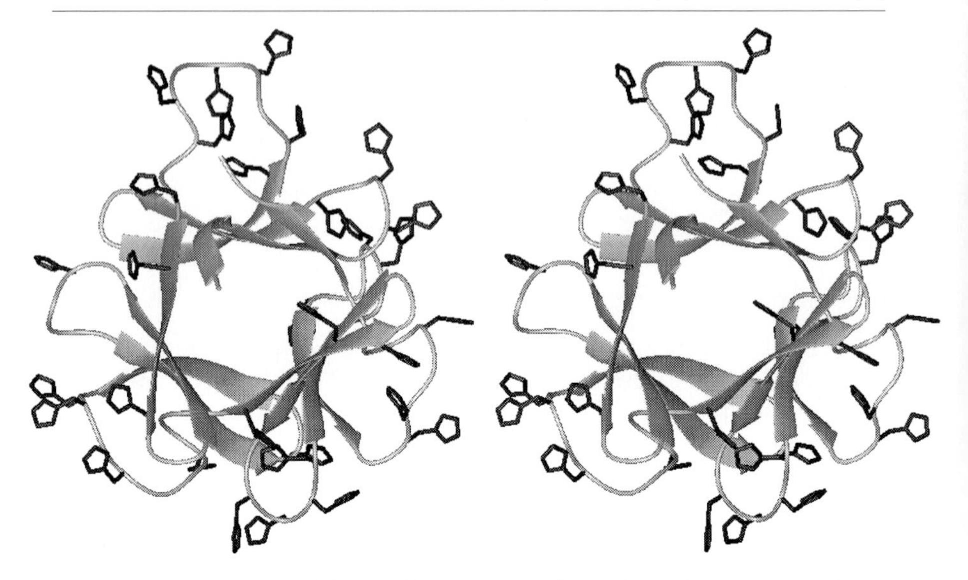

FIG. 1. NMR structure of hisactophilin I. The stereoimage shows the barrel formed by β sheets and the position of histidine residues on the outside of the molecule.[8] (Courtesy Tadeusz Holak, MPI für Biochemie, Martinsried.)

chapter is based on data obtained with cells of *Dictyostelium discoideum*, but should be easily adapted to the specific requirements of other cells.

Structure of Hisactophilins and Binding to Acidic Lipids *in Vitro*

Isoform I of hisactophilins (HsI) was originally isolated from the eukaryotic microorganism *D. discoideum* on the basis of its actin-binding activity, which proved to be strongly pH dependent.[6,7] This 13.5-kDa protein contains 31 histidine residues clustered along the polypeptide chain. Determination of the hisactophilin I structure by nuclear magnetic resonance (NMR) revealed a molecule with a three-fold axis of symmetry (Fig. 1). Antiparallel β strands form the backbone of the molecule, and the histidine clusters are exposed in flexible loops on its surface.[8] Hisactophilin II (HsII) differs only

[6] M. Schleicher, G. Gerisch, and G. Isenberg, *EMBO J.* **3,** 2095 (1984).

[7] J. Scheel, K. Ziegelbauer, T. Kupke, B. M. Humbel, A. A. Noegel, G. Gerisch, and M. Schleicher, *J. Biol. Chem.* **264,** 2832 (1989).

[8] J. Habazettl, D. Gondol, R. Wiltscheck, J. Otlewski, M. Schleicher, and T. A. Holak, *Nature* (*London*) **359,** 855 (1992).

slightly in its sequence from isoform I. It contains 35 histidine residues in its 13.7-kDa polypeptide chain. Both isoforms are myristoylated in *Dictyostelium* cells at their N-terminal glycine residues,[9] and thus differ in their lipid-binding properties from bacterially expressed hisactophilins.[10]

Studies of the binding of hisactophilins to lipid vesicles and monolayers have provided evidence *in vitro* of a pH switch mechanism in the interaction of hisactophilins with negatively charged membranes.[11,12] Because the cytosolic phase of the plasma membrane is negatively charged, these results are applicable to the interaction of hisactophilins *in vivo*.[13] The two isoforms of hisactophilins differ slightly in their calculated isoelectric points, which are p*I* 7.3 for HsI and p*I* 7.5 for HsII. Accordingly, slight differences in the binding of hisactophilins I and II to monolayers of negatively charged lipids have been observed.[11] At pH >7.0 the hisactophilins dissociate from the membranes as a result of electrostatic repulsion.[11]

Protein surface charges calculated from the NMR structure of hisactophilin I for different pH values are illustrated in Fig. 2. The strong pH dependence of the binding of hisactophilins to acidic lipids is caused by the large number of histidine residues exposed on their surface, resulting in steep changes in the net charges of hisactophilins on variation of the pH between pH 6.5 and 7.5. Contribution of the N-terminal myristoyl residue to the membrane association constant is deduced from the binding of hisactophilins to uncharged lipid monolayers.[11]

Shuttling of Hisactophilin–Green Fluorescent Protein Fusions between Cytoplasm and Plasma Membrane

Hisactophilin Fusion Constructs

The sequence of HsI is obtained under accession number JO4472, and that of HsII under U13671. The hisactophilin I and II genes each contain one intron.[14] Fusions of hisactophilin II to the N terminus of GFP are

[9] F. Hanakam, C. Eckerskorn, F. Lottspeich, A. Müller-Taubenberger, W. Schäfer, and G. Gerisch, *J. Biol. Chem.* **270,** 596 (1995).

[10] A. Behrisch, C. Dietrich, A. A. Noegel, M. Schleicher, and E. Sackmann, *Biochemistry* **34,** 15182 (1995).

[11] F. Hanakam, G. Gerisch, S. Lotz, T. Alt, and A. Seelig, *Biochemistry* **35,** 11036 (1996).

[12] C. Naumann, C. Dietrich, A. Behrisch, T. Bayerl, M. Schleicher, D. Bucknall, and E. Sackmann, *Biophys. J.* **71,** 811 (1996).

[13] J. A. F. Op den Kamp, *Annu. Rev. Biochem.* **48,** 47 (1979).

[14] M. Stoeckelhuber, A. A. Noegel, C. Eckerskorn, J. Koehler, D. Rieger, and M. Schleicher, *J. Cell Sci.* **109,** 1825 (1996).

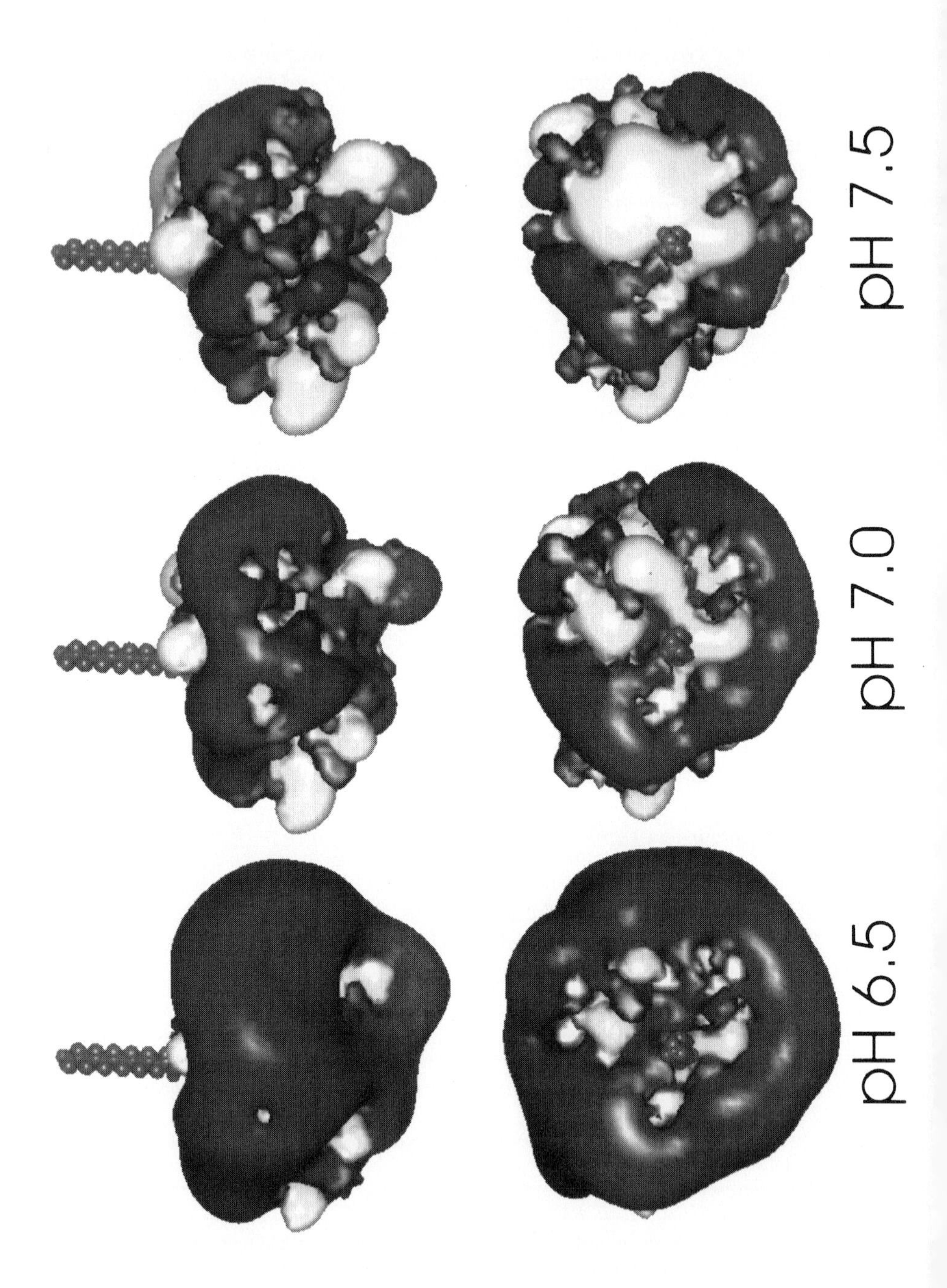
pH 6.5
pH 7.0
pH 7.5

suited for studies of pH-dependent membrane binding *in vivo*.[15] Although hisactophilin is smaller than GFP, it determines the behavior of the fusion protein. A $[(\text{Gly-Gly-Ser})_4]_n$ spacer can be inserted between the hisactophilin and GFP moieties ($n = 1$ to 3 has been tested), but for the interactions described here it proved to be unnecessary.

When the fusions are expressed in *D. discoideum* under the control of the actin 15 promoter the cells become strongly fluorescent. If a contribution of the myristoyl residue to membrane binding is to be avoided, Gly-2 can be converted into alanine (the N-terminal Met-1 is lacking in the mature protein). The glycine-to-alanine conversion at position 2 does not prevent hisactophilin from binding to the plasma membrane *in vivo*, and does not significantly alter the response to pH changes.

Hisactophilin Translocation on pH Changes in Cytoplasm

To change the cytoplasmic pH, cells are subjected to a buffer below pH 7.0, i.e., one containing a weak acid such as propionic acid (pK_d of 4.88), or to a buffer above pH 7.0, containing ammonia or a primary amine. The undissociated forms of these compounds are capable of diffusing through the plasma membrane and, by partial dissociation within the cells, of altering the pH of the cytoplasm.[16]

In buffers at physiological pH values, between pH 6.0 and 8.0, the hisactophilin–GFP fusions are evenly distributed within the cytoplasmic space. Extensive redistribution of the fusion proteins between cytoplasm and plasma membrane is elicited by lowering the pH in the cytoplasm (Fig. 3A). On the addition of 10 m*M* propionic acid in pH 6.0 buffer, the cytoplasm is largely depleted of the protein, which accumulates at the plasma membrane and translocates to the nucleus. This relocalization is reversed on increasing the pH by adding 5 m*M* ammonium chloride and changing the buffer pH to >7.0 (Fig. 3A and B). N-terminally myristoylated hisactophilin as well as nonacylated Ala-2 variants show this pH-dependent relocalization.

[15] F. Hanakam, R. Albrecht, C. Eckerskorn, M. Matzner, and G. Gerisch, *EMBO J.* **15,** 2935 (1996).

[16] K. Inouye, *Biochim. Biophys. Acta* **1012,** 64 (1989).

FIG. 2. Calculated distribution and density of surface charges of hisactophilin II at three different pH values. Dark areas indicate an isocharge surface of $+1kT/e$; light areas indicate an isocharge surface of $-kT/e$. *Top row*: Side view, with the myristoyl residue drawn in a upright direction. *Bottom row*: Top view of the molecule. The electrostatic surface shows a strong increase in negative charges with a rise in pH, within the physiological range of pH 6.5 to 7.5.[11] The surface charges were calculated by DelPhi (Insight II, Biosym, San Diego, CA).

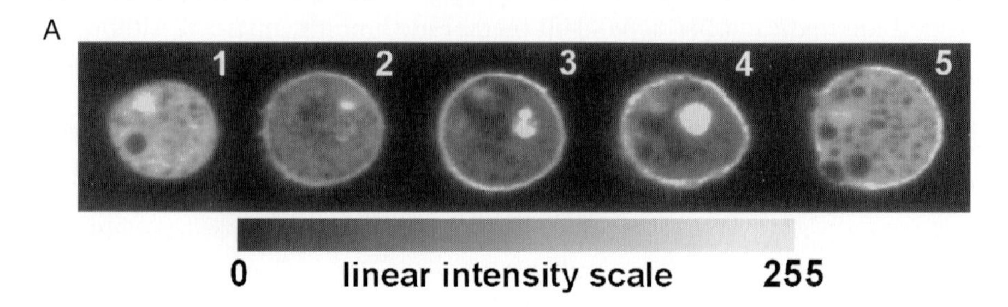

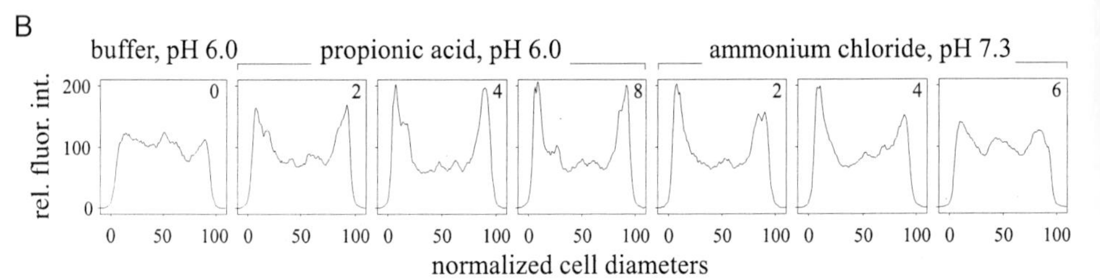

FIG. 3. pH-induced intracellular redistribution of hisactophilin II–GFP. (A) A single cell was incubated with 20 m*M* MES buffer, pH 6.0, without additions (1), and thereafter with the same buffer plus 10 m*M* propionic acid adjusted to pH 6.0 (2 and 3), which was finally replaced by Tris buffer with 10 m*M* ammonium chloride, pH ~7.3 (4 and 5). [Data from F. Hanakam *et al.* (1996).[15]] (B) Time course of the reversible translocation of hisactophilin II–GFP. As in (A), 10 m*M* propionic acid was added, without changing the pH, to cells incubated in 20 m*M* MES buffer, pH 6.0. Subsequently, the solution was partially replaced by ammonium chloride in 20 m*M* Tris buffer (final concentration, 5 m*M* at pH 7.3). Numbers at the curves correspond to minutes after the addition of propionic acid or ammonium chloride. Confocal line scans were passed through cells in such a way that nuclei were not crossed. Data are based on profiles of fluorescence intensities determined in the same set of 10 cells for all panels. Cell diameters are normalized to 100, using custom-made software,[18] and relative fluorescence intensities are expressed in arbitrary units.

Acquisition of Quantitative Data

To quantitate the changes in the ratio of membrane-bound to cytosolic hisactophilin–GFP in response to pH shifts, the profile of fluorescence intensities along an adjustable line through the cell is recorded by digital imaging. A charge-coupled device (CCD) camera in combination with a conventional fluorescence microscope can be used to assign gray levels to fluorescence intensities. However, line scans obtained with a confocal fluorescence microscope are preferable, because they enable one to determine the light emitted from defined volume elements. The spatial resolution is limited by optical constraints. With the best Neofluar 100/1.3 objectives available, the theoretical limit of resolution is generally given by the point spread function as 0.23 μm in the x and y axes, and 0.62 μm in the z axis.[17]

Individual cells differ in size, shape, and GFP fluorescence intensity. Normalization of these parameters is therefore necessary to allow averaging of the profiles of fluorescence intensities over a cell population. First the cell diameter, and then the integral of fluorescence intensity over the diameter of each cell, is normalized (Fig. 3B). Software developed for that purpose including detailed description can be obtained from Albrecht.[18]

Potential Applications of Techniques Described

Hisactophilins have been isolated from *Dictyostelium*, a microorganism living in soil that is exposed in its habitat to rapidly varying pH conditions. Hisactophilins may have a function specific to these conditions.[14] There are no homologous proteins known from cells of higher animals that are capable of establishing pH homeostasis in their extracellular fluid space. As pH-sensing devices hisactophilin cDNA sequences may be transiently or permanently expressed in other organisms. Mammalian cells are capable of transcribing *Dictyostelium* cDNAs cloned into appropriate expression vectors, and of translating the messages into functional proteins. For instance, a fusion of *Dictyostelium* actin with GFP incorporates into actin filament structures of mouse fibroblasts and into the trails of actin formed during the intracellular movement of *Listeria monocytogenes*.[19]

The methods used to monitor the redistribution of hisactophilin may be adapted to studies *in vivo* of the translocation of other pH-dependent proteins tagged with GFP. The association of a number of actin-binding proteins with cellular structures is strongly pH dependent.[20–22] For instance, α-actinin and EF1α bind tightly to actin filaments under slightly acidic conditions, with a marked decline in affinity at increasing pH.[23,24]

With their N-terminal myristoyl residues, hisactophilins resemble a series of mammalian proteins whose association with membranes is based

[17] S. W. Hell and E. H. K. Stelzer, *in* "Handbook of Biological Confocal Microscopy" (J. B. Pawley, ed.), 2nd Ed., pp. 347–354. Plenum Press, New York, 1995.

[18] R. Albrecht, Program Linefit at http://www.biochem.mpg.de/~albrecht.

[19] A. Choidas, A. Jungbluth, A. Sechi, J. Murphy, A. Ullrich, and G. Marriott, *Eur. J. Cell Biol.* **77,** 81 (1998).

[20] M. Hawkins, B. Pope, S. K. Maciver, and A. G. Weeds, *Biochemistry* **32,** 9985 (1993).

[21] J. M. Schmidt, R. M. Robson, J. Zhang, and M. H. Stromer, *Biochem. Biophys. Res. Commun.* **197,** 660 (1993).

[22] B. T. Edmonds, J. Murray, and J. Condeelis, *J. Biol. Chem.* **270,** 15222 (1995).

[23] J. Condeelis and M. Vahey, *J. Cell Biol.* **94,** 466 (1982).

[24] G. Liu, J. Tang, B. T. Edmonds, J. Murray, S. Levin, and J. Condeelis, *J. Cell Biol.* **135,** 953 (1996).

on a combination of hydrophobic interactions mediated by the fatty acid and charge interactions involving the protein moiety.[25–28] The hydrophobic interactions of the myristoyl residue play only a supporting role, so that alteration of the protein moiety switches the protein from its membrane-bound to a soluble state, analogous to the action of the pH switch in hisactophilins. Binding of the myristoylated alanine-rich C kinase substrate L_1 (MARCKS L_2) to the plasma membrane is regulated by phosphorylation, which compensates the charges of basic amino acid residues that are located in the N-terminal region of the polypeptide chain. In the unphosphorylated state, MARCKS interacts with the cytoplasmic membrane face.[29–31] Membrane binding of recoverin in retinal cells is similarly regulated by a Ca^{2+} switch.[32,33] If GFP fusion constructs of these protein were to become available, their shuttling could be studied using similar optical techniques and data acquisition procedures as outlined here for the pH-controlled translocation of hisactophilins.

[25] A. M. Schultz, L. E. Henderson, and S. Oroszlan, *Annu. Rev. Cell Biol.* **4,** 611 (1988).
[26] D. A. Towler, J. I. Gordon, S. P. Adams, and L. Glaser, *Annu. Rev. Biochem.* **57,** 69 (1988).
[27] M. D. Resh, *Cell* **76,** 411 (1994).
[28] G. Milligan, M. Parenti, and A. I. Magee, *Trends Biochem. Sci.* **20,** 181 (1995).
[29] M. Thelen, A. Rosen, A. C. Nairn, and A. Aderem, *Nature (London)* **351,** 320 (1991).
[30] J. Kim, P. J. Blackshear, J. D. Johnson, and S. McLaughlin, *Biophys. J.* **67,** 227 (1994).
[31] J. T. Seykora, M. M. Myat, L. A. H. Allen, J. V. Ravetch, and A. Aderem, *J. Biol. Chem.* **271,** 18797 (1996).
[32] T. Tanaka, J. B. Ames, T. S. Harvey, L. Stryer, and M. Ikura, *Nature (London)* **376,** 444 (1995).
[33] J. B. Ames, R. Ishima, T. Tanaka, J. I. Gordon, L. Stryer, and M. Ikura, *Nature (London)* **389,** 198 (1997).

[8] Measuring Protein Degradation with Green Fluorescent Protein

By STEPHEN R. CRONIN and RANDOLPH Y. HAMPTON

Introduction

Many cellular proteins are selectively degraded in order to control protein quality or quantity. Protein degradation can be highly selective, such that single point mutations or changes in physiological signals can drastically change the degradation rate of a given protein.[1–4] It is now

[1] A. Finger, M. Knop, and D. H. Wolf, *Eur. J. Biochem.* **218,** 565 (1993).

0076-6879/99 $30.00

clear that the high specificity of protein degradation is brought about by a substantial and growing battery of proteins that are committed to targeting other proteins for degradation.[5–8] However, there remain many questions about the number and type of such factors, how they affect and control degradation, and how the numerous and disparate degradation pathways are coordinated and regulated in an intracellular milieu that is rich with proteins undergoing drastically different degradative fates, often in the same aqueous compartment.

The study of protein degradation has, until now, usually required some sort of invasive strategy for examining the half-life of a given degradation substrate, or the steady state level of the substrate, which will vary with changes in degradation rate. Typical assays of protein stability capitalize on specific antibodies to immunoblot or immunoprecipitate the protein under study (e.g., see Refs. 1–4). In either case, the cells expressing the protein must be lysed in order to use these approaches. Immunological approaches can sometimes be used to assess the level of a protein *in situ* and so assess changes in the pool owing to changes in degradation,[9] but even in such cases, fixation of the cells and immunostaining procedures could hardly be considered noninvasive.

Fusion Proteins as Degradation Substrates

Degradation determinants on proteins are often modular and can function autonomously when included in novel, engineered proteins.[10,11] Numerous laboratories have capitalized on this feature by making a fusion gene that will express an easily measured reporter function, such as β-galactosidase, fused to a degradation-determining portion of a protein of interest. This approach allows the levels of the new degradation substrate, and the effects of altering degradation on these levels, to be conveniently measured by assaying the reporter activity that is now subject to the degradation

[2] E. Tsuji, Y. Misumi, T. Fujiwara, N. Takami, S. Ogata, and Y. Ikehara, *Biochemistry* **31,** 11921 (1992).

[3] R. Y. Hampton and J. Rine, *J. Cell Biol.* **125,** 299 (1994).

[4] J. Roitelman and R. D. Simoni, *J. Biol. Chem.* **267,** 25264 (1992).

[5] M. Hochstrasser, *Curr. Opin. Cell Biol.* **7,** 215 (1995).

[6] R. J. Deshaies, *Curr. Opin. Cell Biol.* **7,** 781 (1995).

[7] J. M. Huibregtse, M. Scheffner, S. Beaudenon, and P. M. Howley, *Proc. Natl. Acad. Sci. U.S.A.* **92,** 2563 (199995).

[8] R. Y. Hampton, R. Gardner, and J. Rine, *Mol. Biol. Cell* **7,** 2029 (1996).

[9] F. J. Stafford and J. S. Bonifacino, *J. Cell Biol.* **115,** 1225 (1991).

[10] M. Glotzer, A. W. Murray, and M. W. Kirschner, *Nature (London)* **349,** 132 (1991).

[11] M. Hochstrasser and A. Varshavsky, *Cell* **61,** 697 (1990).

process of interest. However, in most cases, the measurement of such reporter genes still requires lysis of the cells in which they are expressed.

The use of green fluorescent protein (GFP) allows a novel refinement of the reporter approach in the study of protein degradation. The idea is the same as with previous reporters. GFP is expressed as part of a protein that undergoes degradation, and so undergoes the same degradation as the authentic fusion partner. However, now the amount of the resulting protein can be directly measured in living cells by virtue of the fluorescence from the GFP moiety. In theory, this approach allows a totally noninvasive and integrated observation of protein degradation. Our interest is focused on the use of the GFP reporter in genetic studies of protein degradation.[12] However, the example that we describe in the following sections can in all likelihood be generalized to many other experimental circumstances.

HMG-CoA Reductase

HMG-CoA reductase (HMGR) is a key enzyme in the mevalonate pathway from which cholesterol and many other essential molecules are synthesized.[13] The degradation of HMGR is regulated by the mevalonate pathway as a mode of feedback control When flux through the mevalonate pathway is high, HMGR degradation is fast. When flux is slowed, as for instance when cells are treated with the HMGR inhibitor lovastatin, HMGR degradation is slowed. HMGR is an integral membrane protein of the endoplasmic reticulum (ER), and HMGR degradation occurs without exit from the ER. Many molecular details about the degradation mechanism and the manner is which HMGR degradation is coupled to the mevalonate pathway remain to be discovered.

Regulated degradation of HMGR has been conserved between yeast and mammals[3,14,15] In yeast, the Hmg2p isozyme undergoes regulated ER degradation with many features similar to that observed in mammals.[3,12,16,17] This conservation has allowed us to launch a genetic analysis of yeast Hmg2p regulated degradation with the goal of understanding the molecular mechanisms operating in yeast, including those in common with mammals.[8] Our studies have revealed a set of genes required for the degradation of Hmg2p, called *HRD* genes (pronounced *herd*, for Hmg-CoA reductase degradation). Loss-of-function *hrd* mutants such as *hrd1-1* or *hrd1*Δ fail to

[12] R. Y. Hampton, *Curr. Opin. Lipidol.* **9,** 93 (1998).
[13] R. Hampton, D. Dimster-Denk, and J. Rine, *Trends Biochem. Sci.* **21,** 140 (1996).
[14] P. A. Edwards, S. F. Lan, and A. M. Fogelman, *J. Biol. Chem.* **258,** 10219 (1983).
[15] K. T. Chun and R. D. Simoni, *J. Biol. Chem.* **267,** 4236 (1992).
[16] R. Gardner, S. Cronin, B. Leader, J. Rine, and R. Y. Hampton, *Mol. Biol. Cell* **9,** 2611 (1998).
[17] R. Y. Hampton and H. Bahkta, *Proc. Natl. Acad. Sci. U.S.A.* **94,** 12944 (1997).

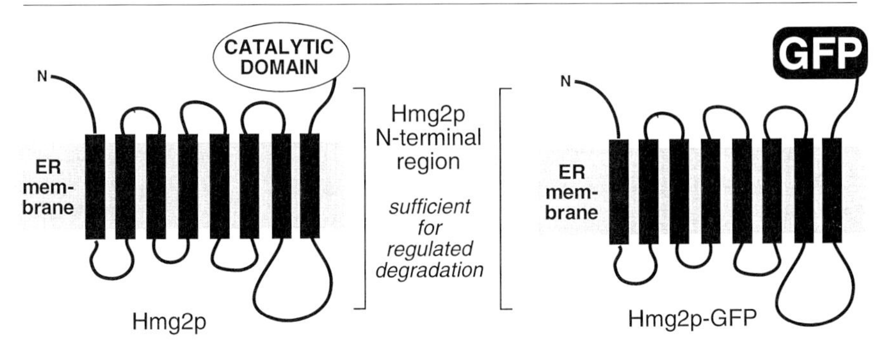

FIG. 1. Hmg2p and the fluorescent reporter Hmg2p–GFP. Hmg2p is anchored in the ER membrane of yeast by the hydrophobic N terminus. The N-terminal anchor is responsible for regulated degradation. The C-terminal catalytic domain is attached to the membrane anchor by a poorly conserved linker region. Hmg2p–GFP, which still undergoes regulated degradation, has the authentic catalytic domain replaced by the GFP reporter. The *HMG2::GFP* fusion gene was produced by fusing the region encoding GFP to the *HMG2* coding region at codon 669.[18] The S65T bright mutation was then introduced by subcloning that portion of the S65T *GFP* coding region into the *HMG2::GFP* gene (see text).[8]

degrade Hmg2p. In the context of this chapter, the known regulatory behavior of Hmg2p degradation and the degradative phenotypes of *hrd* mutants provided strict criteria for the validity of using the GFP reporter as part of a novel fusion protein, Hmg2p–GFP, in the study of Hmg2p protein degradation. Our studies have indicated that the GFP faithfully reports the regulated degradation of Hmg2p, and so use of the Hmg2p–GFP fusion allows a variety of approaches that would otherwise not be possible. More generally, these studies indicate that the GFP reporter could have broad use in the study of protein degradation.

Hmg2p–GFP Reporter Protein

HMGR has a modular structure. The C-terminal portion of the protein provides the essential catalytic activity for which the protein is named. In contrast, the N-terminal, membrane-anchoring portion has no role in catalysis, but anchors the protein to the ER membrane, and is responsible for regulated degradation of the protein (Fig. 1). Accordingly, in both mammals and yeast, engineering an HMGR coding region with the catalytic region of the coding region replaced with a novel 3′ end encoding an enzyme (His4cp in yeast,[3] and β-galactosidase in mammals[15]) results in a fusion protein that still undergoes regulated degradation. GFP was tested as an *in vivo* reporter of Hmg2p degradation by similarly producing an *HMG2* coding region with the 3′, catalytic region codons replaced with the entire *GFP* coding region. The original *HMG2::GFP* coding region expressed the Hmg2p–GFP reporter protein, with the N-terminal 669 amino acids fused to the GFP protein, and a small number of amino acids between the two coding regions from the polylinker from which the GFP was cloned[18] (Fig. 1).

The first Hmg2p–GFP optical reporter protein contained the wild-type GFP as cloned from jellyfish.[18] We have subsequently recloned the *HMG2::GFP* plasmids to include the brighter variant GFP with an S65T mutation[8,16] (the S65T numbering refers to the GFP-coding region considered alone, since that is the way that the mutation is referred to in most cases). Use of the S65T mutant results in an Hmg2p–GFP with approximately six or seven times the brightness of the original GFP.[19] The S65T mutation was introduced into our original fusion gene by subcloning the *Msc*I/*Sal*I fragment of the S65T GFP-coding region from the Clontech (Palo Alto, CA) pS65T-C1 plasmid into the corresponding sites of the original *HMG2::GFP* coding region. The *Msc*I site cuts the *GFP* coding region 8 codons before the S65 position, making this approach generally useful for introducing the bright T65 mutation (or any other downstream mutations from other *GFP* genes) into a preexisting wild-type *GFP* coding region by nonmutagenic cloning techniques. Both the original and the brighter Hmg2p–GFP fusion proteins behave identically as degradation substrates. For the remainder of this work, *HMG2::GFP*, and Hmg2p–GFP, will refer to the gene and proteins, respectively, with the S65T mutation in the GFP.

In our studies, the promoter and plasmid type have been chosen to ensure that the level of Hmg2p–GFP protein is entirely controlled by protein degradation. The *HMG2::GFP* coding region is expressed from a strong, constitutive glyceraldehyde-3-phosphate dehydrogenase (GAPDH) promoter. In this context Hmg2p–GFP synthesis is constant, and unaffected by regulatory perturbations of the mevalonate pathway (such as addition of lovastatin; see below) that control the Hmg2p degradation rate.

Furthermore, expression plasmids with the *HMG2::GFP* coding region are always integrated into the yeast genome in single copy, so that the dose of the gene is not subjected to the variability that autonomously replicating yeast plasmids can display (e.g., between one and four copies per cell for ARS/CEN plasmids). These measures keep the expression of Hmg2p–GFP constant and stable among individual cells and across physiological or

[18] R. Y. Hampton, A. Koning, R. Wright, and J. Rine, *Proc. Natl. Acad. Sci. U.S.A.* **93,** 828 (1996).

[19] R. Heim, D. C. Prasher, and R. Y. Tsien, *Proc. Natl. Acad. Sci. U.S.A.* **91,** 12501 (1994).

genetic conditions that affect degradation. Accordingly, the fluorescence intensity is directly controlled by the degradation rate of Hmg2p–GFP.

We have extensively compared the degradation of the Hmg2p–GFP reporter protein with that of authentic Hmg2p. In every way tested, the fluorescent reporter protein displays degradation behavior that mirrors the parent protein. The degradation of Hmg2p–GFP depends on *HRD* genes.[8,16] Also, Hmg2p–GFP degradation is regulated by signals from the mevalonate pathway.[16–18] For example, inhibitors of early pathway enzymes such as the HMGR inhibitor lovastatin slow the degradation of the reporter protein. Finally, mutations in the N-terminal region of Hmg2p that alter the degradation cause identical alterations when present in Hmg2p–GFP.[16] The focus of this chapter is to describe the techniques that are uniquely possible with the optical reporter. The features of Hmg2p–GFP degradation (*HRD* dependence, physiological regulation) are incorporated in the examples of data collected using the techniques uniquely permitted by the GFP reporter.

Direct Visual Examination of Hmg2p–GFP

A strain of yeast expressing the Hmg2p–GFP reporter protein can be directly examined by fluorescence microscopy[8,16,18] (Fig. 2). The Hmg2p–GFP is distributed in a manner identical to Hmg2p.[18,20] At the levels attained with the strong GAPDH promoter, the ER is proliferated into stacks and whorls of membrane similar to the structures caused by high levels of Hmg2p.[18] The cells are typically grown and examined in yeast minimal medium, but the log-phase fluorescence is quite independent of the exact medium or circumstances of growth. Hmg2p–GFP fluorescence can be observed with no preparation or fixation. The pictures shown in Fig. 2 were taken of samples removed from experimental cultures and placed directly on a microscope slide. In this particular case, a Nikon Optiphot II microscope with an Episcopic fluorescence attachment EFD-3 was used. The standard B-2A filter block, usually used for detection of the fluorescein moiety, allows excitation and filtration of the fluorescence signal (B-2A features: 450- to 490-nm excitation filter; 505-nm dichroic mirror; 520-nm barrier filter). Although we employ this standard filter block, various companies now offer filter blocks optimized for GFP excitation and emission. The light source is a mercury vapor lamp provided with the unit. The main point for direct microscopic examination is that the microscope and optics are all standard tools for fluorescence microscopy, and any of the numerous microscopes typically employed for this purpose will work as

[20] A. J. Koning, C. J. Roberts, and R. L. Wright, *Mol. Biol. Cell* **7,** 769 (1996).

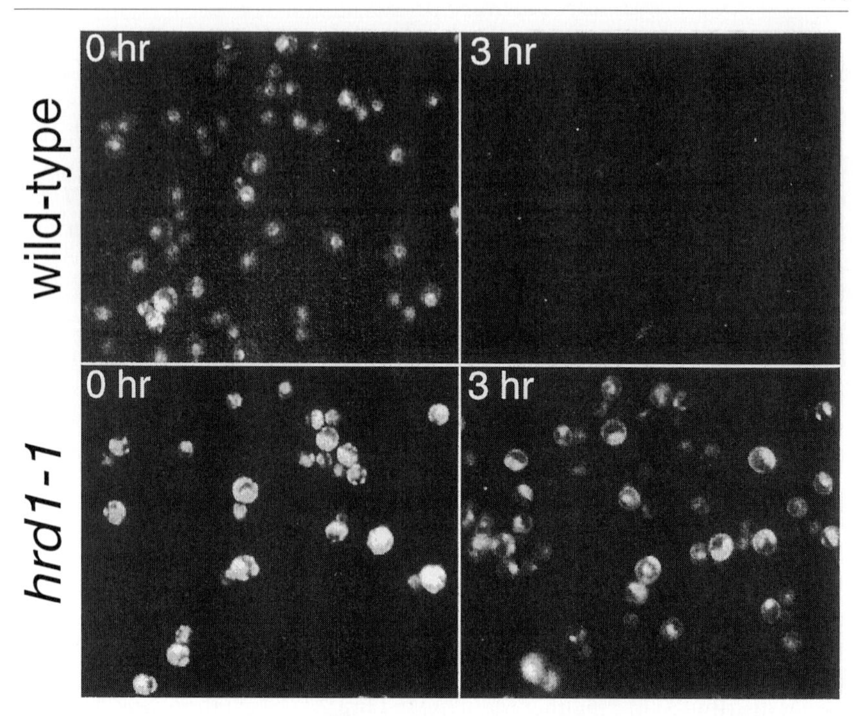

FIG. 2. Direct observation of Hmg2p–GFP degradation by fluorescence microscopy. A wild-type (top row) or degradation-deficient (bottom row) *hrd1-1* strain expressing Hmg2p–GFP were treated with cycloheximide and examined by fluorescence microscopy immediately or after a 3-hr degradation period. The wild-type strain showed substantial loss of Hmg2p–GFP fluorescence. The dark, 3-hr panel from the wild-type strain represents an identical exposure of a similar number of cells photographed in the 0-hr panel. The loss of fluorescence is not observed in the degradation-deficient *hrd1-1* strain (bottom row). All panels were obtained by CCD and frame integrator, and printed on a thermal printer as described.[8,16] All panels were captured and printed using identical settings, and so can be compared for fluorescence intensity. Note that the initial level of fluorescence in the degradation-deficient *hrd1-1* strain (bottom left) is brighter than that of the otherwise identical wild-type strain (top left).

well. The more important issues for the success and/or utility of direct examination concern the actual features and expression levels of the fusion protein under study (see Caveats and Considerations).

Figure 2 shows a typical experiment demonstrating direct examination of Hmg2p–GFP as a degradation substrate. When protein synthesis is blocked Hmg2p–GFP degradation continues and the cellular levels of the reporter protein drop as a function of the degradation rate.[18] Cells expressing the Hmg2p–GFP reporter were treated with cycloheximide (50 μg/ml) to block protein synthesis, and examined immediately or after a 3-hr incubation period to allow protein degradation to proceed (Fig. 2). This experiment was run with otherwise identical strains that had a normal *HRD1* gene (wild type; Fig. 2, top row), or the *hrd1-1* mutation (*hrd1-1*; Fig. 2, bottom row) that renders cells severely compromised for Hmg2p degradation. The top row in Fig. 2 shows a substantial reduction in the fluorescence intensity of the *HRD1* cells after a 3-hr degradation period, caused by loss of the Hmg2p–GFP protein. The bottom row in Fig. 2 shows the results of the identical experiment when the *hrd1-1* mutation is present. The *hrd1-1* mutation strongly blocks the degradation of Hmg2p–GFP, preventing loss of fluorescence during the 3-hr period. We have also used direct observation of Hmg2p–GFP degradation as one way of showing how zaragozic acid hastens the degradation of the reporter by increasing the endogenous signal for degradation.[17]

To observe degradation by direct visual inspection, the production of the fluorescent reporter must be halted, in this case by global cessation of protein synthesis with cycloheximide. While Hmg2p degradation is not adversely affected by cycloheximide, it is possible that the degradation of a different fluorescent substrate could be affected by the drug. In that circumstance, other approaches can be employed to observe the degradation of the entire pool of protein, including removal of a required amino acid to halt protein synthesis,[17] or possibly use of a regulated promoter to stop expression of the fluorescent reporter.

Direct fluorescence microscopy can also be used to observe changes in the steady state levels of Hmg2p–GFP caused by changes in degradation. As an example, the initial steady state level of Hmg2p–GFP in the otherwise isogenic *hrd1-1* strain in Fig. 2 is clearly brighter (bottom left versus top left), owing to stabilization of the protein. Similarly, treatment of strains expressing Hmg2p–GFP with zaragozic acid (ZA) results in a new steady state level of the protein, and thus cellular fluorescence brought about by the effect of ZA to hasten degradation (S. Cronin, H. Bhakta, and R. Hampton, unpublished observation, 1998). Importantly, using the steady state level as a readout of changes in degradation requires that the rate of production of the protein from the fusion gene be unaffected by alterations in the degradation process brought about by genetic or pharmacological means.

Flow Microfluorimetry and Fluorescence-Activated Cell Sorting Analysis of Degradation

Flow microfluorimetry (FMF) is the technique whereby individual cells are measured for fluorescence while passing through a chamber in a laminar sheath of buffer solution. Fluorescence-activated cell sorting (FACS) devices perform FMF (analysis) on a population to query the fluorescence of individual cells, and then a subsequent collection step (sorting) of the individual cells according to designated optical properties ascertained by the FMF. Typically one will sort the most or least fluorescent cells from the bulk population. Because GFP reports fluorescence in living cells, FACS sorting of cells expressing a degraded GFP fusion could be used to isolate mutants with altered degradation of fluorescent reporter protein, and hence atypical fluorescence from the bulk population. FACS sorters can typically process 1000–3000 cells per second, making it entirely resonable to use this approach in the collection of degradation mutants. Most of our work has been in the use of FMF as an analytical tool, and that is the emphasis in the remainder of this chapter. However, the FMF data indicate that the effects of perturbing degradation on the fluorescence histograms are significant enough to allow enrichment of appropriate mutants by FACS.

FMF results are usually plotted as a histogram, showing individual cell fluorescence on the horizontal axis and number of cells with a given level on the vertical axis (Fig. 3). The histogram represents the range and distribution of fluorescences present in the population of cells under analysis. Typical histograms are plotted using the logarithm of fluorescence (in arbitrary units) and represent the collected data from 5000–20,000 cells. Yeast cells expressing Hmg2p–GFP can be directly analyzed by FMF from living cultures with no fixation or preparation. Furthermore, since tens of thousands of cells are typically analyzed, the amount and density of living culture required for analysis can be very small, usually less than 50 μl (a typical early log-phase culture has a cell density on the order of 2×10^6 cells/ml), and often far less than would be amenable to more standard modes of protein analysis. Use of FMF to assess cellular fluorescence is more quantitative and faster than direct microscopic examination, although less informative about cellular localization.

We have used FMF to study the effect of altering degradation on the steady state levels of Hmg2p–GFP,[16] caused by either pharmacological or genetic means. In the top row ("steady state" in Fig. 3) are fluorescence histograms of yeast expressing Hmg2p–GFP under various circumstances that affect degradation of the reporter. The left panel in the top row demonstrates the effect of physiological regulation of Hmg2p–GFP degradation on the steady state fluorescence ("lovastatin treatment" in Fig. 3). When the mevalonate pathway is slowed with a small dose (25 μg/ml) of the HMGR inhibitor lovastatin, the degradation of Hmg2p–GFP is slowed. (The HMGR activity is being provided in this strain by a separately expressed HMGR.) The superimposed, steady state fluorescence histograms of this strain grown for 3 hr in the absence or presence of lovastatin ("no drug" and "lova," respectively, in Fig. 3) show the expected rightward shift

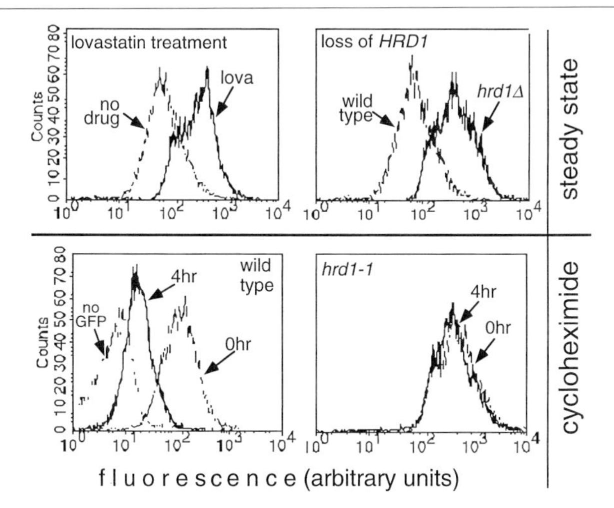

FIG. 3. Use of flow microfluorimetry (FMF) to evaluate steady state (top row) and time-dependent (bottom row) fluorescence of strains expressing the Hmg2p–GFP reporter. *Top row:* Effect of physiological or genetic alterations of Hmg2p–GFP on steady state levels of cellular fluorescence. *Left:* The effect of lovastatin treatment on steady state fluorescence. Cells expressing the Hmg2p–GFP protein were grown for 3 hr in the presence or absence of lovastatin (25 μg/ml) and then subjected to FMF to examine individual cellular fluorescence. The lovastatin slows degradation of Hmg2p–GFP and so causes an increase in the steady state levels of cellular fluorescence, as indicated by a rightward shift of the fluorescence histogram. *Right:* Growing cultures of two strains—one with the normal (wild-type) allele and the other with a null allele (*hrd1*Δ) of the *HRD1* gene—were subjected to FMF. The degradation-deficient *hrd1*Δ strain is significantly more fluorescent as indicated by the histogram shift to the right. *Bottom row:* Direct observation of time-dependent loss of cellular fluorescence due to degradation of the Hmg2p–GFP reporter. *Left:* A strain with a normal *HRD1* gene (wild type) was treated with cycloheximide and examined immediately after treatment, or following a 4-hr degradation period. Degradation of the Hmg2p–GFP protein results in a lowering of cellular fluorescence and a leftward shift of the FMF histogram. The histogram of an identical strain, except that it has no *HMG2::GFP* expression plasmid, is also included to show cellular autofluorescence (peak marked *no GFP*). *Right:* The same experiment run with an identical strain, except that it has an *hrd1-1* allele rendering it degradation deficient.

of the drug-treated cells to brighter fluorescence. The right panel in the top row shows the effect of a genetic deficiency in degradation on Hmg2p–GFP fluorescence. Otherwise identical strains with a normal or null allele of *HRD1* ("wild type" and "*hrd1*Δ," respectively, in Fig. 3) were compared by FMF. The degradation-deficient *hrd1*Δ strain has significantly higher steady state Hmg2p–GFP fluorescence, as indicated by the shift of the histogram to the right when overlaid with that from the otherwise identical strain with a normal *HRD1* gene.

FMF can also be used to observe directly the degradation of the reporter protein after cessation of protein synthesis. When protein synthesis is halted with cycloheximide in strains expressing Hmg2p–GFP, the fluorescent reporter is degraded, and cellular fluorescence drops accordingly. The decline in fluorescence can be monitored by FMF after various degradation periods have been allowed to pass, also shown in Fig. 3. The bottom row ("cycloheximide" in Fig. 3) shows the effect of a 4-hr incubation with cycloheximide on Hmg2p–GFP in a wild-type strain (Fig. 3, left panel) or in a degradation-deficient *hrd1-1* strain (Fig. 3, right panel). In each panel, the initial and 4-hr histograms are superimposed to show the effects of inhibition of protein synthesis. Cycloheximide causes a leftward shift in the wild-type strain fluorescence histogram to lower mean cell fluorescence (Fig. 3, bottom row, "wild-type"). This shift is totally dependent on having a normal *HRD1* gene, and so is completely abrogated in the otherwise identical *hrd1-1* strain that cannot degrade Hmg2p–GFP (Fig. 3, bottom row, "*hrd1-1*").

In testing a new GFP fusion for utility in FMF or FACS, it is important to evaluate cellular autofluorescence. This is best done by assaying otherwise identical strains (or cells of whatever sort being tested) that do not express the GFP fusion being measured. As an example, the histogram of the parent strain to those in Fig. 3, but not expressing Hmg2p–GFP, has been included with the histograms of the wild-type strain in the bottom panel (Fig. 3, bottom row, "no GFP").

An FMF (or FACS) apparatus represents a fairly large investment, so the best strategy to explore using this powerful tool is to find a local laboratory that employs one regularly. Fortunately this is usually quite easy, since most immunology laboratories, and many cell cycle laboratories, own such a device. The histograms shown in Fig. 3 were acquired on a Becton Dickinson (Mountain View, CA) FACScalibur flow microfluorimeter, using fluorescence settings appropriate for GFP (488-nm excitation, 512-nm emission). Each histogram represents individual fluorescence from 10,000 cells. The samples were run directly into the flow system from the experimental cultures in each case. Data were next graphically organized into overlaid histograms with the graphical software used to run the device. Although not the focus of our work, the data collected by FMF are amenable to detailed statistical analysis, and could be used to evaluate protein degradation in quantitative detail.

Direct Examination of Hmg2p–GFP in Living Colonies

We are particularly interested in using yeast to perform a complete genetic analysis of Hmg2p regulated degradation. A significant part of our effort is devoted to obtaining yeast mutants that are deficient in various

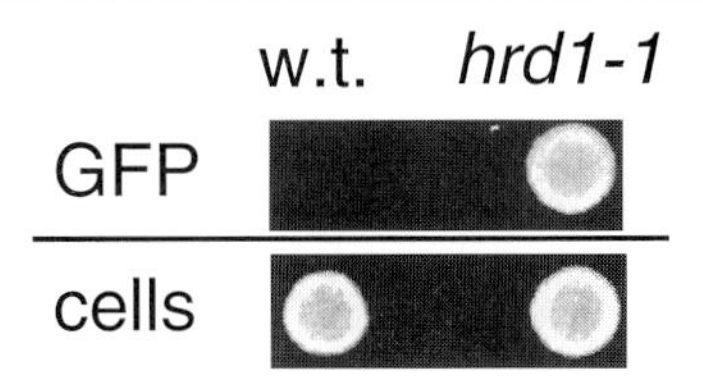

FIG. 4. Colony fluorescence of Hmg2p–GFP. Patches of normal (w.t.) or degradation-deficient (*hrd1-1*) strains expressing Hmg2p–GFP were grown on supplemented minimal agar medium,[8] and subjected to illumination and filtration as described in text to observe green fluorescence (*top row*). The two identical patches were also photographed with room light and no filtration to show the cell density of each patch (*bottom row*). Each patch was prepared by spotting a 50-μl suspension of $\sim 1 \times 10^5$ cells onto the agar medium, and allowing growth for 2 days at 30°.

aspects of Hmg2p regulated degradation.[12] The data presented in Figs. 2 and 3 show that the fluorescence of Hmg2p–GFP in a strain of yeast is affected in a predictable manner by alterations in the degradation of the protein, either by genetic or pharmacological means. Because the fluorescence can be observed in living yeast cells, we developed a method to evaluate the GFP fluorescence in living yeast colonies on agar plates, in order to facilitate screening for mutants with abnormal levels of Hmg2p–GFP due to alterations in regulated degradation.

To examine fluorescence of colonies on agar plates, they must be illuminated with intense light at 488 nm (the absorbance maximum for S65T GFP). This is accomplished with a narrow bandpass filter custom made from Omega Optical (Brattelboro, VT; www.omegafilters.com) with maximal transmittance at 488 nm, and an added blocking filter to remove unwanted red and higher wavelength light from the excitation source. The filter is square in shape with 5-cm sides, the size of a standard photographic slide, allowing it to be placed in the slide holder of a Kodak (Rochester, NY) Carousel 4400 projector. With an FMS bulb, the light from the projector has ample intensity in the 488-nm range, which when passed through the in-place filter results in a field of intense blue light sufficient in size to illuminage multiple agar plates simultaneously. The plates are examined on a bench top in a dark room. The green fluorescence of the illuminated colonies can be directly seen by examination with a long bandpass filter (Kodak Wratten gelatin filter, 75 × 75 mm, No. 12) that removes the substantial contribution of the reflected blue excitation light. The long bandpass filter was purchased at a local camera store and can be taped to a clear face mask or over the lenses of safety goggles.

Figure 4 shows patches of cells grown on agar medium, and expressing the Hmg2p–GFP reporter, that are normal ("w.t.") or degradation deficient (“*hrd1-1*”). The top row in Fig. 4 (“GFP”) shows the green fluorescence of each patch, photographed by illuminating the agar plate with the blue-filtered slide projector, and placing the long bandpass filter in front of a Polaroid camera lens. The bottom row in Fig. 4 (“cells”) represents the same pair of patches photographed with normal white light and no filtration, demonstrating that the density of cells in the two patches is the same. The degradation-deficient strain (“*hrd1-1*” in Fig. 4) is far brighter than the wild-type strain.

The extreme difference in the fluorescence signal compared with what one would expect from the intensity of the wild-type versus *hrd1-1* strains in Fig. 2 is due to a convenient feature of Hmg2p degradation that we have exploited in numerous ways. The agar on which these patches are grown is a typical minimal medium made with yeast nitrogen base, glucose, and the supplements that are required by the auxotrophies of the strain, including the amino acid methionine. On this medium, the cells deplete the amino acid before the glucose, and so cease making protein. Degradation, however, still proceeds, and so the normal strain that can degrade Hmg2p–GFP grows very dark. In contrast, the degradation-deficient (*hrd1-1*) strain cannot degrade the Hmg2p–GFP, and so the patch remains bright despite a similar cessation of Hmg2p–GFP synthesis. Thus, by simply allowing patches to grow for a day or two on supplemented minimal medium, we can score the propensity of that strain to degrade Hmg2p–GFP.

The ease of whole colony analysis offered by the optical colony assay has allowed us to devise numerous, previously infeasible genetic and molecular biological screens relevant to Hmg2p degradation. The protocols are far more straightforward than previous methods to assay protein levels as a gauge of degradation, such as colony immunoblotting. Thus, when an optical reporter works in colony assays such as this, it should certainly be considered as a phenotype for genetic analysis. We are currently looking for mutants deficient in Hmg2p–GFP degradation, mutants deficient in regulation of Hmg2p–GFP stability, variations of the Hmg2p–GFP sequence itself that affect regulated degradation of the protein, high-copy plasmids that alter normal regulated degradation, and mammalian genes that alter degradation- or complement-recessive yeast mutants. Although we use yeast in our current studies, it is likely the same ideas can be applied to the genetics of degradation in other organisms as well.

Caveats and Considerations

The preceding data demonstrate a single example of using GFP as a reporter function in the study of degradation. The majority of people interested in this technique will want to apply it to proteins other than yeast

Hmg2p. With this in mind, there are some considerations and caveats that should be entertained by those who want to harness this powerful tool. These are not to dissuade future fluorophiles from attempting to use GFP, but rather to help allow the greatest chance of success in the exploration of this approach.

GFP fluorogenesis rate versus fusion degradation rate. GFP in its final folded form is fluorescent. The acquisition of this structure, or fluorogenesis, involves both folding and an autooxidative covalent reaction with molecular oxygen. The rate of this process can vary with the type of GFP empolyed,[19] the availability of O_2, and most likely cellular factors that influence protein processing. Many of the mutants that are brighter than the original (and now infrequently used) jellyfish GFP acquire the correct autofluorescent structure fairly rapidly. The steady state fluorescence of a degraded protein with a GFP fusion will be influenced by the rate at which the protein becomes fluorescent. If the degradation rate of the fusion protein is much faster than the rate of GFP fluorogenesis, then only a small fraction of the fusion molecules will survive long enough to become fluorescent. Of course, fluorogenesis of GFP also has its own half-life, so this level of consideration is by no means rigorous. Nevertheless, a degradation rate that is significantly faster than the rate at which the GFP reporter becomes fluorescent will cause diminution of the steady state signal below that predicted from considering the simple effect of degradation on final steady state level of the reporter alone.

On the other hand, the requirement that GFP be fully folded to be optical can be useful in degradation studies. For example, in the case of Hmg2p–GFP, we have shown that the drug zaragozic acid (ZA, an inhibitor of squalene synthase) can stimulate the degradation of Hmg2p and Hmg2p–GFP.[17,18] Even in the absence of protein synthesis, the addition of ZA to fluorescent cells causes rapid loss of signal that is dependent on the *HRD* genes. This effect indicates that in all likelihood the *HRD*-encoded degradation machinery can seize and destroy previously made, fully folded, and previously stable Hmg2p–GFP.

Is loss of fusion GFP fluorescence due to degradation? In the examples herein, we show that various physiological or genetic manipulations have effects on Hmg2p–GFP signal levels and loss rates consistent with bona fide, relevant degradation of the optical reporter. However, these optical effects do not alone confirm that processive degradation is occurring. It is important to verify that the fusion protein is actually undergoing degradation, as opposed to unfolding or cleavage or other unknown processes that at the optical level would be indistinguishable from degradation. This is best done by performing traditional biochemical degradation assays on the fusion itself (examples in Refs. 1–4, 7–11, and 14–18). The most common assay of degradation is the pulse–chase protocol, in which cells are labeled with radioactive amino acids (the pulse step, usually with radioactive methionine), after which incorporation is blocked by addition of excess unlabeled amino acid (the chase), and the protein under study is immunoprecipitated, resolved on sodium dodecyl sulfate–polyacrylamide gels, and autoradiographed to ascertain the rate at which the pulsed sample disappears. Another common assay, similar to that used in Fig. 2, is a "cycloheximide chase," in which protein synthesis is stopped with the drug cycloheximide, and the degradation of the protein of interest is determined by immunoblotting. In either case, antibodies against the fusion protein are needed. Suitable antibodies may be already available against a portion of the authentic protein of interest that is included in the optical reporter. If not, the GFP moiety can also be used for immunodetection or precipitation, either with anti-GFP antibodies made by individual investigators, or with those obtained from a growing number of commercial sources. The original description of Hmg2p–GFP included biochemical analysis of regulated degradation of the new fusion with techniques previously employed in the studies of authentic Hmg2p.[18]

Does the GFP fusion protein go down the same degradation pathway? It is not a given that addition of a novel, folded, 20-kDa domain to a protein will preserve the recognition processes that allow the recognition and degradation of the original protein. If possible, ascertain enough features of the mechanism of degradation to verify that a novel optical reporter behaves in a way that is similar to the authentic protein of interest. Such tests could include examining the effects of physiological perturbations that alter degradation rate, measuring stability of the fusion in degradation-deficient mutants (such as *hrd1-1*), and examining the effects of degradation-altering in-cis mutations when introduced into the reporter. Of course, if one knew everything possible about the degradation mechanism of the original protein, the utility of the optical degradation reporter might be small. Thus, in developing the GFP reporter as a discovery tool, one should verify that the optical protein behaves as expected from information obtained through more invasive and cumbersome methods. Then new discoveries made with the more facile optical approach can be confirmed in traditional studies on the actual protein of interest.

Where to put the GFP moiety? GFP is a remarkably flexible fusion partner. It can be put at the N terminus of proteins, the C terminus (as in Hmg2p–GFP), or even in the middle of some protein sequences. What is most important is to use as much information about the degradation of the protein as possible to inform the placement of the *GFP* coding region, and if possible, to try several different fusions at the early stages of developing the optical approach in order to identify the most useful version.

Which GFP to use? There are a growing number of novel sequence variants of the original GFP protein that can be applied to a given research problem. Variations include altered spectral properties, differing folding rates in particular organisms, and coding regions optimized to express a given version optimally in a desired organism, and even a version that is itself degraded rapidly to allow more dynamic analysis of changes in promoter and translational activity. It is safe to say that the generation of novel variations of the original GFP is far from complete, perhaps including the development of completely novel proteins that function in the same manner. Numerous companies now offer a variety of such variations of the DNA and protein sequence of GFP. Some of these companies include Clontech (www.clontech.com), and Packard (Downers Grove, IL; www.packardinst.com), to name two. It is advisable to perform an Internet search to mesh specific needs with the constantly growing availability of custom GFPs. It also can be useful to take advantage of the GFP newsgroup (for information, e-mail biosci-server@net.bio.net) to investigate the newest developments and to interact with many experienced researchers.

Acknowledgments

The authors thank Dr. Robert Rickert (UCSD Department of Biology) for the use of the FACScalibur flow microfluorimeter, software, and color printer, Dr. Michael Coady (University of Montreal) for valuable suggestions about the colony fluorescence assay, and the members of the Hampton laboratory for continuous effort, interest, and interaction. R.Y.H. also thanks L. Rampulla for consultations without charge, and supportive field work. Supported by AHA Grant 96013020, NIH Grant DK5199601, and a Searle Scholarship.

[9] Studying Nuclear Receptors with Green Fluorescent Protein Fusions

By GORDON L. HAGER

Introduction

Subcellular trafficking of nuclear receptors is an important component in the biological control of these regulatory molecules. The steroid receptors form one major group of the nuclear receptor family, and are the best understood in terms of subcellular distribution. Three distribution patterns have been described classically. The glucocorticoid receptor (GR) is located in the cytoplasm in the absence of ligand and translocates to the nucleus when occupied by ligand.[1,2] The GR seems to be unique in its exclusive cytoplasmic compartmentalization. A fraction of the progesterone receptor (PR) is also present in the cytoplasm, but most of this receptor is found in the nucleus under ligand-free conditions.[3–5] The thyroid receptor (TR) and retinoic acid receptor (RAR) subfamily,[6] as well as the estrogen receptor (ER),[7–9] are usually described as bound to the nucleus both in the presence and absence of hormone.

These views of receptor localization are derived from two experimental approaches, subcellular fractionation and immunolocalization. Both of these approaches require major disruption of cellular architecture: either fixation of cells to render epitopes accessible to large antibody complexes, or complete separation of cellular compartments. While data gathered using these techniques are frequently in agreement on the subcellular distribution for a given receptor, there are frequent discrepancies. For example, the mineralocorticoid receptor (MR) has been variously reported as cytoplasmic and nuclear,[10,11] or exclusively in the nucleus.[12] Likewise, using fixed cell preparations, the GR has been reported to be exclusively in the nucleus.[13]

Some confusion regarding intracellular distribution of macromolecules undoubtedly arises from artifacts inherent in these two approaches. Breakage of cells for preparation of subcellular compartments inevitably disrupts the integrity of these compartments, and may lead to inappropriate localization of complexes. Similarly, fixation of cells can either destroy native

[1] D. B. Mendel, E. Orti, L. I. Smith, J. Bodwell, and A. Munck, *Prog. Clin. Biol. Res.* **322,** 97 (1990).

[2] W. B. Pratt, *Prog. Clin. Biol. Res.* **322,** 119 (1990).

[3] A. Guiochon-Mantel, K. Delabre, P. Lescop, M. Perrot-Applanat, and E. Milgrom, *Biochem. Pharmacol.* **47,** 21 (1994).

[4] J. F. Savouret, M. Perrot-Applanat, P. Lescop, A. Guiochon-Mantel, A. Chauchereau, and E. Milgrom, *Ann. N.Y. Acad. Sci.* **684,** 11 (1993).

[5] A. Guiochon-Mantel, H. Loosfelt, P. Lescop, S. Christin-Maitre, M. Perrot-Applanat, and E. Milgrom, *J. Steroid Biochem. Mol. Biol.* **41,** 209 (1992).

[6] C. Song, R. A. Hiipakka, J. M. Kokontis, and S. Liao, *Ann. N.Y. Acad. Sci.* **761,** 38 (1995).

[7] S. Yamashita, *Histol. Histopathol.* **13,** 255 (1998).

[8] J. Gorski, J. D. Furlow, F. E. Murdoch, M. Fritsch, K. Kaneko, C. Ying, and J. R. Malayer, *Biol. Reprod.* **48,** 8 (1993).

[9] G. L. Greene and M. F. Press, *J. Steroid Biochem.* **24,** 1 (1986).

[10] M. Lombes, N. Farman, M. E. Oblin, E. E. Baulieu, J. P. Bonvalet, B. F. Erlanger, and J. M. Gasc, *Proc. Natl. Acad. Sci. U.S.A.* **87,** 1086 (1990).

[11] H. Sasano, K. Fukushima, I. Sasaki, S. Matsuno, H. Nagura, and Z. S. Krozowski, *J. Endocrinol.* **132,** 305 (1992).

[12] P. T. Pearce, M. McNally, and J. W. Funder, *Clin. Exp. Pharmacol. Physiol.* **13,** 647 (1986).

[13] M. Brink, B. M. Humbel, E. R. de Kloet, and R. van Driel, *Endocrinology* **130,** 3575 (1992).

macromolecular interactions, or mask the epitopes for antibodies directed against purified proteins; both problems would lead to incorrect identification of subcellular location. With the cloning and characterization of the green fluorescent protein (GFP), direct labeling of proteins in living cells is now feasible. Intracellular studies with these chimeras would avoid the major artifacts associated with cell fractionation and immunolocalization in fixed cells.

Successful tagging of the glucocorticoid receptor was first reported by Umesono and colleagues for a truncated version of the receptor,[14] and by Hager and co-workers[15] for the full-length GR. An extensive analysis of the full-length GFP–GR fusion chimera[15] indicated that its transactivation potential and ligand specificity were similar to that of the normal receptor, indicating that the GFP fusions would be highly useful reagents for the study of intracellular receptor trafficking.

Tagging Nuclear Receptors with Green Fluorescent Protein

GFP–fusion chimeras for several of the major nuclear receptor groups have now been reported (see Fig. 1). An example of subcellular localization studied with a GFP–GR fusion is given in Fig. 2. This receptor is localized uniquely in the cytoplasm in the absence of ligand, then undergoes complete nuclear translocation after addition of dexamethasone (dex).[15] Furthermore, as noted in Htun *et al.*,[15] there appears to be an association of the receptor with a fibrillar network in the cytoplasm. Once the receptor has moved into the nucleus, there is also a discrete intranuclear organization of the receptor. A strong focal distribution of the receptor is observed when the GR is activated by an agonist such as dexamethasone.[15] When induced with an antagonist,[15,16] uniform nuclear localization is observed. There are several important implications from these results. First, subnuclear distribution of steroid receptors suggests the existence of specific intranuclear targets. Second, correct recognition of these targets is dependent on the nature of the activating ligand. Finally, the interrelationship between these focal structures and receptor regulatory elements in chromatin is poorly understood, but now becomes an important area of investigation.

Although C-terminal GFP chimeras for several other cellular proteins

[14] H. Ogawa, S. Inouye, F. I. Tsuji, K. Yasuda, and K. Umesono, *Proc. Natl. Acad. Sci. U.S.A.* **92**(25), 11899 (1995).

[15] H. Htun, J. Barsony, I. Renyi, D. J. Gould, and G. L. Hager, *Proc. Natl. Acad. Sci. U.S.A.* **93,** 4845 (1996).

[16] G. L. Hager, C. L. Smith, G. Fragoso, R. G. Wolford, D. Walker, J. Barsony, and H. Htun, *J. Steroid Biochem. Mol. Biol.* **65,** 125 (1998).

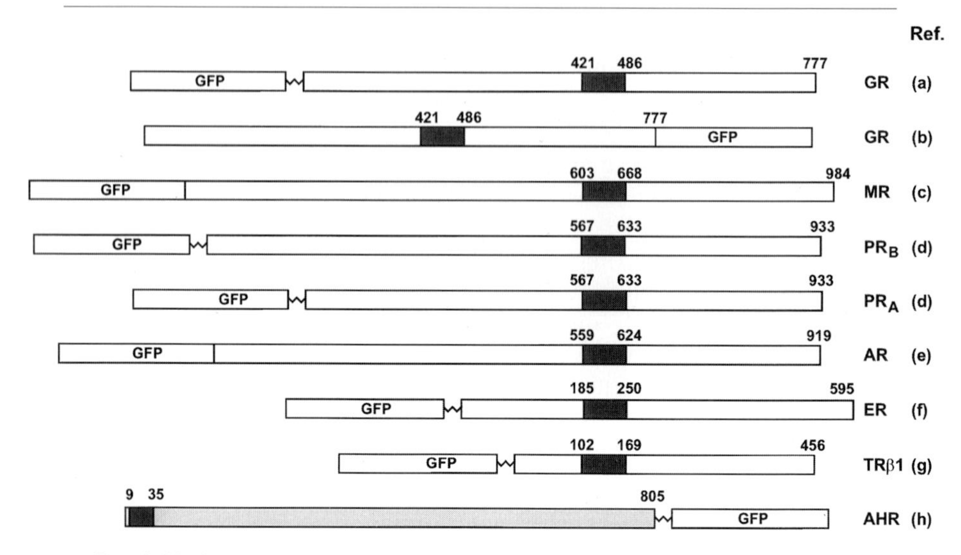

FIG. 1. Nuclear receptor–GFP fusions that have been characterized as of this writing. Schematic presentations of the receptors show the DNA-binding domains (solid fill) and position of the GFP tag. Wavy lines between the GFP domain and the receptor indicate the presence of a peptide linker (see text). Citations for these studies are as follows: a [H. Htun, J. Barsony, I. Renyi, D. J. Gould, and G. L. Hager, *Proc. Natl. Acad. Sci. U.S.A.* **93,** 4845 (1996)], b [K. L. Carey, S. A. Richards, K. M. Lounsbury, and I. G. Macara, *J. Cell Biol.* **133**(5), 985 (1996)], c [G. Fejes-Toth, D. Pearce, and A. Naray-Fejes-Toth, *Proc. Natl. Acad. Sci. U.S.A.* **95,** 2973 (1998)], d [C. S. Lim, C. T. Baumann, H. Htun, W. Xian, M. Irie, C. L. Smith, and G. L. Hager, *Mol. Endocrinol.* in press (1999)], e [V. Georget, J. M. Lobaccaro, B. Terouanne, P. Mangeat, J. C. Nicolas, and C. Sultan, *Mol. Cell. Endocrinol.* **129,** 17 (1997)], f [H. Htun, L. T. Holth, D. Walker, J. R. Davie, and G. L. Hager, *Mol. Biol. Cell* in press (1999)], g [X. Zhu, J. A. Hanover, G. L. Hager, and S.-Y. Cheng, *J. Biol. Chem.* **273**(42), 27058 (1998)], h [C. Y. Chang and A. Puga, *Mol. Cell. Biol.* **18,** 525 (1998)].

have been described, most of the nuclear receptors have been prepared as N-terminal additions (Fig. 1). One exception is a C-terminal GR–GFP fusion described by Macara and colleagues.[17] Using this reagent, these authors studied the involvement of RAN/TC4 GTPase in receptor translocation. A second C-terminal example is provided by the work of Chang and Puga,[18] who utilized a C-terminal fusion to characterize intracellular distribution of the aromatic hydrocarbon receptor (AHR), and identify constitutive nuclear localization signals. Although ligand binding and transactivation features of these receptors were not completely characterized,

[17] K. L. Carey, S. A. Richards, K. M. Lounsbury, and I. G. Macara, *J. Cell Biol.* **133**(5), 985 (1996).

[18] C. Y. Chang and A. Puga, *Mol. Cell. Biol.* **18,** 525 (1998).

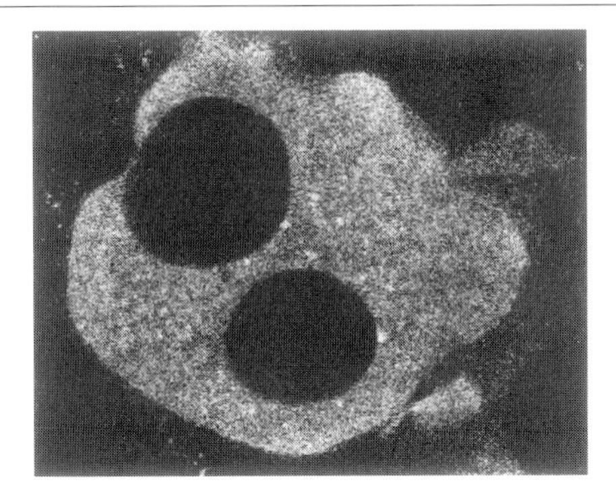

FIG. 2. Intracellular distribution of the glucocorticoid receptor, utilizing GFP–GR. Expression of GFP–GR is shown in a murine C127 cell line. This form of GFP–GR [H. Htun, J. Barsony, I. Renyi, D. J. Gould, and G. L. Hager, *Proc. Natl. Acad. Sci. U.S.A.* **93,** 4845 (1996)] contains the green fluorescent protein fused to the C656G mutant glucocorticoid receptor, which is approximately 20-fold more sensitive to ligand [P. K. Chakraborti, M. J. Garabedian, K. R. Yamamoto, and S. S. J. Simons, *J. Biol. Chem.* **266,** 22075 (1991)]. Cells shown here were untreated with ligand. Images from living cells expressing GFP–GR were taken with a Leica confocal laser scanning microscope, with fluorescent excitation produced by the 488-nm line of a krypton–argon laser, and using a ×100 oil-phase objective. The image from uninduced cells shows complete cytoplasmic localization of the receptor; the receptor in hormone-stimulated cells is found in the nucleus.[15]

these examples suggest that C-terminal labeling also yields a functional receptor.

Preliminary findings on the intracellular distribution of several other members of the major receptor classes are now available. Fejes-Toth *et al.*[19] labeled the MR at the N terminus and studied cytoplasmic/nuclear partitioning of this receptor. Results from this living cell investigation indicated a complex distribution pattern, with the unliganded receptor observed both in the nucleus and the cytoplasm. A similar pattern of distribution was found for the B form of the progesterone receptor by Lim *et al.*[20] These investigators found PR-B in both compartments in the unliganded state. These two receptors (PR and MR) seem to fall in a separate class of receptors that cycle between nucleus and cytoplasm in the absence of hormone.

Georget *et al.*, also using an N-terminal fusion, examined trafficking for the androgen receptor (AR) and found a distribution similar to that for

[19] G. Fejes-Toth, D. Pearce, and A. Naray-Fejes-Toth, *Proc. Natl. Acad. Sci. U.S.A.* **95,** 2973 (1998).

[20] C. S. Lim, C. T. Baumann, H. Htun, W. Xian, M. Irie, C. L. Smith, and G. L. Hager, *Mol. Endocrinol.* in press (1999).

GR.[21] That is, nonstimulated cells contained the AR uniquely in the cytoplasm, and hormone activation led to rapid nuclear translocation. For the thyroid receptor (TR), use of an N-terminal GFP fusion led to the surprising observation that a major subpopulation of cells contained significant amounts of the GFP–TR in the cytoplasm,[22] and this receptor moved to the nucleus in a hormone-dependent manner. This last result is controversial, given that the TR has classically been described as constitutively present in the nucleus. The findings with GFP–TR suggest that accepted models for intracellular distribution of the nuclear receptors may be incomplete, and in some cases in error. Furthermore, the intranuclear focal localization observed with many of the GFP-tagged receptors indicates a level of intranuclear organization not addressed in current paradigms of receptor function.

One final point should be stated concerning the fusion between GFP and a target receptor. The crystal structure of GFP[23] indicates the protein has a rather rigid globular domain. The potential exists that fusion of GFP directly into the open reading frame of a given protein may disrupt an essential feature of the target structure. M. Moser (personal communication, 1996) first suggested that isolation of the GFP moiety from the candidate protein with a peptide linker could minimize the impact of the GFP globular domain. This feature was first incorporated in GFP–GR,[24] and subsequently included in the design of several tagged receptors[18,20,22,25] (linker indicated in Fig. 1 by a wavy line between the receptor and GFP). A comprehensive study to determine whether inclusion of this linker serves to preserve native structure of the target protein has not been performed. However, inclusion of such a linker is a relatively simple modification, and can easily be included in the chimeric design.

Characterization of Fusion Chimera

Prior to use of a given receptor–GFP fusion for studies on intracellular distribution, it is imperative that the basic parameters of the chimeric receptor be verified. The size of the fusion should conform to the predicted

[21] V. Georget, J. M. Lobaccaro, B. Terouanne, P. Mangeat, J. C. Nicolas, and C. Sultan, *Mol. Cell. Endocrinol.* **129,** 17 (1997).

[22] X. Zhu, J. A. Hanover, G. L. Hager, and S.-Y. Cheng, *J. Biol. Chem.* **273**(42), 27058 (1998).

[23] M. Ormo, A. B. Cubitt, K. Kallio, L. A. Gross, R. Y. Tsien, and S. J. Remington, *Science* **273,** 1392 (1996).

[24] H. Htun, D. Walker, and G. L. Hager, *in* "Structure, Motion, Interaction and Expression of Biological Macromolecules" (R. H. Sarma and M. H. Sarma, eds.), p. 157. Adenine Press, Schenectady, New York, 1998.

[25] H. Htun, L. T. Holth, D. Walker, J. R. Davie, and G. L. Hager, *Mol. Biol. Cell* in press (1999).

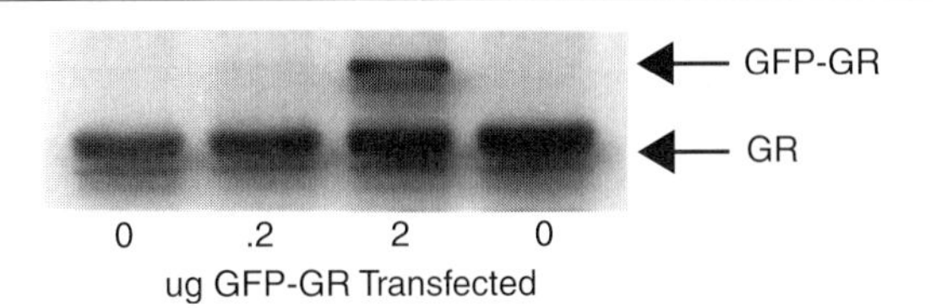

FIG. 3. Expression of the GFP–GR fusion protein. Western blot analysis shows detection of endogenous GR and transfected GFP–GR. Extracts from untreated cells, or from cells transfected with the indicated amounts of GFP–GR DNA, were treated with the BuGR 2 monoclonal GR antibody [B. Gametchu and R. W. Harrison, *Endocrinology* **114,** 274 (1984)] and subjected to ECL (Amersham) analysis. [Reproduced with permission from H. Htun, J. Barsony, I. Renyi, D. J. Gould, and G. L. Hager, *Proc. Natl. Acad. Sci. U.S.A.* **93,** 4845 (1996).]

molecular weight, and should be confirmed by Western blot analysis. Many of the chimeric proteins described to date include an epitope tag, such as the hemagglutinin (HA) antigen, that facilitates discrimination between the fusion chimera and the endogenous, native receptor.[20,22,24,25] To be considered a useful reagent, the fusion chimera should obviously be expressed as a relatively stable protein of the predicted size, with little evidence of breakdown products. As an example of a well-behaved chimera, a Western analysis of GFP–GR is shown in Fig. 3.

The GFP chimera should also be characterized for its transactivation potential with an appropriate hormone-responsive promoter and reporter. This analysis will demonstrate that the basic transactivation domains of the fusion protein are intact and functional. It is useful in this analysis to evaluate the chimeric receptor with both agonists and antagonists, to ensure insofar as possible that the tagged receptor retains proper ligand dependency. An evaluation of GFP–GR transactivation potential and ligand specificity is presented in Fig. 4. It is clear that the tagged receptor retains full transactivation potential, and also maintains a ligand specificity similar to that of the unsubstituted receptor.

The fusion chimera should be tested for ligand affinity by carrying out a dose–response analysis in an *in vivo* response system. The chimeric receptor should manifest a K_d for ligand affinity similar to that measured for the wild-type receptor. Finally, electrophoretic mobility shift analysis (EMSA) with the cognate DNA response element will demonstrate that the fusion maintains native DNA recognition specificity.

Choice of Fluorescent Protein to Be Utilized for Receptor Labeling

Essentially all of the first-generation nuclear receptor–GFP fusions were prepared with the high-efficiency S65T variant of GFP described by Heim

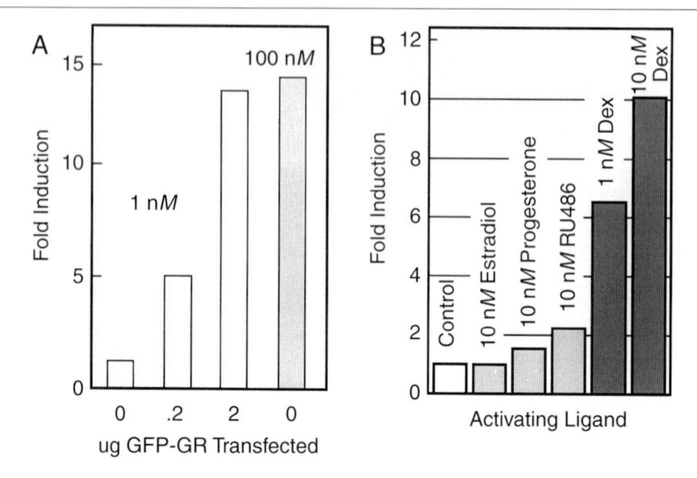

FIG. 4. Transactivation properties of GFP–GR. (A) The response of a transient MMTV reporter plasmid to GFP–GR in 1471.1 cells is shown for 1 n*M* dex (which activates only the GFP–GR fusion; open bars) and 100 n*M* dex (which activates both GFP–GR and the endogenous GR; closed bar). Cell line 1471.1 is described in Archer *et al.* [T. K. Archer, M. G. Cordingley, V. Marsaud, H. Richard-Foy, and G. L. Hager, *in* "Proceedings: Second International CBT Symposium on the Steroid/Thyroid Receptor Family and Gene Regulation" (J. A. Gustafsson, H. Eriksson, and J. Carlstedt-Duke, eds.), p. 221. Birkhauser-Verlag, Berlin, 1989]. (B) Ligand specificity of GFP–GR is presented for activation of endogenous MMTV-LTR-CAT sequences present in cells. Ligand identity and the respective concentrations are displayed above the bars. [Reproduced with permission from H. Htun, J. Barsony, I. Renyi, D. J. Gould, and G. L. Hager, *Proc. Natl. Acad. Sci. U.S.A.* **93,** 4845 (1996).]

et al.[26] This mutation renders the GFP protein stable at mammalian cell growth temperatures, resulting in a much more efficient chromophore. This form of the protein enables the preparation of receptor chimeras that are easily visualized in single-cell analysis.

There are now many color variants of GFP available, permitting for the first time the design of multiple expression experiments in real time in living cells. The first variants available had an excitation spectrum shifted into the ultraviolet (UV) range, and emitted in the visible blue range. These early blue variants were strongly susceptible to photobleaching, and marginally useful for real-time visualization. These forms of GFP were also susceptible to improper folding when fused to target proteins, resulting both in low fluorescence efficiency, and inappropriate localization in subcellular compartments. This last point is, of course, a serious problem.

Fortunately, several stable color-shifted variants with high quantum

[26] R. Heim, A. B. Cubitt, and R. Y. Tsien, *Nature* (*London*) **373,** 663 (1995).

TABLE I
EXCITATION/EMISSION MAXIMA FOR GREEN FLUORESCENT PROTEIN VARIANTS

GFP variant	Excitation maximum (nm)	Emission maximum (nm)	Ref.[a]
GFP–S65T	488	511	1
EGFP	490	507	2
YGFP	513	527	3
BGFP	387	450	4
GFP–W7	433	473	5

[a] References: (1) A. B. Cubitt, R. Heim, S. R. Adams, A. E. Boyd, L. A. Gross, and R. Y. Tsien, *Trends Biochem. Sci.* **20,** 448 (1995); (2 and 3) "Living Colors User Manual." PT2040-1. Clontech Laboratories, Palo Alto, California, 1997; (4) R. H. Stauber, K. Horie, P. Carney, E. A. Hudson, N. I. Tarasova, G. A. Gaitanaris, and G. N. Pavlakis, *Biotechniques* **24**(3), 462 (1998); (5) R. Heim and R. Y. Tsien, *Curr. Biol.* **6,** 178 (1996).

yield are now available (see Table I for an overview of emission/excitation spectra for variants in current use). Pavlakis and colleagues have described a blue GFP that is relatively resistant to photobleaching, and bright enough for abundant proteins.[27] The W7 variant[28] is one of the most stable blue-shifted forms, with an excitation maximum at 433 nm. Enhanced green forms of GFP are commercially available from Clontech (Palo Alto, CA). Also, a "yellow" version (YFP) of the molecular has appeared. Although the excitation/emission spectra for this variant are not widely separated from those of the GFP spectra, it is in fact possible to discriminate between YFP and GFP expressed in the same cell by the use of appropriate instrumentation. Baumann *et al.*[29] describe the use of a new-generation confocal microscope (Leica, Exton, PA) to visualize YFP and GFP separately. This instrument features a continuously adjustable monochromator on the emission side that permits the selection of any desired emission window.

These reagents now provide a powerful new approach in real-time imaging. Multiple protein species can be labeled in different colors, and visualized separately when expressed in the same cell. For example, a given nuclear receptor and a suspected cofactor can be examined for subcellular distribution when coexpressed. In principal, a great deal of experimentation is now possible regarding the distribution, colocalization, and interaction of all the nuclear receptors, cofactors, and other interacting species.

[27] R. H. Stauber, K. Horie, P. Carney, E. A. Hudson, N. I. Tarasova, G. A. Gaitanaris, and G. N. Pavlakis, *Biotechniques* **24**(3), 462 (1998).
[28] R. Heim and R. Y. Tsien, *Curr. Biol.* **6,** 178 (1996).
[29] C. T. Baumann, C. S. Lim, and G. L. Hager, *J. Histochem. Cytochem.* **46**(9), 1073 (1998).

Mode of Expression

Almost all experimentation to date with nuclear receptor GFP chimeras has been performed by transient introduction of the GFP receptor. While this is the mode most readily available, there are potentially serious drawbacks to this method of expression. First, it is now clear that transiently expressed steroid receptors often function aberrantly in gene activation. Smith, Hager, and colleagues have described pronounced differences between the activation potentials of the transient and endogenous progesterone receptor.[30–32] Simons *et al.* also describe major shifts in the dose–response curve for transient versus endogenous receptors.[33] Second, transient transfection often produces high levels of expression per cell, and these levels may induce inappropriate localization of the receptor chimera, or improper interaction with other molecules.

For these reasons, stable introduction of a receptor chimera into a cell line of choice is a much preferred method for study of receptor behavior in that cell. This will likely have three positive effects: the levels of expression per cell are more likely to be in the normal physiological range, all cells in the population will have a similar complement of chimeric receptor, and the stably expressed receptor has a better chance of forming appropriate associations with other components of multiprotein complexes in which receptors usually reside.

Well-characterized systems for the controlled expression of introduced chimeras are also now available. The tetracycline-inducible expression system[34] has been widely used, and has been found to work well with GFP–GR.[35] This approach provides several additional advantages. Levels of receptor expression can be regulated to whatever level is desired, and subcellular trafficking can be observed after expression of the receptor is induced.

Instrumentation

Much of the preliminary characterization for subcellular distribution of a new receptor chimera can be carried out with a standard epifluorescence

[30] C. L. Smith, T. K. Archer, G. Hamlin-Green, and G. L. Hager, *Proc. Natl. Acad. Sci. U.S.A.* **90,** 11202 (1993).

[31] G. L. Hager, T. K. Archer, G. Fragoso, E. H. Bresnick, Y. Tsukagoshi, S. John, and C. L. Smith, *Cold Spring Harbor Symp. Quant. Biol.* **58,** 63 (1993).

[32] C. L. Smith and G. L. Hager, *J. Biol. Chem.* **272**(44), 27493 (1997).

[33] D. Szapary, M. Xu, and S. S. Simons, Jr., *J. Biol. Chem.* **271,** 30576 (1996).

[34] M. Gosen and H. Bujard, *Proc. Natl. Acad. Sci. U.S.A.* **89,** 5547 (1992).

[35] D. Walker, H. Htun, and G. L. Hager, unpublished observations (1998).

microscope. These instruments are now widely available, and when equipped with a high-quality oil emersion lens (×60–100) provide excellent images with sufficient resolution for initial experimentation. For higher resolution studies, however, particularly in attempts to define intranuclear receptor trafficking patterns, a state-of-the-art laser scanning confocal microscope is indispensable. Furthermore, when attempting to discriminate between multiple chromophores, the excitation wavelength selections available on these modern instruments are invaluable.

Newer technology promises to bring more sophistication to the analysis of receptor function with GFP chimeras. Collection of three-dimensional images through "optical deconvolution" or "extensive reassignment of light" is rapidly emerging as an alternative to confocal imaging. Deconvolution offers the significant advantage of utilizing essentially all of the fluorescence emitted from a sample, whereas confocal imaging is only about 1% efficient. This is a major advantage when using chromophores excited in the UV band of the spectrum, because these wavelengths can be quite damaging to cells. The even more recent development of "two-photon" imaging also promises to expand the boundaries of possible experimentation.

Future Potential

Studies on the distribution and trafficking of nuclear receptors using chimeras with the green fluorescent protein and its color variants are in their infancy. Two general points have emerged. First, addition of the large GFP peptide to most nuclear receptors does not seriously impact activities of the receptors, insofar as these properties have been examined. Indeed, in this laboratory we have now tagged eight different receptors (the GR, ER, PR, TR, AHR, PPAR [peroxisome proliferator receptor] α, β, and γ); all of these molecules retain relatively normal transcriptional activation efficiencies and ligand-binding profiles. The fusion chimeras are also efficiently expressed as appropriately sized polypeptides with little evidence of degradation. Second, these reagents clearly provide a powerful new tool with which to examine receptor trafficking and subcellular distribution in living cells. In particular, new insights into the intranuclear distribution of the nuclear receptors have already emerged with the introduction of this technology.

Several applications are obvious for the immediate future. Real-time observation of intracellular receptor movement is now possible. This opens a new approach for investigations into several difficult issues such as receptor cycling and cytoplasmic/nuclear shuttling. The use of fluorescence recovery after photobleaching (FRAP) will permit a direct approach to issues such as receptor cycling off and on chromatin targets. One major avenue of investigation will involve studies of the colocalization of receptors, coactivators, and transcription factors. These reagents will provide a major new tool in exploring architecture of the interphase nucleus, which is poorly understood. The potential exists to detect direct molecular interaction between receptors and interacting factors by the use of fluorescence resonance energy transfer (FRET), or fluorescence correlation spectroscopy. FRET analysis for energy transfer between GFP variants has been reported.[28] It is not yet clear whether these sophisticated approaches for the detection of direct molecular interactions will emerge as a major tool in the study of nuclear receptors. Finally, the ability to monitor receptor function in living cells in response to ligand stimulation provides new opportunities to characterize ligand effects on compartmentalization, and to conduct searches for new classes of agonists and antagonists. In summary, the application of GFP methodology has enormous implications in the study of nuclear receptor biology.

Section II

Localization of Molecules

[10] Localization of Calmodulin in Budding Yeast and Fission Yeast Using Green Fluorescent Protein

By MARK R. FLORY and TRISHA N. DAVIS

Introduction

Calmodulin is a small, multifunctional, eukaryotic protein that reversibly binds Ca^{2+}. Studies in a variety of organisms indicate that calmodulin performs a broad range of functions encompassing both Ca^{2+}-dependent and Ca^{2+}-independent activities. The Ca^{2+}-dependent functions of calmodulin have been studied in a variety of organisms. In smooth muscle cells, Ca^{2+}–calmodulin facilitates muscle contraction by activating myosin light chain kinase.[1] In mammalian skeletal muscle and liver, Ca^{2+}–calmodulin induces glycogenolysis via activation of phosphorylase kinase.[2] Finally, Ca^{2+}–calmodulin regulates motility in *Paramecium*.[3] The genetic analyses afforded by fungal systems shed additional light on calmodulin function. Calmodulin is required for viability in the budding yeast *Saccharomyces cerevisiae*,[4] the fission yeast *Schizosaccharomyces pombe*,[5] and the filamentous fungus *Aspergillus nidulans*.[6] In *S. cerevisiae*, Ca^{2+} binding to calmodulin is dispensable for vegetative growth[7] but is required for survival in mating pheromone and for maintenance of ion homeostasis. In the latter case, Ca^{2+}–calmodulin activates both Ca^{2+}–calmodulin-dependent kinase and the Ca^{2+}–calmodulin-dependent phosphatase calcineurin.[8–10] By contrast, a mutant *A. nidulans* calmodulin that cannot bind Ca^{2+} does not support cell cycle progression,[11] while *S. pombe* calmodulin must retain at

[1] J. T. Stull, *in* "Calmodulin" (P. Cohen and C. B. Klee, eds.), pp. 91–122. Elsevier Science Publishers, Amsterdam, 1988.

[2] P. Cohen, *in* "Calmodulin" (P. Cohen and C. B. Klee, eds.), pp. 123–144. Elsevier Science Publishers, Amsterdam, 1988.

[3] J. A. Kink, M. E. Maley, R. R. Preston, K.-Y. Ling, M. A. Wallen-Friedman, Y. Samai, and C. Kung, *Cell* **62,** 165 (1990).

[4] T. N. Davis, M. S. Urdea, F. R. Masiarz, and J. Thorner, *Cell* **47,** 423 (1986).

[5] T. Takeda and M. Yamamoto, *Proc. Natl. Acad. Sci. U.S.A.* **84,** 3580 (1987).

[6] C. D. Rasmussen, R. L. Means, K. P. Lu, G. S. May, and A. R. Means, *J. Biol. Chem.* **265,** 13767 (1990).

[7] J. R. Geiser, D. van Tuinen, S. E. Brockerhoff, M. M. Neff, and T. N. Davis, *Cell* **65,** 949 (1991).

[8] K. W. Cunningham and G. R. Fink, *Mol. Cell. Biol.* **16,** 2226 (1996).

[9] M. J. Moser, J. R. Geiser, and T. N. Davis, *Mol. Cell. Biol.* **16,** 4824 (1996).

[10] T. C. Pozos, I. Sekler, and M. S. Cyert, *Mol. Cell. Biol.* **16,** 3730 (1996).

[11] A. R. Means, M. F. VanBerkum, I. Bagchi, K. P. Lu, and C. D. Rasmussen, *Pharmacol. Ther.* **50,** 255 (1991).

METHODS IN ENZYMOLOGY, VOL. 302
0076-6879/99 $30.00

least one functional Ca^{2+}-binding site for cell growth.[12] In *A. nidulans,* Ca^{2+}–calmodulin is required for activation of both calcineurin and Ca^{2+}–calmodulin-dependent protein kinase.[13] Similar essential targets of Ca^{2+}–calmodulin in *S. pombe* have not yet been identified.

In addition to genetic and biochemical approaches, localization studies have proved a powerful tool for probing calmodulin function in budding yeast and fission yeast. In budding yeast, calmodulin localizes to two different subcellular sites. At each of these sites, calmodulin performs an essential function independent of Ca^{2+} binding. First, calmodulin localizes to the yeast centrosome or spindle pole body (SPB). At the SPB, calmodulin binds the coiled-coil protein Spc110p.[14–16] The interaction between calmodulin and Spc110p is essential for proper assembly of spindle pole components[17] and for the integrity of the spindle during mitosis.[17–21] Second, calmodulin localizes to cortical actin patches and, during cytokinesis, to the neck region.[22] Calmodulin functions at these locations to facilitate polarized growth.[18,23] This localization is mediated by the unconventional type V myosin Myo2p,[22] which colocalizes with calmodulin to sites of cell growth.[22,24–26] In *S. pombe,* calmodulin similarly localizes to the SPB and to sites of polarized growth. Furthermore, fission yeast strains with temperature-sensitive mutations in the calmodulin gene demonstrate defects in spindle architecture and chromosome segregation similar to those observed for conditional *S. cerevisiae* calmodulin mutants.[27] Despite these similarities, corresponding targets of calmodulin activity in *S. pombe* have not yet been identified.

Although localization studies have been a powerful starting point for understanding calmodulin function in budding yeast and fission yeast, localization of calmodulin using traditional methods involving cell fixation and

[12] M. J. Moser, S. Y. Lee, R. E. Klevit, and T. N. Davis, *J. Biol. Chem.* **270,** 20643 (1995).
[13] C. D. Rasmussen, K. P. Lu, R. L. Means, and A. R. Means, *J. Physiol. Paris* **86,** 83 (1992).
[14] J. R. Geiser, H. A. Sundberg, B. H. Chang, E. G. Muller, and T. N. Davis, *Mol. Cell. Biol.* **13,** 7913 (1993).
[15] J. V. Kilmartin, S. L. Dyos, D. Kershaw, and J. T. Finch, *J. Cell Biol.* **123,** 1175 (1993).
[16] D. A. Stirling, K. A. Welch, and M. J. Stark, *EMBO J.* **13,** 4329 (1994).
[17] H. A. Sundberg, L. Goetsch, B. Byers, and T. N. Davis, *J. Cell Biol.* **133,** 111 (1996).
[18] T. N. Davis, *J. Cell Biol.* **118,** 607 (1992).
[19] J. V. Kilmartin and P. Y. Goh, *EMBO J.* **15,** 4592 (1996).
[20] D. A. Stirling, T. F. Rayner, A. R. Prescott, and M. J. R. Stark, *J. Cell Sci.* **109,** 1297 (1996).
[21] G. H. Sun, A. Hirata, Y. Ohya, and Y. Anraku, *J. Cell Biol.* **119,** 1625 (1992).
[22] S. E. Brockerhoff, R. C. Stevens, and T. N. Davis, *J. Cell Biol.* **124,** 315 (1994).
[23] Y. Ohya and D. Botstein, *Science* **263,** 963 (1994).
[24] S. E. Brockerhoff and T. N. Davis, *J. Cell Biol.* **118,** 619 (1992).
[25] S. H. Lillie and S. S. Brown, *J. Cell Biol.* **125,** 825 (1994).
[26] G. H. Sun, Y. Ohya, and Y. Anraku, *Protoplasma* **166,** 1625 (1992).
[27] M. J. Moser, M. R. Flory, and T. N. Davis, *J. Cell Sci.* **110,** 1805 (1997).

immunostaining has been confounded by several factors. As discussed above, the multiple functions of calmodulin result in localization of calmodulin to many sites within the same cell. In *S. cerevisiae,* fixation conditions that allow visualization of calmodulin at the SPB eliminate the ability to observe calmodulin at sites of polarized growth, and vice versa.[14,16,17,24,26] While calmodulin may be observed at both sites simultaneously in a fixed *S. pombe* cell, it is often difficult to resolve these multiple localization patterns. Localization to the SPB occurs along longitudinal axes within the cell while localization to sites of polarized growth is generally confined to the cortex. Thus, these two patterns can sometimes be separated in different focal planes. However, localization of calmodulin to the SPB is often obscured by an overlapping signal from calmodulin localized to one of the numerous actin patches accompanying cell wall growth.[28,29] Furthermore, immunolocalization of mutant proteins is often further hindered because the mutant proteins have lost antigenicity or are expressed at a reduced level.[30] Finally, comparison of immunolocalization patterns of homologous proteins in different organisms is often biased by the process of fixation, as different fixation regimes are almost always required for effective immunostaining in different organisms.

To overcome the problems inherent in localizing calmodulin in fixed yeast cells, we have taken advantage of the ability to easily create functional protein fusions to calmodulin. Calmodulin and calmodulin-like proteins are often fused to other proteins to facilitate characterization of calmodulin-regulated proteins. In these cases, calmodulin fusions have been successfully used to detect proteins in electrophoretic gels or proteins transferred to membranes using both radioactive and nonradioactive detection systems. These assays employ fusions of calmodulin to biotin,[31] protein A,[32] glutathione-*S*-transferase,[33] and a hemagglutinin epitope.[34] More recently, the green fluorescent protein (GFP) of the jellyfish *Aqueoria victoria* was fused to calmodulin and calmodulin-like proteins, creating powerful reagents for both biochemical assays and also for subcellular localization studies. Fusions of GFP(S65T) to human calmodulin and to the human calmodulin-like protein (CLP) have been successfully used as probes in far-Western blotting

[28] M. Flory and T. Davis, unpublished observation (1998).

[29] J. A. Waddle, T. S. Karpova, R. H. Waterston, and J. A. Cooper, *J. Cell Biol.* **132,** 861 (1996).

[30] T. Davis, unpublished observation (1997).

[31] M. L. Billingsley, J. W. Polli, K. R. Pennypacker, and R. L. Kincaid, *Methods Enzymol.* **184,** 451 (1990).

[32] D. A. Stirling, A. Petrie, D. J. Pulford, D. T. W. Paterson, and M. J. Stark, *Mol. Microbiol.* **6,** 703 (1992).

[33] R. Fischer, Y. Wei, and M. Berchtold, *Biotechniques* **21,** 292 (1996).

[34] G. Szymanska, M. B. O'Connor, and C. M. O'Connor, *Anal. Biochem.* **252,** 96 (1997).

for the detection of calmodulin- and CLP-binding proteins.[35] Of even greater significance, GFP fusions eliminate the problems of fixation by allowing localization studies to be performed in live cells. For example, fusions of GFP to the small calcium-binding protein centrin clearly label the mitotic spindle poles in live cultured cells.[36]

Given the confounding factors inherent to fixation of yeast cells, fusions to GFP have proved to be a powerful tool for localization studies in both the budding yeast *S. cerevisiae* and the fission yeast *S. pombe.* General methods have previously been described for the construction, preparation, and visualization of GFP fusion proteins in budding yeast[37] and fission yeast.[38] Here we describe the construction, preparation, and visualization of fusions between GFP and calmodulin in budding yeast and fission yeast. Our experience from work with GFP–calmodulin fusions in two yeasts provides a sound framework for others planning to use GFP to localize proteins in yeast. Multiple methods for both construction and observation of fusions are discussed, and we also provide strategies for visualization of GFP fusions that localize to more than one subcellular site. We find that localization using GFP–calmodulin fusion proteins in live yeast cells complements and extends findings from procedures requiring fixation and immunostaining of cells. GFP–calmodulin fusion proteins permit simultaneous observation of calmodulin at multiple sites in single cells. These multiple localization patterns may be differentiated from one another on the basis of their kinetics and position relative to the cell cortex. Moreover, in budding yeast, use of GFP–calmodulin fusion proteins allows localization of a mutant calmodulin that cannot be localized using immunostaining of fixed cells. Finally, use of GFP–calmodulin fusion proteins allows comparison of calmodulin localization in budding yeast and fission yeast without the biases of cell fixation.

Fusion of Green Fluorescent Protein to Calmodulin in Budding Yeast and Fission Yeast

Budding Yeast

In constructing fusions to GFP, we have found it useful to include a flexible glycine–alanine spacer between GFP and the protein of interest.

[35] N. Garamszegi, Z. P. Garamszegi, M. S. Rogers, S. J. DeMarco, and E. E. Strehler, *Biotechniques* **23,** 864 (1997).

[36] R. A. White and J. L. Salisbury, Regulation of centrosome protein targeting. *In* "ASCB/EMBO/H. Dudley Wright Conference: Centrosomes and Spindle Pole Bodies." University of California, Santa Cruz, 1997.

[37] S. L. Shaw, E. Yeh, K. Bloom, and E. D. Salmon, *Curr. Biol.* **7,** 701 (1997).

[38] K. Nabeshima, S. Saitoh, and M. Yanagida, *Methods Enzymol.* **283,** 459 (1997).

This linker likely allows the relatively large GFP moiety more thermodynamic degrees of freedom. This increases the likelihood that the fusion protein will be functional by allowing the two halves of the fusion to fold properly and by reducing the ability of GFP to inhibit interactions with target proteins. We have incorporated this Gly–Ala spacer into functional GFP fusions with a broad range of yeast proteins. These include small proteins such as calmodulin[27] and thioredoxin,[39] as well as much larger proteins such as Spc110p,[40] Myo2p,[41] and the SPB component Spc97p.[42] In all of these cases, the function of the GFP fusion was demonstrated by showing that the fusion can support growth of a strain deleted for the corresponding wild-type gene. The fact that smaller proteins such as calmodulin and thioredoxin are still functional even when fused to GFP is particularly remarkable, as GFP is roughly twice the size of these small proteins. However, we caution that while a GFP fusion may be able to substitute for the wild-type protein under most conditions, the fusion may not functionally mimic the wild-type protein in all circumstances. For example, in budding yeast, fusion of GFP(S65T) to calmodulin exacerbates the mutant phenotype of certain conditional *spc110* alleles despite the fact that this fusion fully complements a chromosomal deletion of the wild-type calmodulin gene.[40]

In budding yeast, we find that GFP(S65T)–calmodulin fusions can be readily visualized when expressed from centromeric (CEN) plasmids under control of the native *CMD1* promoter. CEN plasmids may be stably maintained at a low copy number in a culture through selection for an auxotrophic marker on the plasmid.[43] However, a small fraction of cells expressing GFP fusions from CEN plasmids in this manner show either no fluorescent signal or localization of GFP throughout the cell. A majority of cells show appropriate localization patterns. Two mutant calmodulin alleles, *cmd1-6* and *cmd1-1,* were also fused to *GFP(S65T)* on centromeric plasmids. However, only fusion of *GFP(S65T)* to *cmd1-6* could functionally replace a deletion of *CMD1,* indicating that fusion of *GFP(S65T)* to the temperature-sensitive *cmd1-1* allele is a lethal mutation.

The most consistent localization patterns of S65T-GFP–calmodulin fusions across a population of live budding yeast cells were obtained with single-copy GFP(S65T)–calmodulin fusions integrated into the genome. Two methods, utilizing either a cloning-based strategy or a polymerase chain reaction (PCR)-based strategy, can be used to integrate GFP fusions in the budding yeast genome. We used the following cloning-based method

[39] L. Schoening and E. Muller, personal communication (1996).

[40] M. Moser and T. Davis, unpublished observation (1994).

[41] R. Stevens and T. Davis, unpublished observation (1997).

[42] T. Nguyen and T. Davis, unpublished observation (1996).

[43] P. Hieter, D. Pridmore, J. H. Hegemann, M. Thomas, R. W. Davis, and P. Philippsen, *Cell* **42,** 913 (1985).

to integrate GFP–calmodulin fusions at the calmodulin (*CMD1*) locus in the budding yeast genome. Initially, a plasmid was constructed that contains the *CMD1* promoter and *GFP*(*S65T*), fused in-frame to calmodulin (*GFP*(*S65T*)::*CMD1*), a stop codon, and the *CMD1* terminator. A linear DNA fragment from this plasmid containing the *CMD1* promoter, *GFP*(*S65T*)::*CMD1,* a stop codon, and the *CMD1* terminator was transformed into a strain containing a chromosomal deletion of *CMD1.* This strain is normally maintained at room temperature by a plasmid containing the temperature-sensitive calmodulin mutant allele *cmd1-1.*[18] The *cmd1-1* allele allows growth at 21° but not at 37°. Taking advantage of the temperature-sensitive nature of the *cmd1-1* allele, we isolated cells that had integrated the *GFP*(*S65T*)::*CMD1* linear DNA fragment into the genome by selection for cells that could grow at 37° following transformation. Proper integration at the *CMD1* locus was confirmed by Southern blot, PCR, and DNA sequence analysis.[27]

If the GFP is to be fused to the C terminus of the protein, we recommend the use of a convenient integrative PCR-based method for integration.[44] This method allows construction of single-copy C-terminal GFP fusion proteins integrated into the genome under control of the native promoter. This procedure demands no plasmid construction but does require the synthesis of two long (~80-mer) gene-specific primers. One gene-specific primer contains sequence encoding the C terminus of the protein of interest while the second primer contains sequence flanking the 3′ end of the coding sequence of interest. These primers are used to amplify the GFP tag and a selectable marker from a plasmid cassette. The primers are designed such that the resulting linear PCR product encodes the C terminus of the protein of interest fused in-frame to GFP, a stop codon, a selectable marker, and sequence flanking the 3′ end of the coding sequence of interest. This linear product is then transformed into either a haploid or diploid yeast strain, where it is integrated into the genome by homologous recombination. Integration at the correct locus is confirmed by diagnostic PCR and Southern blotting. Cassettes are available for GFP tagging utilizing resistance to G418 or, alternatively, histidine prototrophy as conferred by the *his5*$^+$ gene of *S. pombe.* The *his5*$^+$ gene of *S. pombe* complements mutations in the *HIS3* gene of *S. cerevisiae.* Use of the *his5*$^+$ gene of *S. pombe* allows selection without the use of a *S. cerevisiae* selectable marker, which can promote integration at genomic sites away from the locus of interest. Finally, while this method employs no flexible linker between GFP and the protein of

[44] A. Wach, A. Brachat, C. Alberti-Segui, C. Rebischung, and P. Philippsen, *Yeast* **13,** 1065 (1997).

interest, we have successfully used integrative PCR strategies to create functional GFP fusions to two yeast spindle pole components.[28,45]

Fission Yeast

In *S. cerevisiae,* episomal expression of GFP–calmodulin fusions is useful, whereas in *S. pombe* localization patterns are too inconsistent when expressing GFP–calmodulin fusions from multicopy plasmids. A significant fraction of cells showed no GFP fluorescence, while other cells exhibited bright fluorescence across the entire cytoplasm and nucleus.[40] A likely explanation for this variability is the relatively large range in copy number inherent to most noncentromeric plasmids. *Schizosaccharomyces pombe* centromeres are too large to be accommodated on plasmids and therefore cannot be used to maintain plasmids at low copy number.[46] However, *S. pombe* plasmids have been developed that allow a low copy number to be maintained either by a highly efficient, autonomously replicating sequence element on the plasmid[47] or by experimental manipulation of G418 concentration.[48] In some instances the elevated gene dosage conferred by high copy number plasmids in fission yeast provides an advantage. For example, when the spindle pole body (SPB) component Sad1p[49] is fused to GFP and expressed from a multicopy plasmid, visualization of both the SPB and the nuclear envelope is possible. Although not necessarily a reflection of the true *in vivo* localization of Sad1p, the additional visualization at the nuclear envelope facilitates characterization of conditional mutants such as the temperature-sensitive topoisomerase mutant *top2-191*.[38]

As with budding yeast cells, the most consistent localization patterns of GFP–calmodulin fusions across a population of live fission yeast cells were obtained with single-copy GFP–calmodulin fusions integrated into the genome. Cloning-based strategies or PCR-based strategies similar to those described above for *S. cerevisiae* can be used to integrate GFP fusions in the fission yeast genome. We used the following cloning-based method to integrate GFP–calmodulin and S65T-GFP–calmodulin fusions at the calmodulin (*cam1*$^+$) locus in the fission yeast genome. A strain was created that contains a chromosomal deletion of the *cam1*$^+$ open reading frame (ORF) but that is viable when carrying a plasmid encoding vertebrate calmodulin. Vertebrate calmodulin can functionally replace *S. pombe* cal-

[45] T. Nguyen and T. Davis, unpublished observation (1998).

[46] Y. Chikashige, N. Kinoshita, Y. Nakaseko, T. Matsumoto, S. Murakami, O. Niwa, and M. Yanagida, *Cell* **57,** 739 (1989).

[47] C. Brun, D. D. Dubey, and J. A. Huberman, *Gene* **164,** 173 (1995).

[48] H. Tohda, H. Okada, Y. Giga-Hama, H. Okayama, and H. Kumagai, *Gene* **150,** 275 (1994).

[49] I. Hagan and M. Yanagida, *J. Cell Biol.* **129,** 1033 (1995).

modulin.[12] This strain was transformed with an uncut plasmid containing the *cam1*$^+$ promoter, *GFP,* or *GFP(S65T)* fused in-frame to calmodulin, a stop codon, the *cam1*$^+$ terminator, and finally a selectable marker conferring leucine prototrophy. Because the chromosomal *cam1*$^+$ ORF in this strain is absent, integration of *GFP*- or *GFP(S65T)*–calmodulin at the *cam1* locus, driven by homology in the *cam1* promoter and terminator sequences, was observed to occur at a low frequency. Leucine prototrophs that had lost the plasmid encoding vertebrate calmodulin and that had integrated the *GFP-cam1*$^+$ or *S65T-GFP-cam1*$^+$ sequences at the *cam1* locus were thus identified. Proper integration was confirmed by Southern blot analysis.[27] As an alternative to controlling expression from the native promoter, similar strategies are available for construction of single-copy GFP fusions integrated into the genome with expression controlled by either the constitutive *ADH1* promoter[50] or the thiamine-regulatable *nmt1* promoter.[38,51]

Similar to the synthetic lethality produced by fusion of GFP to the budding yeast mutant calmodulin protein Cmd1-1p, creation of an integrated, single-copy fusion of GFP to the conditional fission yeast calmodulin mutant CamE14p was not possible. Localization of CamE14p by immunofluorescence using anti-calmodulin antibodies was not informative, as only a broad pattern of fluorescence was seen in the nucleus.[52] Therefore, an attempt was made using the cloning-based method described above to create fusions of GFP(S65T) to the temperature-sensitive *S. pombe* calmodulin mutant protein Cam1E14p. However, we were unable to isolate single-copy integrants. Southern blot analysis showed that the leucine prototroph strains we did isolate contained multiple copies of the *GFP(S65T)::camE14* plasmid integrated into the genome. The GFP(S65T)–calmodulin in these cells was overexpressed at levels 15 times higher than the wild-type GFP fusion, as measured by fluorescence-activated cell sorting (FACS). As a consequence, GFP(S65T)–Cam1E14p fusion localized uniformly throughout the cells at both permissive and nonpermissive temperatures, as did plasmid-encoded fusion protein. Thus it was concluded that fusion of GFP(S65T) to the temperature-sensitive calmodulin encoded by *cam1-E14* is a lethal mutation that can be rescued only by overexpression.[27]

Derivatives of the integrative PCR-based method described above for use in budding yeast[44] are now available for integration of GFP fusions

[50] P. R. Russell and B. D. Hall, *J. Biol. Chem.* **258,** 143 (1983).
[51] K. Maundrell, *J. Biol. Chem.* **265,** 10857 (1990).
[52] M. Flory and T. Davis, unpublished observation (1996).

into the fission yeast genome.[53] We now favor the integrative PCR approach, as it reduces the amount of plasmid construction required. Plasmid cassettes are available for the construction of both single-copy C-terminal GFP fusions under control of the native promoter and single-copy N-terminal GFP fusions under control of the *nmt1 promoter*.[51] Selection is based on a marker conferring resistance to G418.

Preparation of Live Budding Yeast Cells and Live Fission Yeast Cells for Microscopy

For preparation of budding yeast, an equal volume (~3 μl) of a culture containing a fusion of GFP to calmodulin grown in yeast extract–peptone–dextrose (YPD) medium at 21° or 30° was mixed with synthetic dextrose complete (SDC) medium containing 1% low melting temperature agarose on a microscope slide. For preparation of fission yeast, an equal volume (~3 μl) of a culture containing a fusion of GFP or GFP(S65T) to calmodulin grown in yeast extract-liquid (YEL) medium at 21° or 30° was mixed with supplemented Edinburgh Minimal Medium 2 (EMM2) medium containing 37° low melting temperature agarose (LMA) on a microscope slide. For both budding and fission yeast samples, a 22 × 22 mm coverslip was then placed on top of the mixture, which was then allowed to solidify at room temperature under an ~50-g weight for 10 min. Cells prepared in this manner could be viewed for up to 2 hr, at which point the cells begin to demonstrate a diffuse and blotchy localization of GFP. Individual cells could be subjected to continuous excitation for 30 sec before the GFP signal began to bleach. Therefore, excitation was removed except when needed to identify, focus, and photograph cells.[27,52]

For rapid scanning of cells requiring less than 15 min, the SDC–LMA or EMM2–LMA may be omitted. A small amount (~5 μl) of liquid SDC or EMM2 medium containing cells is placed on a slide and covered with a 22 × 22 mm coverslip. No weight need be applied to the sample. Beyond 15 min, samples begin to dry out and the GFP localization becomes diffuse and blotchy. This method is less desirable for photographs, as the cells are not held stationary.[52] Alternatively, a procedure involving mounting of live budding yeast cells onto slabs of 25% gelatin containing minimal medium is particularly suited to time-lapse photography.[37] For examination of high-density cells in the same focal plane, fission yeast cells containing GFP fusions may be cultured between agarose and a concanavalin A-coated

[53] J. Bahler, J. Q. Wu, M. S. Longtine, N. G. Shah, A. McKenzie III, A. B. Steever, A. Wach, P. Philippsen, and J. R. Pringle, *Yeast* **14,** 943 (1998).

coverslip. Finally, to reduce the significant fluorescent background inherent to agarose, fission yeast cells may be cultured and visualized in a microwell at the bottom of a 35-mm plastic culture dish.[38]

Microscopy for Observing Green Fluorescent Protein–Calmodulin Fusions

We viewed GFP–calmodulin fusions in yeast cells using a Zeiss Axioplan (Carl Zeiss, Thornwood, NY) microscope with a ×100 1.4 objective and an optivar set at 1.25. The microscope was equipped with a 100-W mercury lamp as the light source. The filter set used was the XF14 (425DF45, 475DCLPO2, 535DF55) (Omega Optical, Brattelboro, VT). Images were captured using an Imagepoint cooled CCD video camera (Photometrics, Tuscon, AZ) fitted to the microscope in conjunction with IP laboratory software (Signal Analytics Vienna, VA). Fusions containing the S65T mutation could always be easily seen by eye, whereas fusions without this mutation were more easily viewed using the CCD camera with increased gain and exposure time. To visualize and photograph cells containing GFP–camodulin fusions without the S65T mutation, we employed laser scanning confocal microscopy. In this case, cells were visualized with an MRC600 laser scanning confocal microscope (Bio-Rad, Hercules, CA) on 10% power with a ×60 objective. In fission yeast, GFP–calmodulin localization patterns assessed by confocal laser scanning microscopy were identical to that of GFP(S65T)–calmodulin observed using fluorescence microscopy. Time-lapse images were taken with exposures collected manually at ~2-min intervals. Time-lapse images were then compiled into a movie format using QuickTime software. All other images were prepared for publication with Photoshop 4.0 (Adobe Systems, San Jose, CA).[27]

Three-dimensional deconvolution fluorescence microscopy may provide the most informative subcellular localization of GFP fusion proteins in the future. This method generates a three-dimensional representation of localization patterns within the volume of a yeast cell. In this method, several two-dimensional images are collected serially at different focal planes through the sample. These images are then compiled into a three-dimensional image using computerized deconvoluting techniques that minimize optical distortion from out-of-focus information.[54,55]

[54] D. A. Agard, Y. Hiraoka, P. Shaw, and J. W. Sedat, *Methods Cell. Biol.* **30,** 353 (1989).

[55] Y. Hiraoka, J. R. Swedlow, M. R. Paddy, D. A. Agard, and J. W. Sedat, *Semin. Cell Biol.* **2,** 153 (1991).

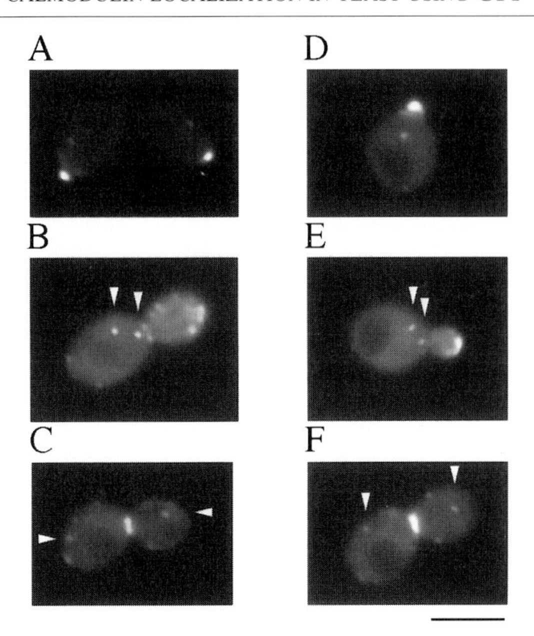

FIG. 1. Localization of GFP(S65T)–Cmd1p (A–C) and GFP(S65T)–Cmd1-6p (D–F) to the SPB of *S. cerevisiae* as visualized by fluorescence microscopy. GFP–calmodulin localization to the SPB (arrowheads). Bar: 5 μm. [Reproduced from M. J. Moser, M. R. Flory, and T. N. Davis, *J. Cell Sci.* **110,** 1805 (1997), with permission from the Company of Biologists, Ltd.]

Observation of Green Fluorescent Protein–Calmodulin Fusions at Spindle Pole Body and Sites of Polarized Growth in Live Cells

Localization to Spindle Pole Body

In both live budding yeast cells and live fission yeast cells, GFP(S65T)–calmodulin is found to localize to the spindle pole body (SPB). In *S. cerevisiae,* unbudded cells and cells with small buds each contain one dot of fluorescence, as expected for either a single SPB or for two SPBs in side-by-side configuration (Fig. 1A). Cells with a medium bud contain two small fluorescent dots near the neck region of the cell as expected for a protein at the poles of a short spindle (Fig. 1B). In a large budded cell, both mother and daugher contain a dot of GFP–Cmd1p, presumably corresponding to their single SPB (Fig. 1C).[27] This localization pattern agrees with localization patterns determined by both immunofluorescence microscopy[14,16] and by immunoelectron microscopy.[17,56,57] The Ca^{2+}-binding mutant GFP–

[56] A. Spang, K. Grein, and E. Schiebel, *J. Cell Sci.* **109,** 2229 (1996).
[57] H. Sundberg and T. Davis, unpublished observation (1995).

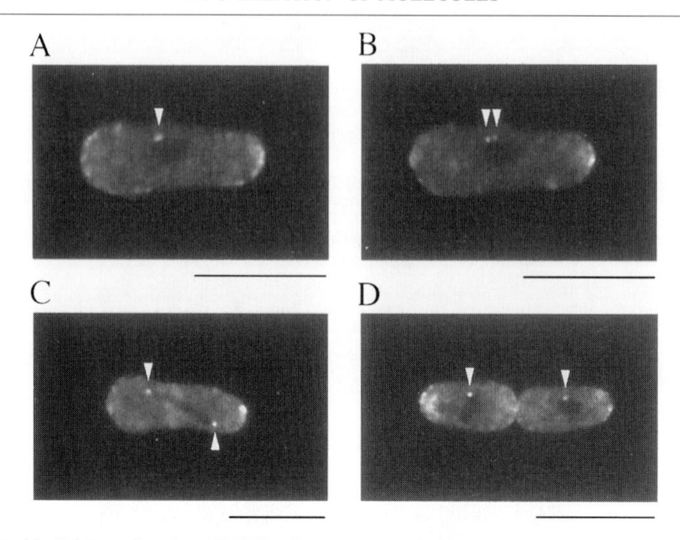

FIG. 2. (A–D) Localization of GFP–Cam1p to the SPB of *S. pombe* as visualized by laser scanning confocal microscopy. Note that the focal plane optimal for visualization of the SPB (arrowheads) results in poor visualization of the cortical cytoskeleton. Bars: 5 μm. [Reproduced from M. J. Moser, M. R. Flory, and T. N. Davis, *J. Cell Sci.* **110,** 1805 (1997), with permission from the Company of Biologists, Ltd.]

Cmd1-6p fusion had a pattern of localization to the SPB identical to that of the wild-type fusion protein (Fig. 1D–F). This is the first demonstration that a mutant calmodulin defective in binding Ca^{2+} localizes to the SPB. Methods involving fixation of cells and immunostaining did not allow observation of this localization.[27]

In *S. pombe,* small fluorescent dots behaved exactly as previously demonstrated for both the budding and fission yeast SPBs.[58–60] In cells early in their cell cycle as judged by their small size, a single spot of fluorescence was observed at the edge of the nucleus (Fig. 2A). The dot of fluorescence elongated and eventually divided (Fig. 2B). Later the separation between the dots increased (Fig. 2C), with one spot of the pair eventually being distributed to each new daughter cell following mitosis and cytokinesis (Fig. 2D). Notably, calmodulin was not seen at the additional MTOCs that nucleate microtubules near the septum in *S. pombe* cells following mitosis. This localization pattern matches those determined by immunofluorescence

[58] B. Byers, *in* "The Molecular Biology of the Yeast *Saccharomyces*—Life Cycle and Inheritance" (J. N. Strathern, E. W. Jones, and J. R. Broach, eds.), pp. 59–96. Cold Spring Harbor Laboratory Press, Cold Spring Harbor, New York, 1981.

[59] I. M. Hagan and J. S. Hyams, *J. Cell Sci.* **89,** 343 (1988).

[60] E. K. McCully and C. F. Robinow, *J. Cell Sci.* **9,** 475 (1971).

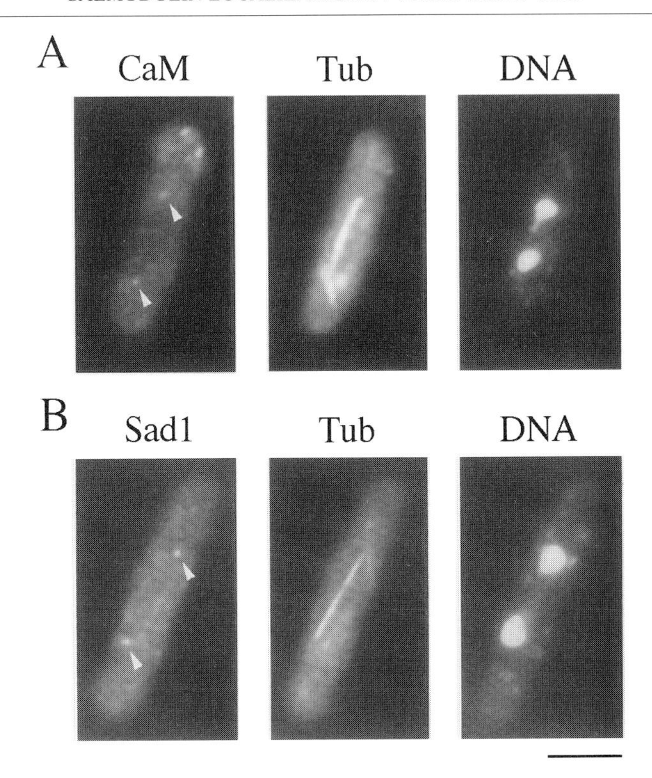

FIG. 3. Immunolocalization of calmodulin and Sad1p at the ends of the *S. pombe* spindle (Tub). A wild-type culture of *S. pombe* (*cam1*$^{+}$) grown at 30° to logarithmic phase in YEL supplemented with adenine, leucine, and uracil was prepared for immunofluorescence as described by Moser *et al.*[27] Arrowheads indicate Cam1p or Sad1p localization to the SPB. Bar: 5 μm. [Reproduced from M. J. Moser, M. R. Flory, and T. N. Davis, *J. Cell Sci.* **110,** 1805 (1997), with permission from the Company of Biologists, Ltd.]

using antibodies against Cam1p, against tubulin, and against a known SPB component, Sad1p[49] (Fig. 3). A QuickTime movie of the GFP–Cam1p behavior at the SPB can viewed at http://weber.u.washington.edu/~moser/GFP-Cam1p.mov.[27]

Localization to Sites of Polarized Growth

In both live budding yeast cells and live fission yeast cells, GFP(S65T)–calmodulin also localizes to sites of polarized growth. In *S. cerevisiae,* unbudded cells and cells with small buds each have a bright patch of fluorescence at the nascent bud site or the small bud (Fig. 1A). Cells with a medium bud contain patches of fluorescence that appear in the bud similar to the

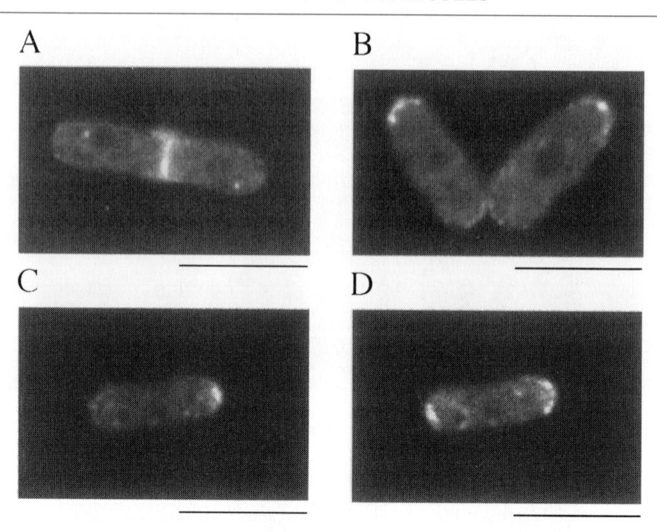

FIG. 4. Localization of GFP–Cam1p to sites of polarized growth in *S. pombe* as visualized by laser scanning confocal microscopy. (C) and (D) are images of the same cell taken 9 min apart. Note that the focal plane optimal for visualization of the cortical cytoskeleton results in poor visualization of the SPB. Bars: 5 μm. [Reproduced from M. J. Moser, M. R. Flory, and T. N. Davis, *J. Cell Sci.* **110,** 1805 (1997), with permission from the Company of Biologists, Ltd.]

pattern seen for the actin capping protein, Cap2p.[29] In a large budded cell, a bright ring appears at the neck between mother and daughter (Fig. 1C).[27] This localization pattern is in complete agreement with studies using immunofluorescence microscopy,[14,16] which indicated localization of calmodulin to areas of cell growth overlapping sites of actin patches.[24,26] This localization pattern has also been confirmed by immunoelectron microscopy.[17,56,57] The Ca^{2+}-binding mutant GFP–Cmd1-6p fusion had a pattern of localization to the sites of cell growth identical to that of the wild-type fusion protein (Fig. 1D–F).

Sites of polarized growth labeled by GFP–calmodulin are also observed in *S. pombe.* In fission yeast cells undergoing septation and cytokinesis, bright fluorescence was found on both sides of the growing septum dividing the cell (Fig. 4A). In a newly divided cell immediately following cytokinesis, fluorescence was seen to move from the septum to the opposite or "old" end of the cell (Fig. 4B and C). The old end corresponds to the former end of the mother cell and is in opposition to the end directly created by septation. *Schizosaccharomyces pombe* cells are known to grow only from the old end just after division.[61] Later the fluorescence redistributed to

[61] J. M. Mitchison and P. Nurse, *J. Cell Sci.* **75,** 357 (1985).

both ends of the growing cell (Fig. 4D). This bipolar localization coincides with an event in the *S. pombe* cell cycle known as NETO, or new end take off, the time when both ends of the cell begin to grow.[61] These observations support the localization patterns determined by immunofluorescence using antibodies against Cam1p in cells fixed by a modified combined aldehyde method.[27]

Multiple Localization Patterns of Green Fluorescent Protein–Calmodulin: Temporal and Spatial Differentiation

It is of great convenience to simultaneously observe the multiple localization patterns of calmodulin in a single living yeast cell. In *S. cerevisiae,* the simultaneous localization of calmodulin at both the SPB and sites of polarized growth in a single cell is only possible using GFP(S65T)–calmodulin fusions. Previous methods employing fixation showed calmodulin at only one place or the other.[16,17,26,62,63] Furthermore, in both live budding yeast and live fission yeast cells expressing GFP–calmodulin, differences in the kinetics of the multiple localization patterns allow clear differentiation between localization to the SPB and to sites of polarized growth. Separation of SPBs from the center to the ends of individual cells embedded in LMA takes approximately 20 min at room temperature in both budding and fission yeast.[27] This is in marked contrast to GFP–calmodulin at sites of polarized growth, which can be seen to undergo rapid and extensive remodeling in seconds.[27,29] Thus, GFP–calmodulin in live cells observed at the relatively stable SPBs was easily distinguished from GFP–calmodulin at areas of polarized growth, which rapidly transited along the cell periphery. Furthermore, in *S. pombe* observation of dynamic events such as NETO are facilitated by the ability to monitor GFP–calmodulin in single cells transiting an entire cell cycle. NETO is much more difficult to recognize in fixed cells, where asymmetric staining patterns resembling the initial part of NETO often result from incomplete fixation and/or permeabilization of cells.[52]

In both yeasts, these two localization patterns can also be separated using spatial criteria. SPBs marked by GFP–calmodulin fusions separate along the interior longitudinal axis of the cell. Therefore, the SPB fluorescence is most clearly seen in a focal plane that bisects the center of the cell. By contrast, GFP–calmodulin at sites of polarized growth are generally confined to the cortical periphery of the cell. Thus, the localization to sites

[62] S. E. Brockerhoff, C. G. Edmonds, and T. N. Davis, *Protein Sci.* **1,** 504 (1992).
[63] J. R. Geiser, Ph.D. Thesis, University of Washington, Seattle (1993).

of polarized growth is most clearly seen when the focus is in a plane that is level with the cortex of the cell. In this focal plane, localization of GFP–calmodulin at the SPBs is often indistinct. This is particularly true in *S. pombe* (Fig. 4B–D). Furthermore, use of GFP–calmodulin fusions in live cells precludes the need to fix and permeabilize cells, methods often necessary for labeling of a SPB epitope but that frequently distort the cortical shape of the cell. For example, permeabilization of *S. pombe* cells with subsequent fixation in methanol/acetone allows robust localization of SPBs using anti-calmodulin antibodies, but the processed cells ultimately appear round rather than elliptical. Finally, fixation in both yeasts often eliminates useful subcellular reference points such as the nucleus, which remain intact in live cells containing GFP fusions.[52]

Visualization of Calmodulin–Green Fluorescent Protein in Fixed Yeast Cells

Using immunofluorescence techniques, GFP(S65T) fusions may be visualized in *S. cerevisiae* cells fixed on ice for 15 min in 3.7% formaldehyde. Both polyclonal (Santa Cruz, Santa Cruz, CA) and monoclonal (Clontech, Palo Alto, CA) antibodies against the GFP protein are now available for localization of GFP fusion proteins by immunofluorescence microscopy. This is especially useful for purposes of colocalization, as cells fixed in this manner may be stained with multiple sets of antibodies for immunofluorescence microscopy. This procedure has proven effective for colocalization of spindle pole components. For example, costaining with a pool of mouse-derived mixed monoclonal antibodies[64] and a rabbit-derived polyclonal antibody specific for GFP (Santa Cruz) effectively colocalizes Spc98p with GFP(S65T)–Spc97p in cells fixed as described above.[42]

Attempts were made to visualize GFP–calmodulin fusions in *S. pombe* cells fixed using a modified combined aldehyde procedure.[27] However, as these GFP fusions did not contain the enhancing S65T mutation, the fluorescence intensity from GFP substantially quenched following fixation. Only a faint background localization to the mitochondria was detectable. However, useful protocols are available for visualization of the fluorescent signal from GFP(S65T) fusions in fission yeast cells fixed with methanol or glutaraldehyde.[38] Immunofluorescence can then be performed for purposes of subcellular colocalization.

Acknowledgments

This work was supported by National Institutes of Health Grants T32 GM-07270 (to M.R.F.) and GM-40506 (to T.N.D.).

[64] M. P. Rout and J. V. Kilmartin, *J. Cell Biol.* **111,** 1913 (1990).

[11] Analysis of Microtubule Organization and Dynamics in Living Cells Using Green Fluorescent Protein–Microtubule-Associated Protein 4 Chimeras

By KEITH R. OLSON and J. B. OLMSTED

Introduction

The inherent dynamic instability of microtubules is essential for a variety of cellular processes ranging from cell division and cell motility to establishment of cell shape and polarity. Even subtle perturbation of microtubule dynamics during mitosis can result in mitotic defects and cell death. Evidence has established a link between regulation of cellular kinases, microtubule dynamics, and cell cycle progression.[1] Thus, detailed *in vivo* analysis of the cellular proteins and events involved in regulation of microtubule dynamics is necessary for a better understanding of the cell biology involved in processes such as cellular morphogenesis, mitosis, progression through the cell cycle, and apoptosis

Prior studies of microtubule dynamics in living cells have utilized either video-enhanced differential interference contrast (DIC) microscopy or microinjection of purified tubulin subunits that have been covalently modified with fluorescent molecules.[2,3] This chapter describes an alternative method for *in vivo* analysis of microtubule dynamics following transfection of mammalian expression constructs encoding fusion proteins of the green fluorescent protein (GFP) and MAP4, a ubiquitously expressed cellular protein that localizes to the microtubule cytoskeleton.

MAP4 is a member of a family of microtubule-associated proteins (MAPs) which include the well-studied neuronal proteins MAP2 and tau.[4,5] MAP proteins are thought to be regulators of microtubule function in living cells. Like other family members, MAP4 associates specifically with microtubules both *in vitro* and *in vivo*, and is able to promote *in vitro*

[1] P. R. Andreasson and R. L. Margolis, *J. Cell Biol.* **127,** 789 (1994).
[2] L. Cassimeris, N. K. Pryer, and E. D. Salmon, *J. Cell Biol.* **107,** 2223 (1998).
[3] P. Sammak and G. G. Borisy, *Nature (London)* **332,** 720 (1988).
[4] A. Matus, MAP 2. *In* "Microtubules" (K. Roberts and J. S. Hyams, eds.), pp. 155–166. Wiley-Liss, New York, 1994.
[5] E. M. Mandelkow, O. Schweers, G. Drewes, J. Biernat, N. Gustke, B. Trinczek, and E. Mandelkow, *Ann. N.Y. Acad. Sci.* **777,** 96 (1996).

0076-6879/99 $30.00

microtubule assembly.[6] However, MAP4 is unusual in its nearly ubiquitous expression in both differentiated and proliferating adult mammalian tissues. It also differs from the neuronal MAPs in that it has no adverse affect on microtubule organization when exogenously expressed in tissue culture cells.[7,8]

Heterologous expression of either MAP2 or tau proteins stabilizes microtubules through a reduction in dynamic instability, thereby altering microtubule organization and promoting the formation of microtubule bundles.[9] This property limits the usefulness of these proteins as markers for normal microtubule behavior. In contrast, overexpression of the full-length MAP4 protein does not induce microtubule rearrangements or affect microtubule stability in response to treatment with nocodazole, a microtubule-depolymerizing drug.[7,8] These properties make MAP4 tagged with green fluorescent protein (GFP) suitable as an *in vivo* indicator of microtubule organization and dynamics.

GFP is a 27-kDa protein originally isolated and cloned from the jellyfish *Aequorea victoria* and it possesses properties that facilitate its use as a fluorescent reporter protein.[10,11] When illuminated with blue light, GFP emits green light at a maximum of 510 nm in the absence of added cofactors. The fluorescence produced has limited phototoxicity and negligible photobleaching over extended periods of constant illumination with either a mercury light source or the 488-nm emission of an argon laser. GFP has now been fused to a variety of proteins and utilized for cell biological studies in living cells with no apparent adverse effects on normal protein function, indicating that it is an innocuous reporter. As outlined elsewhere in this volume, these properties make GFP useful for a variety of *in vivo* analyses.

In the following sections we describe methods and present data illustrating the utility of three GFP–MAP4 fusion proteins for *in vivo* analysis of microtubule organization and dynamics. One chimera was generated using the full-length MAP4 protein to which GFP was fused at the amino terminus. Amino-terminal fusions of GFP and the P4-3 spectral mutant of GFP

[6] J. B. Olmsted and H. Murofushi, MAP 4 (MAP U). *In* "Guidebook to the Cytoskeletal and Motor Proteins" (T. Kreis and R. Vale, eds.), pp. 115–116. Oxford University Press, Oxford, 1993.

[7] K. R. Olson, J. R. McIntosh, and J. B. Olmsted, *J. Cell Biol.* **130,** 639 (1995).

[8] S. Barlow, M. L. Gonzalez-Garay, R. R. West, J. B. Olmsted, and F. Cabral, *J. Cell Biol.* **126,** 1017 (1994).

[9] E. Mandelkow and F.-M. Mandelkow, *Curr. Opin. Cell Biol.* **7,** 72 (1995).

[10] D. C. Prasher, V. K. Eckenrode, W. W. Ward, F. G. Prendergast, and M. J. Cormier, *Gene* **111,** 229 (1992).

[11] M. Chalfie, *Photochem. Photobiol.* **62,** 651 (1995).

were also engineered using a fragment of MAP4 comprising the complete basic microtubule-binding domain of MAP4 (BD).[12] These fusion proteins are referred to as GFP–BD and GFPP43–BD chimeras, respectively. The effects of some of these chimeric proteins as well as other subdomains of MAP4 on microtubule organization have been described previously.[7]

Engineering Green Fluorescent Protein–Microtubule-Associated Protein 4 Expression Constructs

Construction of pRc/CMV-GFP

The basic procedures used for generation of mammalian expression constructs for expression of GFP–MAP4 and GFP–BD have been described previously (Olson *et al.*[7]), and are outlined in more detail below. Generation of a mammalian expression vector expressing GFP was accomplished by polymerase chain reaction (PCR) amplification of the coding region from the pGFP 10.1 plasmid DNA template (generously provided by M. Chalfie, Columbia University).[13] PCR amplification employed *Taq* DNA polymerase and standard small-scale reaction conditions with oligodeoxynucleotide (ODNs) that included engineered *Hin*dIII or *Not*I restriction enzyme recognition sites. The primer sequences used for amplification were

GFP5′Hind: GGATAAGCTTATGACTAAAGAAGAAGAACTTTT
GFP3′Not: GGATGCGGCCGCTTTGTATAGTTCATCCATGCC

PCR amplimers were separated on a 1.2% agarose gel, excised from the gel, and purified using a GeneClean kit (Bio101, La Jolla, CA). Purified DNA fragments were then digested with the appropriate restriction enzymes and ligated into the pRc/CMV mammalian expression vector (Invitrogen, San Diego, CA). The resultant DNA construct was called pRc/CMV-GFP and utilized the constitutively active cytomegalovirus (CMV) promoter to drive GFP expression in a variety of liposome-transfected mammalian tissue culture cells (3T3, BHK, CHO, Cos-7, Hela, L, LLCPK, PAM, and PtK). This vector and its derivatives also contain sequences for drug resistance to ampicillin and neomycin, allowing both plasmid amplification in bacteria and stable selection of mammalian cell lines following transient transfection.

To generate fusion proteins with the P4-3 mutant version of GFP, the PCR primers described above were used to amplify the coding region of the mutant GFP from a bacterial expression vector (plasmid generously provided by R. Tsien, UCSD).[12] The PCR product was then gel purified,

[12] R. Heim and R. Y. Tsien, *Curr. Biol.* **6,** 178 (1996).
[13] M. Chalfie, Y. Tu, G. Euskirchen, W. W. Ward, and D. C. Prasher, *Science* **263,** 802 (1994).

restriction digested, and ligated into similarly digested pRc/CMV to generate the pRc/CMV-P43GFP expression vector. This and all other DNA vectors were sequenced to confirm insert identity and orientation.

Both the pRc/CMV-GFP and pRc/CMV-P43GFP constructs expressed proteins that distributed uniformly throughout the cytoplasm and nucleus of the cell (data not shown; see Fig. 2 in Olson *et al.*[7]).

Generation of Green Fluorescent Protein–Microtubule-Associated Protein 4 Constructs

For fusion of wild-type and mutant GFPs to either full-length MAP4 or the basic domain (BD) alone of the protein, MAP4 sequences were PCR amplified using DNA primers with engineered restriction sites. In all instances, the sense primer contained a 5′ *Not*I restriction site, followed by an extra nucleotide to maintain the reading frame, and sequence specific to the region of MAP4 being amplified. The antisense primer contained a 5′ *Apa*I restriction site for subcloning of PCR amplimers into pRc/CMV-GFP plasmid digested with *Not*I and *Apa*I enzymes, a stop codon for efficient translation termination, and sequence specific to MAP4 for amplification of the appropriate portion of the cDNA. The specific primer sequences were

5′ MAP4: GTAAGCGGCCGCAATGGCCGACCTCAGTCTTGTG
3′ MAP4: GAGGGCCCGCTTGTCTCCTGGATCTGGCTGTC
5′ BD: GGATGCGGCCGCCATGTCCCGGCAAGAAGAAGC
3′ BD: TAGGGCCCTGCAGGAAAGTGGCCAACGTTATC

All DNA vectors were sequenced over the fusion site to confirm insert identity and maintenance of the proper reading frame. The chimeric proteins produced by these expression constructs are referred to as GFP–MAP4, GFP–BD, or GFPP43–BD.

Preparation and Imaging of Cells Containing Green Fluorescent Protein–Microtubule-Associated Protein Chimeras

Transfection

Each of the expression vectors was introduced into a wide range of tissue culture cells via liposome-mediated transfection.[7] The more efficient transfections occurred with highly purified DNA prepared over an ion-exchange column (Qiagen, Chatsworth, CA), Lipofectamine transfection reagent (GIBCO-BRL, Gaithersburg, MD), and a protocol supplied by the manufacturer of these liposomes. We found that transfection efficiencies were highly variable (typically 1–33% transfected cells) and dependent on cell type and the individual lots of liposomes. Optimal transfection efficiencies were obtained with cells grown on glass coverslips in 35-mm dishes using 1.2 μg of column-purified plasmid DNA and 5–7 μl of Lipofectamine reagent in a total volume of 1.2 ml of Opti-MEM serum-free culture medium (GIBCO-BRL). Microtubule patterns in transient transfectants were best visualized by allowing cells to grow and recover at 37° under conditions of 95% air and 5% CO_2 for 24–36 hr in standard growth medium (F-12) supplemented with 10% fetal calf serum (FCS). Although GFP expression could be detected as early as 12 hr after initial DNA uptake, increased levels of protein expression were apparent after longer incubation times. As previously outlined, transiently transfected cells could be used for the selection of stable cells lines using the neomycin resistance conferred by the pRc/CMV vector.[7] The ability to generate these cell lines indicated that the GFP–MAP4 chimera had no adverse effects on cell growth or survival.

Cell Preparation for Microscopy

Microscopic observation of live cells was accomplished by inverting coverslips onto a hanging drop of growth medium (~150 μl) on depression slides (18-mm diameter, 0.5 mm deep; Fisher Scientific, Pittsburgh, PA). After coverslips were mounted and blotted dry, they were sealed using a 1:1:1 (w:w:w) mixture of petroleum jelly, lanolin, and paraffin wax (Valap) that had been liquefied on a hot plate for 1–2 min. These slides could be stored for short periods of time in the cell incubator. In some cases, cells were kept warm (33–37°) during observation by using an air curtain positioned near the microscope stage.

The expression of the GFP chimeras could also be visualized following fixation and mounting of coverslips onto standard microscope slides with 25 μl of Gelvatol solution [350 mg of Gelvatol (Monsanto, St. Louis, MO) solubilized in 3 ml of distilled water and 1.5 ml of glycerol]. Fixation methods that we found optimal for visualization of GFP include (1) fixation in 100% methanol at −20° for 7 min followed by three washes in 1× phosphate-buffered saline (PBS), pH 7.2; or (2) fixation at room temperature for 20 min in a solution of paraformaldehyde–glutaraldehyde [2% paraformaldehyde, 0.1% glutaraldehyde, 0.1 *M* piperazine-*N*,*N*′-bis(2-ethanesulfonic acid) (PIPES), 1 m*M* $MgSO_4$] followed by three washes in PHEM buffer, pH 6.9 [60 m*M* PIPES, 2.5 m*M* *N*-2-hydroxyethylpiperazine-*N*′-2-ethanesulfonic acid (HEPES), 2 m*M* $MgSO_4$, 10 m*M* EGTA]. Cells fixed by either method could be extracted (e.g., with 1% Triton X-100) and signal would be maintained. An example of a mouse L cell expressing the GFP–BD chimera that had been fixed with aldehydes and extracted is shown in Fig.

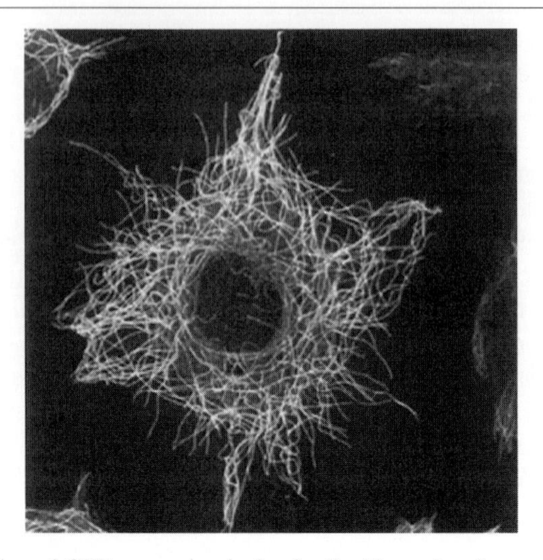

FIG. 1. Detection of GFP expression in fixed cells. Mouse L cells were transfected with the GFP–BD expression construct and incubated for 36 hr prior to aldehyde fixation and extraction. This confocal micrograph is a flat 3D projection generated from a *Z* series collected through the entire cell volume.

1. Well-preserved microtubule arrays labeled with GFP–BD are readily observed. The ability to visualize GFP following fixation facilitates dual-label analyses in which procedures for the second label require fixation and extraction (e.g., immunofluorescence).

Imaging of Green Fluorescent Protein-Expressing Cells

All of the microscopy described in this chapter was carried out using a Leica ×100 oil immersion lens (NA 0.7–1.4) mounted on a Leica DM-IRB inverted microscope that was equipped with Leica TCS-NT hardware and software for laser-scanning confocal microscopy. Cells transfected with GFP–MAP4 expression vectors were identified prior to laser scanning using a standard fluorescein long-pass filter set and illumination with a 50-W mercury arc lamp. These cells were then laser scanned and analyzed via confocal microscopy using the 488-nm emission wavelength of an argon laser, an RSP500 reflection short-pass mirror, and BP525/50 bandpass filter. For analysis of microtubule organization and generation of three-dimensional (3D) projections from confocal *Z* series, images were typically obtained using medium scan speed and were averaged over four to eight sequential scanning intervals. This resulted in a total laser illumination time per optical section of approximately 6–12 sec, depending on the number of averages and a total laser exposure per cell that varied with number of optical sections acquired. The argon laser was typically used at a constant attenuated power to enhance cell survival, and any adjustments necessary to account for differences in GFP expression level were made by altering the sensitivity of the photomultiplier tube detector (electronic gain). Confocal *Z* series were then collected through the entire cell volume and used to generate single-angle projections. The number of sections acquired was adjusted so that the thickness of the sections (*Z* resolution) for each cell matched pixel resolution in the *XY* dimension. This configuration approximated a cubic voxel and gave optimal 3D resolution.

One advantage of using GFP in studies requiring prolonged image acquisition is shown in Fig. 2A. A histogram in which integrated fluorescence intensity per cell area is plotted versus time measured in seconds illustrates that the fluorescence of GFP (in this case linked to MAP4) is extremely stable. These data were generated using Metamorph image analysis software (Universal Imaging, Brandywine, PA) to measure integrated pixel intensity within a predefined area of the cell drawn directly on each image within a time series. In this case, the cell had been illuminated for 30–60 sec before acquisition was initiated. At the baseline laser power used in this example, fluorescent signal remained constant, and the cell could be analyzed with continuous laser illumination for more than 10 min.

GFP has the unusual property of undergoing photoactivation prior to reaching the maximal fluorescent signal intensity (Fig. 2B). Unlike the time course illustrated in Fig. 2A, the images used for this analysis were collected with minimal illumination prior to the onset of laser scanning and image acquisition. GFP fluorescence increases linearly during the first few image intervals, and then reaches a stable plateau. This photoactivation property of GFP is useful for distinguishing GFP fluorescence from overlapping background fluorescence, as the latter undergoes rapid photobleaching.

Morphological Phenotypes of Constructs

Analysis of single-angle 3D projections of live cells expressing either GFP–MAP4 or GFP–BD chimeras demonstrates that the microtubule arrays are organized in readily distinguishable patterns in these two types of transfectants. Cells expressing GFP–MAP4 (Fig. 3A) have typical radial microtubule arrays organized from a single centrosome, while cells expressing GFP–BD (Fig. 3B) contain microtubules that are rearranged into thick microtubule bundles that have no apparent centrosomal origin. In previous experiments, treatment of cells expressing either GFP–MAP4 or GFP–BD with the microtubule-depolymerizing drug nocodazole suggested that the microtuble rearrangements induced by GFP–BD were the result of en-

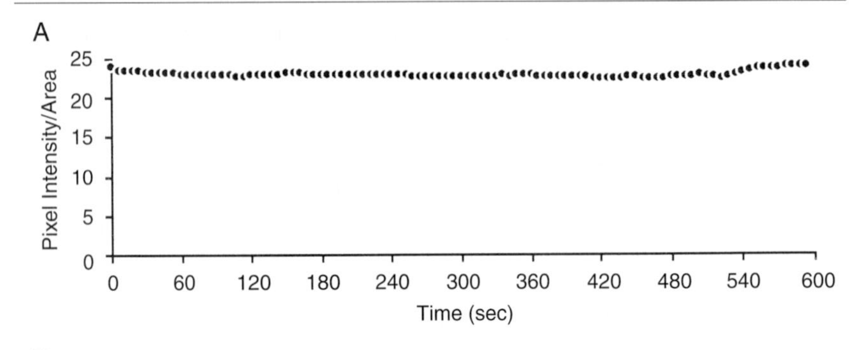

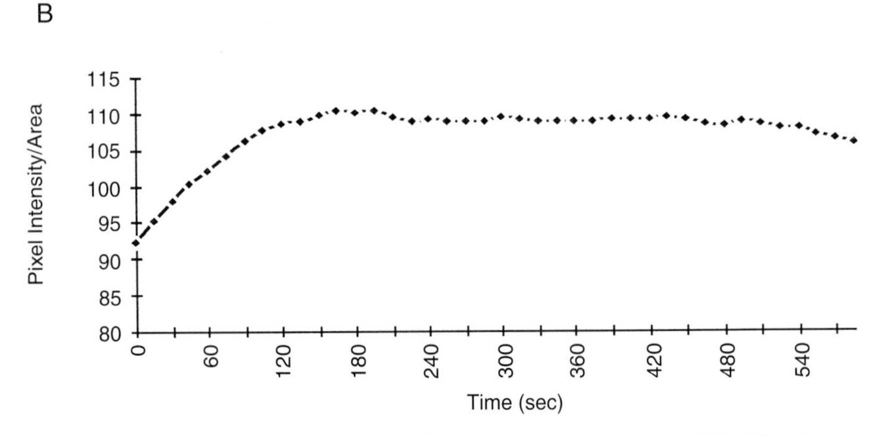

FIG. 2. Properties of GFP fluorescence. (A) A BHK cell expressing the GFP–BD chimera was examined for ~1 min, and images were then acquired at 6-sec intervals for 10 min. The integrated pixel intensity per cell area was plotted versus time, demonstrating that GFP signal intensity remains constant throughout the time course. (B) A BHK cell expressing GFP–BD was rapidly identified by conventional fluorescence microscopy, and then images were collected at 15-sec intervals. Pixel intensity was recorded as in (A). Note the increase in signal intensity on irradiation that then reaches a stable plateau.

hanced microtubule stability.[7] We therefore used confocal microscopy of cells expressing these chimeras to analyze microtubule dynamics directly.

Analysis of Microtubule Dynamics with Green Fluorescent Protein–Microtubule-Associated Protein Constructs

Microtubules are dynamic structures that undergo frequent and rapid transitions between periods of growth and shrinkage.[14] To monitor these

[14] T. Mitchison and M. Kirschner, *Nature* (*London*) **312,** 237 (1984).

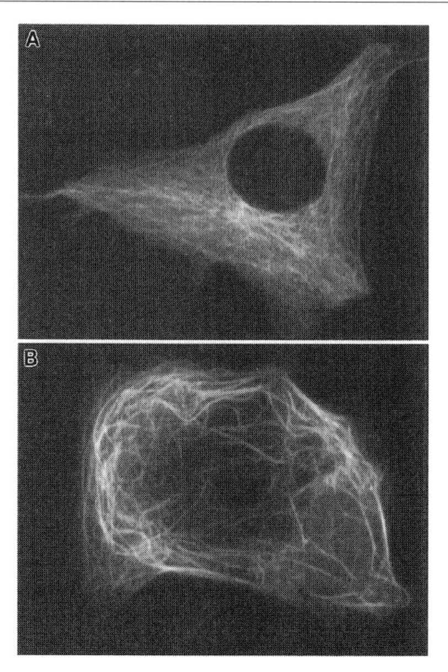

FIG. 3. Comparison of microtubule organization in live cells using GFP–MAP4 and GFP–BD chimeras. 3T3 cells were transfected with GFP–MAP4 or GFP–BD and incubated for 36 hr prior to analysis by confocal microscopy. These images are 3D projections tilted at an 8° angle from center. Both chimeric proteins label microtubules evenly throughout the cells and are excluded from nuclei. Microtubule arrays show a typical radial distribution of individual microtubules emanating from the centrosome in cells expressing GFP–MAP4 (A), whereas cells expressing GFP–BD (B) have disrupted microtubule organization and contain microtubule bundles.

changes, individual images must be acquired in a interval that is long enough to maximize image resolution, yet short enough to ensure that the majority of microtubule length changes are measured. For this type of analysis, image resolution is particularly important for accurate estimation of the position of the microtubule end.

Acquisition of a Time Series

A region of the cell near the periphery is magnified using an electronic zoom to help identify microtubules that can be distinguished from one another and readily traced. Once an appropriate focal plane has been selected, a predefined number of images is acquired within this Z plane. The scan speed of the laser is selected to maximize image resolution while minimizing image acquisition time. In our analyses of microtubule dynamics using the Leica confocal microscope, an image acquisition time of 4 sec

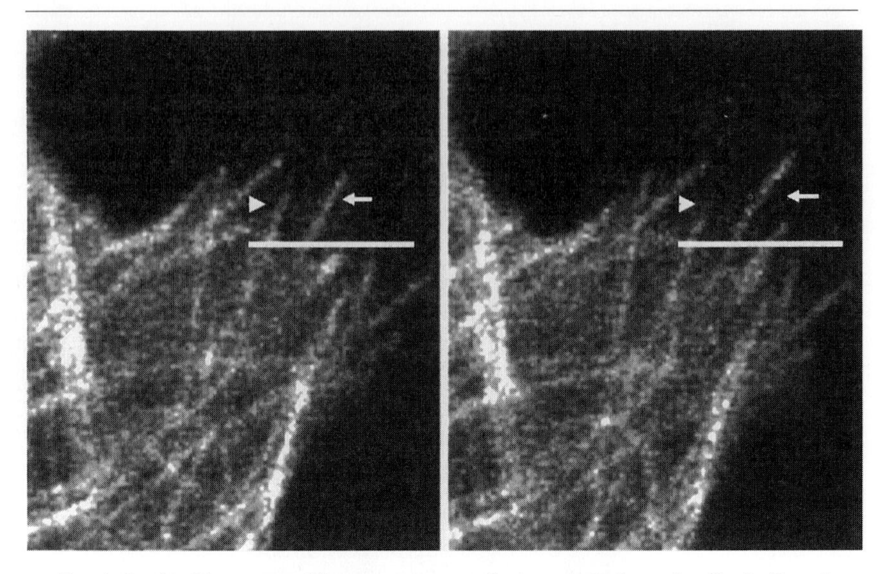

FIG. 4. Confocal images used for measurement of microtubule dynamics. Single Z sections are shown for two sequential images separated by 8 sec in a time series analysis of cells expressing GFP–MAP4. The white line serves as a fiducary mark for microtubule measurement. The arrowhead shows a microtubule that shrinks during the time interval, and the arrow indicates the position of a microtubule that grows.

was achieved using the slow2 scan mode and images were collected at this time interval for a total of 140 sec. As shown in Fig. 4, the image resolution obtained with these settings permits changes in microtubule ends to be readily visualized in two images within a time series that were acquired 8 sec apart. One potential limitation in the number of time points acquired derives from data storage. For example, the studies described here for a 140-sec time course generated 36 individual data files, resulting in a file series of approximately 9.5 MB for each microtubule to be analyzed.

Microtubule Tracing

For quantification of changes in microtubule length, a time course series of images is converted into a stack file format using Metamorph image analysis software. This file format allows one to scroll through the image series, using a simple window-based application, and position text or drawings that are maintained at a constant position on each image of the series. Stack files can also be used to create video sequences of the entire time course, allowing identification of individual microtubules that are suitable for tracing. Only microtubules that were clearly visible at the cell periphery for over half the time course were used for analysis of microtubule dynamics.

An example of an image used for tracing is shown in Fig. 4. The microtubule marked with the arrowhead has decreased in length during the time interval, whereas the microtubule marked by an arrow has increased in length. Using Metamorph software, a line that serves as a fiduciary mark is drawn on an image at a position perpendicular to the microtubule(s) being analyzed, and this line is then reproduced throughout the image stack. The cursor-driven software is used to trace the microtubule along its length from the fiduciary mark to the microtubule end for each of the 36 frames in the video sequence. A value representing pixel distance for the microtubule tracing is exported to a Microsoft Excel spreadsheet and used to determine actual micron changes in microtubule length, using calibrations obtained from the confocal software. Pixel-to-micron calibrations are determined for individual time series based on the image size (512 $\times$ 512 pixels), the lens objective ($\times$100), and the extent of electronic zoom.

Microtubule Life History Plots

An example of the data derived from tracings of an individual microtubule from a CHO cell expressing GFP–MAP4 are shown in Table I. The general characteristics of microtubule dynamics are then determined by plotting the changes in microtubule length measured in microns (Table I, column 5) versus time measured in seconds (Table I, column 4). Examples of four separate microtubule life histories obtained from analyses of cells expressing either GFP–MAP4 or GFP–BD are illustrated in Fig. 5. These plots indicate that microtubules in cells expressing GFP–BD are essentially static compared with those in cells with GFP–MAP4, and show little overall length change and few examples of microtubule shrinkage. This behavior indicates that in the absence of flanking acidic sequences present in the full-length protein, the MAP4 basic domain is able to suppress microtubule dynamic instability. This perturbation in dynamics is likely to account for the altered microtubule organization in cells expressing this chimera. Unlike GFP–BD, GFP–MAP4 does not dramatically alter microtubule dynamic instability and typical microtubule arrays are observed in cells expressing this chimera.

Calculation of Microtubule Dynamics Parameters

The data shown in Table I and Fig. 5 can be further analyzed for calculation of the parameters of microtubule dynamics that are used to

TABLE I
ANALYSIS OF MICROTUBULE DYNAMICS[a]

Image	Plane	Distance a.u.	Time (sec)	Length (μm)	Change (μm)	p, g, s[b]	r, c[b]	Growth (μm)	Shrinkage (μm)
CMP14	1	44.6945	0	4.4695					
	2	40.5262	6	4.0526	−0.41683	s			−0.4168
	3	48.7347	12	4.8735	0.82085	g	r	0.8209	
	4	43.9616	18	4.3962	−0.4773	s	c		−0.4773
	5	49.5782	24	4.9578	0.56166	g	r	0.5617	
	6	39.6989	30	3.9699	−0.98793	s	c		−0.9879
	7	49.0626	36	4.9063	0.93637	g	r	0.9364	
	8	49.5974	42	4.9597	0.05348	p			
	9	48.1457	48	4.8146	−0.14517	p			
	10	32.45	54	3.245	−1.56957	s	c		−1.5696
	11	42.0857	60	4.2086	0.96357	g		0.9636	
	12	46.3767	66	4.6377	0.4291	g	r	0.4291	
	13	32.0707	77	3.2071	−1.4306	s	c		−1.4306
	14	33.4306	78	3.3431	0.13599	p			
	15	34.2053	84	3.4205	0.07747	p			
	16	38.6333	90	3.8633	0.4428	g	r	0.4428	
	17	38.0278	96	3.8028	−0.06055	p			
	18	28.3196	102	2.832	−0.97082	s	c		−0.9708
	19	37.1616	108	3.7162	0.8842	g	r	0.8842	
	20	23.0217	114	2.3022	−1.41399	s	c		−1.414
	21	23.087	120	2.3087	0.00653	p			
	22	17	126	1.7	−0.6087	s	c		−0.6087
	23	17.0294	132	1.7029	0.00294	p			
	24	26.1725	138	2.6173	0.91431	g		0.9143	
	25	33.3386	144	3.3339	0.71661	g		0.7166	
	26	39.91	150	3.991	0.65714	g	r	0.6571	
	27	24.0832	156	2.4083	−1.58268	s	c		−1.5827
	28	22.8035	162	2.2804	−0.12797	p			
	29	24.8428	168	2.4843	0.20393	g		0.2039	
	30	28.8498	174	2.885	0.4007	g	r	0.4007	
							Total:	7.9312	−9.4584
							Average:	0.6609	−1.0509
							SD:	0.2519	0.46968
							SE:	0.0727	0.15656

[a] Data generated from the analysis of a single microtubule in a CHO cell expressing GFP–MAP4.
[b] p, Pause; g, growth; s, shrinkage; res, rescue; c, catastrophe.

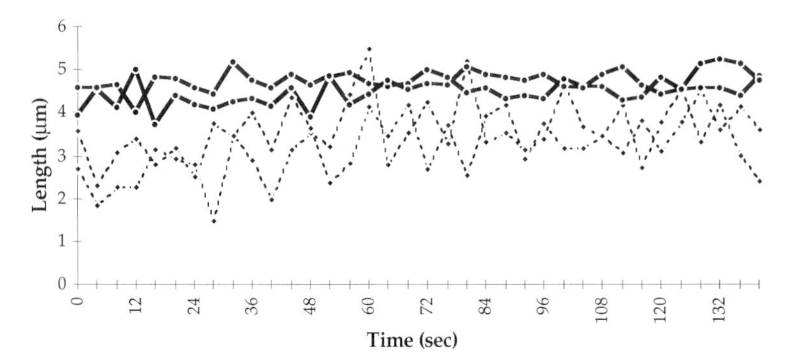

FIG. 5. Microtubule life history plots. Individual microtubules visualized using either GFP–BD (solid lines) or GFP–MAP4 (dashed lines) were imaged under constant laser illumination at 4-sec intervals for 140 sec. Microtubule life histories were generated by plotting the values for microtubule length versus time for each interval in the time series. Note that microtubules labeled with GFP–MAP4 show characteristic dynamic behavior while those labeled with GFP–BD are essentially static.

characterize populations of microtubules.[14,15] These parameters include, but are not limited to, a measure of the percentage of time microtubules spend in growth, shrinkage, and pause, the mean rates of microtubule growth and shrinkage, the average total growth and shrinkage distances, the frequencies of rescue and catastrophe, the average rescue and catastrophe distance, and calculation of an overall measure of total microtubule dynamics termed *dynamicity*. Examples of values obtained for these parameters from populations of microtubules analyzed in cells expressing either GFP–BD or GFP–MAP4 are shown in Table II.

Comparison of the dynamicity values obtained for microtubules decorated with either GFP–MAP4 or GFP–BD confirms previous findings from the microtubule life histories that the GFP–BD chimera inhibits dynamic instability. Nearly all of the parameters of microtubule dynamics are affected in a statistically significant manner by expression of the GFP–BD chimera. These data indicate that the microtubule reorganizations induced by GFP–BD are caused by a global effect on microtubule dynamics leading to increased microtubule stability and the formation of microtubule bundles. In contrast, cells expressing full-length GFP–MAP4 show dynamic parameters that are generally comparable to those reported previously in analyses of similar cell types in which microtubules were visualized by microinjection

[15] E. Shelden and P. Wadsworth, *J. Cell Biol.* **120,** 935 (1993).

TABLE II
PARAMETERS OF MICROTUBULE DYNAMICS: GFP–BD VERSUS GFP–MAP4[a]

Parameter	MAP (17 MTs, 8 cells)	BD (4 MTs, 1 cell)
Average percent pause:	14.3 ± 1.6	36.4 ± 2.7[b]
Average percent growing:	44.8 ± 1.2	34.3 ± 4.5[c]
Average percent shrinking:	40.9 ± 1.1	29.3 ± 2.9[b]
Average growth rate (μm/sec):	0.19 ± 0.01	0.11 ± 0.01[c]
Average shrinkage rate (μm/sec):	0.21 ± 0.01	0.10 ± 0.01[b]
Average growth/4 sec (μm):	0.76 ± 0.05	0.44 ± 0.06[c]
Average shrinkage/4 sec (μm):	0.83 ± 0.04	0.41 ± 0.04[b]
Average growth distance (μm):	11.82 ± 0.70	5.22 ± 0.84[b]
Average shrinkage distance (μm):	11.93 ± 0.68	4.13 ± 0.32[b]
Average net change (μm):	−0.10 ± 0.18	1.09 ± 0.66
Frequency of rescue (1/sec):	0.182 ± 0.007	0.192 ± 0.012
Frequency of catastrophe (1/sec):	0.190 ± 0.006	0.233 ± 0.010[c]
Average rescue length (μm):	1.04 ± 0.06	0.54 ± 0.08[c]
Average catastrophe length (μm):	1.06 ± 0.07	0.43 ± 0.07[b]
Dynamicity (μm/sec):	0.165 ± 0.009	0.008 ± 0.012[b]

[a] Statistical significance calculated using Student's t test. MAP, Microtubule-associated protein; BD, binding domain; MT, microtubule.
[b] $p < 0.001$.
[c] $0.01 < p < 0.001$.

of rhodaminated tubulin.[15] These data indicate that exogenous expression of MAP4 results in little perturbation of existing microtubule organization or behavior.

Additional Applications of Green Fluorescent Protein–Microtubule-Associated Protein 4 Chimeras

Dual-Labeling Studies

GFP–MAP4 can be used as a convenient marker for microtubules, thus facilitating dual-labeling experiments in which microtubules are tracked relative to other organelles or cellular structures. For example, the emission spectrum of GFP readily allows covisualization of GFP–MAP4 fusion proteins in both fixed and live cells with rhodamine-labeled proteins, Texas Red-labeled proteins, and other fluorophores that emit light in the blue, orange, and red spectra. We have also labeled organelles and proteins in living cells expressing GFP–MAP4 using the DNA-specific dye 4′,6-

diamidino-2-phenylindole (DAPI; Sigma, St. Louis, MO), the mitochondrial marker MitoTracker Red (Molecular Probes, Eugene, OR), and rhodamine-labeled tubulin, and have also labeled fixed cells with antibodies to tubulin and other cellular proteins.[7] An example of this approach is outlined below for the covisualization of microtubules tagged with GFP–MAP4 and mitochondria in living cells.

For *in vivo* labeling with MitoTracker Red, a 1 m*M* stock solution made in dimethyl sulfoxide (DMSO) was diluted 1 : 1000 into the culture medium of cells transfected with GFP–MAP4 and cells were observed 5–60 min after further incubation at 37°. MitoTracker Red dye was maximally excited by the 568-nm emission of a krypton–argon laser. Covisualization of the two fluorescent signals was performed using a double-dichroic mirror and two photomultiplier tube detectors. GFP signal was collected as described earlier, while MitoTracker Red illumination was passed through an LP590 long-pass filter prior to detection at the second photomultiplier tube. As shown in Fig. 6, images of single optical sections obtained from cells containing GFP–MAP4 and MitoTracker Red demonstrate that mitochondrial distribution parallels that of microtubules and extends from the perinuclear area of the cell to the periphery. These patterns are similar to those observed using double-immunofluorescence analysis of fixed samples, and indicate that exogenous GFP–MAP4 expression does not overtly perturb mitochondrial distribution.[16]

Analysis of Microtubule Organization Using GFPP43–BD

A number of spectral variants of GFP have now been generated that emit at either lower or higher wavelengths than wild-type GFP. We examined the feasibility of imaging microtubules using the P4-3 mutant of GFP that emits blue light maximally at 445 nm.[12] Cells transfected with this construct were imaged by confocal microscopy using a water-cooled argon UV laser as the source for the 363-nm excitation wavelength of this GFP mutant. As shown in Fig. 7A, cells expressing the GFPP43–BD fusion protein are relatively dim compared with those containing GFP–BD, but microtubule bundles and altered microtubule organization are detected that are similar to those observed previously with GFP–BD. While these data indicate that the GFPP43–BD fusion protein can be used as a reporter for microtubule organization, there are some limitations in the overall utility of the GFP mutant. As shown in Fig. 7B, a major drawback of this mutant GFP is that signal intensity decreases rapidly on irradiation. This rapid photobleaching limits the usefulness of the P4-3 mutant for analyses

[16] L. B. Chen, *Methods Cell Biol.* **29,** 103 (1989).

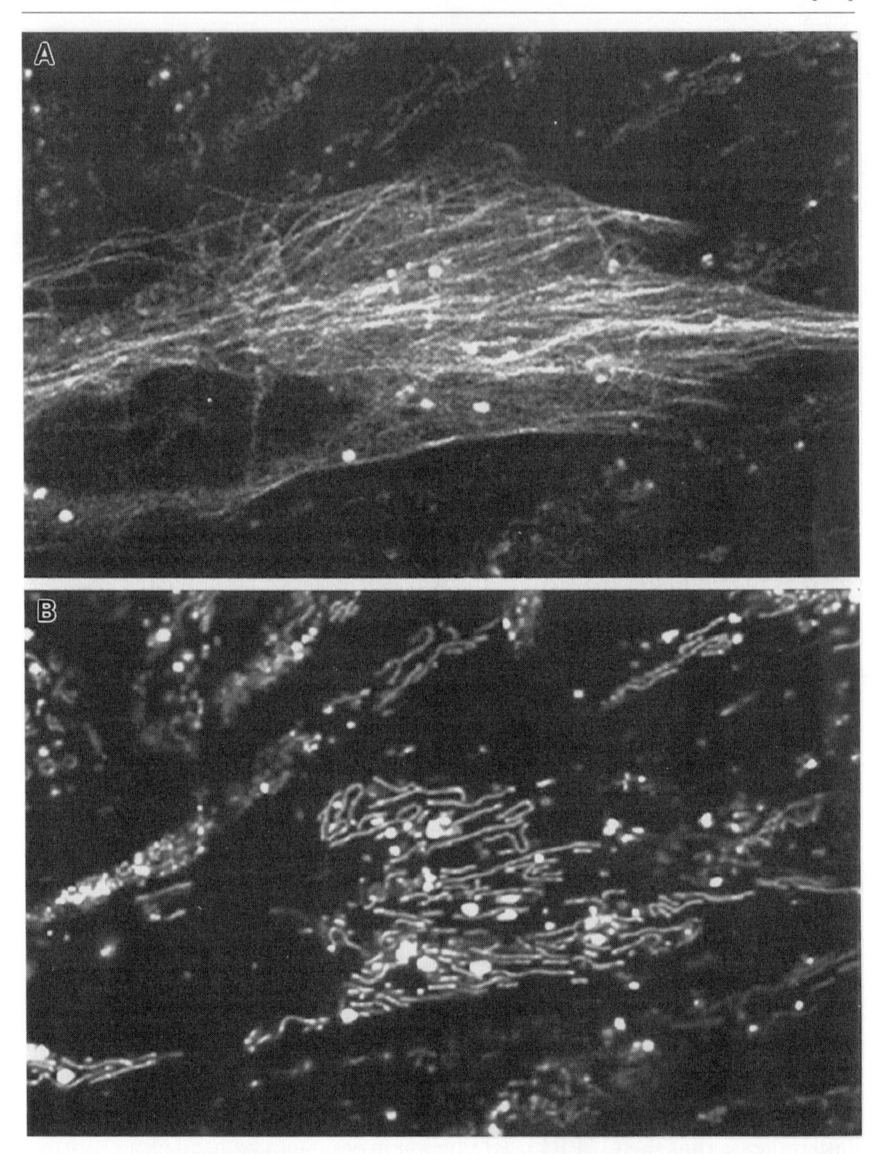

FIG. 6. Confocal analysis of GFP–MAP4 and mitochondrial distribution. BHK cells were transiently transfected with GFP–MAP4 and incubated for 24 hr prior to *in vivo* labeling with MitoTracker Red for 30 min. Live cells were then simultaneously analyzed for GFP–MAP4 fluorescence (A) and MitoTracker Red signal (B) using confocal microscopy. Alignment of mitochondria along microtubules is not affected by exogenous GFP–MAP4 expression.

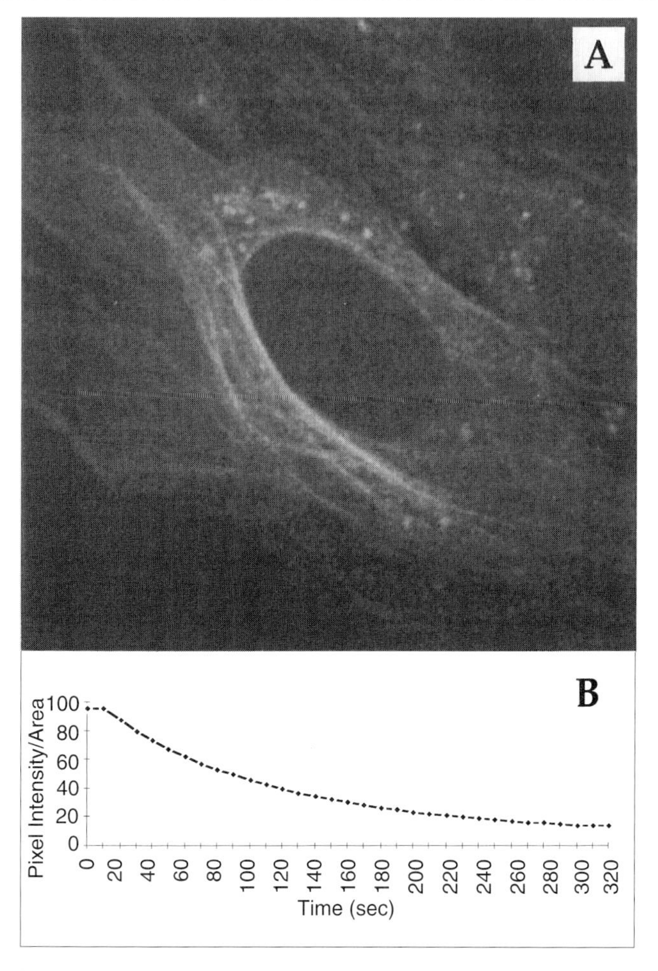

FIG. 7. Confocal analysis using the P4-3 mutant of GFP. BHK cells were transiently transfected with the GFPP43–BD expression construct and analyzed using confocal microscopy following 40 hr of incubation. (A) A single optical section of a cell imaged using irradiation from the UV argon laser. Cells expressing GFPP43–BD contain microtubule bundles similar to those previously observed following expression of GFP–BD, but the signal is much weaker. (B) Histogram plotting fluorescence intensity of GFPP43–BD per cell area versus time. This GFP spectral variant has decreased photostability compared with wild-type GFP.

of microtubule dynamics or other processes in which extended periods of observation are required for sufficient data collection. However, this mutant can still be employed for dual-label studies in which two proteins are tagged with two different spectral variants of GFP. The P4-3 GFP mutant has also been used for analyses of protein–protein interaction using fluorescence resonance energy transfer, and we are currently utilizing the GFPP43–BD chimera in conjunction with other GFP-labeled proteins to investigate this phenomenon.[12,17]

Concluding Remarks

Traditional methods of measuring microtubule dynamics *in vivo* have focused on analysis of fluorescently derivatized tubulin subunits or visualization of microtubules by video-enhanced differential interference contrast microscopy. The methods described here allow noninvasive analysis of microtubule dynamics following transfection of cells with expression vectors encoding GFP–MAP4 fusion proteins. GFP permits extended time course analyses using confocal microscopy without detectable phototoxicity or photobleaching. These properties proved to be particularly useful for the analysis of microtubule dynamics, generating high-quality images over long time periods with a relatively short image acquisition time of 4 sec. The narrow emission spectrum of GFP further broadens its general utility for covisualization with other fluorescently labeled proteins, or chimeric proteins incorporating different spectral mutants of GFP.[12]

Acknowledgment

This work was supported by Grants NIH GM22214 and NSF MCB9603730 to J.B.O.

[17] R. D. Mitra, C. M. Silva, and D. C. Youvan, *Gene* **173,** 13 (1996).

[12] Trafficking of the Androgen Receptor

By VIRGINIE GEORGET, BÉATRICE TEROUANNE, JEAN-CLAUDE NICOLAS, and CHARLES SULTAN

Introduction

The androgen receptor (AR) belongs to the superfamily of nuclear receptors characterized by a common structure and mechanism of action.[1,2] The AR is an androgen-dependent transcription factor composed of four domains: the amino-terminal transcription activation domain; the DNA-binding domain, which interacts with a specific DNA sequence called the androgen-responsive element (ARE); the hinge region, which includes the nuclear localization signal (NLS); and the carboxy-terminal ligand-binding domain. For its transcriptional activity, the AR needs to bind the androgen, pass the nuclear membrane, and interact as a dimer with the ARE to trigger a cascade of transcriptional events.

The analysis of the subcellular localization of the steroid receptors has usually been performed by immunotechniques (immunocytochemistry or immunohistochemistry). It is generally acknowledged that the estrogen receptor (ER) and the progesterone receptor (PR) are predominantly nuclear, with a continuous shuttle between the nucleus and cytoplasm.[3] The intracellular localization of the mineralocorticoid receptor (MR), the glucocorticoid receptor (GR), and the AR are more controversial.[3] According to the immunostaining protocol, these receptors have been described as being either in the cytoplasm or in the nucleus in the absence of ligand, and exclusively in the nucleus after incubation with ligand. These techniques require the fixation and permabilization of cells, which can lead to artifacts in the pattern of subcellular localization. Moreover, the AR can be in different states, i.e., associated with the heat shock proteins in an unliganded form or associated with DNA or transcription factors in the liganded form. The accessibility of the epitope to antibodies may vary for these different forms and this could induce artifactual results in the immunostaining. Having considered all the limits of immunocytochemistry, we developed a model using a chimera of AR fused to the green fluorescent protein (GFP). This

[1] R. M. Evans, *Science* **240,** 889 (1988).

[2] D. Mangelsdorf, C. Thummel, M. Beato, P. Herrlich, G. Schütz, K. Umesono, B. Blumberg, P. Kastner, M. Mark, P. Chambon, and R. M. Evans, *Cell* **83,** 835 (1995).

[3] A. Guiochon-Mantel, K. Delabre, P. Lescop, and E. Milgrom, *J. Steroid Biochem. Mol. Biol.* **56,** 1 (1996).

0076-6879/99 $30.00

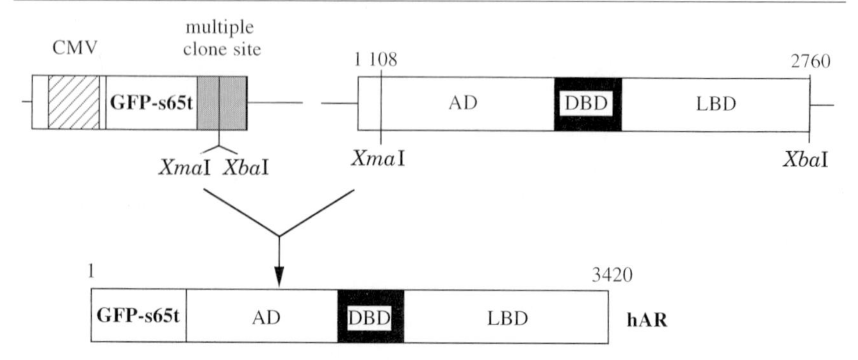

FIG. 1. Construction of GFP expression plasmids and structure of GFP fusion proteins. Shown is the schematic structure of pC1-S65T-GFP, encoding a full-length red-shifted variant of GFP, and the schematic structure of the human AR cDNA with its three domains: the activation domain (AD), the DNA-binding domain (DBD), and the ligand-binding domain (LBD). The amino-terminal truncated hAR cDNA (*Xma*I–*Xba*I) was fused with GFP-S65T.

fluorescent reporter permitted the visualization of the AR in living transfected cells.[4]

We first verified that the fusion protein (GFP–AR) conserved the functional characteristics of the AR. We demonstrated the advantages of this GFP–AR tool versus immunodetection. The intracellular dynamics of the AR were evaluated and quantified in living cells, which suggested some applications of the GFP–AR model, such as anti-androgen screening and androgen insensitivity study.

Plasmid Construction and Functional Properties of Green Fluorescent Protein–Androgen Receptor

Construction

Two different GFPs are used: wild-type GFP contained in pC1-GFP (Clontech, Palo Alto, CA) and GFP-S65T, a red-shifted variant of GFP in which a mutation in the chromophore has been introduced, contained in pC1-S65T-GFP (Clontech); but only GFP-S65T is fused to the AR (Fig. 1). To digest pC1-S65T-GFP with *Xba*I, the vector is transformed in DM1 cells (*dam*-host) (GIBCO-BRL, Cergy-Pontoise, France) to demethylate the *Xba*I site. The AR cDNA is isolated from pCMV5-hAR by *Xma*I–*Xba*I

[4] V. Georget, J. Lobaccaro, B. Térouanne, P. Mangeat, J. Nicolas, and C. Sultan, *Mol. Cell. Endocrinol.* **129,** 17 (1997).

digestion and ligated at the cognate sites of pC1-S65T-GFP digested with *Xma*I and *Xba*I. In pGFP-AR, GFP-S65T is fused to the amino terminal of a human AR (hAR) that lacks the first 36 amino acid residues. The ligation product is transformed in DH5α *Escherichia coli* cells and positive clones are selected by neomycin antibiotics. Constructions are verified by enzymatic digestion and then sequenced in ligation fragments to verify the correct reading frame.

Functional Properties of Green Fluorescent Protein–Androgen Receptor: Comparison with Androgen Receptor

The protein expression, the androgen-binding characteristics, and the transcriptional activity capacity of the GFP–AR are evaluated by transient transfection experiments and cotransfection assays.[4] Mammalian cells expressing GFP exhibit stronger fluorescence when grown at 30°.[5] All experiments are performed at 30 and 37°.

Materials. The COS-7 cell line is transfected with 10 μg of pCMV5-hAR, pC1-S65T-GFP, or pGFP-AR in dishes with a 10-cm diameter for protein expression detection by Western blot; and with 50 ng of pCMV5-hAR or pGFP-AR with 0.25 μg of pCMV-β-galactosidase per well in 12-well tissue culture dishes for androgen-binding assay in whole cells.

The CV-1 cells are transfected for the transactivation assay with 50 ng of pCMV5-hAR or pGFP-AR with 0.25 μg of pCMV-β-galactosidase and 1 μg of androgen-dependent reporter gene [p-tyrosine aminotransferase–thymidine kinase (TAT-tk) regulating the luciferase gene] per well in 12-well tissue culture dishes.

Methods. Cells are transfected by the calcium phosphate DNA precipitation method 8 hr after trypsinization. The precipitate is removed 12 hr later and cells are cultured in Dulbecco's minimal essential medium (DMEM; GIBCO-BRL) supplemented with penicillin (100 units/ml) and streptomycin (100 μg/ml) in the presence or absence of hormones for 30 hr at 30 or 37°.

WESTERN BLOT. Cells are solubilized in lysis buffer [160 m*M* Tris (pH 6.9), 200 m*M* dithiothreitol (DTT), 4% sodium dodecyl sulfate (SDS), 20% glycerol, 0.004% bromphenol blue] in the presence of protease inhibitors [1 m*M* phenylmethylsulfonyl fluoride (PMSF), 0.05 m*M* leupeptin, 0.01 m*M* pepstatin], boiled for 5 min, and centrifuged at 13,000 *g* for 10 min. The supernatant is subjected to SDS–polyacrylamide gel electrophoresis (SDS–PAGE) and then Western-transferred onto nitrocellulose membranes by electroblotting. The filters are saturated with a 10% milk solution and

[5] J. Pines, *Trends Genet.* **11,** 326 (1995).

incubated either with anti-AR antibody (SpO61)[6] diluted 1:2000 or with anti-GFP antiserum (anti-GFP; Clontech) diluted 1:1000. The filters are then incubated in the presence of a peroxidase-conjugated anti-rabbit IgG diluted 1:5000. Blots are developed using the enhanced chemiluminescence (ECL) detection system (Amersham, les Ulis, France).

ANDROGEN-BINDING ASSAY. The transfected cells are incubated in duplicate for 2 hr at 37° with various concentrations (0.05 to 0.3 nM) of synthetic androgen [^{3}H]R1881 (total binding) and duplicate wells are incubated together with 100 nM unlabeled R1881 (nonspecific binding). The cells are then washed with cold phosphate-buffered saline (PBS) and harvested in lysis buffer [25 mM Tris-H_3PO_4 (pH 7.8), 2 mM DTT, 2 mM EDTA, 1% Triton X-100 and 10% glycerol]. Aliquots are counted for radioactivity and for β-galactosidase activity assay. After subtraction of nonspecific binding from total binding, the dissociation constant (K_d) and the maximum androgen-binding sites (B_{max}) were derived from the Scatchard representation.

TRANSACTIVATION ASSAY. CV-1 cells are cultured for 30 hr at 30 or 37° in serum-free medium with various concentrations of R1881 (3.10^{-12} to 10^{-9} M). Transfected cells are lysed as described above. The luciferase activity is measured by the reaction of lysate with the luciferin solution: 270 μM coenzyme A, 470 μM luciferin, 530 μM ATP, 20 mM Tris-H_3PO_4, 1.05 mM $MgCl_2$, 2.7 mM $MgSO_4$, 0.1 mM EDTA, and 33 mM DTT. Luciferase activity is measured in an LKB (Bromma, Sweden) luminometer and expressed in arbitrary units of luminescence.

Results and Comments. By immunoblot analysis, the GFP–AR was visualized with an anti-AR antibody and an anti-GFP antibody as a protein of 130 kDa. GFP was detected with an apparent molecular mass of 27 kDa and the AR was detected with a mass of 110 kDa. The GFP and GFP–AR were less expressed than the AR, which could be due to the difference in the vectors containing the AR or GFP and the GFP–AR. The use of the two antibodies revealed that the GFP–AR was sensitive to degradation because in both cases smaller bands were observed. The expression level appeared to be identical for cells cultured at 30 and 37° but the GFP–AR was more degraded at 37°.

In transient transfected cells, the androgen-binding affinity of the GFP–AR was conserved in comparison with the wild-type AR, and the B_{max} value was slightly reduced. The GFP–AR was able to transactive the androgen-regulated reporter gene only if the cells were cultured at 30°. This phenomenon has already been reported for the GFP–GR fusion pro-

[6] J. H. van Laar, M. M. Voorhost-Ogink, N. D. Zegers, W. J. A. Boersma, E. Claassen, J. A. G. M. van der Korput, J. A. Ruizeveld de Winter, T. H. Van der Kwast, E. Mulder, J. Trapman, and A. O. Brinkmann, *Mol. Cell. Endocrinol.* **67,** 29 (1989).

tein.[7] The GFP–AR demonstrated an induction luciferase activity with a dose dependence identical to that of the AR, but it reached a reduced maximal activity in comparison with the AR. This partial activity of the GFP–AR can be explained by a low expression of the protein previously observed by Western Blot, by the deletion of amino acids 1–36 of the amino-terminal cDNA of the AR in GFP–AR, or by a cumbersome conformation of the fusion protein, which is less appropriate than the AR to transactivate reporter gene.

The GFP–AR, however, conserves the essential functional characteristics of the AR, such as the androgen-binding characteristics and transactivation capacity. The GFP tag does not modify the properties of the AR and can thus be used to determine the subcellular localization of the AR.

Green Fluorescent Protein–Androgen Receptor Localization by Fluorescence and by Immunodetection

The subcellular localization of the GFP–AR has been determined by detection of GFP fluorescence and by antibodies conjugated to Texas Red [rhodamine isothiocyanate (RITC) filter] within the same cell. Because the reported immunodetection results of AR localization were controversial, we tested different conditions of cell permeabilization for indirect immunofluorescence.

Materials

COS-7 cells: Cultured directly on microscope glass coverslips (20 × 20 mm) and transfected using the calcium phosphate method with 1 μg of pGFP-AR, pC1-GFP, or pC1-S65T-GFP, as described above

Solutions:
- Fixation solution: Paraformaldehyde, 4% in PBS
- Permeabilization and washing solution:
  - In PBS: Saponin (0.05, 0.1, or 0.3%) or Triton X-100 (0.02, 0.1, and 0.3%)-Tween 20 (0.05%)
- Saturation solution: Normal donkey serum, 2.5% diluted in permeabilization solution
- Antibody dilution solution: 0.5% bovine γ-globulin diluted in permeabilization solution
- Antibodies:
  - SpO61, rabbit polyclonal anti-AR antibody[6]
  - Anti-GFP, a rabbit anti-GFP antiserum (Clontech)

[7] H. Ogawa, S. Inouye, F. I. Tsuji, K. Yasuda, and K. Umesono, *Proc. Natl. Acad. Sci. U.S.A.* **92,** 11899 (1995).

Texas Red-conjugated donkey anti-IgG rabbit antibody (Jackson Immunoresearch, West Grove, PA)

Microscopes: Conventional direct epifluorescence microscope with a ×40 objective and a fluorescein isothiocyanate (FITC) filter for GFP detection and an RITC filter for immunofluorescence visualization

Methods

The following protocol was written for immunofluorescence in whole COS-7 cells; appropriate modifications can be adapted to other cell types.

Cells are incubated in the presence or absence of 10^{-6} *M* R1881 for 1 hr at 37°. The cells are then placed on ice for 10 min, washed twice with PBS, and fixed for 1 hr at room temperature in 4% paraformaldehyde. The cells are washed twice with PBS and incubated with various concentrations of saponin or Triton X-100 (to permeabilize the cells) for 30 min at room temperature with gentle agitation. For saturation of nonspecific sites, the cells are treated for 30 min with saturation solution, the saponin or Triton X-100 concentration of which corresponds to the concentration used for permeabilization. The slides are then incubated for 2 hr with either anti-GFP or anti-AR antibody diluted 1:500. After cell washing, the slides are incubated for 2 hr with the secondary antibody diluted 1:100. After another wash, coverslips are mounted on the slides with fluorescent mounting medium (Dako, Carpinteria, CA) and observed under a microscope the next day. To control the specificity of immunostaining, some slides are incubated with just the primary antibody or the secondary antibody.

Results and Comparison of Permeabilization Conditions

The cells transfected with GFP-S65T or its corresponding fusion protein (GFP–AR) were easily identified by GFP fluorescence. The level of GFP fluorescence differed from that of cellular autofluorescence, and the transfected cells corresponded to approximately 10% of the COS-7 cells. Conversely, the fluorescence was not detected in cells expressing wild-type GFP. GFP-S65T was distributed throughout the cytoplasm and nucleus of the COS-7 cells, and this distribution was not modified by addition of androgen. The GFP–AR was localized predominantly in the cytoplasm in the absence of hormone. The intensity of the detected fluorescence depended on the cell depth. The higher perinuclear fluorescence exhibited by some cells was thus due to a thicker cytoplasm around the nucleus rather than to a particular staining pattern. After incubation with 10^{-6} *M* R1881 for 1 hr, the GFP–AR was predominantly nuclear. This revealed the hormone dependence of GFP–AR trafficking.

In immunocytochemistry, the spatial distribution of the AR was strongly affected by the conditions of permeabilization. A pattern of localization identical to GFP fluorescence was obtained when cells were permeabilized with 0.3% saponin or 0.02% Triton X-100 (Fig. 2, see color insert). With these concentrations, the GFP–AR was cytoplasmic in the absence of hormone and nuclear after hormonal incubation. Control experiments were performed with transfected cells expressing the AR, and the same cellular distribution was observed. However, as a function of the concentration of permeabilization solution, the spatial distribution differed. Indeed, the cells permeabilized with Triton X-100 (0.1 or 0.3%) exhibited a nuclear localization of immunofluorescence in the absence and in the presence of hormone. A lower concentration of saponin (0.05 or 0.1%) revealed a cytoplasmic immunofluorescence in the absence of hormone, as visualized in the same cell by GFP fluorescence. In the presence of hormone, the GFP–AR was immunodetected in the cytoplasm, whereas the GFP fluorescence was nuclear. This was verified with the anti-GFP and anti-AR antibodies.

Given the results of AR immunodetection, we concluded that GFP was of potentially great utility. Moreover, GFP can be visualized in living cells, which suggests that the localization of the GFP–AR could be studied in terms of its dynamics along with the analysis of the fate of the AR in the same living cell.

Various Fluorescence Microscope Technologies for Green Fluorescent Protein Analysis

Analysis of the subcellular localization of GFP could be done using various fluorescence microscope technologies such as the confocal scanning microscope, the laser scanning cytometer, and the epifluorescence microscope. Each technique presents some advantages and disadvantages for specific GFP applications.

Confocal Scanning Microscope

In confocal scanning microscopes, the object is viewed not as a whole but as a series of points on which both illumination and detection are tightly focused. This avoids the problem often encountered with fluorescence, i.e., the swamping of the in-focus fluorescence signal by noise in the form of fluorescence from out-of-focus portions of the object. In some applications, the confocal microscope produces a dramatic increase in the sharpness and contrast of ultimate images. This technology could be applied to GFP to determine a precise localization of GFP fusion protein. However, the confocal scanning microscope is not adapted for kinetics studies. The time required for the analysis of one cell is too long for the study of a rapid process and it can induce substantial photobleaching of the fluorescence. Therefore, this technology is not adapted for dynamics studies.

Laser Scanning Cytometer

ACAS 570 (adherent cell analysis and sorting; Meridian, Okemos, MI): The ACAS provides fluorescence measurement of structure, function, and response at the single-cell level. Cells of interest are positioned above the objective and the focused laser beam is pulsed as the stage moves to scan the cell. Data are collected at intervals as small as 0.25 μm.

Advantages

Sensitive, quantitative measurement is obtainable with minimal photobleaching

The inverted fluorescence microscope allows maximum flexibility in sample handling, from glass slides to multiwell culture dishes

The motorized *x–y* stage permits storage of cell coordinates for repeated scans of the same cell over time. The ACAS 570 is well adapted for kinetics studies

The ACAS 570 permits the measurement of cellular dynamics, such as receptor mobility, by FRAP (fluorescence redistribution after photobleaching) techniques. FRAP techniques involve photobleaching of the fluorescent molecule in discrete subcellular areas with a high-intensity laser pulse, thereby creating nonfluorescent patches. The rate at which the patches are filled by unbleached molecules from surrounding areas is measured by periodic, low-intensity laser scanning

Disadvantages

The absence of an epifluorescence microscope in parallel with the ACAS 570 leads to some difficulties in detecting rapidly the positive cells in a transient transfection

The time required to scan one cell is too long for visualization of rapid processes

The low resolution and sharpness do not permit identification of organelles in the cell

The ACAS 570 can be used for quantitative kinetics to study only predefined processes

Epifluorescence Microscope

For most GFP applications, epifluorescence microscopes are sufficient. A microscope coupled to a camera is a good system for analysis of GFP,

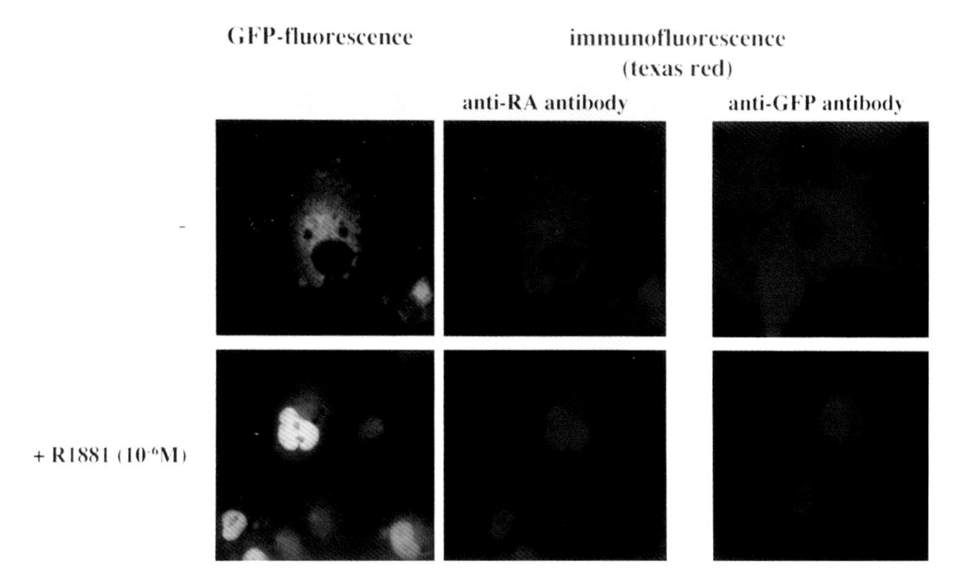

FIG. 2. Comparison of GFP fluorescence and immunofluorescence. The subcellular localization of the GFP–AR in COS-7 cells was analyzed by GFP fluorescence and by immunofluorescence within the same cell. Some cells were incubated with 10^{-6} *M* R1881 at 37° for 1 hr before immunostaining. The cells were permeabilized with 0.3% saponin and incubated with either anti-GFP or anti-AR antibody. The same cell was observed with an FITC filter for GFP fluorescence and with an RITC filter for immunodetection of Texas Red-conjugated antibody. The same results were obtained when cells were permeabilized with 0.02% Triton X-100.

providing an excellent compromise between sufficient sensitivity, good resolution, simplicity of operation, and high rapidity of image acquisition. The images can be quantified with NIH Image software. This microscopic system is well adapted for kinetics study.

Of these three technologies that we have tested, we prefer the epifluorescence microscope and have used it to study GFP–AR kinetics in COS-7 cells.

Intracellular Dynamics and Quantification of Green Fluorescent Protein–Androgen Receptor in Living Cells

In transfected cells, the dynamics of the GFP–AR in the presence of androgen can be evaluated in the same living cell by epifluorescence microscopy and, after quantification of fluorescence, in the various cellular compartments.

Quantification of Green Fluorescent Protein–Androgen Receptor Dynamics

Materials. We use an inverted epifluorescence microscope (Diaphot 200; Nikon; Tokyo, Japan) with ×10, ×20, and oil immersion ×60 differential interference contrast (DIC) objectives. This microscope is coupled to a CCD camera (Night Owl; EGG, Berthold, Germany) which permits image acquisition and processing with Winlight software (EG&G).

Methods. COS-7 cells are cultured in 2 × 2 cm^2 Lab-Tek chamber slides (Nunc, Naperville, IL) and transfected with 2 μg of p-GFP-AR. These chamber slides have the depth of a coverslip and thus the cells can be observed with the oil immersion ×60 objective, with good resolution and low fluorescence noise. The medium is replaced with phenol red-free DMEM before microscopy, because the phenol red exhibits a high background fluorescence. The transfected cells, easily and rapidly distinguished, are maintained at 37° during the kinetics studies. A single cell is recorded with the CCD camera with a 0.1-sec time acquisition in DIC visible light and a 1-sec time acquisition in fluorescence. This image corresponds to time zero of the kinetics. The medium is removed and replaced by 1 ml of medium with androgen. The same cell is observed and recorded every 15 min (an observation every 5 min could have damaged the cell after several observations).

All of the images recorded with Winlight software were saved as TIFF images for quantification by NIH Image software. We measured the fluorescence of the nuclear area and of the total cellular area, using the same surface for each image. Thus, the intensities of pixels were summed within

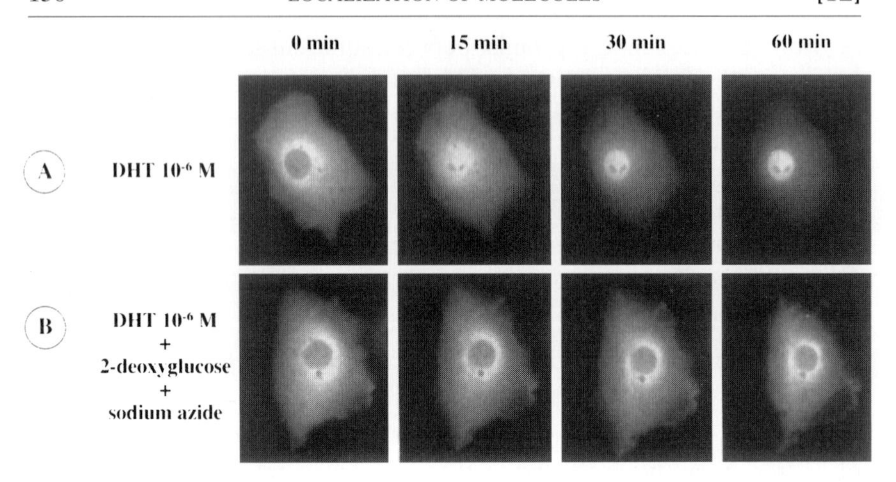

FIG. 3. Dynamics of GFP–AR translocation induced by DHT. COS-7 cells expressing the GFP–AR were analyzed directly by epifluorescence microscopy. Living cells were observed and recorded after 15, 30, and 60 min of incubation with 10^{-6} *M* DHT alone (A) or with inhibitors of ATP synthesis (B).

the nuclear and total cellular areas and corrected for background fluorescence. The percentages of nuclear fluorescence compared with the total cellular fluorescence are calculated and represented on graphs as a function of time in minutes.

Results and Comments. Cytoplasmic fluorescence was observed without hormone, as previously described, after immunostaining. The cells were then incubated with the natural androgen dihydrotestosterone (DHT). After 15 min of incubation at 37° with 10^{-6} *M* DHT, the cytoplasmic fluorescence diminished, whereas the nuclear fluorescence increased (Fig. 3A). After 60 min, the fluorescence signal was predominantly nuclear (85% of total receptors were nuclear), although a complete translocation was never observed. All of the cells in the chamber exhibited the same kinetics. In addition, we determined the energy dependence of AR nuclear import, as has been previously described for the PR.[8] Cells were preincubated for 1 hr with inhibitors of ATP synthesis (sodium azide, 10 m*M*) in glucose-minus DMEM supplemented with 6 m*M* 2-deoxyglucose. With the epifluorescence microscope, we followed the fate of the AR after addition of 10^{-6} *M* R1881. While the GFP–AR entered rapidly into the nucleus under normal conditions, the GFP–AR in the presence of the inhibitor of ATP synthesis remained cytoplasmic after 1 hr (Fig. 3B). This suggested that

[8] A. Guiochon-Mantel, P. Lescop, S. Christin-Maitre, H. Loosfelt, M. Perrot-Applanat, and E. Milgrom, *EMBO J.* **10,** 3851 (1991).

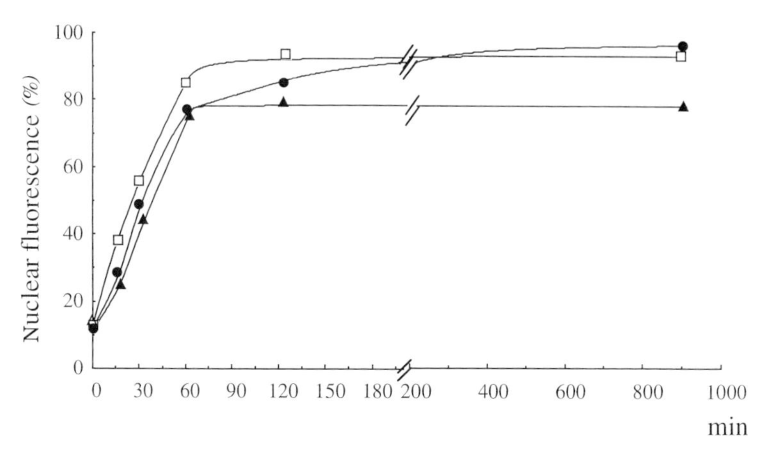

FIG. 4. Quantification of nuclear trafficking of the GFP–AR in the presence of various concentrations of DHT (10^{-6}, 10^{-7}, and 10^{-9} *M*). The nuclear import of the GFP–AR was analyzed in transfected cells and images were quantified by NIH Image software. For each set of conditions, the intensities of pixels were summed within the individual nuclei and the total cellular areas and corrected for background fluorescence. The percentages of nuclear fluorescence were calculated and pooled for each point.

AR nuclear transport is an energy-dependent process. We also tested DHT at various concentrations (10^{-9}, 10^{-7}, and 10^{-6} *M*). The initial rate of nuclear import, evaluated by quantification, was identical with different concentrations (Fig. 4).

GFP is an effective reporter, providing good temporal and spatial resolution in living cells. The GFP–AR model appears to be a new tool for studying AR translocation.

Applications of Green Fluorescent Protein–Androgen Receptor

Screening of Antiandrogens

The antiandrogens are antihormones that repress the action of androgens. The binding of antiandrogen to the AR induces an abnormal conformation of the AR that could block the mechanism of AR action at different steps, such as the dissociation of heat shock proteins, nuclear import, dimerization, and the interaction with DNA or transcription factors.[9] We hypothe-

[9] C. W. Kuil and A. O. Brinkmann, *Eur. Urol.* **29**(Suppl. 2), 78 (1996).

sized that the antiandrogen that would be able to block a preliminary step such as nuclear import could be an efficient antiandrogen with a reduced agonist activity. The antiandrogens are classified in two groups: the steroidal antiandrogens (R2956 and cyproterone acetate) and the nonsteroidal antiandrogens (OH-flutamide, inocoterone, bicalutamide or casodex, and nilutamide or anandron).

We tested the ability of each antiandrogen alone to induce AR nuclear import and, in competition with androgen, their ability to block the nuclear translocation induced by androgen. In parallel, we evaluated the agonistic and antagonistic activities of each antiandrogen by cotransfection experiments.

Methods. Transcriptional activity is evaluated as previously described with CV-1 cells transfected with the GFP–AR and an androgen-dependent reporter gene. Cells are incubated with various concentrations of antiandrogens in the presence or absence of R1881 at 10^{-10} M, the concentration at which the maximal luciferase induction was obtained. The incubation of antiandrogen alone determined the agonistic activity and, in competition with androgen, the antagonistic property.

Concerning nuclear trafficking, COS-7 cells were transfected with the GFP–AR and incubated with 10^{-6} M antiandrogen alone or in competition with DHT at 10^{-9} M, a concentration at which nuclear import was efficient.

Results and Comments. Most of the antiandrogens exhibited agonistic activity (inocoterone, cyproterone acetate, R2956, and OH-flutamide); bicalutamide displayed low agonistic activity and nilutamide had no agonistic activity even at 10^{-6} M. These last two are the most efficient antiandrogens. Only nilutamide (10^{-6} M) did not induce AR nuclear transfer and it was able to block the transfer normally induced by androgen (Fig. 5).[10] This reflects and explains its antagonistic activity. Regarding bicalutamide, it quite likely blocks the mechanism of AR action in a step that follows translocation.

The GFP–AR is a new tool for screening for new antagonists that are able to block nuclear transport, which would provide more potent antiandrogens in the treatment of endocrine disease. We next applied the GFP–AR model in the study of environmental antiandrogens.

Application to Natural Androgen Receptor Mutants Described in Patients with Androgen Insensitivity Syndrome

Defects in the *AR* gene can cause androgen insensitivity syndromes, with a wide spectrum of phenotypes ranging from the complete female

[10] V. Georget and C. Sultan, in preparation (1999).

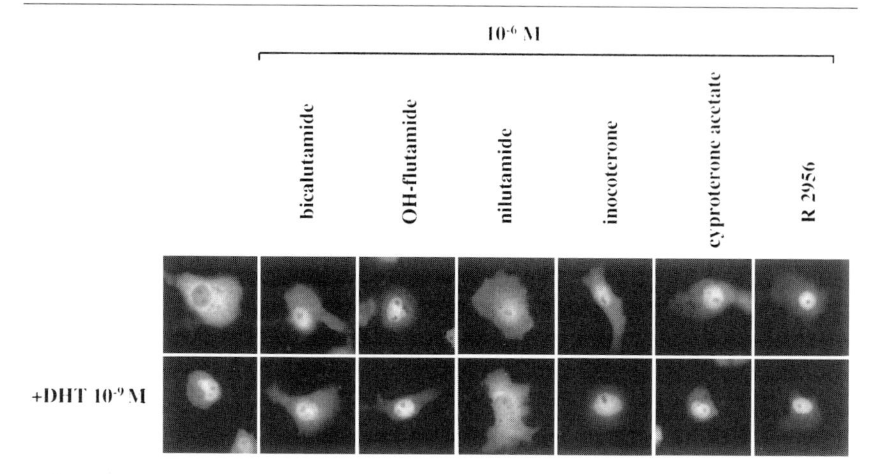

FIG. 5. Nuclear import of the GFP–AR in the presence of antiandrogens. GFP–AR-transfected cells were incubated with various steroidal and nonsteroidal antiandrogens at 10^{-6} *M* alone or in competition with 10^{-9} *M* DHT. After 1 hr, cells were observed by epifluorescence microscopy.

phenotype in 46,XY subjects [complete androgen insensitivity syndrome (CAIS), grades 6 and 7] to partial androgen resistance [or partial androgen insensitivity syndrome (PAIS), grades 1–5].[11] Abnormalities of the *AR* gene have been identified in the eight exons of *AR* and can vary from complete or partial gene deletions to, most often, single base mutations.[12] A single amino acid substitution in the AR protein may thus partially or completely affect the mechanism of AR action. Like many investigators, we have evaluated the activity of these mutants by *in vitro* techniques, previously described, to establish the relationship between the genetic mutation identified in the *AR* gene and the phenotype observed in the patient. This is not always easily done for the partial form of androgen resistance. The GFP–AR model has been applied to AR mutants to evaluate the trafficking of two mutants in the ligand-binding domain (LBD) and to determine the usefulness of this method in the investigation of PAIS.

Methods. Two mutations detected in patients with PAIS, substitutions

[11] C. A. Quigley, A. Debellis, K. B. Marschke, M. K. Elawady, E. M. Wilson, and F. S. French, *Endocr. Rev.* **16,** 271 (1995).

[12] B. Gottlieb, H. Lehvaslaiho, L. K. Beitel, R. Lumbroso, L. Pinsky, and M. Trifiro, *Nucleic Acids Res.* **26,** 234 (1998).

R840C (grade 3)[13,14] and G743V (grade 4 or 6) (personal data; and see Ref. 15), and one mutation detected in CAIS, L707R,[16] which is used as "negative control," are recreated in the *AR* gene sequence cloned in pCMV5 by site-directed oligonucleotide mutagenesis. The construction of the AR mutants fused to GFP is then obtained in the same manner as for the GFP–AR. The expression level, the androgen-binding characteristics, and the transactivation properties are evaluated as previously described in the section concerning the wild-type GFP–AR. The intracellular dynamics of the GFP–AR mutants were evaluated in transfected COS-7 cells incubated with various concentrations of DHT. For each concentration, real-time kinetics are studied with the epifluorescence microscope and images are quantified with NIH Image software, as described above.

Results and Comments. In transient transfection assays, with an identical expression level, GFP–AR-G743V and GFP–AR-R840C displayed decreased androgen-binding affinity compared with the GFP–AR and were fully effective in transactivation, but only at high concentration. This explained the abnormal function of the AR in these patients but did not permit differentiation in terms of the severity of phenotype. Moreover, the androgen-binding characteristics revealed a faster dissociation of androgen for GFP–AR-G743V than for GFP–AR-R840C. As expected, we observed no androgen binding by and no transcriptional activity of GFP–AR-L707R.

The nuclear trafficking of the GFP–AR mutants was evaluated in terms of two parameters: the rate of nuclear transfer and the maximal amount of receptors imported into the nucleus. At 10^{-6} *M* DHT, the GFP mutants associated with PAIS entered the nucleus in a fashion similar to that of the wild-type GFP–AR, whereas GFP–AR-L707R remained cytoplasmic (Fig. 6). Concerning exclusively the mutants associated with PAIS, the rate and maximal degree of nuclear import at 10^{-7} *M* DHT were both reduced, even more so for GFP–AR-G743V. The difference between mutants was more pronounced at 10^{-9} *M* DHT because GFP–AR-G743V entered the nucleus with even slower kinetics.[17]

We observed that the nuclear transfer capacities of these LBD mutants

[13] L. K. Beitel, E. P. Kazemi, M. Kaufman, R. Lumbroso, A. M. DiGeorge, D. W. Killinger, M. A. Trifiro, and L. Pinsky, *J. Clin. Invest.* **94,** 546 (1994).

[14] C. L. Bevan, B. B. Brown, H. R. Davies, B. A. J. Evans, I. A. Hughes, and M. N. Patterson, *Hum. Mol. Genet.* **5,** 265 (1996).

[15] R. Nakao, T. Yanase, Y. Sakai, M. Haji, and H. Nawata, *J. Clin. Endocrinol. Metab.* **77,** 103 (1993).

[16] S. Lumbroso, J. M. Lobaccaro, V. Georget, J. Leger, N. Poujol, B. Terouanne, D. Evainbrion, P. Czernichow, and C. Sultan, *J. Clin. Endocrinol. Metab.* **81,** 1984 (1996).

[17] V. Georget, B. Térouanne, S. Lumbroso, J. C. Nicolas, and C. Sultan, *J. Clin. Endocrinol. Metab.* **83,** 3597 (1998).

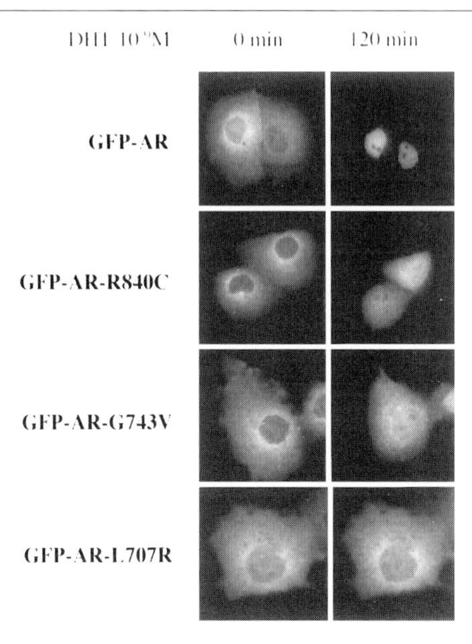

FIG. 6. Quantification of the trafficking of the GFP–AR and GFP–AR mutants associated with PAIS (GFP–AR-R840C and GFP–AR-G743V) or CAIS (GFP–AR-L707R). Transfected COS-7 cells were incubated with 10^{-9} *M* DHT for 2 hr. The fusion proteins were detected in live cells using an epifluorescence microscope coupled to a CCD camera.

are in correlation with the severity of the phenotype and, thus, the GFP–AR model could be a useful complementary tool to understanding the phenotype–genotype relationship of AR function in patients with AIS.

In conclusion, the GFP–AR model has permitted analyses of the dynamics of GFP–AR trafficking in naive cells. Nuclear transfer could also be evaluated in androgen target cells, such as prostatic cells.

[13] Use of Green Fluorescent Protein for Visualization of Cell-Specific Gene Expression and Subcellular Protein Localization in *Bacillus subtilis*

By CHRIS D. WEBB and ORNA RESNEKOV

Introduction

The application of the green fluorescent protein (GFP) to studies of cell biology in *Bacillus subtilis* has opened up exciting new areas of research. An important attribute of GFP is that it can be used in live *B. subtilis* cells as a reporter for gene expression, protein localization, or the analysis of cell morphology. In addition, protein fusions to GFP can be used to screen mutants rapidly or to observe dynamic processes such as chromosome segregation with relative ease. Extensive treatment of cells is not necessary; under most circumstances one simply places a drop of culture on a glass slide for immediate viewing under a microscope.

In *B. subtilis,* folding of wild-type GFP and maturation of the chromophore take less than 40 min at 37°C.[1] Therefore, detection of a signal from a fusion protein follows soon after expression of the gene fusion. The time between expression and the detection of fluorescence could conceivably be decreased with new mutant forms of *gfp*. *B. subtilis* cells are relatively small (1–2 μm in length and 0.5 μm in diameter) and yet with the appropriate tools for visualization, fluorescence from fusion proteins can be observed in whole cells and even at specific subcellular locations. In most instances (see Table I and references therein) the joining of the *gfp* coding sequence to the coding sequence of a gene of interest has not had a deleterious effect on either the function of the protein being studied or its pattern of localization.[1–6] In a few cases temperature sensitivity has been reported.[4,7]

Studies in *B. subtilis* have made use of the unique features of GFP to

[1] C. D. Webb, A. Decatur, A. Teleman, and R. Losick, *J. Bacteriol.* **177,** 5906 (1995).
[2] F. Arigoni, K. Pogliano, C. D. Webb, P. Stragier, and R. Losick, *Science* **270,** 637 (1995).
[3] P. Glaser, M. E. Sharpe, B. Raether, M. Perego, K. Ohlsen, and J. Errington, *Genes Dev.* **11,** 1160 (1997).
[4] W. M. Huang, J. L. Libbey, P. Van der Hoeven, and S. X. Yu, *Proc. Natl. Acad. Sci. U.S.A.* **95,** 4652 (1998).
[5] D. C.-H. Lin, P. A. Levin, and A. D. Grossman, *Proc. Natl. Acad. Sci. U.S.A.* **94,** 4721 (1997).
[6] O. Resnekov, S. Alper, and R. Losick, *Gene Cells* **1,** 529 (1996).
[7] P. A. Levin, personal communication (1998).

0076-6879/99 $30.00

TABLE I
PUBLISHED GENE FUSIONS TO *gfp*

Gene fused to *gfp* coding sequence[a,e]	Function of wild-type protein	Localization of fluorescence in *Bacillus* cells	Controlling transcription factor, if known	Codons present in gene fusion to *gfp*[b]	*gfp* used[c]	Ref.[d]
cotE	Coat protein	In the mother cell surrounding the forespore	σ^E	1–171	wt	1
cotE	Coat protein	Mother cell specific (high level)	σ^E	0 (transcriptional fusion)	wt	1
csfB	Unknown	Forespore specific (low level)	σ^F	1–50	wt	1
dacF	Putative penicillin-binding protein	Forespore specific	σ^F and σ^G	1–33	S65T or Y66H (BFP)	2
divIVA	Cell division protein	Old, nascent, and future division sites		1–164	S65T	3
gerE	Sporulation-specific transcriptional activator	Mother cell specific	σ^K	1–51	wt	1
gyrA	Subunit of gyrase	Nucleoid associated		1–821	F99S, M153T, V163A	4
gyrB	Subunit of gyrase	Nucleoid associated		1–638	F99S, M153T, V163A	4
hbs	Histone-like protein	Nucleoid associated		1–92	wt	5
parC	Required for chromosome segregation. Subunit of topoisomerase IV	Entire cell, with brighter signals at cell poles		1–806	F99S, M153T, V163A	4
parE	Required for chromosome segregation. Subunit of topoisomerase IV	Entire cell		1–655	F99S, M153T, V163A	4

(*continued*)

TABLE I (*continued*)

Gene fused to *gfp* coding sequence[a,e]	Function of wild-type protein	Localization of fluorescence in *Bacillus* cells	Controlling transcription factor, if known	Codons present in gene fusion to *gfp*[b]	*gfp* used[c]	Ref.[d]
spo0J	Chromosome partition protein	Nucleoid associated, close to the cell poles	σ^A	1–282	S65T, V163A	6
spo0J	Chromosome partition protein	Nucleoid associated, close to the cell poles	σ^A	1–282	S65T	7
spoIIE	Phosphatase	Both potential polar division sites	σ^A	1–826	wt	8
spoIIE (*B. megaterium*)	Putative phosphatase	Sites of potential septum assembly near the cell poles		?–582	wt	9
spoIIGB	Sporulation-specific transcription factor	Mother cell compartment and septal membranes	σ^A	1–55	wt	10
P_{dacF}-*spoIIGB*	Sporulation-specific transcription factor	Forespore specific	σ^F or σ^G	1–55	wt	10
spoIIQ	Required for engulfment	Forespore specific (high level)	σ^F	1–63 (transcriptional fusion)	S65T	11, 12
spoIVA	Needed for assembly of the spore coat	In the mother cell as a shell around the forespore in the presence of the wild-type protein	σ^E	1–492	S65T	2
spoIVFB	Protease/regulator involved in the activation of a sporulation-specific transcription factor	Mother cell side of the sporulation septum and later surrounding the forespore	σ^E	1–283	F64T	13

P_{xylA}-*spoIVFB*	Protease/regulator involved in the activation of a sporulation-specific transcription factor	Not published		1–283	F64T	14
sspE-2G	Small acid-soluble protein	Forespore specific (high level)	σ^F and σ^G	1–21	wt	1
P_{spoVG}-*gfp-lacI*	LacI is a repressor of the *E. coli* lactose operon when bound to the lactose operator	Lactose operator on chromosome	σ^H		S65T	15
P_{veg}-*gfp-lacI*	LacI is a repressor of the *E. coli* lactose operon when bound to the lactose operator	Lactose operator on chromosome	σ^A		S65T	1

[a] Natural promoters were used unless otherwise indicated. All studies were performed using *B. subtilis* genes in *B. subtilis,* except for one case in which a gene from *B. megaterium* was studied in *B. megaterium.*

[b] All are translational fusions, except for one of the *cotE-gfp* gene fusions and the *spoIIQ-gfp* gene fusion. All are fusions of the *gfp* coding sequence to the 3′ end of the gene of interest, except for the *gfp-lacI* fusions. The codons that are included in the gene fusions are noted.

[c] wt, Wild type; BFP, blue fluorescent protein.

[d] References: (1) C. D. Webb, A. Decatur, A. Teleman, and R. Losick, *J. Bacteriol.* **177,** 5906 (1995); (2) P. J. Lewis and J. Errington, *Microbiology* **142,** 733 (1996); (3) D. H. Edwards and J. Errington, *Mol. Microbiol.* **24,** 905 (1997); (4) W. M. Huang, J. L. Libbey, P. Van der Hoeven, and S. X. Yu, *Proc. Natl. Acad. Sci. U.S.A.* **95,** 4652 (1998); (5) P. Kohler and M. A. Marahiel, *J. Bacteriol.* **179,** 2060 (1997); (6) D. C.-H. Lin, P. A. Levin, and A. D. Grossman, *Proc. Natl. Acad. Sci. U.S.A.* **94,** 4721 (1997); (7) P. Glaser, M. E. Sharpe, B. Raether, M. Perego, K. Ohlsen, and J. Errington, *Genes Dev.* **11,** 1160 (1997); (8) F. Arigoni, K. Pogliano, C. D. Webb, P. Stragier, and R. Losick, *Science* **270,** 637 (1995); (9) I. Barak, J. Behari, G. Olmedo, P. Guzman, D. P. Brown, E. Castro, D. Walker, J. Westpheling, and P. Youngman, *Mol. Microbiol.* **19,** 1047 (1996); (10) J. Ju, T. Luo, and W. G. Haldenwang, *J. Bacteriol.* **179,** 4888 (1997); (11) J.-A. Londono-Vallejo, C. Frehel, and P. Stragier, *Mol. Microbiol.* **24,** 29 (1997); (12) P. Stragier, personal communication (1997); (13) O. Resnekov, S. Alper, and R. Losick, *Gene Cells* **1,** 529 (1996); (14) O. Resnekov and R. Losick, *Proc. Natl. Acad. Sci. U.S.A.* **95,** 3162 (1998); (15) C. D. Webb, A. Teleman, S. Gordon, A. Straight, A. Belmont, D. C.-H. Lin, A. D. Grossman, A. Wright, and R. Losick, *Cell* **88,** 667 (1997).

[e] Note added in proof: Four new publications describing gene fusions to *gfp* have appeared: R. A. Britton, D. C. Lin, and A. D. Grossman, *Genes Dev.* **12,** 1254 (1998); L. J. Wu, A. Feucht, and J. Errington, *Genes Dev.* **12,** 1371 (1998); P. Fawcett, A. Melnikov, and P. Youngman, *Mol. Microbiol.* **28,** 931 (1998); K. P. Lemon and A. D. Grossman, *Science* **282,** 1516 (1998).

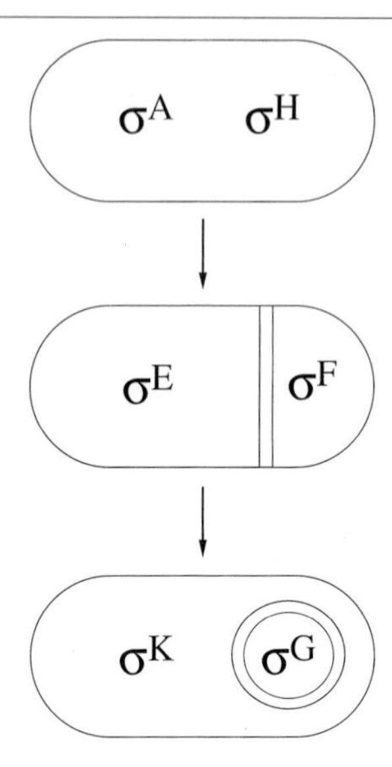

FIG. 1. Schematic illustration of the sequential activation of transcription factors and the morphological changes associated with the developing cell during sporulation. At the initiation of sporulation two transcription factors, σ^H and σ^A, are active. These transcription factors are also active during vegetative growth. After an asymmetric division, the forespore (the smaller compartment) and mother cell (the larger compartment) lie side by side within the sporangium. Gene transcription is governed by σ^F in the forespore and by σ^E in the mother cell [P. Margolis, A. Driks, and R. Losick, *Science* **254,** 562 (1991)]. In the next morphological stage (when the forespore is wholly contained within the mother cell) gene transcription is principally controlled by σ^G in the forespore and by σ^K in the mother cell [P. Stragier and R. Losick, *Annu. Rev. Genet.* **30,** 297 (1996)]. Cell wall material is not shown in this diagram; lines indicate membranes. For a complete review of the developmental pathway of sporulation and the regulation of compartment-specific gene expression, see Stragier and Losick [*Annu. Rev. Genet.* **30,** 297 (1996)].

complement other techniques in addressing the following questions during growth and sporulation (for an overview of sporulation see Fig. 1, Losick and Stragier,[8] and Shapiro and Losick[9]): (1) How are proteins localized in cells and where do specific proteins localize within cells? (2) How is a protein complex assembled in a cell? (3) How are chromosomes segregated to daughter cells? (4) Does the chromosome have a specific architecture? (5) What is the mechanism by which the forespore compartment is engulfed by the mother cell? (6) How is compartment-specific gene expression established and maintained? (7) What governs the bacterial cell cycle and cell

[8] R. Losick and P. Stragier, *Nature* (*London*) **355,** 601 (1992).
[9] L. Shapiro and R. Losick, *Science* **276,** 712 (1997).

division? (8) What are the components, complexes, and control points of specific signal transduction pathways? (see Table I).

This chapter is intended as a resource for those using GFP in their studies of *B. subtilis.* We are assuming that the reader has constructed a *B. subtilis* strain containing either a transcriptional or translational fusion to *gfp.* We have not made a recommendation as to which *gfp* variant to choose, but we have listed the different variants of *gfp* that have been used successfully in *B. subtilis* in Table I. We have also not included basic methods for working with *B. subtilis.* A recommended starting point for such methods is Harwood and Cutting.[10]

Equipment for Observing Green Fluorescent Protein Fluorescence

The following is a list of the items required to view and record fluorescence from GFP in *B. subtilis:*

- Fluorescence microscope, including
  - Magnification lens, ×100 [1.3–1.4 numerical aperture (NA); higher NA is advantageous]
  - Filter sets [for fluorescein isothiocyanate (FITC) and 4′,6-diamidino-2-phenylindole (DAPI); see Tips for Maximum Fluorescence]
  - Camera [35 mm or cooled charge-coupled device (CCD)]; Provia Fujichrome film is recommended for slides, owing to the higher saturation of green in the film
  - Condenser for phase-contrast and/or Nomarski optics
  - Glass slides [multitest slide, 15 wells per slide, 4-mm diameter (ICN Pharmaceuticals, Irvine, CA) or a standard slide]
  - Dark room

To observe and record the fluorescence dynamically from a GFP fusion protein one needs (see Webb *et al.*[11]) the following:

- Cooled CCD digital camera, color or black and white: Considerations for choosing a camera include resolution, sensitivity, and cost
- Software for camera control and image capturing: This software needs to be flexible and to have adequate technical support for maintenance and upgrades. Another important aspect of the software is the ability to overlay images (useful when comparing the images between two different colored labels). MetaMorph software (Universal Imaging, West Chester, PA) meets these requirements

[10] C. R. Harwood and S. M. Cutting, "Molecular Biological Methods for *Bacillus.*" John Wiley & Sons, Chichester, UK, 1990.

[11] C. D. Webb, P. Graumann, J. A. Kahana, A. Teleman, P. Silver, and R. Losick, *Mol. Microbiol.* **28,** 883 (1998).

Automatic shutter to minimize exposure of samples to excitation light

The easiest way of recording results and preparing figures is to work with digital images. For this, one needs the following:

Computer for capturing and storing images: A compact disk (CD) writer is an inexpensive and useful way to store large numbers of digital images

Software for the preparation of finished figures or slides, such as Adobe (Mountain View, CA) Photoshop and Illustrator or Deneba (Miami, FL) Canvas

Slide scanner for the conversion of 35-mm slides into digital format

Photographic quality printer with high resolution, such as the Tektronix (Wilsonville, OR) dye-sublimation printers. An inexpensive alternative is a late-model ink jet printer

Video graphics array (VGA)-to-video converter for the transfer of digital movies to video tape: This is useful for presentations

In the following section, we describe the sample preparation steps needed for optimum detection of GFP.

Preparation of Specimens for Viewing

Growth Conditions

The conditions under which strains are grown affect the intensity of the fluorescent signal. As noted in some studies, lower temperatures tend to result in a brighter signal of GFP.[1,12] This is probably due to effects on the folding and maturation of the GFP chromophore. Several mutants of GFP are now available that may overcome this problem by allowing for a rapid or stable formation of the chromophore at higher temperatures.[13,14] If these mutants are not available, simply growing strains at 30° or even 25° increases the intensity of the fluorescent signal.

One idiosyncrasy of the use of GFP in *B. subtilis* is that the signal intensity can vary from experiment to experiment. This seems to be more apparent in experiments using vegetatively growing cells as opposed to cells that are sporulating. The explanation for this observation is unclear and may simply be related to the particular colony that was selected from a plate. Therefore, it is advisable to monitor several different colonies for

[12] J. Ju, T. Luo, and W. G. Haldenwang, *J. Bacteriol.* **179,** 4888 (1997).

[13] J. A. Kahana and P. A. Silver, *in* "Current Protocols in Molecular Biology" (F. M. Ausubel, R. Brent, R. E. Kingston, D. E. Moore, J. G. Seidman, J. A. Smith, and K. Struhl, eds.), p. 9.6.13. John Wiley & Sons, New York, 1996.

[14] Y. Kimata, *Methods Enzymol.* **303,** [31], 1999 (this volume).

GFP fluorescence. Another variable that can affect the signal strength is the medium in which the strains are grown. For fusion proteins that are synthesized early in sporulation (such as SpoIIE–GFP) sporulation in DS medium[10] results in a brighter fluorescent signal than sporulation by the resuspension method.[15]

Sample Preparation

To observe optimum fluorescent signals, care must be taken in the sample preparation. If a colored medium is used to grow cells (such as LB or DS medium), the cells should be washed [e.g., in phosphate-buffered saline (PBS)] prior to slide preparation and viewing. These media contain ingredients that fluoresce and interfere with the observation of low levels of GFP. An alternative is to fix the cells.

The fixation of cells allows for the viewing of multiple samples from strains at one session on the microscope. We find that fixation does not affect the pattern or intensity of fluorescence compared with cells that are not fixed.[6,16] For fixation, cells are processed as for immunofluorescence,[17,18] using a final concentration of 2.5% paraformaldehyde–0.005% glutaraldehyde in 30 m*M* Na_3PO_4 (pH 7.4) for 15 min at room temperature, and then transferred to ice for 30 min. Subsequently they are washed in PBS three times and resuspended in GTE buffer [50 m*M* glucose, 10 m*M* EDTA, 20 m*M* Tris (pH 7.5)]. Cells can be viewed immediately or kept at 4° for as long as 2 months.

Slide Preparation

A prerequisite for GFP detection is the use of nonfluorescent colorless glass slides. Each new box of slides should be checked to ensure that the glass has no color, since the authors have found that even with the same brand and type of slides, box-to-box variation is common. The color can best be seen by looking at the edges of the slides when they are packed together in a box. The use of a slide with green-tinted glass obscures the fluorescence from GFP and should be avoided.

We use two ways to position cells directly on the slides. In the first method one places 3 μl of a cell suspension on a glass slide and positions a coverslip on top of the sample. To minimize fluid flow, a folded tissue

[15] C. D. Webb, unpublished observations (1995).

[16] C. D. Webb, A. Teleman, S. Gordon, A. Straight, A. Belmont, D. C.-H. Lin, A. D. Grossman, A. Wright, and R. Losick, *Cell* **88,** 667 (1997).

[17] E. J. Harry, K. Pogliano, and R. Losick, *J. Bacteriol.* **177,** 3386 (1995).

[18] K. Pogliano, E. Harry, and R. Losick, *Mol. Microbiol.* **18,** 459 (1995).

paper is placed on top of the coverslip and a flat object is used to press down firmly and evenly on the coverslip (a coverslip box is useful for this purpose). This has the effect of removing excess fluid and causing a tight seal between the coverslip and the slide. As long as no dirt or other contaminants are present, a minimum of fluid flow should be observed and the cells are ready for viewing.

An alternative method involves the use of poly-L-lysine (0.1%; Sigma, St. Louis, MO) to adhere the cells to the slide. For this purpose we recommend the use of multiwell slides from ICN Pharmaceuticals. It is helpful to prepare the slides in a clean petri dish, using the top of the petri dish as a cover to avoid accumulating dirt in the wells. Add 10 μl of poly-L-lysine to each well on the slide, and let the solution sit for 2 min. Gently aspirate the solution without touching the bottom of the wells and allow the remaining poly-L-lysine to air dry completely. The glass in the wells will appear opaque when dry. Add 10 μl of a fixed cell suspension in GTE (see above) to each well and let sit for 2 min. Cells will settle and stick to the poly-L-lysine-treated glass. (Nonfixed cells can also be used in this procedure, but must be used immediately.) Gently remove the 10 μl from each well of the slide and wash each well once with 10 μl of PBS. Allow the cells to air dry completely to ensure their adherence to the slide. Rehydrate the cells by adding 10 μl of PBS in each well, and let sit for 2 min. Remove the PBS, equilibrate each well in SlowFade equilibration buffer (Molecular Probes, Eugene, OR) for 2 min, add SlowFade (in PBS–glycerol) to each well, and cover the slide with a coverslip. After cleaning the edges and the bottom of the slide, the cells are ready for viewing. It is convenient to store individual slides in petri dishes at −20°.

Agarose Pads on Slides

Placing a sample of a cell suspension on a pad of agarose immobilizes the cells and allows one to observe GFP fluorescence dynamically during growth. A 1–1.5% agarose (SeaKem; FMC Bioproducts, Philadelphia, PA) solution should be made in minimial medium. If different brands or types of agarose are used, a different percentage may be needed to achieve the best consistency. An agarose pad that is too thin is difficult to handle. Melt the agarose so that the solution is hot. Place about 0.8 ml of the hot solution onto a glass slide, and quickly place a second slide on top, taking care to avoid formation of bubbles. The second slide causes an even layer of agarose to be formed. After a few minutes, when the agarose layer has cooled, lift the two slides from the bench. First, remove the excess agarose from the edges, using a razor blade. Then carefully slip one glass slide away from the other without tearing the agarose (this takes some practice!). Soon after separation of the slides, and before the agarose pad dries out, place 3 μl of the cell sample on the agarose pad and position a 22-mm^2 coverslip on top of the sample. Using a sharp razor blade, carefully remove excess agarose from all four sides (any excess may prevent a tight seal in the next step). This can then be sealed using melted Valap (a mixture, 1:1:1 by weight, of petroleum jelly, lanolin, and paraffin wax). Dip a cotton swab in the melted Valap, and rapidly swipe the Valap across one edge of the coverslip. Repeat for each side of the coverslip. Avoid spreading excess Valap on the top of the coverslip, as it can mix with the lens immersion oil and have a deleterious effect on the quality of the image. Once the coverslip is sealed the slide is ready for the microscope.

Double Labeling of Cells

Double labeling can be useful to label the perimeter of the cell, the DNA, or a second protein in addition to the protein fusion of interest. The choice of which second label to use (if at all) needs to be made once the intensity of the fluorescent signal of the protein fusion is determined. If the protein fusion has an intense fluorescent signal then the range of choice is much greater. Whatever second label is chosen, once at the microscope the fluorescent image from the GFP fusion should be recorded first because it tends to fade with time.

The cell wall and septa of cells can be visualized using phase contrast or differential interference contrast.[1,6,11,16] The authors have also used a fluorescently labeled lectin[16] (wheat germ agglutinin–fluorescein; Molecular Probes). This staining is strong and is yellow-green in color. Therefore the authors visualized the GFP-tagged protein fusion by immunofluorescence using a secondary antibody coupled to a red fluorophore (Cy-3; Jackson ImmunoResearch, West Grove, PA).

To label DNA, propidium iodide (PI, which labels both RNA and DNA red; Molecular Probes), 4′,6-diamidino-2 phenylindole (DAPI, which labels DNA blue; Molecular Probes) and 7-amino-actinomycin D (which labels DNA red; Molecular Probes) have been used. In the authors' experience PI does not work well with weak GFP signals because a band pass filter set must be used that significantly decreases the observed GFP signal (see below). DAPI is probably the label of choice. In the case of one protein fusion, SpoIIE–GFP, DAPI has been reported to interfere with the fluorescent signal from GFP.[19] One published report demonstrated the use of 7-amino-actinomycin D to label DNA while visualizing the chromosome partition protein Spo0J–GFP.[5] Spo0J–GFP accumulates to relatively high levels in cells.

[19] N. King, personal communication (1998).

To label a second protein in addition to the protein fusion of interest the authors would have liked to use a GFP variant with a different spectral emission, such as the blue fluorescent protein (BFP). Using one available BFP variant (Y66H) has proved to be impractical owing to cell autofluorescence at the emission peak of the protein.[20] New variants of GFP or modified filter sets need to be found that would make this possible in *B. subtilis.* In the interim, experiments that combine the immunofluorescence of one protein and direct visualization of a protein fusion to GFP are feasible as long as the protein fusion is relatively highly expressed.[21]

Tips for Maximum Fluorescence of Green Fluorescent Protein

Whether fluorescence of a protein fusion to GFP is visible depends primarily on the cellular abundance and the effective concentration of the protein in question. Therefore if a protein fusion accumulates to low levels but is concentrated in a small area of the cell, it may be easily visible. The same level of a fusion protein distributed throughout the cell cytoplasm may not be possible to detect. In the next sections we describe several tips for detecting fluorescence from weakly expressed *gfp* fusions.

Equipment

Several features of the equipment play a critical role in the detection of fluorescence from GFP. These include the filter set, the lamp, and the optics of the microscope. Use of the appropriate filter set is most important. For example, the FITC long pass (emission wavelengths, 515 nm and longer) allows for the passage of yellow light, which increases the intensity of the observable signal. When a band pass filter set (emission wavelengths, 515 to 550 nm) is used the intensity of the signal decreases significantly owing to the more narrow spectrum of emitted light. Another useful feature of having a long pass filter set is that it is possible to distinguish by eye the fluorescence due to GFP from the more yellow autofluorescence of the cell. Other filter sets are available that are optimized for GFP. However, we have tested one such set and found that the image did not improve significantly (HiQ GFP longpass; Chroma, Brattleboro, VT). Although the intensity of the signal increased, the overall brightness of the field increased as well. Many new filter sets have been developed and should be tested with *B. subtilis.*

[20] P. J. Lewis and J. Errington, *Microbiology* **142,** 733 (1996).
[21] C. D. Webb and O. Resnekov, unpublished observations (1997).

Another factor that plays an important role in the signal is the lamp. Performance varies with the brand and model of bulb; we found that the 100-W Osram (HB103 W/2, Danvers, MA) mercury bulb works well. As the age of the bulb increases, the signal intensity from protein fusions to GFP decreases significantly. Therefore, the bulb should be replaced after about 170 hr of use, especially if one is dealing with a low-intensity signal. It is important to align the lamp properly so that even and bright illumination can be seen across the field of view. It is possible to enhance a signal by not aligning the bulb for even illumination but by focusing all of the light on a smaller area in the field of view. In fact, the Olympus (Norwood, MA) BX60 microscope has a conversion lens that focuses the light into a smaller area (Cat. No. U-UCV). The obvious drawbacks of this are the smaller viewing area and nonuniform field illumination. However, with the Olympus these drawbacks are not noticeable at the higher magnifications (×100 or ×60 objective lens) that are needed for work with *B. subtilis.*

The other factor that affects the viewing of fluorescence from GFP is the optics of the microscope itself. What may be adequate for viewing larger specimens in other organisms with highly abundant signals may not be sufficient for the detection of low-intensity signals in cells as small as *B. subtilis.* We favor the Olympus BX60 microscope. Other microscopes may be used, and several published reports used a Zeiss (Thornwood, NY) Axiovert or Axiophot microscope.[12,20,22,23] The performance of the various brands of microscopes is dependent on the optics of the lenses and the directness of the excitation and emission light paths. For maximum signal, the camera should be configured so that the light travels in the most direct and shortest path. No unnecessary mirrors or lenses should be inserted between the sample and the camera. For many applications most optimized microscopes will be adequate. However, when the level of abundance of GFP is low, a signal detectable in the Olympus BX60 may not be visible in other microscopes. Whatever microscope one chooses it should be optimized for viewing of protein fusions to GFP (see above and Table II). Initially it may be useful to include a strain that contains a well-characterized abundant fusion protein, such as CotE–GFP (see Table I).

Other Tips and Tricks

In addition to having optimal equipment and growing the cells under particular conditions, other steps can be taken to enhance the fluorescent signal from GFP. One trick that has immediate results is to expose the

[22] I. Barak, J. Behari, G. Olmedo, P. Guzman, D. P. Brown, E. Castro, D. Walker, J. Westpheling, and P. Youngman, *Mol. Microbiol.* **19,** 1047 (1996).

[23] P. Kohler and M. A. Marahiel, *J. Bacteriol.* **179,** 2060 (1997).

TABLE II
CHECKLIST FOR OPTIMUM SIGNAL

Item	Check for:
Microscope	Model and brand; some configurations may not be best for low-intensity signals
Mercury lamp	Age and alignment; should be 100 W, new, and properly aligned
Light path	No unnecessary filters or glass; an optivar, neutral density filters, or Nomarski prisms should not be in the light path; distance light needs to travel should be minimized
Filter set	Proper excitation and emission wavelengths; should be an FITC long-pass filter
Lenses	Dirt and mixed oil; should be clean and have a high NA (1.3 or higher)
Immersion oil	Nonfluorescent; should be labeled as such
Glass slides	Nonfluorescent; should have no color when examined on edge
Media	Any DS or LB media; should be replaced with PBS or another colorless medium

cells to ultraviolet (UV) light (excitation wavelengths, 360–370 nm) for a few seconds just before viewing GFP with the FITC filter set. (The source of the UV light can be from a filter set that is used for viewing DAPI.) This temporarily increases the brightness of fluorescence from GFP, and it works well with GFP in *B. subtilis.*[1] If this is combined with a low amount of phase-contrast light the outline of the cells can also be seen. One curious feature that has been noted with JH642-based strains, but not with PY79-derived strains, is that after exposure to UV light the cells exhibit extremely bright autofluorescence when viewed with the FITC filter set.[24] This can be useful for observing the boundaries of the cell but may interfere with weak GFP signals.

Another technique that aids in increasing the fluorescent signal from GFP is to fix the samples with either formaldehyde (see above) or treatment of the cells with sodium azide.[6,22] After either procedure, the samples should be left at 4° for several hours. The GFP signal increases in brightness even though the cells have been fixed or treated with sodium azide. This is presumably because a significant amount of GFP was made prior to fixation or treatment with azide, and the increase in signal intensity reflects the maturation of the chromophore. An easy alternative to both of these procedures is simply to leave the flask of cells on the bench for a few hours without fixation prior to slide preparation. This also results in an increase in the fluorescent signal from GFP.

For fusion proteins expressed during vegetative growth, another way

[24] C. D. Webb and F. Arigoni, personal communication (1995).

to increase the signal is to allow the cells to reach stationary phase. Like fixation or treatment with azide, this increases the signal, presumably by allowing the GFP to "catch up" in its folding and maturation of the chromophore. A dilution of the cells into fresh medium results in immediate growth for a subset of the population.

Constructs that are Campbell-like insertions into the chromosome can be grown on successively higher concentrations of drug.[25] This results in amplification of the plasmid in the chromosome, thereby adding to the overall concentration of the fusion protein in the cell and increasing the fluorescent signal accordingly.

What should one do if no signal is observed with a newly constructed fusion to GFP? First, as mentioned above, it is useful to check the equipment by examining the fluorescence from a control strain containing a well-characterized gene fusion that exhibits a high level of fluorescence (e.g., *cotE–gfp;* see Table I). Second, make sure that all of the equipment is in optimal working order. We have provided a checklist (Table II) for help with this step. Third, one should try the tips described in this section. Fourth, one should check that the fusion is expressed (and if one is looking at a protein fusion, that it is in frame) by doing a Western blot (see below). Fifth, one should think about the linker between the coding region of the protein of interest and GFP. As a result of cloning strategies, many of the protein fusions created have linker regions of several amino acids. In any particular case, it is possible that the linker region is needed or may be detrimental to the fluorescence of GFP. And, finally, it is possible that the use of antibodies against GFP with fluorescent secondary antibodies may improve the signal, as described in the next section.

Use of Green Fluorescent Protein Antibodies

Immunofluorescence[17,18] with antibodies directed against GFP and a secondary antibody coupled to a fluorophore has been used in double-labeling experiments or when the inherent fluorescence of a protein fusion has been difficult to detect.[16,26] The authors have successfully used Clontech product 8361-2 (a polyclonal serum) at a dilution of 1 : 2000 to 1 : 4000. This approach results in a considerably stronger signal than direct visualization of the fluorescence due to GFP. The signal is also specific and results in little background fluorescence. It does, however, require fixation of the cells and titration of the primary antibody. The dilutions suggested above will be useful for most fusion proteins. Another variable that may affect the fluorescent signal is the concentration of the fixative. The authors have

[25] A. Driks and R. Losick, *Proc. Natl. Acad. Sci. U.S.A.* **88,** 9934 (1991).

[26] O. Resnekov and R. Losick, personal communication (1996).

found that titration of glutaraldehyde (suggested range, from 0 to 0.25% final concentration) is critical to seeing an intense signal.[16,26]

Antibodies specific to GFP have also been used to detect fusion proteins in cell extracts by Western blot analysis.[4,6,12,27] This can be useful to check the expected size, the cellular abundance, and the modification or stability of a fusion protein. Both monoclonal antibodies[12] and polyclonal antibodies[6,26] from Clontech have been used in this type of analysis. The authors have found that the polyclonal antibodies produced a stronger signal with less background than did the monoclonal antibodies when analyzing extracts prepared from growing or sporulating cells from *B. subtilis.*[28]

Past, Present, and Future Explorations

Unpublished Observations

In contrast to the many successful experiments using GFP in *B. subtilis* (see Table I), other endeavors have been less fruitful. The authors attempted to use GFP as a reporter of temporal gene expression levels in an analogous fashion to experiments with β-galactosidase. A spectrofluorimeter was used to measure the levels of fluorescence in either extracts from or whole sporulating cells that were expressing GFP under the control of the *cotE* promoter. The level of fluorescence from the *B. subtilis* cells was not sufficiently high to be useful even though the *cotE* promoter is highly expressed during sporulation. The fluorescence from a control strain of *Escherichia coli* cells with GFP under the control of the T7 promoter was several orders of magnitude brighter than the background fluorescence, while the fluorescence from the *B. subtilis* cells was at most twofold higher than background.[15]

We have not been able to see green *B. subtilis* colonies on plates when viewed with a hand-held UV light source.[28] Presumably the production of the protein fusions that we have worked with has not been high enough. However, we have detected fluorescence on plates using a control strain of *E. coli* cells with *gfp* under the control of the T7 promoter.[28,29]

Finally, the later stages of sporulation have not proved to be amenable to the use of protein fusions to GFP. Only a few examples have been tried but they have either not been as successful as was expected (in terms of the strength of the fluorescent signal),[1] or they have not exhibited fluorescence at all.[29] Especially troublesome have been promoters that are under

[27] O. Resnekov and R. Losick, *Proc. Natl. Acad. Sci. U.S.A.* **95,** 3162 (1998).
[28] O. Resnekov, unpublished observations (1996).
[29] C. D. Webb, unpublished observations (1996).

the control of σ^G. As Table I shows, no examples of a fluorescent fusion protein in the forespore whose gene is solely under the control of σ^G have been published. These difficulties could be attributable to a decrease in pH in the developing forespore at the fourth hour of sporulation,[30] the oxygen content of cells at the late stages of sporulation, the metabolic activity of the developing sporangium, or the inherent phase brightness of the developing forespore late in sporulation.

Future Questions to Be Addressed with Green Fluorescent Protein in Bacillus subtilis

From Table I and the references therein it is obvious that the use of GFP in *B. subtilis* research has opened up exciting new avenues of study. It is likely that the use of GFP and other fluorescent proteins to address questions related to cell biology will grow increasingly common. We have compiled a list of areas in which GFP could prove to be a useful tool. Last, we describe new technologies that could be combined with and expand the use of GFP in *B. subtilis.*

- Identification of mutants with altered morphology
- Study of germination
- Membrane biogenesis and trafficking
- Engulfment
- Flagella, chemotaxis
- Competence and DNA uptake
- Cell movement
- Signal transduction systems
- Protein or RNA trafficking between compartments
- Visualizing bacterial cytoskeletal elements
- Protein orientation in membranes
- Compartment-specific gene expression
- Purification of protein complexes by use of the GFP tag
- Chromosome architecture
- Colocalization of proteins
- Teaching tool
- Environmental studies, by monitoring bacteria released to the environment
- Screening for proteins that show interesting patterns of localization (i.e., transposon-generated protein fusions)
- Easy scoring of populations by microscope or separation of subpopulations by fluorescence-activated cell sorting (FACS)

[30] N. G. Magill, A. E. Cowan, M. A. Leyva-Vazquez, M. Brown, D. E. Koppel, and P. Setlow, *J. Bacteriol.* **178,** 2204 (1996).

Spectral variants of GFP are useful to determine the subcellular localization of multiple proteins that one suspects either colocalize or localize to discrete sites within a cell. One technical problem that arises when using this technique in *B. subtilis* with a conventional fluorescence microscope is that *B. subtilis* cells are small and therefore the imaging requirements are near the limits of the resolving power of a light microscope. A newer technique has been described that may help with this type of problem. Fluorescence imaging microspectrometry (FIMS) combines the spatial resolution of a standard microscope with the spectral resolution of a conventional fluorimeter.[31] This allows for the determination of the exact spectral characteristics of each pixel in an image and therefore permits visualization of the relative locations of two or more GFP variants or any other fluorophores.

Two other techniques that help increase the resolution of a microscope might also be tried with *B. subtilis* cells. These are exhaustive photon realignment (EPR) and two-photon microscopy. EPR increases the signal-to-noise ratio by computationally reassigning photons in an image to the proper plane of focus.[32] The resulting image is much cleaner (due to the absence of unfocused light) and can even be converted to a pseudo-three-dimensional image. Two-photon microscopy uses two different laser sources, oriented perpendicular to each other, each emitting one-half of the wavelength of light required for excitation.[33] This results in a higher resolution image because the required energy for excitation can occur only at the intersection of the two laser beams.

Studies that involve tracking a process require multiple images from the same sample over time. With a conventional fluorescence microscope such studies are limited by photobleaching of the fluorescence from GFP and the physiological effects of repeated exposure to the light, which can kill the cells.[11] These difficulties can be minimized by the use of a technique called total internal reflection (TIR) or evanescent microscopy. Use of a TIR microscope results in a strong signal-to-noise ratio. The technique is minimally invasive because the excitation energy required is low and therefore physiological effects on the cell and photobleaching of GFP fluorescence are minimized.[34] TIR has also been used to observe two differently

[31] D. C. Youvan, W. J. Coleman, C. M. Silva, J. Petersen, E. J. Bylina, and M. M. Yang, Biotechnology et alia <www.et-al.com> **1,** 1 (1997).

[32] Q. Sun and W. Margolin, *J. Bacteriol.* **180,** 2050 (1998).

[33] K. D. Niswender, S. M. Blackman, L. Rhode, M. A. Magnuson, and D. W. Piston, *J. Microsc.* **180,** 109 (1995).

[34] N. L. Thompson and B. C. Lagerholm, *Curr. Opin. Biotechnol.* **8,** 58 (1997).

colored fluorophores simultaneously, so it may also be useful for tracking proteins labeled with spectral variants of GFP.[35]

We envision the use of GFP and other fluorescent proteins in *B. subtilis* research as leading to further interesting questions and novel forms of technology. We eagerly await the results of such studies.

Acknowledgments

We thank A. Driks, A. Hofmeister, and N. King for critical review of the manuscript. We also thank E. Angert, P. Graumann, and P. A. Levin for helpful comments, T. Wilson for help with the spectrofluorimeter experiments, J. Kahana for help with the time-lapse and slide preparation techniques, and P. A. Silver for collaboration. O.R. thanks B. Weisberg and the National Institute of Child Health and Human Development for hospitality during the writing of this manuscript. C. D. W. was a predoctoral fellow of the National Science Foundation and is currently a postdoctoral fellow of Associated Western Universities (Salt Lake City, UT) at DARPA. Some of the work described in this chapter was supported by a Claudia Adams Barr Investigator Award to P. A. Silver and an NIH Grant GM18568 to R. Losick.

[35] K. Saito, M. Tokunaga, A. H. Iwane, and T. Yanagida, *J. Microsc.* **188,** 255 (1997).

[14] Signaling, Desensitization, and Trafficking of G Protein-Coupled Receptors Revealed by Green Fluorescent Protein Conjugates

By LARRY S. BARAK, JIE ZHANG, STEPHEN S. G. FERGUSON, STEPHANE A. LAPORTE, and MARC G. CARON

Introduction

G protein-coupled receptors (GPCRs) are components of signal transduction cascades that regulate the physiological responses to a wide variety of hormones and neurotransmitters.[1,2] Included among the many processes GPCRs regulate are cardiac and vascular tone, odorant perception, diges-

[1] S. Watson and S. Arkinstall, "The G-Protein Linked Receptor Facts Book." Academic Press, San Diego, California, 1994.

[2] E. M. Ross, *in* "Goodman and Gilman's The Pharmacological Basis of Therapeutics," p. 33. Pergamon Press, New York, 1990.

0076-6879/99 $30.00

tion, and central nervous system transmission.[3–5] Approximately 200 G protein-coupled receptors have been identified and it is estimated that hundreds to thousands more may exist.[5,6] Because these receptors respond to a diverse array of ligands and are coupled to multiple types of G proteins and second-messenger pathways, the characterization of newly identified GPCRs represents a formidable challenge.[7,8] However, biochemical studies of rhodopsin and the β_2-adrenergic receptor (β2AR) suggest that in addition to the many common properties that exist in these signaling mechanisms, GPCRs also follow a common paradigm to dampen or terminate their signaling activity. This phenomenon, which is referred to as *desensitization*, involves at least two distinct types of proteins, G protein-coupled receptor kinases (GRKs) and arrestins.[9,10] In particular, β-arrestins not only bind to GRK-phosphorylated β2ARs and uncouple them from G proteins, but they also initiate agonist-mediated β2AR sequestration (internalization) and resensitization of its signaling ability. Owing to the difficulties in obtaining sufficiently large quantities of purified GPCRs, GRKs, and β-arrestins for *in vitro* studies, and the difficulties in observing these proteins in live cells, data concerning the cell biology of their interactions are lacking. A means to follow the real-time behavior and to assess the activity of GPCRs, GRKs, and β-arrestins in living cells should have a profound impact for understanding the regulation of GPCR signal transduction, and has become a major goal of our laboratory. Therefore we have constructed and characterized green fluorescent protein (GFP) conjugates of GPCRs, GRKs, and β-arrestins and have demonstrated their utility in real-time studies of GPCR signal transduction in living cells.

Properties of G Protein-Coupled Receptors, G Protein-Coupled Receptor Kinases, and β-Arrestins

All GPCRs are structurally similar. They contain seven membrane-spanning domains connected by alternating intracellular and extracellular

[3] P. Nef, *Receptors Channels* **1,** 259 (1993).

[4] B. Alberts, D. Bray, A. Johnson, J. Lewis, M. Raff, K. Roberts, and P. Walter, *in* "Essential Cell Biology" (E. Lawrence and V. Neal, eds), p. 481. Garland Publishing, New York, 1998.

[5] M. E. Presti and J. D. Gardner, *Am. J. Physiol.* **264**(1), G399 (1993).

[6] D. Larhammar, A. G. Blomqvist, and C. Wahlestedt, *Drug Design Discov.* **9,** 179 (1993).

[7] S. Cotecchia, B. K. Kobilka, K. W. Daniel, R. D. Nolan, E. Y. Lapetina, M. G. Caron, R. J. Lefkowitz, and J. W. Regan, *J. Biol. Chem.* **265,** 63 (1990).

[8] R. D. Blitzer, G. Omri, M. De Vivo, D. J. Carty, R. T. Premont, J. Codina, L. Birnbaumer, S. Cotecchia, M. G. Caron, and R. J. Lefkowitz, *J. Biol. Chem.* **268,** 7532 (1993).

[9] R. Sterne-Marr and J. L. Benovic, *Vitam. Horm.* **51,** 193 (1995).

[10] M. J. Lohse, J. L. Benovic, J. Codina, M. G. Caron, and R. J. Lefkowitz, *Science* **248,** 1547 (1990).

loops, extracellular amino termini, cytoplasmic carboxy termini, and ligand-binding sites that can be transmembrane or extracellular. Through interactions with one of several classes of heterotrimeric G proteins at their cytoplasmic domains, GPCRs may be coupled to any one of several second-messenger systems.[9]

GPCRs are homologously desensitized by only a limited number of GRKs and arrestin proteins. GRKs compose a family of serine–threonine kinases with significant sequence homology.[9,11,12] The phosphorylation sites for these kinases are contained in the C-terminal tails and intracellular cytoplasmic loops of the receptors. Whereas several sites on a particular receptor may be phosphorylated by a single GRK, agonist-mediated phosphorylation of a single site may be sufficient to initiate homologous receptor desensitization.[13,14] GRKs have a variable tissue distribution. GRK1 (rhodopsin kinase) is restricted to the visual system and pineal gland. GRK2 (β-adrenergic receptor kinase 1) is most abundant in brain, heart, lung, and white blood cells, and has a critical role in normal cardiac physiology.[9,15,16] Transgenic mice overexpressing GRK2 demonstrate decreased cardiac contractility, whereas mice expressing a single copy of the GRK2 gene or a GRK2 inhibitor protein show enhanced contractility.[17–19] Of the remaining GRKs, GRK3 (β-adrenergic receptor kinase 2) is found in the olfactory system, brain, spleen, heart, and lungs; GRK4 in the testes, GRK5 in the heart and lungs, and GRK6 in a broadly distributed tissue pattern.[9] Four different arrestins have been identified and characterized: they are visual arrestin, cone arrestin, β-arrestin 1, and β-arrestin 2. Visual arrestins are predominantly found in the eye. β-Arrestin 1 and 2 are widely distributed and especially elevated in brain, heart, lungs, and blood cells.[9]

Little is known about the dynamic regulation of β-arrestins in tissue. The binding of β-arrestin 1 and 2 to the β2AR has been well studied *in*

[11] K. Palczewski and J. L. Benovic, *Trends Biochem. Sci.* **16,** 387 (1991).

[12] S. S. G. Ferguson, L. S. Barak, J. Zhang, and M. G. Caron, *Can. J. Physiol. Pharmacol.* **74,** 1095 (1996).

[13] H. Ohguro, K. Palczewski, L. H. Ericsson, K. A. Walsh, and R. S. Johnson, *Biochemistry* **32,** 5718 (1993).

[14] Z. L. Fredericks, J. A. Pitcher, and R. J. Lefkowitz, *J. Biol. Chem.* **271,** 13796 (1996).

[15] G. Kong, R. Penn, and J. L. Benovic, *J. Biol. Chem.* **269,** 13084 (1994).

[16] J. L. Arriza, T. M. Dawson, R. B. Simerly, L. J. Martin, M. G. Caron, S. H. Snyder, and R. J. Lefkowitz, *J. Neurosci.* **12,** 4045 (1992).

[17] W. J. Koch, H. A. Rockman, P. Samama, R. A. Hamilton, R. A. Bond, C. A. Milano, and R. J. Lefkowitz, *Science* **268,** 1350 (1995).

[18] W. J. Koch, B. E. Hawes, J. Inglese, L. M. Luttrell, and R. J. Lefkowitz, *J. Biol. Chem.* **269,** 6193 (1994).

[19] M. Jaber, W. J. Koch, H. Rockman, B. Smith, R. A. Bond, K. K. Sulik, J. Ross, Jr., R. J. Lefkowitz, M. G. Caron, and B. Giros, *Proc. Natl. Acad. Sci. U.S.A.* **93,** 12974 (1996).

vitro.[20] However, their abilities to interact with many other GPCRs in cells have been only recently demonstrated by our laboratory, using a β-arrestin 2–GFP conjugate.[21] Splice variants of β-arrestin 1 and 2 also exist, but it is unknown how this affects their biochemistry. The regulation of β-arrestin activity does seem to be dependent on GPCR type, GRK activity, their own phosphorylation by protein kinase C (PKC), and by their intracellular compartmentalization. For example, even though β-arrestins are generally cytosolic, they have been found concentrated at neuronal synapses with GRKs.[12,22]

GRK- and β-arrestin-mediated desensitization of GPCRs has been shown to require agonist occupancy of the receptor, in contrast to second messenger-mediated phosphorylation of receptors by protein kinase A (PKA) or PKC. As a result of GRK phosphorylation, receptor affinity for β-arrestins can increase more than 10-fold, as was demonstrated *in vitro* for the equimolar binding of the β2AR to β-arrestin 2 or in whole cells for the ability of β-arrestin to enhance receptor internalization.[20,23] For a functional interaction to occur between predominantly cytoplasmic β-arrestins and membrane-localized GPCRs, one of them must change its cellular compartmentalization. Assessment of the kinetics and magnitude of this redistribution would provide measures of both receptor activation and desensitization.

GRK phosphorylation followed by β-arrestin binding contributes to both β2AR desensitization and initiation of sequestration, a process leading to receptor resensitization.[24–26] Agonist-dependent interaction of β-arrestin with the receptor serves as the switch that initiates receptor migration to a clathrin-coated pit, and once receptors are assembled at coated pits they can be internalized. This paradigm may be common to different types of GPCRs, because many of them rapidly sequester in response to agonist. Therefore, the direct regulation of sequestration by β-arrestin or the ability of a particular receptor to be phosphorylated by a GRK or to interact with a β-arrestin may contribute substantially to its rate of resensitization.

[20] V. V. Gurevich, S. B. Dion, J. J. Onorato, J. Ptasienski, C. M. Kim, R. Sterne-Marr, M. M. Hosey, and J. L. Benovic, *J. Biol. Chem.* **270,** 720 (1995).

[21] L. S. Barak, S. S. G. Ferguson, J. Zhang, and M. G. Caron, *J. Biol. Chem.* **272,** 27497 (1997).

[22] H. Attramadal, J. L. Arriza, C. Aoki, T. M. Dawson, J. Codina, M. M. Kwatra, S. H. Snyder, M. G. Caron, and R. J. Lefkowitz, *J. Biol. Chem.* **267,** 17882 (1992).

[23] S. S. G. Ferguson, W. E. Downey, A. M. Colapietro, L. S. Barak, L. Menard, and M. G. Caron, *Science* **271,** 363 (1996).

[24] L. S. Barak, L. Menard, S. S. Ferguson, A. M. Colapietro, and M. G. Caron, *Biochemistry* **34,** 15407 (1995).

[25] S. Pippig, S. Andexinger, K. Daniel, M. Puzicha, M. G. Caron, R. J. Lefkowitz, and M. J. Lohse, *J. Biol. Chem.* **268,** 3201 (1993).

[26] S. S. Yu, R. J. Lefkowitz, and W. P. Hausdorff, *J. Biol. Chem.* **268,** 337 (1993).

Moreover, receptor resensitization should require β-arrestin dissociation from the receptor. It has been suggested that receptor dephosphorylation occurs in internalized acidic endosomes prior to receptor recycling to the plasma membrane, so that it would be expected that β-arrestin dissociation from the receptor probably occurs prior to these events.[27] Where the process occurs and at which step it takes place should be readily discernable, if the cellular dynamic behaviors of receptors and β-arrestins can be simultaneously followed. While all GPCRs may not internalize through clathrin-coated pits, activation of all GPCRs so far tested promote the association of β-arrestin.[21,28]

The use of N-terminal, epitope-tagged GPCRs has greatly facilitated the study of sequestration. Tagged GPCRs can be specifically labeled and visualized by immunofluorescence using monoclonal antibodies directed against these accessible surface epitopes. Flow cytometry measurements of the time-dependent agonist-mediated redistributions of various GPCRs suggest that they do not all internalize by the same pathway. β2ARs apparently sequester via a clathrin-dependent mechanism whereas the angiotensin II type 1A receptor internalizes despite blockade of clathrin-mediated endocytosis.[28] While experiments indicate that β_2-adrenergic sequestration absolutely requires β-arrestin binding and that GRK-mediated phosphorylation is only a facilitative step, the degree to which many other GPCRs are regulated by these mechanisms remains unclear.[28] For the β2AR the interdependent contributions of GRK phosphorylation and β-arrestin binding to sequestration are evident from the behavior of the phosphorylation-impaired Y326A-β2AR mutant. In HEK-293 cells under conditions of endogenous GRK expression this mutant receptor sequesters poorly, but its sequestration can be enhanced to levels of wild-type receptor by the overexpression of GRK2.[24,29,30] Thus determination of the time-dependent redistributions of receptors, GRKs, and β-arrestins would provide significant insight into the biology and kinetics of signal transduction regulation.

Attempts to study GPCRs, GRKs, and β-arrestins in live cells have been limited by the comparatively low endogenous expression levels of most GPCRs and the permeability barrier provided by the cell plasma membrane. Transfection of cells with GPCR cDNA may increase GPCR membrane expression many fold and render receptors visible using immu-

[27] K. M. Krueger. Y. Daaka, J. A. Pitcher, and R. J. Lefkowitz, *J. Biol. Chem.* **272,** 5 (1997).

[28] J. Zhang, S. S. G. Ferguson, L. S. Barak, L. Menard, and M. G. Caron, *J. Biol. Chem.* **271,** 18302 (1996).

[29] L. S. Barak, M. Tiberi, N. J. Freedman, M. M. Kwatra, R. J. Lefkowitz, and M. G. Caron, *J. Biol. Chem.* **269,** 2790 (1994).

[30] S. S. G. Ferguson, L. Menard, L. S. Barak, W. J. Koch, A. M. Colapietro, and M. G. Caron, *J. Biol. Chem.* **270,** 24782 (1995).

nofluorescence. However, these techniques cannot guarantee the permanent labeling of receptors or provide easy access to intracellular antigens. Thus the behaviors of intracellular signal transduction proteins have generally remained out of reach to real-time fluorescence microscopy. The use of green fluorescent protein (GFP) technology may provide insight into how GPCR signal transduction regulatory proteins function *in vivo*. Our experience in this area is described in the following sections.

Design of Green Fluorescent Protein-Conjugated G Protein-Coupled Receptors, G Protein-Coupled Receptor Kinases, and β-Arrestins

Ideally, the properties of GFP-conjugated proteins should recapitulate the biochemical and biophysical characteristics of the unconjugated forms. Our experiences in creating and characterizing adrenergic receptor mutants indicate that their N and C termini are the two most promising positions for GFP conjugation, because the transmembrane domains and loops generally govern receptor conformation and provide binding sites for ligands and associated signal transduction proteins. Our decision to place GFP at the C terminus of the β_2-adrenergic receptor was motivated by two further considerations. First, the β2AR has a relatively long cytoplasmic tail, and minor truncation of this tail has relatively little effect on receptor biochemistry. Second, additions of certain epitopes at the receptor N terminus have been demonstrated to reduce receptor expression drastically. However, this problem was circumvented, at least for the Flag epitope, by placing before it a cleavable signal peptide, and perhaps a similar strategy could be considered for N-terminal GFP conjugation.[31] Using a C-terminal conjugation strategy and site-directed mutagenesis, we have constructed GFP variants of the β2AR, the Y326A-β2AR, and the angiotensin II type 1A receptor.

GFP variants have been engineered that possess spectral properties significantly different from the original GFP isolated from the jellyfish *Aequorea victoria*.[32,33] These modified GFPs have been engineered for enhanced brightness, visible light excitation maxima, and red-shifted emission spectra.[34] In this chapter protein-conjugated GFP refers to the S65T variant, which is at least 10–20 times brighter than the original form, has a 480-nm excitation maximum, and can be excited and observed with filter sets optimized to view fluorescein fluorescence.

[31] X. M. Guan, T. S. Kobilka, and B. K. Kobilka, *J. Biol. Chem.* **267,** 21995 (1992).

[32] D. C. Prasher, *Trends Genet.* **11,** 320 (1995).

[33] D. C. Prasher, V. K. Eckenrode, W. W. Ward, F. G. Prendergast, and M. J. Cormier, *Gene* **111,** 229 (1992).

[34] R. Heim and R. Y. Tsien, *Curr. Biol.* **6,** 178 (1996).

GFP was conjugated to the C terminus of an N-terminal epitope-tagged (hemagglutinin, HA) β2AR. The terminal stop codon of the receptor was replaced with a *Sal*I restriction site by directed mutagenesis. The resulting receptor cDNA was ligated in frame between the *Sac*I/*Sal*I polylinker sites of a eukaryotic GFP expression vector obtained from Clontech (Palo Alto, CA), which we had modified to express S65T-GFP (at the time an S65T version was unavailable).[21] However, numerous GFP expression vectors containing mutated and enhanced GFP variants have since become commercially available.

The proliferation of GFP expression vectors has decreased the need for complicated cloning strategies. Therefore, the design of GFP fusion proteins should depend predominantly on two factors: (1) which end of the protein is a more important determinant of the behavior or activity that needs to be preserved, and (2) which biological functions of the protein can be readily assessed. We followed this strategy for creating GFP-conjugated GRK2 (βARK1). GRK2 can be divided into three functional regions. The N-terminal third binds cellular components, the middle third constitutes the catalytic domain, and the C-terminal third segment interacts with G protein βγ subunits. Of the six known GRKs, only GRK2 and GRK3 contain similar C-terminal βγ-binding domains, and the interaction of GRK2 with G protein βγ subunits constitutes an important regulatory step in GPCR phosphorylation and desensitization. The majority of GRK2 is cytosolic, so that phosphorylation of the receptor by GRK2 (or GRK3) requires a functional translocation of kinase to the plasma membrane directed by G protein βγ subunits and phospholipids.[12] Because this function of GRK2 can be readily assessed, GFP was added to the C terminus of GRK2 in order to avoid modifying the presumed receptor-binding domain, and in a manner similar to that described above using the same S65T-GFP expression vector. Interestingly, $G_{\beta\gamma}$ binding is not critical for the kinase activity of other GRKs, because GRKs 4–6 are either palmitoylated or contain a polybasic domain at their C termini that allows them to associate with the plasma membrane. In such a case GFP conjugation might be more judiciously performed at the N-terminal portion of the protein.

Arrestin protein binding is the terminal step in the homologous desensitization of GPCRs. The crystal structure of visual arrestin predicts that the receptor-binding domain is located in the first quarter of the molecule[35] These data agree with a conclusion from mutagenesis studies, that the receptor-binding site of β-arrestin 2 is in its N-terminal domain.[9] Moreover, our findings that C-terminal GFP constructs of β-arrestin 1 and β-arrestin

[35] J. Granzin, U. Wilden, H. W. Choe, J. Labahn, B. Krafft, and G. Buldt, *Nature* (*London*) **391,** 918 (1998).

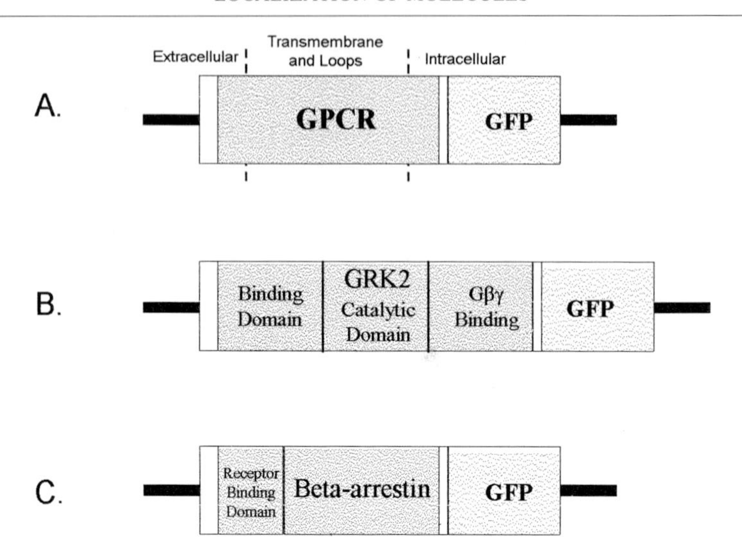

FIG. 1. Linear models indicating the relative sizes of the GFP conjugates of the (A) β2AR, (B) G protein-coupled receptor kinase 2, and (C) β-arrestin.

2, conjugated as described above with S65T-GFP, are biologically active also support this notion.

Characterization of Green Fluorescent Protein-Conjugated G Protein-Coupled Receptors, G Protein-Coupled Receptor Kinases, and β-Arrestins

GFP conjugates of signal transduction proteins should be characterized using known biochemical or functional assays prior to their use in addressing new biological problems. Our experiences with the three types of signal transduction proteins suggest that their conjugation to GFP probably does not significantly change their biochemical properties. Although only minor differences have been observed between the conjugated and native proteins, this conclusion applies only to those properties that have been examined. This is surprising in light of the relatively large, although smaller, size of the GFP in relation to the native proteins (Fig. 1). However, the crystal structure of GFP suggests a compact tertiary structure.[36] Thus, a sufficiently long linker between GFP and its conjugate may be appropriate to minimize potential changes in the biochemical properties of the chimeric protein.

[36] M. Ormo, A. B. Cubitt, K. Kallio, L. A. Gross, R. Y. Tsien, and S. J. Remington, *Science* **273,** 1392 (1996).

The biochemistry of the β2AR has served as a paradigm for other GPCRs, and its characterization also provides insight into how the GFP conjugates should be evaluated. At a minimum the following assays should be performed to assess receptor pharmacology, antagonist binding, agonist binding, second-messenger coupling (in this case to adenylyl cyclase); and if possible receptor phosphorylation. Little difference was seen between the wild-type receptor and the β2AR–GFP conjugate (β2AR–GFP) evaluated for these properties. Saturation binding measurements using ^{125}I-labeled pindolol showed that the dissociation constants for both were equal to 70 ± 10 p*M*. Isoproterenol, a pure βAR agonist with nanomolar affinity, displayed the same ability to compete the radiolabeled antagonist (pindolol) binding from the wild-type receptor and β2AR–GFP. Moreover, the ability of β2AR–GFP to couple to G protein, and activate the adenylyl cyclase, was equal to or slightly greater than that of wild-type β2AR. These data indicate that the GFP group does not substantially change the receptor conformation responsible for ligand binding or significantly interfere with receptor coupling to G protein. The ability of β2AR–GFP to be phosphorylated was also directly evaluated by measuring the agonist-mediated incorporation of ^{32}P into the protein. β2AR–GFP was maximally phosphorylated in response to agonist to approximately 80% of that observed with wild-type β2AR. When tested in the paradigm of agonist-promoted receptor internalization, which is dependent on both receptor phosphorylation and β-arrestin binding, β2AR–GFP sequestered as well as the wild-type receptor. Flow cytometry measurement showed that the decrease in surface receptor after 30 min of agonist exposure in HEK-293 cells was 57 ± 5% for wild-type receptor and 62 ± 11% for β2AR–GFP.[37]

Fewer tools or assays are available to characterize GRKs and arrestins as opposed to GPCRs because comparatively less is known about their cell biology and biochemistry. This has somewhat simplified our evaluations of GRK2–GFP and β-arrestin–GFP (βarr–GFP). GRK2–GFP was characterized by its ability to enhance the internalization of the sequestration-impaired Y326A-β2AR (i.e., phosphorylate the mutated receptor) in HEK-293 cells and to promote $G_{\beta\gamma}$-mediated phosphorylation of rhodopsin *in vitro*.[38] Each of these functions might be expected to be impaired more readily than other activities of GRK2, because these functions of GRK2 depend on C-terminal motifs in close proximity to the GFP moiety. Nevertheless, we demonstrated that in HEK-293 cells, overexpression of GRK2–GFP promotes sequestra-

[37] L. S. Barak, S. S. G. Ferguson, J. Zhang, C. Martenson, T. Meyer, and M. G. Caron, *Mol. Pharmacol.* **51,** 177 (1997).

[38] J. A. Pitcher, R. A. Hall, Y. Daaka, J. Zhang, S. G. Ferguson, S. Hester, S. Miller, M. G. Caron, R. J. Lefkowitz, and L. S. Barak, *J. Biol. Chem.* **273,** 12316 (1998).

tion of the Y326A-β2AR nearly as well as wild-type GRK2. In addition, cell homogenates containing GRK2–GFP, when combined with purified rhodopsin and varying concentrations of $G_{\beta\gamma}$ subunits, were able to phosphorylate rhodopsin 60% as well as wild-type GRK2. These results indicate that while the functional kinase activity of the GRK2–GFP–$G_{\beta\gamma}$ complex is mildly impaired, the ability of GRK2–GFP to interact with GPCRs remains.

βarr2–GFP function was characterized using a sequestration assay based on the observation that β2ARs sequester poorly in COS-7 cells as a result of the relatively low levels of endogenous β-arrestins.[39] Overexpression of wild-type β-arrestin 1 or 2 generally increases β2AR sequestration three- to fourfold, and overexpression of β2AR2–GFP was observed to increase it to a similar degree. These data suggest that addition of GFP to the C terminus of β-arrestin does not interfere with its receptor binding and trafficking functions.[21] In contrast, addition of the Flag epitope slightly proximal to the C terminus of β-arrestin 2 resulted in a β-arrestin construct unable to enhance β2AR sequestration normally (our unpublished data). Both the amino and carboxy termini of β-arrestin contain regulatory domains, so that it is unexpected, considering the relatively small size of the Flag epitope compared with GFP, that βarr2–GFP is even functional. This retention of function supports the observation that the GFP moiety is noninteracting and also suggests a general approach for the C-terminal modification of arrestins. Additional studies have since demonstrated that an N-terminal βarr2–GFP conjugate is also functional.

Instrumentation and Methodology

Once the GFP conjugates have been characterized their cell biology can be evaluated using conventional optical methods. We have observed β2AR–GFP fluorescence by flow cytometry, fluorescence spectrophotometry, conventional fluorescence microscopy, and laser scanning confocal microscopy. Many of our current experiments now measure the time-dependent redistribution of GFP-conjugated receptors or β-arrestins in live cells in response to agonists, and our best results have been acquired using the confocal microscope.

While we have developed permanent clones expressing GFP-conjugated receptors, GRKs, and β-arrestins, most experiments are performed using live HEK-293 or COS-7 cells that are transiently transfected with the appropriate cDNA. The advantage of a transient expression system rests in its versatility and the variety of conditions that can be explored within a given experiment. For these experiments cells are plated at 50% confluence in

[39] L. Menard, S. S. G. Ferguson, J. Zhang, L. Fang-Tsyr, R. J. Lefkowitz, M. G. Caron, and L. S. Barak, *Mol. Pharmacol.* **51,** 800 (1997).

100-mm tissue culture dishes on day 0. On day 1, the cDNA of interest in a GFP expression plasmid is transfected into the cells by coprecipitation with calcium phosphate. In this method, 5–10 μg of plasmid cDNA is resuspended in 450 μl of 18-MΩ water to which 50 μl of 2.5 *M* calcium chloride is added. Five hundred microliters of 2× Hanks' balanced salt solution is then added dropwise to this mixture. The resulting suspension is then added dropwise over the cells, which are now usually between 60 and 80% confluent. On day 2 the medium is replaced with fresh medium, and the cells are returned to the incubator for 24–48 hr for use on days 3 or 4. For confocal microscopy transfected cells are plated into 35-mm dishes (MatTek, Ashland, MA) containing centered, 1-cm wells formed from glass coverslip-sealed holes. For live cell experiments cells are placed in minimal essential medium supplemented with 20 m*M* *N*-2-hydroxyethylpiperazine-*N'*-2-ethanesulfonic acid (HEPES). The experiments are performed with a Zeiss (Thornwood, NY) LSM-410 confocal microscope using the time series function. A baseline scan of GFP fluorescence is performed at 488 nm. Ligand is then added in volumes of 50 to 100 μl (5–10% of the total volume), and scans of the same field are typically made every 30 sec to 2 min. Fluorescence images acquired on the Zeiss LSM-410 confocal microscope are analyzed using the LSM-image software or transferred to Adobe Photoshop (Adobe, Mountain View, CA).

Cell Distribution of β2AR–GFP and Y326A-β2AR–GFP

GFP technology offers significant advantages over conventional fluorescence techniques for the study of the cellular regulation of GPCRs. These advantages include the following:

1. Receptors or proteins are stoichiometrically labeled by GFP with a ratio of 1:1.
2. Every GFP-tagged molecule is potentially visible despite its intracellular location.
3. The labeling and visualization of intracellular proteins does not require permeabilization or extraction of the cell membrane, thus preventing the introduction of fixation artifacts.
4. The same set of live cells can be observed over extended intervals (days) since GFP-labeled protein is a product of cellular synthesis.

Figure 2 shows video images acquired on a Leitz (Deerfield, IL) model DM fluorescence microscope connected to an Optronics (Goleta, CA) VI-470 CCD video camera system and a Sony (New York, NY) model UP-5600 color video printer with a UPK-5502SC digital interface board. The panels display images of 4% paraformaldehyde-fixed HEK-293 cells that permanently express either β2AR–GFP (Fig. 2A–D, H) or the seques-

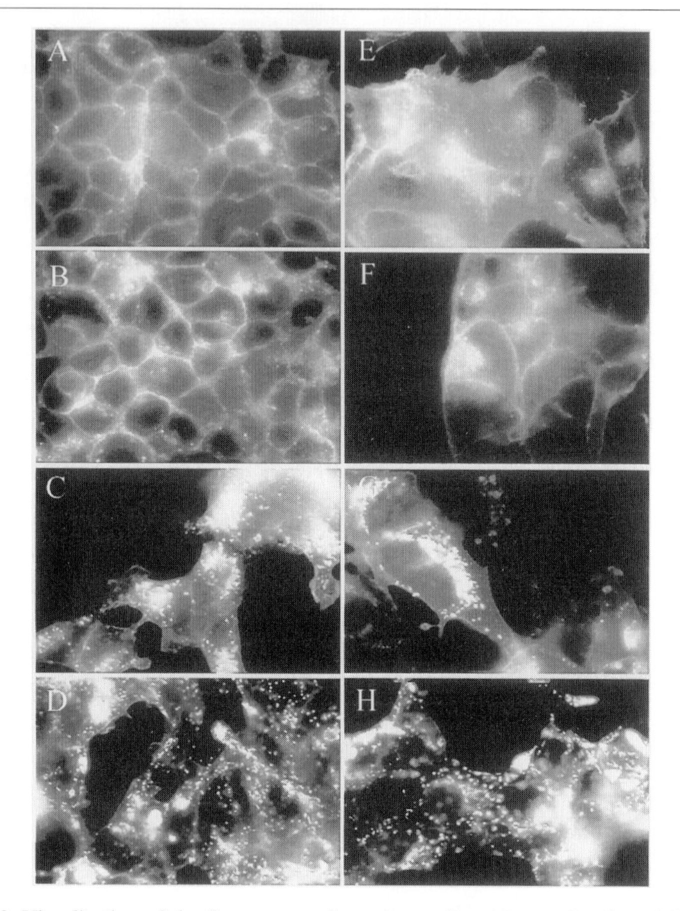

FIG. 2. Visualization of the time course of agonist-mediated internalization of the β2AR and the Y326A-β2AR. Permanent clones of HEK-293 cells containing each receptor were treated as described in text and visualized using standard fluorescence microscopy. Receptors can be observed in at least three different cellular compartments (plasma membrane, endosomal, and perinuclear) in this series of experiments. (See text for details.)

tration-impaired and phosphorylation-impaired Y326A-β2AR–GFP conjugate (Fig. 2E–G). These cells were exposed to saturating concentrations of the agonist isoproterenol at 37° for various periods of time in order to examine receptor sequestration and downregulation. β2AR–GFP should respond to the agonist in a manner similar to unconjugated β2AR, as discussed previously. Figure 2A shows the plasma membrane distribution of β2AR–GFP in cells never exposed to agonist. In these permanent clones receptor fluorescence is homogeneous and uniform from cell to cell. In addition, there is a paucity of intracellular fluorescence. In contrast, cells exposed only to isoproterenol for 15 min (Fig. 2C) have large amounts of internalized receptors. These cells demonstrate a loss of plasma membrane staining at their periphery, and internalized receptors appear as bright patches and white dots in the cell interior. The cells in Fig. 2B appear roughly similar to those in Fig. 2A. However, they were first exposed to 20 μM isoproterenol for 15 min, washed, and then incubated with saturating concentrations of the antagonist propranolol for 1 hr. Cells in Fig. 2D and H were treated for 3 hr with either agonist followed by antagonist (Fig. 2D) or agonist alone (Fig. 2H). In contrast to results shown in Fig. 2B, in which sequestered β2AR–GFP returns to and remains on the cell surface in the presence of antagonist, a large intracellular receptor population persists in Fig. 2D. Moreover, these receptors have the same perinuclear appearance as those in Fig. 2H, which are treated with agonist only. The failure of the GFP-conjugated receptors to return to the cell surface in the long-term presence of agonist and their aggregated perinuclear distributions are most consistent with receptor downregulation. In contrast to β2AR–GFP, Y326A-β2AR–GFP does not sequester. Figure 2E–G shows cells permanently expressing Y326A-β2AR–GFP after 0 min, 15 min, and 3 hr of exposure to the agonist isoproterenol. No internalization of receptor has occurred after 15 min, in agreement with previous sequestration measurements of the Y326A-β2AR.[29] However, after 3 hr of exposure to isoproterenol the distribution of the Y326A-β2AR–GFP as seen in Fig. 2G is similar to that of β2AR–GFP (Fig. 2D and H). The cells in Fig. 2G show a considerable loss of staining along the plasma membrane. Furthermore, perinuclear aggregates of receptors are now apparent. These experiments demonstrate the utility of GFP-conjugated GPCRs for studying receptor behavior. Moreover, the results suggest that sequestration is not a prerequisite for downregulation.

Determination of Receptor Activation by Green Fluorescent Protein-Conjugated G Protein-Coupled Receptor Kinase and β-Arrestin

Phosphorylation by GRKs followed by β-arrestin binding is perhaps a common step in the homologous desensitization of most GPCRs. These processes should occur within seconds of receptor activation, and should coincide with the translocation of GRK and β-arrestin from the cytosol to the plasma membrane. We have attempted to verify by confocal microscopy in live HEK-293 cells that translocations of GRK2–GFP and βarr2–GFP do indeed occur. Translocation should be reflected as an increase in plasma membrane fluorescence that coincides with a decrease in fluorescence from the cell interior. An optimal signal-to-noise ratio will result following agonist treatment if the cytoplasmic fluorescence signal is reduced to levels of cellular autofluorescence and the plasma membrane now contains all of the GFP-conjugated protein. These limits can be practically approached by increasing receptor expression (usually greater than 1 pmol/mg of cell protein) with respect to the amount of expressed GFP-conjugated GRK or β-arrestin protein, and by using cells with relatively small diameters. For example, HEK-293 cells produce better fluorescence signals than the larger COS-7 cells for the same amount of GFP expression. Our attempts to monitor GRK2–GFP translocation under these conditions have failed to demonstrate any persistent GRK2–GFP translocation to membrane-bound β2ARs or angiotensin II type 1A receptors within 10–15 sec of agonist exposure, although preliminary experiments have indicated that a sustained translocation to the substance P receptor can be observed within that time frame. Most unexpected, however, was the novel observation made with GRK2–GFP that GRK2 colocalizes with tubulin in mitotic spindles.[38]

The translocation and binding of βarr2–GFP to an agonist-activated receptor is a more persistent process, and is consistent with the slower time course observed for receptor sequestration. The agonist-mediated translocation of βarr2–GFP is shown in Fig. 3. Figure 3A is a confocal scan of HEK-293 cells that have not been agonist treated and that express receptors at the plasma membrane and βarr2–GFP in the interior of the cell. The nucleus excludes βarr2–GFP and so appears dark, whereas the cytosol, containing the vast majority of βarr2–GFP, appears white (fluorescent). Figure 2B shows the redistribution of βarr2–GFP 90 sec following the addition of agonist. The majority of GFP fluorescence now appears at the edge of the cell as a result of βarr2–GFP binding to plasma membrane-localized receptor. Depending on the experimental conditions, βarr2–GFP redistribution can easily occur within 10–15 sec. However, because images result from the averaging of between 4 and 16 scans that require approximately 1 sec each to obtain, it is unlikely that this technique can resolve cellular processes that occur in time intervals of less than 1 to 3 sec. Controlling variables such as the temperature or coexpression of GRKs can decrease or accelerate the rate of βarr–GFP translocation.

Analysis of β-Arrestin 2–Green Fluorescent Protein Translocation that Occurs in Single Cells

In this section we describe how numerical and kinetic data can be obtained from confocal images using the scans presented in Fig. 3. Many computer programs such as NIH Image are available over the Internet or commercially to analyze these types of images. Figure 4 demonstrates a simple type of analysis. Histograms of the average fluorescence at a point, along a line or in a defined area, can be made using the imaging software

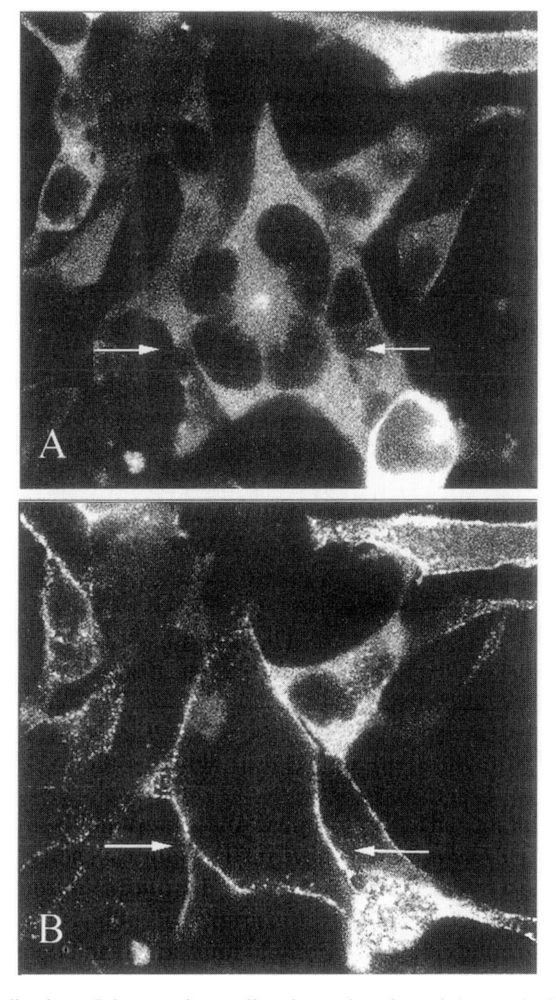

FIG. 3. Visualization of the agonist-mediated translocation of β-arrestin 2–GFP to membrane-bound receptors. (A) Image before agonist is administered; (B) 90 sec after agonist is added. In these confocal images the disappearance of GFP fluorescence from the cytosol and the increase in GFP fluorescence at the plasma membrane (arrows) is a measure of receptor activation, receptor desensitization, and the recompartmentalization of β-arrestin 2.

available with the Zeiss LSM-410 microscope. Histogram trace A in Fig. 4 corresponds to a line drawn through the lower portion of the cell in Fig. 3A and trace B corresponds to a similar line drawn through the image of the agonist-treated cell in Fig. 3B. It is apparent that the decrease in fluorescence in the cell interior (Fig. 3B) following agonist treatment is reflected by the concave appearance of the corresponding histogram (Fig.

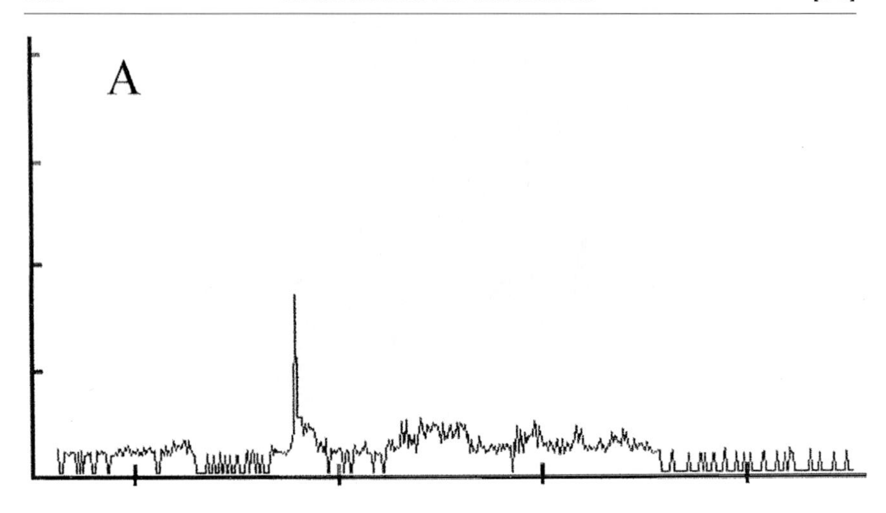

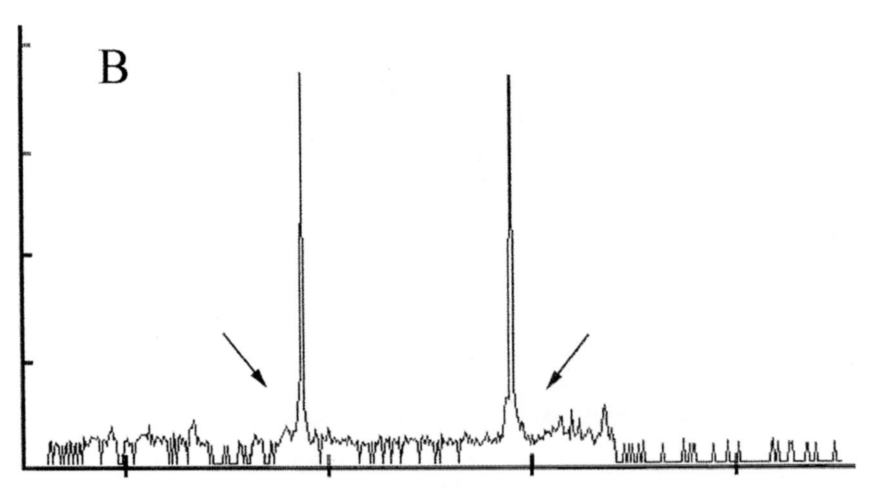

FIG. 4. Measurement of the distribution of β-arrestin 2–GFP fluorescence. The relative fluorescence intensity of β-arrestin 2–GFP along a line drawn through the cells before and after agonist exposure is proportional to the height of the curves at each point. Note that the concave shape of the curve in (B) corresponds to the movement of β-arrestin 2–GFP to the plasma membrane.

4B). Similarly, the two peaks in the lower histogram reflect the increase in fluorescence at the cell membrane (Fig. 3B).

As a result of the stoichiometric labeling of βarr2–GFP, the measured fluorescence signal should be proportional to the amount of activated receptor. Background noise is now mainly a result of cell autofluorescence rather than of nonspecific or nonuniform probe labeling as is often the case with conventional fluorescence. Therefore, we can use translocation of βarr2–GFP as a measure of any process that changes or affects the population of activated receptors at the cell membrane. For example, the pharmacology of an unknown agonist can be determined by integration of the fluorescence originating from the plasma membrane of a single cell. Known increases in the extracellular agonist concentration over time would produce measurable changes in the magnitude of membrane-translocated βarr2–GFP. A plot of agonist concentration versus plasma membrane fluorescence intensity would thus yield a dose–response curve. We have performed this experiment for the β2AR and the agonist isoproterenol, and the results are in agreement with direct radioligand binding data (our unpublished data).

Limitations

Study of the biology of β-arrestins and GRKs in live cells had been all but impossible prior to the generation of the corresponding GFP conjugates. Thus, there are only limited data concerning the effects of β-arrestins on GPCRs other than the β2AR and the m_2-muscarinic receptor. Although in its infancy, this new GFP technology will contribute immensely to the study of the biochemistry and cell biology of GPCRs. While the technology will circumvent and solve many practical problems, it will still suffer some of the same limitations that currently confine fluorescence measurements in general: the resolving power of the optics, cell autofluorescence, and fluorescence collected from out-of-the-plane of focus.

Cellular autofluorescence, which diminishes with longer wavelength excitation, perhaps remains the greatest impediment for obtaining larger signal-to-noise ratios. For most experiments we obtain signal-to-noise ratios of 5–10 to 1, and can easily discriminate probe distribution from other cell fluorescence. Under the best of circumstances signal-to-noise ratios of 50 to 1 have been measured. In either case, these signals are more than large enough to measure probe kinetics. Moreover, the dynamic range of the signal is well within the 8-bit range (256-gray scale) of most imaging systems. Further improvements will perhaps require GFP variants that are excited by green light and fluoresce in the red portion of the spectrum.

The relative expression levels and biological fidelity of the transfected proteins are other important factors limiting signal quality. For example, the magnitude of the βarr–GFP fluorescence signal at the plasma membrane increases with increasing amounts of plasma membrane-localized receptor. Therefore, for the purpose of measuring processes such as receptor activation, more receptor means larger signals until some other component of the signal transduction system becomes limiting, such as the amount of G protein available to produce high agonist-affinity receptor complexes. At the least, we have been able to measure translocation in cells containing 50,000–100,000 receptors (i.e., with 0.5–1 pmol/mg cell protein). The sensitivity is somewhat improved since the receptors are measured at the edge of the cell, reducing the autofluorescence. Sensitivity can also be improved by using smaller cells with the same receptor expression. However, the ultimate goal should be to make measurements at endogenous receptor expression levels, which are at least 5- to 10-fold lower.

Each GFP-conjugated protein utilized in our laboratory undergoes either a biochemical or functional characterization. However, an unappreciated biochemical or biological property of the GFP conjugate can still be affected. For example, the G protein-related signal transduction properties of the β2AR are essentially unaffected by conjugation of GFP at its C terminus. However, properties that require a free C-terminal motif might well be affected. Therefore, for proteins such as β-arrestin, construction and characterization of both N-terminal and C-terminal GFP conjugates should perhaps be undertaken in spite of theoretical considerations. Similar reasoning applies when only segments rather than full-length constructs of these proteins are expressed conjugated to GFP. While it is unlikely that new binding specificity will be created by conjugation of a protein to GFP, it is not unlikely that existing ones will be affected, and the degree to which that occurs must be determined in order for the construct to be useful for a particular assay.

Conclusion

GFP technology will significantly change the way GPCR signal transduction proteins are studied in cells. In this chapter we have presented the methodology to construct and characterize GPCRs, GRKs, and β-arrestins conjugated to GFP. In addition, we have demonstrated that their behaviors can be studied in living cells in real time. Even though this technology is in its infancy, at the least it will contribute to the study of GPCR-based diseases, the search for new classes of drugs, and the identification of ligands that bind to orphan receptors.

Acknowledgments

This work was supported by National Institutes of Health Grants HL 03422 (to L.S.B.) and NS 19576 (to M.G.C.), a Heart and Stroke Foundation of Ontario Grant NA-3349 to (S.S.G.F.), a Heart and Stroke Foundation of Canada Grant (to S.A.L.), and an unrestricted Neuroscience Award from Bristol-Myers Squibb and an unrestricted grant from Zeneca Pharmaceutical Co. (to M.G.C.). M.G.C. is an investigator of the Howard Hughes Medical Institute.

[15] Use of Green Fluorescent Proteins Linked to Cytoskeletal Proteins to Analyze Myofibrillogenesis in Living Cells

By GUISSOU A. DABIRI, JOSEPH C. AYOOB, KENAN K. TURNACIOGLU, JEAN M. SANGER, and JOSEPH W. SANGER

Introduction

The coupling of fluorescent probes to cytoskeletal molecules inside cells allows the labeled molecules to be seen in live cells. With video and digital cameras and image-processing software,[1-3] the dynamic distribution of the proteins and the structures they form can be followed in single living cells during activities such as myofibrillogenesis,[4] cell division,[5] and cell locomotion[6]; whereas previously, similar information had to be reconstructed from a series of separate cells fixed sequentially during a process with clearly identifiable time points such as chromosome position in cell division.

Cultured muscle cells have provided many insights into myofibril formation,[7-9] but they lack the characteristic morphological shape reflective of the stage of myofibril development, and they do not develop with sufficient synchrony in culture to correlate time with stage of myofibril formation. Time-lapse observations of single myocytes in a culture offer a way of following the stepwise assembly of myofibrils.[4] To visualize muscle as it forms in live cells, myofibrillar proteins can be coupled to a fluorescent tag and trace amounts microinjected into cells that are forming myofibrils. By

[1] R. D. Allen, *Annu. Rev. Biophys. Chem.* **14,** 265 (1986).
[2] S. Inoue and K. R. Spring, "Video Microscopy: The Fundamentals," pp. 1–741. Plenum Press, New York, 1997.
[3] D. L. Taylor and Y. Wang (eds.), *Methods Cell Biol.* **30,** 1 (1989).
[4] J. M. Sanger, B. Mittal, M. B. Pochapin, and J. W. Sanger, *J. Cell Biol.* **102,** 2053 (1986).
[5] J. M Sanger, B. Mittal, J. S. Dome, and J. W. Sanger, *Cell Motil. Cytoskel.* **14,** 201 (1989).
[6] D. L. Taylor, P. A. Amato, K. Luby-Phelps, and P. McNeil, *Trends Biochem. Sci.* **9,** 88 (1984).
[7] H. Holtzer, J. M. Marshall, and H. Finck, *J. Biophys. Cytol.* **3,** 705 (1957).
[8] J. W. Sanger, B. Mittal, and J. M. Sanger, *Cell Motil.* **4,** 405 (1984).
[9] D. Rhee, J. M. Sanger, and J. W. Sanger, *Cell Motil. Cytoskel.* **28,** 1 (1994).

0076-6879/99 $30.00

injecting two proteins, each labeled with a different dye, or by injecting a fluorescent protein and exposing the cell to a vital dye, two probes can be correlated with one another in a living cell.[4] These studies are technically difficult, however, and only those myofibrillar proteins that can be isolated and labeled with a fluorescent dye can be used for injection. Furthermore, fluorescently labeled proteins slowly turn over, leading to a loss of signal over time.

The development of plasmids designed to couple the cDNA for green fluorescent protein (GFP) to the cDNA encoding another protein[10] has made it possible for muscle proteins to be visualized in live cells.[11–14] With this approach the roles of selected proteins can be examined during the process of myofibril formation in living, transfected cells. Because GFP contains a stable fluorophore[14,15] and the GFP-linked proteins are continually synthesized in the transfected cells, the fluorescent signal can be followed over many hours and days provided overexpression of the protein does not affect the cells adversely. Transfected cells could also be microinjected with rhodamine-labeled proteins to follow the assembly of two different proteins into myofibrils. Alternatively, the development of variants of GFP that fluoresce at other wavelengths will allow several different sarcomeric proteins to be followed concurrently in the same living cell. The hardware currently available for imaging, including cameras, filters, and digital storage devices, and the software for image processing and analysis[2,3,16] make it feasible to localize and measure fluorescently tagged molecules and study their dynamics with time-lapse observations of live cells.

GFP-linked probes for proteins of the actin–myosin cytoskeleton of muscle can serve as markers for the assembly of thin and thick filaments into organized sarcomeric arrays.[9,12] In this chapter, we present methods for the culturing of embryonic chick cardiac and skeletal muscle cells, and for transfecting the primary cultures of muscle cells with a plasmid encoding the actin-binding protein, α-actinin, fused to green fluorescent protein. Also

[10] M. Chalfie, Y. Tu, G. Euskirchen, W. W. Ward, and D. C. Prasher, *Science* **263,** 802 (1994).

[11] K. K. Turnacioglu, B. Mittal, G. A. Dabiri, J. M. Sanger, and J. W. Sanger, *Mol. Biol. Cell* **8,** 705 (1997).

[12] G. A. Dabiri, K. K. Turnacioglu, J. M. Sanger and J. W. Sanger, *Proc. Natl. Acad. Sci. U.S.A.* **94,** 9493 (1997).

[13] D. Auerbach, B. Rothen-Ruthishauser, S. Bantle, M. Leu, E. Ehler, D. Helfman, and J.-C. Perriard, *Cell Struct. Funct.* **22,** 139 (1997).

[14] M. Ormo, A. B. Cubitt, K. Kallio, L. A. Gross, R. Y. Tsien, and S. J. Remington, *Science* **273,** 1392 (1996).

[15] F. Yang, L. G. Moss, and G. N. Phillips, *Nature Biotechnol.* **14,** 1246 (1996).

[16] Y. Wang and D. L. Taylor (eds.), *Methods Cell Biol.* (Special Issue) **29,** 1 (1989).

described are the conditions for time-lapse observations of the appearance of α-actinin in premyofibrils at the spreading edges of cardiomyocytes and subsequently in Z bands in mature myofibrils.

Methods

Linking Green Fluorescent Protein to Sarcomeric Protein

Choosing Which End of the Protein, N or C Terminus, to Link to Green Fluorescent Protein. The green fluorescent protein molecule contains about 240 amino acid residues folded into a specific tertiary structure, required for its autofluorescence.[14,15] To the surprise of many of its users, the large size of the GFP has usually not affected the function of the fusion protein.[17–20] Nevertheless, there is some evidence that proximity of the GFP to the functional domain of the protein can interfere with normal localization of the protein in the cell. When muscle cells were transfected with a plasmid in which GFP was coupled to the N terminus of α-actinin, the region of the molecule containing the actin-binding domain, only diffuse fluorescence was seen in the cells; whereas when linked to the C terminus of α-actinin, the GFP probe localized in the Z bands of both cardiac and skeletal muscle cells (Figs. 1 and 2),[12] presumably through binding to the titin and actin filaments embedded there.[21] A similar effect of GFP placement was seen with fragments of titin coupled to GFP[11,22]: the N terminus of titin is embedded in the Z bands, and when GFP was linked to the N terminus of the titin fragments the fluorescent constructs were unable to target to the Z band, but when GFP was linked to the C terminus of the titin fragments they did target to the Z band, as expected.

The possible inhibitory steric effect of GFP on the activity of proteins can also be minimized by inserting a linker of several or many amino acids between the GFP and the protein. A linker consisting of a minimum of 10 alanines was needed between yeast actin (Act1p) and GFP for clear localization in cortical actin patches without background fluorescence.[23] If the actin was linked directly to GFP or with a linker of only four alanines,

[17] T. Stearns, *Curr. Biol.* **5,** 262 (1995).
[18] G. Gerisch, R. Albrecht, C. Heizer, S. Hodgkinson, and M. Maniak, *Curr. Biol.* **5,** 1280 (1995).
[19] R. Rizzuto, M. Brini, P. Pizzo, M. Murgia, and T. Pozzan, *Curr. Biol.* **5,** 635 (1995).
[20] S. L. Moores, J. H. Sabry, and J. A. Spudich, *Proc. Natl. Acad. Sci. U.S.A.* **93,** 443 (1996).
[21] M. Gautel, B. Goulding, B. Bullard, K. Weber, and D. O. Furst, *J. Cell Sci.* **109,** 2747 (1996).
[22] K. K. Turnacioglu, B. Mittal, G. A. Dabiri, J. M. Sanger, and J. W. Sanger, *Cell Struct. Funct.* **22,** 73 (1997).
[23] T. Doyle and D. Botstein, *Proc. Natl. Acad. Sci. U.S.A.* **93,** 3886 (1996).

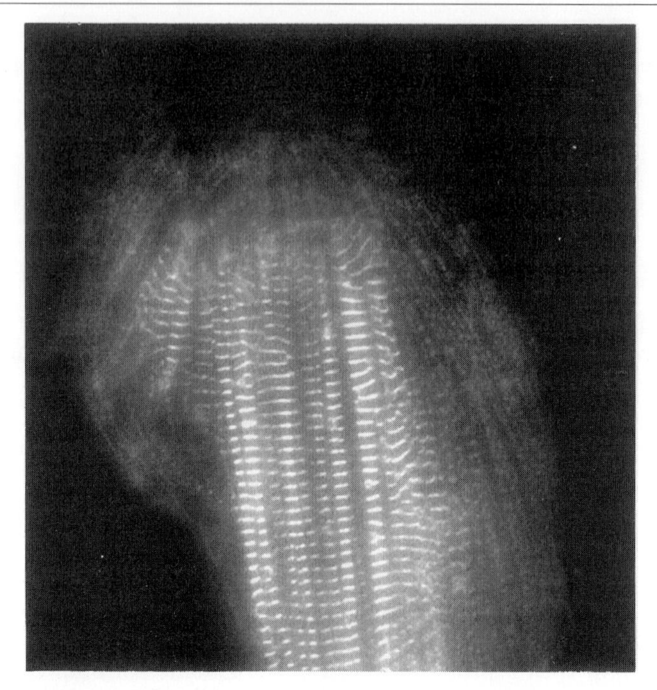

FIG. 1. Embryonic chick cardiomyocyte transfected with a GFP plasmid encoding α-actinin. Note the concentration of GFP–α-actinin in the Z bands in this living cell. The distances between two adjacent Z bands (2 μm) mark the limit of one sarcomere.

there was a high background fluorescence. *Dictyostelium* actin linked directly to GFP incorporates into the thin filaments of the cells, but for actin–myosin interaction to occur a linker of five amino acids was required between the GFP and the N terminus of the actin monomer.[24] We have found that GFP linked directly to two cDNA-encoded z-repeat domains of the Z-band fragment of titin (40 to 50 amino acids per z-repeat) targeted to the Z band but also exhibited diffuse cytoplasmic fluorescence. However, when a portion of the I-band region of titin, which does not target to the Z band, was inserted as a linker between the z-repeats and GFP, the z-repeats targeted to the Z band exclusively.[25] This use of linkers between GFP and small domains of a protein may provide a means for determining

[24] M. Westphal, A. Jungbluth, M. Herdecker, B. Muhlbauer, C. Heizer, J.-M. Schwartz, G. Marriot, and G. Gerisch, *Curr. Biol.* **12,** 176 (1997).

[25] K. K. Turnacioglu, B. Mittal, J. C. Ayoob, J. M. Sanger, and J. W. Sanger, *Mol. Biol. Cell* **8,** 375a (1997).

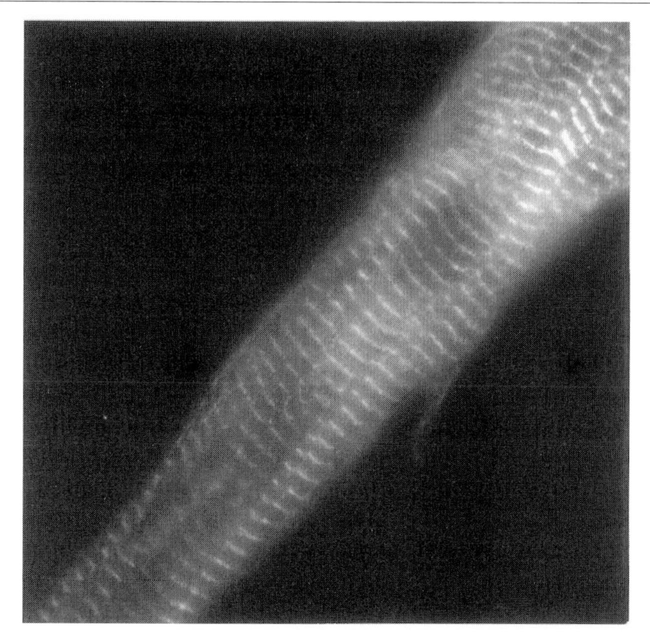

FIG. 2. Embryonic chick myotube transfected with a GFP plasmid encoding α-actinin. The distances between the adjacent Z bands in this living skeletal muscle cell is aobut 2 μm.

which part(s) of a protein are capable of targeting the protein to specific sites in living cells.

Working with Green Fluorescent Protein Vectors. The commercially available GFP vectors [e.g., Clontech (Palo Alto, CA), Packard (Downers Grove, IL), and Quantum (Montreal, Quebec, Canada)] have been engineered with start and stop codons for expression of the GFP cDNA. The stop codon in the cDNA encoding the protein of interest must be removed when that cDNA is fused to the 5′ end of the GFP sequence. Start codons may need to be introduced when working with cDNAs encoding protein fragments. Standard polymerase chain reaction (PCR) techniques can be used for the addition or elimination of start and stop codons. At the same time, convenient restriction sites may be introduced as well for cloning purposes.

Isolation of Embryonic Muscle Cells from Avian Heart and Breast Tissue

Avian embryonic muscle cells grown in culture have a number of advantages for studying the formation of myofibrils. The embryonic cardiomyocytes grow in tissue culture as flat cells that are ideal for examination of myofibrillogenesis by phase-contrast and epifluorescence microscopy.

However, these cultures are not a homogeneous population of cells, but a mixture of cardiomyocytes, fibroblasts, and epithelial cells. In contrast to the flat cardiomyocytes, myoblasts isolated from avian embryonic breast muscles grow and fuse with other myoblasts to form thick, round myotubes containing several layers of myofibrils. With these cells, confocal microscopy allows the myofibrils to be viewed in a series of single focal planes without out-of-focus fluorescence interfering. An advantage of skeletal muscle cultures is that they can be prepared with few contaminating fibroblasts. Addition of DNA inhibitors to cultures of myoblasts after they have ceased dividing and have begun to fuse will inhibit fibroblast division.

Fertilized eggs (obtained from a registered USDA dealer) are the starting material for cultures of cardiac and skeletal muscle cells. The eggs are best kept in a slowly rocking, humidified egg incubator at 37° until use. A less expensive alternative is to use a Styrofoam incubator with a heating element and a rotating egg holder (Ward's Biology, Rochester, NY). The cultured muscle cells isolated from the embryos should be kept in a tissue culture incubator at 37° in 5% CO_2. A pan containing distilled water is placed in the bottom of the incubator to slow the loss of fluid in the culture dishes. The water level of this pan of water should be checked once a week and replenished as necessary.

Tables I and II describe step-by-step procedures for isolating embryonic cardiomyocytes and skeletal myoblasts, respectively. These isolation methods have evolved from procedures described by a number of laboratories working on embryonic cardiomyocytes,[26–29] and skeletal muscle cells.[4,30–32] Table III describes the preparation of embryo extract, a necessary ingredient for the growth of skeletal muscle cells. Embryo extract was prepared according to the procedure of J. Lash (University of Pennsylvania, personal communication) based on a modification of the original description by Chen.[33]

Isolated avian embryonic cardiomyocytes spread and contract intermittently in culture over several days and occasionally undergo cell divisions.[26,27] Many of the dividing cardiomyocytes fail to undergo complete cytokinesis and form binucleated cardiomyocytes. Because the cytoplasm of the cardiomyocytes is phase dense, it is often difficult to see the myofibrils

[26] S. Chacko, *Dev. Biol.* **35,** 1 (1973).
[27] J. W. Sanger, *J. Exp. Zool.* **201,** 463 (1977).
[28] M. W. Cavanaugh, *J. Exp. Zool.* **128,** 573 (1955).
[29] R. L. DeHaan, *Dev. Biol. Suppl.* **2,** 208 (1968).
[30] R. Bischoff and H. Holtzer, *J. Cell Biol.* **36,** 111 (1968).
[31] D. Yaffee, *Proc. Natl. Acad. Sci. U.S.A.* **61,** 477 (1968).
[32] G. D. Fischbach, *Dev. Biol.* **28,** 407 (1972).
[33] J. M. Chen, *Exp. Cell Res.* **7,** 518 (1954).

TABLE I
PREPARATION OF CHICK EMBRYONIC CARDIOMYOCYTES

Necessary materials for dissections and cell isolations:

Heart medium
Two 10-ml tubes of 0.05% trypsin–0.53 m*M* EDTA
Sterile 100- and 35-mm petri dishes
Sterile 100-mm tissue culture dish
A dozen 7-day-old fertilized chick eggs
Sterile forceps and iridectomy knives or razor blades
Dissecting microscope
Waste beaker

To make 100 ml of heart medium:

1. Place 10 ml of fetal bovine serum in the upper chamber of a 200-ml 0.22-μm pore size filter system
2. Fill to total volume of 100 ml with MEM (without L-glutamine)
3. Apply vacuum to filter system

Preparation of cardiomyocytes:

1. Rinse the eggs with a flow of 70% ethanol and let them dry
2. Break open each egg; remove the embryos with sterile tweezers and place them in a sterile 100-mm petri dish
3. Place the petri dish on the stage of a dissecting microscope and observe the chest area of the embryo. The hearts are outside the thoracic cavity at this embryonic stage. Dissect out the beating red hearts and place them in a single pile in a sterile 35-mm petri dish
4. Carefully examine the excised hearts under the dissecting microscope and remove any material adhering to the hearts, e.g., ribs, chest muscle, skin. Cut the group of hearts with fine, sharp iridectomy knives (or razor blades). At the end of the operation, each heart should be in about eight pieces
5. Collect the tissue with 10 ml of the trypsin solution and place the tissue into a 14-ml plastic tube
6. Place the tightly capped tube into a 37° incubator (5% CO_2) for 15 min
7. Invert the tube to disturb the settled cells
8. Let the pieces settle again for about 2 min. Slowly remove the trypsin solution above the pellet with a sterile pipette. This removes most of the nucleated red blood cells
9. Add fresh trypsin (10 ml) to the minced heart peices
10. Place the capped tube for another 15 min in the incubator
11. Spin the tube for 5 min at a setting of 5 in a tabletop centrifuge (we use an IEC Clinical certrifuge)
12. Gently resuspend the pellet in 10 ml of normal heart medium. Carefully triturate the pellets, i.e., draw the pieces up and down until they have separated into cells. Let the larger pieces settle to the bottom of the tube for a few minutes. Carefully remove the cell suspension above the large pieces at the bottom of the tube
13. Preplate the cell suspension on a 100-mm tissue culture dish (8 to 10 ml per dish) for at least 1 hr in the 37° incubator. The cardiac fibroblasts will attach readily to the dish
14. Collect the medium from the preplate dish. Wash this dish lightly several times with the previously collected medium to release any loosely attached cardiac muscle cells

(*continued*)

TABLE I (*continued*)

15. Gently wash the preplate dish with a fresh 10 ml of muscle culture medium. Collect and add this medium to the previous wash collected in step 14
16. Count the collected cells in a hemocytometer
17. Place a sterile glass coverslip in the middle of twelve 35-mm plastic dishes. These glass coverslips are 22 × 22 mm and 1.5 mm thick. The glass coverslips can be made sterile by overnight UV illumination or by flaming after a quick rinse in 70% ethanol. Sterile plastic dishes are also available that have glass bottoms (Mar Tek, Ashland, MA); these are particularly useful for observations of live cells on an inverted microscope
18. Plate the cells onto the 35-mm dishes at a density of 0.5×10^6 cells per dish. Each heart yields approximately 0.5 million cells, sufficient for one coverslip. The cardiac muscle cells attach to the glass surfaces as round balls and gradually spread on the surfaces after 24 hr. The cardiomyocytes are often beating in culture. Fibroblasts will also be present in the cultures. The lack of glutamine in the cultures slows the growth of the fibroblasts. (Inhibitors of DNA synthesis are not used because the embryonic cardiomyocytes can divide and would be affected by these drugs)
19. Feed the cardiac muscle cultures every other day. The muscle cells are healthy for about 1 week under these conditions
20. These primary cultures cannot be subcultured for the propagation of muscle cells. New primary cells will have to be isolated from chick embryos. We prepare cultures of cardiomyocytes once a week. However, these cultures can be subcultured for fibroblasts for several more weeks before their rate of division drops drastically

by phase-contrast microscopy. The mononuclear myoblasts isolated from avian embryonic skeletal muscles will fuse on the second and subsequent days of culture to form long multinucleated myotubes that begin spontaneous contractions approximately 4–5 days after plating. The nuclei in these myotubes do not undergo further mitotic divisions, and there may be as many as several hundred nuclei per myotube. The density of myofibrils is much higher in myotubes than in cardiomyocytes. The alignment of the myofibrils with Z bands in register across the myotube makes it possible to detect sarcomeres by phase-contrast microscopy. This is particularly so in cultures of quail myotubes. Before transfecting cultured muscle cells, one should be sure that the culture conditions produce beating cells with myofibrils resembling those shown in the literature.

Methods of DNA Transfection of Muscle Cells in Culture

The transfection efficiency for primary cardiac muscle cells is low (5 to 10% compared with that for cell lines. A much higher rate of transfection is obtained in cultures of skeletal muscle because transfection occurs before most myoblasts fuse with one another to form myotubes. Even if only a small percentage of fusing cells are transfected , the myotube formed will be transfected.

TABLE II
AVIAN MYOTUBE PREPARATION

Necessary materials:

- Collagen-coated 35-mm dishes (glass bottom or coverslips)
- Avian myotube medium
- Sterilized Sweeney filters
- Sterilized forceps
- Quail or chick eggs, 10 day
- Petri and tissue culture dishes, 100 mm
- Hanks' solution without $CaCl_2$, $MgCl_2$, or $MgSO_4$ (alternate solution: two 10-ml tubes of 0.05% trypsin–0.53 m*M* EDTA)
- Dissecting microscope
- Waste bucket

To make 100 ml of avian myotube medium:

1. Combine 10 ml of chick embryo extract (see Table III for preparation), 10 ml of horse serum, and 1.5 ml of L-glutamine in a 100-ml tube
2. Fill up to 100 ml with MEM (without L-glutamine)

Preparation of avian myotubes:

1. Dishes containing coverslips or glass-bottom dishes should be coated with collagen several hours before the start of the skeletal muscle preparation. The collagen-coated dishes should be placed in a tissue culture hood with the UV light on. The lids of the dishes are removed from the dishes but left in the hood. This allows the collagen-coated dishes to be dried and sterilized at the same time. This is typically done the night before the preparation. However, it can also be done at the start of the work day. The dishes will need several hours to be dried and sterilized
2. Rinse the eggs in the egg carton with 70% ethanol and allow them to dry for about 5 to 10 min
3. Break open each egg into a sterile petri dish; remove the embryo and place it into a separate 100-mm sterile petri dish. Repeat this procedure until all of the embryos are in one dish. If at any time during the egg-cracking procedure a dead or grossly underdeveloped embryo is detected, the embryo is thrown out and a new petri dish is used for the next egg cracking
4. Remove one embryo from the 100-mm petri dish and place it in a sterile 60-mm petri dish. Place the dish on the stage of a dissecting microscope. Decapitate the embryo, and place the embryo on its back. Spread its legs to stabilize the body and gently peel off skin tissue from the chest area
5. Scoop out the breast tissue by pinching the muscle with the curved ends of the forceps. Place the excised tissue into another sterile 35-mm petri dish. Repeat this procedure with all of the embryos. Place all of the dissected muscle tissue in the same 35-mm petri dish. Examine the excised muscle tissue under the dissecting microscope and remove any bones or skin in the muscle sample

(*continued*)

TABLE II (*continued*)

6. Add 10 ml (or 1 ml/embryo) of Hanks' balanced salt solution without $CaCl_2$, $MgCl_2$, or $MgSO_4$ to this dish. (An alternative dissociating solution is the use of trypsin–EDTA. Ten-day-old and younger avian muscle readily disassociates in Ca^{2+}- and Mg^{2+}-free solutions. One must use trypsin solutions to dissociate myoblasts from muscles of older embryos.) Transfer the suspension of tissue into a 14-ml tube
7. Place the tightly capped test tube in a tissue culture incubator (37°, 5% CO_2) for 15 min
8. After 15 min change the Hanks' solution above the muscle that has settled to the bottom of the tube and incubate the sample for another 15 min in the incubator
9. Spin the tube for 5 min at a setting of 6 in a tabletop centrifuge
10. Gently remove the supernatant without disturbing the pellet. Gently resuspend the pellet in 10 ml of avian myotube medium. Let the resuspended pellet stand for 1–2 min to allow any large pieces of tissue to settle
11. Carefully remove the supernatant and filter it through a Sweeney filter, using a 10-ml syringe
12. Preplate the filtrate of cells for 1 hr in a 100-mm tissue culture-grade petri dish to decrease the number of fibroblasts in the preparation
13. Collect the medium from the preplate dish. Wash this dish lightly several times with the previously collected medium to release any loosely attached myoblasts
14. Gently wash the preplate dish with a fresh 10 ml of avian muscle culture medium. Collect and add this medium to the previous wash collected in step 13
15. Count the collected cells in a hemocytometer
16. Plate the cells onto the collagen-coated dishes at a density of 0.5×10^6 cells per dish
17. Change the medium after 2 days of culture in the incubator. Many of the myoblasts will have exited the mitotic cycle by this time. These cells begin to fuse with one another to form elongated multinucleated myotubes. To inhibit the growth of the fibroblasts in these cultures, cytosine arabinoside (ara-C) is added at a concentration of 10^{-5} *M* on day 3 of culture
18. Feed the skeletal muscle cultures every other day. The muscle cells are healthy for about 1 week under these conditions. Myofibrils form in these myotubes on the third day of culture. The contractions of the myotubes begin to pull the cells from the surface of the dish by the end of the week
19. These cultures cannot be subcultured to propagate fresh muscle cells. Primary cultures of avian skeletal muscle cells will need to be prepared again

Lipid-Based Methods. Embryonic chick cardiomyocytes (see Table I) can be optimally transfected with DNA using lipid-based reagents such as LipofectAMINE (GIBCO-BRL, Gaithersburg, MD). Cardiomyocytes isolated from 7-day-old chick embryos should be transfected at 50% confluency (about 0.5×10^6 cells/35-mm dish) after 1 day in culture.[12] For each 35-mm dish of cells to be transfected, plasmid DNA (1–2 μg) is added to 100 μl of serum-free culture medium in one tube (tube A), and LipofectAMINE (10 μl) is added to 100 μl of serum-free medium in a second tube (tube B). The contents of tube A are added to tube B, and the combined mixture is incubated at room temperature for 15 to 30 min following the manufacturer protocol. Serum-free medium (800 μl) is added to the DNA–

TABLE III
CHICK EMBRYO EXTRACT PROTOCOL

Necessary materials:

One hundred twenty 11-day old eggs for 200 ml of chick embryo extract
Two sterile 250-ml stoppered graduated cylinders
Two sterile forceps
Sterile petri dishes
Simms medium
Test tubes, 15 ml
Ethanol (70%)
Waste bucket for eggs

To make Simms medium (1 liter):

1. In a sterile, 1-liter bottle, combin the following:
   1.00 g glucose, 8.00 g NaCl, 0.20 g KCl, 0.15 g $CaCl_2$, 0.20 g $MgCl_2 \cdot 6H_2O$, 0.02 g Na_2HPO_4, 1.00 g $NaHCO_3$
2. Bring the total volume up to 1 liter and sterilize in an autoclave for 15 to 30 min. Store at 4°

Preparation of chick embryo extract:

1. Wash a table and tray with 70% alcohol
2. Spray 70% alcohol on the eggs and let dry. (We find it easy to work with a dozen eggs at a time.)
3. Using sterile technique (an open flame), pour 10 ml of Simms medium into each of three petri dishes under a tissue culture hood. Take one dish to a table in the tissue culture room
4. Break open an egg and empty it into an empty sterile petri dish. Pick the embryo out of the egg by using sterile forceps and place it into the petri dish containing Simms medium
5. Pour out the yolk into the waste bucket without splashing. Break the next egg into the same dish. Repeat the procedure in step 4 until all 12 embryos are collected
6. If an embryo is dead or deformed during the cracking procedure, throw out the entire egg and discard the petri dish. Use a new petri dish for the next egg breaking
7. Take the dish containing the 12 embryos to the hood, decapitate each embryo using tweezers, rinse by placing the embryos into the second dish of Simms, and repeat with the third dish
8. Have two sterile 250-ml stoppered graduated cylinders ready under the hood
9. Take a 20-ml syringe out of the sterile bag by the tip and put the plunger back into the bag
10. Flame the syringe tip lightly
11. Using sterile forceps, put all 12 embryos into the syringe
12. Using the plunger, squeeze the embryos into one of the graduated cylinders
13. With each dozen eggs, use a new syringe. Repeat steps 2 to 13 for each of the remaining nine dozen eggs.
14. Add an equal volume of Simms medium to the embryos in the graduated cylinder and swirl
15. Let the mixture sit for at least 1.5 hr at room temperature
16. Occasionally, swirl the mixture to obtain an even distribution
17. Using sterile technique, aliquot about 10-ml portions into previously labeled 15-ml test tubes
18. Quick-freeze and store in the freezer (−20°). The embryo extract is good for 1–2 years

lipid mixture, and the combined mixture is added to a 35-mm dish of cells that has been rinsed recently with serum-free medium. The cells are incubated with this DNA–lipid mixture for 5–6 hr, following which 1 ml of normal muscle medium with serum is added. The next day the transfection medium is exchanged for fresh heart medium. GFP fluorescence should be detectable between 18 and 48 hr posttransfection. The procedures presented here are optimized for chick embryonic cardiomyocytes transfected with plasmids encoding α-actinin (Fig. 1) or titin fragments. The conditions may have to be varied for different species of cardiac muscle cells, e.g., cats, mice, or rats.

The chick cardiomyocytes transfected with a plasmid encoding α-actinin–GFP exhibited fluorescently labeled Z bands within 24 hr of transfection (Fig. 1). Contraction of these transfected cardiac muscle cells occurred as often as in untransfected cells. Moreover, transfections of the embryonic cardiomyocytes did not interfere with their ability to undergo cell division.[12] When control cardiomyocytes were transfected with the GFP plasmid lacking any insert, the GFP fluorescence was not localized to any part of the myofibril, i.e., there was a diffuse distribution of the fluorescence in the cells. This is an important control that the investigator should use to determine that the GFP is not binding nonspecifically to structures in the cell.

Calcium Phosphate Method. Whereas the lipid method works best for transfections of cardiomyocytes, the calcium phosphate method[34] gave the best transfection results for myotubes.[11] Chicken or quail myotubes, from 10- to 11-day-old eggs (see Table II), are plated at 0.5×10^6 cells/35-mm dish 1 day before transfection. When the cells reach 50–80% confluency on the next day, the medium is exchanged for 2 ml of fresh medium, about 2 hr prior to transfection. The calcium phosphate–DNA precipitate is prepared as described below.

For each 35-mm dish of cells, 2 μg of DNA in 125 μl of sterile distilled water (tube A) is mixed with the contents of tube B, containing 125 μl of 2× HBS (HEPES-buffered saline: 250 m*M* NaCl, 10 m*M* KCl, 1.5 m*M* $Na_2HPO_4 \cdot 2H_2O$, 12 m*M* dextrose, 50 m*M* HEPES). $CaCl_2$ (2 *M*, 15.5 μl) is slowly added to this mixture in a dropwise manner, while vortexing vigorously. The combined three-part mixture in tube B is incubated at room temperature for 30 min. The medium of each dish to be transfected is replaced with 2 ml of fresh skeletal muscle medium. Two hundred microliters of the calcium phosphate–DNA mixture is then added directly to the 2 ml of freshly added culture medium in each dish of cells. Transfection is

[34] J. Sambrook, E. F. Fritsch, and T. Maniatis, "Molecular Cloning: A Laboratory Manual." Cold Spring Harbor Laboratory Press, Cold Spring Harbor, New York, 1989.

permitted to occur overnight at 37°. On the next day, the transfection medium is replaced with fresh culture medium and the cells are allowed to incubate for another 24 hr. This calcium phosphate method has also been used successfully to transfect myotubes with other plasmids.[35]

Using the procedures described above, we have observed a 50% transfection efficiency of myotubes with α-actinin–GFP (Fig. 2). α-Actinin–GFP localizes to the Z bands of these skeletal muscle myofibrils and does not affect the random contractions normally exhibited by cultured myotubes. To confirm that the GFP–α-actinin was in the Z bands, the transfected myotubes were fixed, permeabilized, and stained with anti-α-actinin antibodies. The GFP–α-actinin and the anti-α-actinin antibodies were both localized to the Z bands.

Microinjection Method. Direct microinjection has been used to introduce DNA and/or proteins into specific cells of interest.[11,36–41] This approach allows two different agents to be introduced simultaneously into a single cell,[42] such as an antibody or a rhodamine-labeled protein along with a GFP fusion probe. A 10-fold lower amount of DNA is needed for microinjection purposes (needle concentration is 0.1 μg/ml as opposed to the 1 to 2 μg used in the preceding transfection methods[40]). The stock solution of plasmid DNA is diluted to 0.1 μg/ml with distilled water.

Microscopy of Living Green Fluorescent Protein-Transfected Cells

Culture Conditions. The great advantage of transfecting cells with green fluorescent protein fusions is the ability to analyze the distribution of the molecules in living cells. We have found that it is most convenient to do this on an inverted microscope. Cardiac or skeletal myocytes can be grown in 35-mm culture dishes with glass bottoms (MatTek, Ashland, MA), which consist of a glass coverslip glued over a 1-cm circular opening in the bottom dish. We have found that embryonic chick cardiomyocytes and fibroblasts, as well as a number of tissue culture cell lines (PtK2, REF-52, Caco-2, and

[35] T. Schultheiss, Z. Lin, H. Ishikawa, I. Zamir, C. J. Stoeckert, and H. Holtzer, *J. Cell Biol.* **114,** 953 (1991).

[36] M. R. Capecchi, *Cell* **22,** 479 (1980).

[37] B. Mittal, B. A. Danowski, J. M. Sanger, and J. W. Sanger, *Cell Motil. Cytoskel.* **23,** 188 (1992).

[38] M. B. Pochapin, J. M. Sanger, and J. W. Sanger, *Cell Tissue Res.* **234,** 309 (1983).

[39] J. M. Sanger, M. B. Pochapin, and J. W. Sanger, *Cell Tissue Res.* **240,** 287 (1985).

[40] J. M. Sanger, R. Golla, D. Safer, J. K. Choi, K. Yu, J. W. Sanger, and V. T. Nachmias, *Cell Motil. Cytoskel.* **31,** 307 (1995).

[41] J. S. Dome, B. Mittal, M. B. Pochapin, J. M. Sanger, and J. W. Sanger, *Cell Differ.* **23,** 37 (1988).

[42] N. B. Cole, C. L. Smith, N. Sciaky, M. Terasaki, M. Edidin, and J. Lippincott-Schwartz, *Science* **273,** 797 (1996).

LLC-PK), grow readily on these dishes. Skeletal muscle cells need to grow on collagen-coated surfaces. Therefore, we coat the glass in the MatTek dishes with collagen (type I rat tail collagen, 5 mg/ml; Collaborative Biomedical Products, Bedford, MA) at least 1 hr before plating. The transfected cells can be observed with low-power or oil immersion lenses without the need for mounting coverslips in special chambers.

The microscopic stage is maintained at 37° with a heat curtain (ASI 400 air stream incubator; NevTek, Burnsville, VA). To maintain an atmosphere of 5% CO_2 around the culture dish on the stage, we have constructed a homemade chamber fashioned from the bottom of a 100-mm plastic petri dish that is placed over the culture dish. With a heated 22-gauge needle a hole is made in the side of the 100-mm dish, and carbon dioxide is perfused into the petri dish through tubing connecting a tank of 5% CO_2, balanced with air, to a disposable 22-gauge needle inserted into the hole in the petri dish. Plastic wrap (e.g., SaranWrap) is placed over the stage with a hole for the objective lens, and the chamber is inverted over the glass-bottom culture dish on the microscope stage. To minimize the escape of the CO_2 and water vapor, a weight, such as an annular microscope stage insert, is placed over the petri dish chamber to press it onto the plastic wrap. The CO_2 should be bubbled through distilled water to humidify it before delivery to the homemade chamber. We often add drops of distilled water around the culture dish on the plastic wrap surface under the petri dish chamber to ensure high humidity in the chamber, to reduce loss of water from the tissue culture dish during observations over several days. The chamber can be readily lifted to exchange the culture medium in the dish of cells on the microscope. We recommend changing the medium every 12 hr for long-term observations.

Imaging of Cells. The transfected cells can be observed with phase-contrast and epifluorescence optics with pairs of images acquired sequentially with a cooled CCD camera (Photometrics AT 200; Photometrics, Tucson, AZ) controlled by Metamorph software (Universal Imaging, West Chester, PA). It is important in studying living cells to keep the exposure to the light of the mercury lamp as brief as possible to avoid photodamage to the cells. In our experience exposures should not exceed 500 msec for time-lapse sequences. Thick muscle cells require images to be acquired in several focal planes to visualize all of the myofibrils in an area of the cell. In this case a *z*-axis motor (Ludl Electronic Products, Hawthorne, NY) is useful. When fluorescent images of several planes have been collected for one time point, they can be combined digitally to reveal all of the fibrils present. A thin part of a cell containing one layer of myofibrils is easier to analyze, and for cardiomyocytes the thin edge of the muscle cell that is spreading is where myofibrils readily assemble. When the cell remains in position on the stage, it is easier to compare changes by superimposing the images gathered at all of the time points. Image-processing software allows time-lapse recording to be automated and images to be saved to disk. Changes in the intensity of the GFP signal can be measured over time and correlated with changes in structure. Inverting the contrast of the fluorescent image from white to black sometimes improves detection, as can digital sharpening filter.[2]

Fluorescent Staining of Transfected Cells

GFP-transfected muscle and nonmuscle cells can be fixed, permeabilized, and stained with specific antibodies and secondary fluorescent antibodies to determine the distribution of the GFP-tagged sarcomeric protein relative to other sarcomeric proteins in the cell. We have stained transfected muscle cells with antibodies directed against α-actinin, muscle myosin, and nonmuscle myosin IIB.[15,22] GFP-transfected cells can also be fixed and stained with the F-actin-staining agent, rhodamine-labeled phalloidin. The large size of the GFP probe does not interfere with the access of the various anti-sarcomeric antibodies or phalloidin, and although the GFP signal is not as bright paraformaldehyde-fixed cells as in live cells, it usually can be detected.

Problems Encountered in Cells Transfected with Green Fluorescent Protein-Sarcomeric Proteins

Under- and Overexpression. If the fluorescent signal is weak owing to underexpression of the GFP–cytoskeletal protein in transfected cultures, the concentrations of added GFP plasmids should be increased. Normally, 1–2 μg of DNA per 35-mm dish is used to successfully transfect cardiomyocytes and nonmuscle cells.[11,12] A 10-fold lower concentration of DNA is used in microinjection studies.[11] We usually transfect one or two dishes of muscle cells with a plasmid encoding just GFP along with the experimental fusion GFP constructs to check our culturing and transfection results.

On the other hand, overexpression of the GFP–sarcomeric protein may lead to so much unincorporated protein that it is difficult to see the myofibrils. If this happens, the amount of plasmids added to the cultures or the time for transfection can be reduced. Conditions of overexpression can also be used to create a dominant effect to determine the role of certain protein domains in myofibrillar structures. Muscle cells do not retain their myofibrils in the presence of large amounts of a 1.1-kb fragment of the Z-band region of titin linked to GFP.[11] By measuring the GFP fluorescence in different cells and in single cells monitored by time-lapse imaging, we found that cells synthesizing low amounts of the titin fragment were unaffected. We used this dominant negative effect of truncated titin to examine possible roles for titin in myofibril assembly.[39]

Photobleaching. Although the GFP signal is resistant to photobleaching, prolonged exposure to excitation does reduce the fluorescence of the transfected cells and introduce photodamage as well. Therefore, exposure of transfected cells to the excitation light should be minimized to prevent diminution of the fluorescence. We use epifluorescence microscopy to rapidly locate a transfected cell. However, the cell is positioned in the field of view and focused with phase-contrast microscopy before final imaging of the GFP signal. Neutral density filters should, if possible, be inserted in the pathway of the exciting light to diminish its intensity. Our camera exposures are kept as short as possible (less than 1 sec) to prevent photobleaching of the signal. In stark contrast to cells injected with dye-labeled proteins, transfected cells continue to produce fluorescent GFP-fused proteins. We have been able to follow the same transfected muscle cell for more than 2 weeks. On the other hand, deliberate photobleaching of a defined region of a GFP-labeled structure can be analyzed by quantitation of the fluorescence recovery after photobleaching (FRAP) to measure the mobility of the GFP-tagged protein within cellular structures.[42]

Summary

Once the appropriate site has been selected for the attachment of GFP to the sarcomeric protein, it is quite remarkable that the large size of the GFP molecule does not appear to interfere with the localization of the fluorescent sarcomeric proteins into the sarcomeric regions of the myofibrils. A similar approach using truncated parts of sarcomeric proteins linked to GFP should allow studies of the targeting properties of other sarcomeric domains[12,13,25] for localization and assembly studies.

Section III

Special Uses

[16] Fluorescent Proteins in Single- and Multicolor Flow Cytometry

By LONNIE LYBARGER and ROBERT CHERVENAK

Introduction

Advances in green fluorescent protein (GFP) expression technology have fueled the drive to develop efficient and sensitive methods for GFP detection. Flow cytometry, at least in principle, should be applicable to the detection of GFP-derived fluorescence in mixed populations of cells to obtain quantitative and qualitative data regarding those cells. This could include simply monitoring the activity of a single genetic regulatory element that is engineered to control GFP expression, or combining this information with cell surface phenotypic data and/or additional fluorescent protein reporter construct data in multiparameter analyses. Experimental data from several laboratories have now demonstrated that all of these possibilities can be realized with the proper reagents and instrumentation. In large part, this success can be ascribed to the generation of GFP variants that possess spectral properties that are superior to those of wild-type GFP, in terms of flow cytometric detection.

Choice of Fluorescent Proteins

Flow cytometry can be used to detect wild-type GFP expression in mammalian cells subsequent to transfection, using the commonly employed 488-nm wavelength from an argon laser for excitation.[1,2] However, the signal obtained is modest, even when the protein is expressed from relatively powerful promoter elements. Several mutants of GFP, differing from wild-type GFP by one or a few amino acids, are better suited for flow cytometry owing to their increased efficiency of fluorescence and their altered excitation and emission spectra. For single fluorescent protein (FP) detection, one of the green-emitting variants that excites efficiently in the 488-nm (blue light) range is probably a better choice as compared with the ultraviolet (UV)-excitable mutants. Such variants are generally brighter (although

[1] J. Ropp, C. Donahue, D. Wolfgang-Kimball, J. Hooley, J. Chin, R. Hoffman, R. Cuthbertson, and K. Bauer, *Cytometry* **21,** 309 (1995).

[2] L. Lybarger, D. Dempsey, K. J. Franek, and R. Chervenak, *Cytometry* **25,** 211 (1996).

0076-6879/99 $30.00

effective UV-excitable FPs do exist[3,4]) and can be used with the simplest analyzers. These include GFP-S65T,[5] EGFP (enhanced GFP[6,7]), GFPsg25,[4] and EYFP [enhanced yellow fluorescent protein[8,9] (clone 10C variant[8])]. In particular, EYFP (527-nm emission maximum) is attractive because this variant emits bright fluorescence with 488-nm excitation and optimal detection of this fluorescence does not require modification of the cytometer optics,[10] which may be required for optimal detection of the other variants.

Multicolor FP detection has been most frequently accomplished by combining one of the green-emitting variants with a UV-excitable mutant.[4,11,12] Although this approach can be useful, many of the UV mutants are relatively dim in comparison with the blue-excitable forms or are simply not excited well by the UV light emitted from an argon laser. If the cytometer is equipped with both a krypton laser and an argon laser, FPs are available that permit sensitive detection of two distinct regulatory elements.[3] In addition, our laboratory utilizes two blue-excitable variants, EGFP and EYFP, for two-color analyses.[10] Certainly, the choice of FP variants will depend on the experimental system and the available flow cytometry hardware. This chapter focuses on single- and two-color FP analyses using EGFP and EYFP in conjunction with 488-nm stimulation, because fluorescence detection of these two variants can be achieved on virtually any cytometer.

Transfection of Mammalian Cells

Published reports have described successful expression of FPs following transfection of cells via all of the standard transfection methods. Thus, essentially any method should provide cells suitable for flow cytometric

[3] M. T. Anderson, I. M. Tjioe, M. C. Lorincz, D. R. Parks, L. A. Herzenberg, G. P. Nolan, and L. A. Herzenberg, *Proc. Natl. Acad. Sci. U.S.A.* **93,** 8508 (1996).

[4] R. H. Stauber, K. Horie, P. Carney, E. A. Hudson, N. I. Tarasova, G. A. Gaitanaris, and G. N. Pavlakis, *BioTechniques* **24,** 462 (1998).

[5] R. Heim, A. B. Cubitt, and R. Y. Tsien, *Nature* (*London*) **373,** 663 (1995).

[6] B. P. Cormack, R. H. Valdivia, and S. Falkow, *Gene* **173,** 33 (1996).

[7] T.-T. Yang, L. Cheng, and S. Kain, *Nucleic Acids Res.* **24,** 4592 (1996).

[8] M. Ormö, A. Cubitt, K. Kallio, L. Gross, R. Tsien, and S. Remington, *Science* **273,** 1392 (1996).

[9] *CLONTECHniques* **XII,** 10 (1997).

[10] L. Lybarger, D. Dempsey, G. H. Patterson, D. W. Piston, S. R. Kain, and R. Chervenak, *Cytometry* **31,** 1 (1998).

[11] J. Ropp, C. Donahue, D. Wolfgang-Kimball, J. Hooley, J. Chin, R. Cuthbertson, and K. Bauer, *Cytometry* **24,** 284 (1996).

[12] T.-T. Yang, P. Sinai, G. Green, P. A. Kitts, Y. T. Chen, L. Lybarger, R. Chervenak, G. H. Patterson, D. W. Piston, and S. R. Kain, *J. Biol. Chem.* **273,** 8212 (1998).

analysis. With each method, especially transfection with some liposome reagents, it is imperative to carry negative (i.e., mock-transfected) control cells through the experimental procedure, as transfection may alter the light-scattering and autofluorescent properties of the cells and, therefore, affect the subsequent analyses. A procedure is detailed below for electroporation of NIH 3T3 murine fibroblasts using a BTX (San Diego, CA) model T820 Electrosquareporator.

1. Harvest and count cells. It is important that the cells be in the exponential growth phase at the time they are transfected. Resuspend the cell pellet in complete growth medium [Dulbecco's modified Eagle's medium supplemented with glutamine (0.2 m*M*), gentamicin (50 mg/liter), 2-mercaptoethanol (0.05 m*M*), and 10% fetal calf serum] at a concentration of 10^7 cells/ml. Keep the cells on ice.

2. Prepare all reagents. For each transfection, label a 35-mm tissue culture plate and fill each plate with 6 ml of growth medium. If several transfections are to be performed, a six-well plate may be more convenient. Pipette 200 μl (2 million cells) of the cell suspension into a sterile, 4-mm electroporation cuvette for each reaction.

3. Add DNA to each cuvette just prior to electroporation. DNA should be of the highest purity, obtained by CsCl-gradient centrifugation or ion-exchange chromatography. Five to 50 μg of DNA (or of each DNA construct in the case of cotransfection) may be used per reaction, depending on the strength of the promoter in the cells to be transfected. The optimal quantity of DNA will have to be determined for each cell type. DNA should be at a concentration of ≥1 mg/ml. Pipette DNA into the cuvette and mix thoroughly. Remove any bubbles that form and place the cuvette in the electrode chamber.

4. Using the low-voltage (LV) mode, set the appropriate pulse amplitude (voltage), pulse length in milliseconds, and pulse number. For NIH 3T3 cells, we use 240 V, 20 msec, and one pulse, respectively. After the pulse is delivered, remove the cuvette from the chamber.

5. With a Pasteur pipette and bulb, remove the cells from the cuvette and transfer to a culture plate that contains growth medium. Rinse the cuvette (two or three times) with medium to remove all of the cells. There will be cell clumps.

6. Place the cells in an incubator and culture overnight. Analyze the cells for FP expression after 20–24 hr. With most cell lines we have analyzed, maximal expression is reached by 24 hr, and a 48-hr incubation is unnecessary. Harvest the cells and wash in a 3× volume of phosphate-buffered saline (PBS). Use of trypsin to detach adherent NIH 3T3 cells will not affect FP detection, but trypsin can damage cell surface proteins. Therefore, it may

be necessary to detach the cells by more gentle means [incubation in an EDTA solution, (Versene; Life Technologies, Gaithersburg, MD) if the expression of surface molecules is to be monitored in combination with GFP.

7. Resuspend the cell pellets in fluorescence-activated cell sorting (FACS) buffer (PBS–2% newborn calf serum–0.1% sodium azide). Sodium azide is a metabolic inhibitor that is added to prevent the alteration of cell surface phenotype during analysis. If cells are to be placed in culture following analysis, omit sodium azide from the buffer. In any case, keep the cells on ice after they are harvested. If possible, cells should be at a concentration of 5×10^5–2×10^6/ml, with a minimum sample volume of 300 μl.

Stable Fluorescent Protein Expression

For some experimental situations, stable FP expression is required. For example, we have generated stable FP-expressing cells that are useful for exploring critical parameters in FP detection by flow cytometry. This provides an abundant source of FP-positive cells without the need to perform transfections prior to each analysis. Using retroviral vectors, stable FP expression can be readily achieved through drug selection or on the basis of FP-derived fluorescence and cell sorting. Typically, stably expressing cell populations exhibit more uniform levels of expression, albeit lower levels, compared with transiently transfected cells. One problem with transiently transfected cells is that a broad spectrum of fluorescent intensities can be seen, and this range can exceed the capabilities of a 4-log amplifier that is standard on many cytometers.

A brief protocol to obtain stable FP-expressing cells is listed below. In this example, a retrovirus-based vector is utilized that encodes a hygromycin resistance (hygR)/GFP fusion gene. This vector can be used to generate stable cell lines following transient transfection of cells, or following infection of target cells with retrovirus particles derived from the vector.[2] For a detailed discussion of FP expression from retroviral vectors, readers are encouraged to refer to chapters 28 and 30 in this volume.

1. Harvest NIH 3T3 cells 48 hr after infection with retrovirus or transfection with the retroviral construct. Fluorescence microscopic analysis should reveal GFP-positive cells.

2a. *Drug selection:* Split cells 1:4 into normal growth medium that is supplemented with hygromycin B (500 μg/ml; Boehringer Mannheim, Indianapolis, IN). Cell death should become evident 2/3 days after the addition of hygromycin. Dead cells will become rounded and eventually detach from the culture surface. After 1 week, selection will be nearly complete, with virtually all of the viable cells expressing GFP. Infection should yield a stable cell line more quickly than transfection, because every infected cell will possess a stable copy of the provirus. In contrast, only a minor fraction of the transfected cells will harbor integrated copies of the plasmid vector. Therefore, it may take longer with transfection (2–3 weeks) to obtain sufficient numbers of stable clones to perform analyses.

2b. *Cell sorting:* GFP fluorescence can also be used as a means of selection for stable expression. After infection, GFP-positive cells can be isolated via cell sorting and placed back into culture. These cells maintain expression of GFP over time, in a manner nearly identical to that of stable GFP-expressing cells obtained by hygromycin selection.[2] In some cases, additional rounds of cell sorting may be required to select pure populations of GFP-expressing cells.

Single-Color Analysis

Reporter studies that require a single FP are most easily performed using mutants that excite well with 488-nm light from the argon laser, such as EGFP or EYFP. EGFP is widely used in reporter studies and this variant can be detected flow cytometrically using the standard filter configuration for fluorescein detection. However, better signals may be achieved with an emission filter that captures slightly shorter wavelengths. For example, we have found that replacement of the standard 530/30-nm bandpass filter on a FACS Vantage (Becton Dickinson, San Jose, CA) with a 510/20-nm bandpass filter significantly improves the signal-to-noise ratio for GFP-S65T and EGFP detection.[2] Although this is a simple and inexpensive modification, the optics are not easily accessed on some analyzers, especially those certified for clinical analysis. For this reason, EYFP represents a sound alternative to EGFP. EYFP is sufficiently stimulated at 488 nm to produce a signal that is comparable in intensity to that obtained with EGFP.[10] Furthermore, because the emission maximum of EYFP is nearly identical to that of fluorescein, no optical alterations are necessary to achieve efficient fluorescence detection. Regardless of which FP is used, the subsequent analyses are performed in a similar manner.

The FP signal should be treated no differently than signals from typical fluorochromes. A negative control is required, preferably one that has been mock transfected or mock infected in order to set the light scatter and background fluorescence gating. Most of our analyses are performed on a FACS Vantage equipped with an Enterprise (Coherent, Inc., Santa Clara, CA) argon ion laser producing 125 mW of 488-nm light. We have not observed a significant fluctuation in the signal-to-noise ratio at different laser powers. In fact, we have achieved similar results with an EPICS Profile II (Coulter Corp., Miami, FL), which uses a 15-mW argon ion laser.

One important consideration is the intensity of the FP signal. Transient

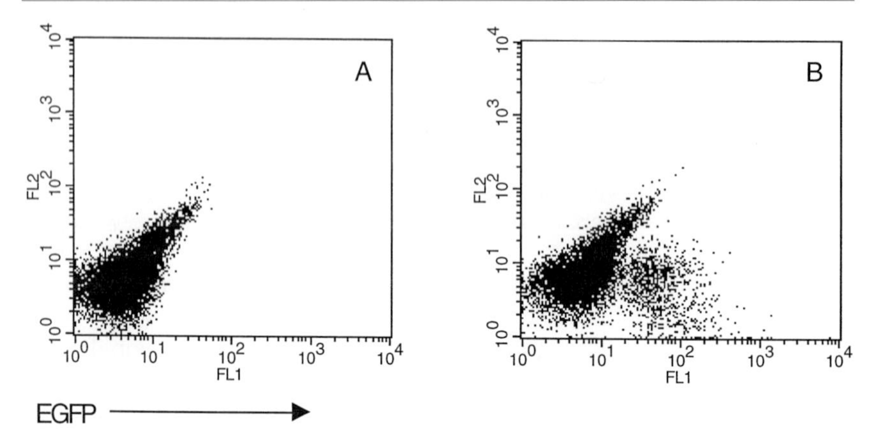

FIG. 1. Two-parameter analysis to distinguish GFP-derived fluorescence from background fluorescence. Murine splenic B cells were stimulated for 48 hr with LPS (25 μg/ml) and then mock infected (A) or infected with EGFP retrovirus (B).[10] Cells were analyzed 48 hr after infection. EGFP fluorescence was measured in FL1 (510/20 bandpass filter) and cellular autofluorescence was monitored in FL2 (585/42 bandpass filter normally used to detect PE). Electronic compensation (FL2 − 18.0% FL1) was applied during analysis.

transfection will generate a population of cells that express widely variable levels of FP. In fact, some vectors that employ strong promoters [such as the cytomegalovirus (CMV) immediate-early promoter/enhancer] will generate FP levels that are detected as off-scale even when the photomultiplier tube (PMT) voltages are adjusted so that the negative control cells are largely off-scale in the negative direction. This is particularly problematic when two-color FP analyses are performed (see below). It must be determined which cells are most important for the experiment, the brightest, the dimmest, or the best average of the population, and the PMT voltages should be adjusted accordingly. Alternatively, less DNA can be used in the transfections to help diminish the signal. The result will be that fewer of the transfected cells will cross the detection threshold (detected as positive), but this should be acceptable for most experiments.

Cells cultured *in vitro* tend to be highly autofluorescent and this fact can obscure relatively weak FP signals. To visualize more clearly low-level FP expression, two-parameter analysis of cells that are singly FP positive can be helpful. An example is shown in Fig. 1. Splenic B cells were stimulated to divide with lipopolysaccharide (LPS) and then infected with a retroviral vector encoding a hygromycin resistance/EGFP fusion protein. By analyzing the EGFP signal (FL1) versus an irrelevant signal in another PMT [FL2, normally used for phycoerythrin (PE) detection] in a two-parameter fashion, it is easier to visualize low-level EGFP fluorescence above the

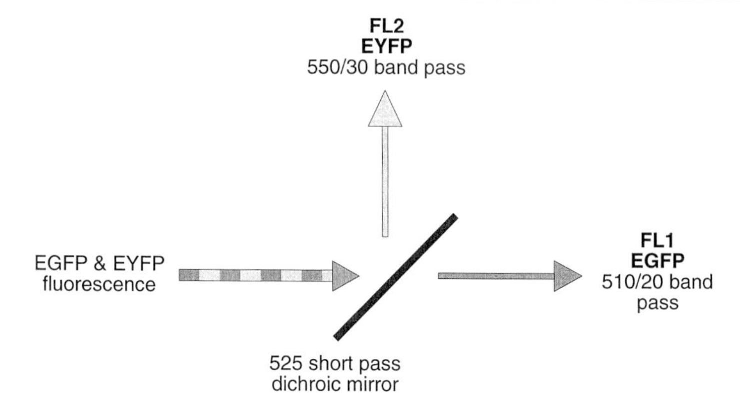

FIG. 2. Optics for dual detection of EGFP and EYFP. In this scheme, fluorescence from both proteins is elicited with 488-nm light and the signals are separated using the indicated filters.

autofluorescence background. As noted in the mock-infected sample (Fig. 1A), LPS treatment causes the cells to become autofluorescent. Because many of these cells are as bright in FL1 as the true EGFP-positive cells (Fig. 1B), two-parameter analysis reveals EGFP-expressing cells that would otherwise go undetected. Here, electronic compensation is needed to subtract the EGFP signal from FL2, because EGFP fluorescence does cross over into the PE filter.

Two-Color Fluorescent Protein Detection

As mentioned previously, various strategies have been developed to permit the detection of multiple FPs. In Fig. 2, a scheme is presented to analyze simultaneously the expression of two distinct FPs, EGFP and EYFP, but many of the general considerations introduced here should be applicable to other systems. Our initial experiments with EYFP revealed that this mutant yields a signal that is comparable in intensity to that of EGFP. For this reason, we sought to determine whether the fluorescence emission from both proteins could be separated, in spite of the similar emission spectra of these proteins. Figure 2 depicts the optical configuration we found to be useful in the simultaneous analysis of EGFP and EYFP fluorescence.[10] (In [17] in this volume,[12a] an application of this detection method is presented.) This scheme requires 488-nm light for excitation and the indicated filters to segregate fluorescence from the two proteins. This type

[12a] L. Lybarger and R. Chervenak, *Methods Enzymol.* **302,** [17], 1999 (this volume).

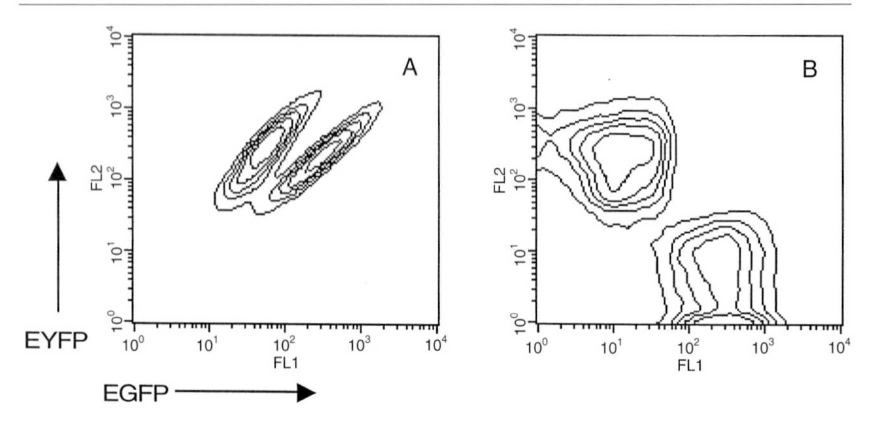

FIG. 3. Resolution of the EGFP and EYFP signals. NIH 3T3 cells were infected with retrovirus vectors encoding either hygR/EGFP or hygR/EYFP fusion proteins, and then completely selected with hygromycin. A 1 : 1 mixture of cells from each population was then analyzed using the optical configuration shown in Fig. 2, both before (A) and after (B) compensation was applied. [Adapted from Lybarger *et al.*[10]]

of analysis should be possible using any analyzer that has optics that are easily modified.

The minimal experimental controls required for each analysis include mock-transfected/infected cells, cells that express only EGFP, and cells that express only EYFP. Mock-transfected cells are used to determine the background fluorescence levels that are detected in each PMT (FL1 and FL2 in Fig. 2). The single-color controls are critical for setting electronic compensation, because crossover of each fluorescent signal into the opposite PMT will occur. For most analyses, compensation levels of $<50\%$ in each direction are required. These levels may vary from cytometer to cytometer, depending on such factors as the sensitivity inherent in each PMT, the exact optical filters that are employed, and the PMT voltages that are applied during analysis.

Figure 3 depicts the analysis of a population of cells that express EGFP or EYFP, both before and after compensation is applied. NIH 3T3 cells were infected with hygR/EGFP or hygR/EYFP retroviruses and completely selected with hygromycin. Each FP-expressing population can be resolved once the compensation is set. In this experiment, each FP was expressed at moderate levels. When much higher levels of FPs are expressed, such as the levels observed following transient transfection, simultaneous detection is still feasible[10] (also see [17] in this volume.[12a]). However, it becomes extremely difficult to determine the appropriate compensation levels when many of the events are off-scale. Therefore, as mentioned for single-color analyses, the DNA quantities used for transfection should be adjusted to obtain reasonable signals.

Three-Color Fluorescent Protein Detection

With the increasing number of available GFP variants, it seems likely that three-color (or more) FP analysis will be possible. Indeed, we have performed three-color detection using the EGFP/EYFP system outlined here in combination with an UV-excitable mutant, enhanced blue fluorescent protein[12] (EBFP). Figure 4 represents analysis of NIH 3T3 cells that were infected with a mixture of hygR/EGFP and hygR/EYFP retroviruses. A pool of stable cell lines was generated, some of which express both proteins (see Fig. 4A and B). Subsequent to transfection of these cells with an EBFP expression construct, signals from each FP can be detected in a fraction of the cells. However, fluorescence is not strongly stimulated from EBFP with UV light from the argon laser. For this reason, EBFP may be most useful as a second FP to use with EGFP in dual-color fluorescence microscopy. By combining EGFP and EYFP fluorescence with other FPs that possess spectra better suited to flow cytometric analysis, effective three-color detection may be possible.

Combining Fluorescent Proteins with Common Fluorochromes

The versatility and power of FP reporter studies can be greatly extended by combining cell surface phenotypic analysis with FP reporter protocols. In theory, FP-derived fluorescence should be compatible with many of the common fluorochromes used in flow cytometry. Fluorescence from the green-emitting variants could be analyzed using the PMT that is normally used to detect fluorescein. However, this precludes the use of fluorescein-labeled probes for analyzing cell surface proteins. If UV excitation is available, especially if the UV and 488-nm lines are not colinear, more possibilities exist through the use of UV-excitable proteins. We have analyzed EGFP fluorescence in conjunction with single-color cell surface staining using phycoerythrin, allophycocyanin, or red-670.[2] In addition, one report describes FP signal detection along with eight-color cell surface staining.[13] Thus, cell surface staining and FP reporter signals can be analyzed simultaneously to produce an additional level of information about the target cells.

[13] M. Roederer, S. De Rosa, R. Gerstein, M. Anderson, M. Bigos, R. Stovel, T. Nozaki, D. Parks, L. Herzenberg, and L. Herzenberg, *Cytometry* **29,** 328 (1997).

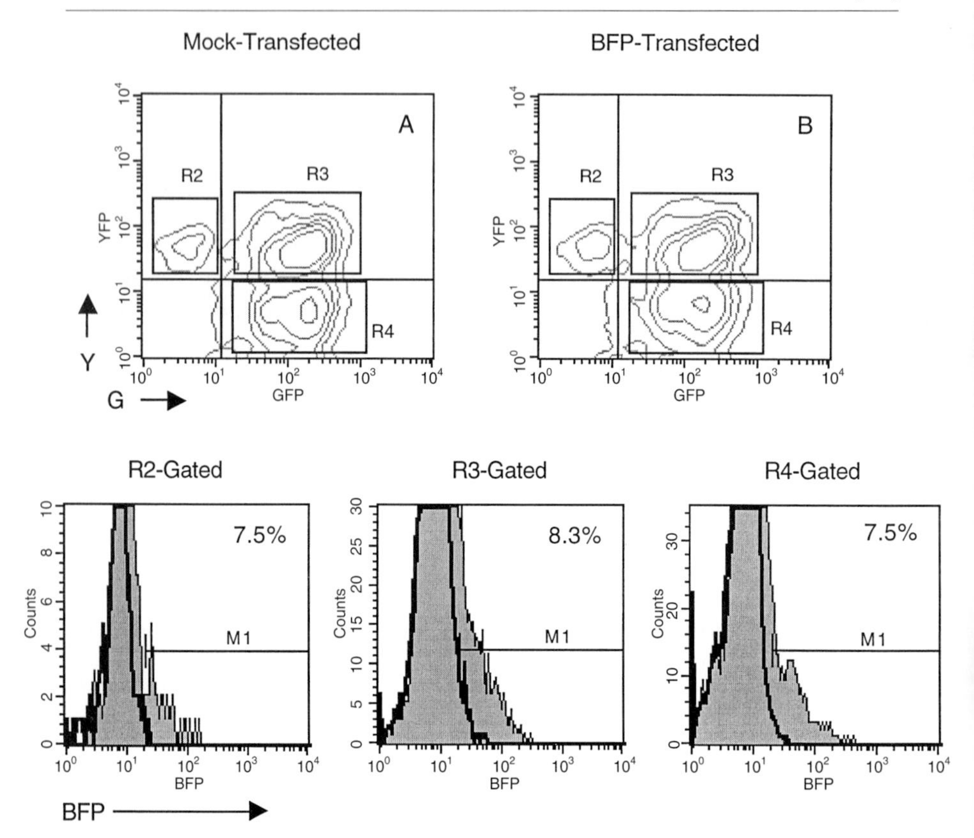

FIG. 4. Two- and three-color analysis. An EGFP/EYFP-expressing cell line was generated by coinfection of NIH 3T3 cells with a mixture of EGFP- and EYFP-encoding retroviruses. This cell line was then either mock transfected (A) or transfected (B) with 30 μg of an EBFP expression construct (pEBFP-C1; Clontech). EBFP fluorescence (380-nm maximum excitation; 440-nm maximum emission) was stimulated with UV light from the argon laser and collected with a 450/65 bandpass filter. EBFP fluorescence is shown in the bottom row of histograms for cells that fall within each indicated region (R1–R3) for the mock-transfected (heavy line) and EBFP-transfected (shaded peak) cells. The percentage of EBFP-transfected cells minus the percentage of mock-transfected cells in region M1 is given.

Undoubtedly, this methodology will become prominent in many future applications that involve FPs.

Conclusions

Flow cytometry represents a high-throughput method to rapidly obtain quantitative data regarding FP-expressing cell populations, and cell sorting may be employed to obtain pure preparations of cells with the desired phenotype. Furthermore, multiple transcriptional elements that are coupled to distinct FPs may be independently analyzed and this information can be interpreted in the context of cell surface phenotypic data. The choice of which FPs, and which expression technolgies, to use will be dictated by the demands of each experimental system. With careful forethought, the available (and yet to be discovered) spectral variants of GFP should meet the requirements of a wide range of experimental designs.

Acknowledgments

The authors are grateful to Deborah Dempsey for flow cytometry assistance and to Dr. Steve Kain (Clontech) and Jim Grobholz (Omega Optical) for providing critical reagents. This work was supported by National Institutes of Health Grant R01 AI31567.

[17] Flow Cytometric Analysis of Transcription: Use of Green Fluorescent Protein Variants to Control Transfection Efficiency

By LONNIE LYBARGER and ROBERT CHERVENAK

Introduction

In the assessment of DNA sequences involved in the control of gene transcription, a common approach has been to transfect cell lines with genetic constructs in which the regulatory sequences under investigation are used to control the expression of a reporter gene. When such analyses involve regulatory sequences whose functions are thought to be cell type or tissue specific, it is often necessary to examine reporter expression in numerous cell lines, representing a variety of cell types, and to compare levels of expression between these cell lines.

One drawback to this method is that individual cell lines tend to vary considerably in terms of the efficiency with which they can be transfected by any of the common techniques. Thus, a relatively low level of expression of a reporter gene in a particular cell type may be due either to the inefficient activation of the regulatory elements in that cell type, or to an inherently low transfection efficiency of the cells under study. A standard method to circumvent this problem is to correct for the transfection efficiency by cotransfection of the cells with a second reporter, such as the β-galactosidase gene. Here, we present a variation of this strategy wherein cotransfection

0076-6879/99 $30.00

of cells with two variants of the green fluorescent protein (GFP) gene makes it possible not only to control for the relative transfection efficiency of individual cell lines, but to examine the level of expression of the reporter genes across the entire population of cells, and at the single-cell level as well. Because flow cytometry is used for the detection of reporter gene expression in this system, the rapid and simultaneous detection of both reporter genes can be accomplished. An example is provided that utilizes this strategy to evaluate the activity of a tissue-specific enhancer in various cell types.

Cytometer Setup

To perform genetic reporter studies, a two-color fluorescent protein (FP) flow cytometric detection scheme is employed.[1] In this scheme, the expression patterns of two GFP variants, enhanced GFP[2,3] (EGFP) and enhanced yellow fluorescent protein[4,5] (EYFP), are monitored simultaneously. For the concurrent analysis of EGFP and EYFP it is necessary to modify the conventional optical setup found on most commercial cytometers. A more lengthy discussion of the considerations involved in implementing these modifications is presented elsewhere (see Ref. 1, and [16] in this volume[5a]). Briefly, however, a generic configuration that works well on most cytometers involves the replacement of three optical filters. First, the optical filter installed in front of the photomultiplier tube (PMT) used to detect fluorescein isothiocyanate (FITC) fluorescence is removed and replaced with a 510/20-nm bandpass filter for the detection of EGFP. Next, the optical filter in front of an adjacent PMT [usually one set up to detect phycoerythrin (PE) fluorescence] is removed and replaced with a 550/30-nm bandpass filter for the detection of EYFP. Finally, the dichroic filter, used to separate the light passed to these two PMTs, is removed and replaced with a 525-nm dichroic filter. Whether this is a shortpass or longpass dichroic element will depend on the exact configuration of the cytometer used.

[1] L. Lybarger, D. Dempsey, G. H. Patterson, D. W. Piston, S. R. Kain, and R. Chervenak, *Cytometry* **31,** 1 (1998).

[2] B. P. Cormack, R. H. Valdivia, and S. Falkow, *Gene* **173,** 33 (1996).

[3] T.-T. Yang, L. Cheng, and S. Kain, *Nucleic Acids Res.* **24,** 4592 (1996).

[4] M. Ormö, A. Cubitt, K. Kallio, L. Gross, R. Tsien, and S. Remington, *Science* **273,** 1392 (1996).

[5] *CLONTECHniques* **XII,** 10 (1997).

[5a] L. Lybarger and R. Chervenak, *Methods Enzymol.* **302,** [16], 1999 (this volume).

DNA Constructs

In our laboratory, constructs have been generated to evaluate the activity of the T cell receptor (TCR) β-chain enhancer[6,7] in various cell lines and data obtained from these constructs are used here for illustrative purposes. A basic EGFP expression vector has been created that contains a hygromycin resistance/EGFP fusion gene[1] plus 3′ processing signals. Distinct regulatory elements have been cloned upstream of this cassette, including the Moloney murine leukemia virus 5′ long terminal repeat (LTR) for constitutive EGFP expression, the TCR β-chain enhancer juxtaposed to the basal promoter elements of the LTR, and the LTR basal promoter elements alone [enhancer(−)]. For constitutive expression of EYFP, a retroviral vector that encodes a hygromycin resistance/EYFP fusion protein[1] is used. These constructs were introduced into cells via electroporation (see [16] in this volume[5a] for a detailed protocol).

Experimental Design

A scheme is presented to compare the activity of a tissue-specific enhancer (here, the TCR β-chain enhancer) between various cell types. In a typical analysis, the following samples would constitute the minimum required for adequate instrument setup and data analysis.

1. Mock-transfected cells: These cells should be subjected to the same procedures that will be used to transfect cells in subsequent samples, but without the addition of EGFP- or EYFP-encoding DNA. These cells are necessary to establish the background fluorescence levels and light-scattering properties of each cell type under investigation. It has been our experience that autofluorescence of mock-electroporated cells is nearly identical to that of control cells; however, cells transfected using some liposome reagents tend to show a dramatic increase in autofluorescene relative to controls, which can sometimes obscure the true fluorescent signals from EGFP or EYFP.
2. Cells transfected with a construct in which the EGFP gene is under the control of promiscuous promoter/enhancer elements: These cells are necessary to set appropriate electronic compensation between the EGFP and EYFP signals.

[6] P. Krimpenfort, R. de Jong, Y. Uematsu, Z. Dembic, S. Ryser, H. von Boehmer, M. Steinmetz, and A. Berns, *EMBO J.* **7,** 745 (1988).

[7] S. McDougall, C. L. Peterson, and K. Calame, *Science* **241,** 205 (1988).

3. Cells transfected with a construct in which the EYFP gene is under the control of promiscuous promoter/enhancer elements: Again, these single-positive cells are necessary for setting signal compensation.

4. Cells cotransfected with the constructs used in samples 2 and 3: These are necessary to establish that cotransfection is working properly and that the construct used to identify the transfected cells is working adequately.

5. Cells cotransfected with a construct in which EGFP expression is under the control of the experimental regulatory sequences (TCR β-chain enhancer) and a construct in which EYFP expression is under the control of the promiscuous promoter/enhancer elements (the construct used in sample 3).

6. Cells cotransfected with an EGFP reference construct and the EYFP constitutive expression construct: In our case, the EGFP reference is a construct that contains only the basal promoter and lacks the TCR β-chain enhancer, referred to as the enhancer(−) construct. This sample is required to determine the basal levels of EGFP expression, and therefore the activity of the basal promoter, in each cell type. Experimental data from all of these samples can be analyzed as described in the next section.

Data and Analysis

Data obtained from the flow cytometric analysis of transiently transfected cells, using the EGFP and EYFP constructs, can be analyzed in several ways, depending on the needs of the investigator and the experimental design. To normalize data with respect to transfection efficiency, the analysis can be focused on those cells that have been detectably transfected by limiting the analysis solely to those cells that express EYFP from the promiscuous retroviral LTR promotor/enhancer–EYFP construct. Thus, the fraction of EGFP-expressing cells can be calculated by determining the percentage of EGFP/EYFP double-positive cells among the total population of EYFP-expressing cells. This simple method can be used to compare the relative activities of two or more EGFP-expressing constructs by comparing, with each construct, the ratio of cells that express the EYFP control to cells that express both constructs. For example, if a given cell line activates an enhancer to be tested, a greater proportion of the transfected cells (EYFP-positive) will be positive for EGFP expression, relative to the enhancer(−).

Although the preceding approach may be suitable for some applications, the overall expression of reporter constructs by a population of cells is a function not only of the number of cells that express the reporter, but of the amount of reporter protein produced by each cell. To deal with this issue, we have devised an analysis that takes into account both the percentage of reporter-positive cells, as well as the mean fluorescence of these cells. To do so, we have introduced a new value, termed *fluorescence units* (FU). In essence, fluorescence units are an indicator of the total amount of a particular FP that is produced in transfected cells. The fluorescence units for a given fluorescent protein are calculated by the following formula:

$$\text{FU} = \%\ \text{FP-positive cells} \times (\text{mean fluorescence of FP-positive cells minus negative gate channel number})$$

Note: The negative gate channel number is determined by setting the lower boundary of the analysis region such that between 0.2 and 0.5% of the mock-transfected cells appear as positive for the FP. Subtraction of this value from the mean fluorescence helps to correct for day-to-day variations in both autofluorescence of the cells and in cytometer setup.

One important advantage this information can provide is the relative efficiency of transfection from one sample to the next. To correct data for transfection efficiency, the fluorescence units of the fluorescent protein under the control of promiscuous promoter/enhancer elements (here, EYFP) can be calculated and normalized to produce a correction factor for transfection efficiency for each cell sample. Calculation of the correction factor simply involves dividing the EYFP fluorescence units of the test sample by the EYFP fluorescence units of the reference sample. The fluorescence units of the fluorescent protein under the control of the experimental regulatory elements (EGFP) can then be adjusted for the efficiency of transfection in the sample by dividing it by the correction factor for that sample.

An example of this type of analysis is illustrated in Fig. 1. Here, ASL1-W cells (murine thymoma) were tested for their ability to activate transcription from the TCR β-chain enhancer. As a reference, EGFP expression from the enhancer(−)–EGFP construct was also monitored. TCR β-chain enhancer-mediated expression is, therefore, defined as an increase in EGFP fluorescence from the enhancer-containing construct versus the promoter-only construct when the EGFP fluorescence units are normalized for transfection efficiency (using the correction factor obtained by comparing EYFP fluorescence units). The plots shown in Fig. 1 were obtained from the analysis of ASL1-W cells approximately 20 hr after transfection. The experimental data are calculated and listed in Table I, along with data obtained from the transfection of NS-1 (murine myeloma) and NIH 3T3 cells (murine fibroblast) with the same constructs. The data indicate that ASL1-W cells strongly activate the enhancer, whereas the other cell lines do not. It is

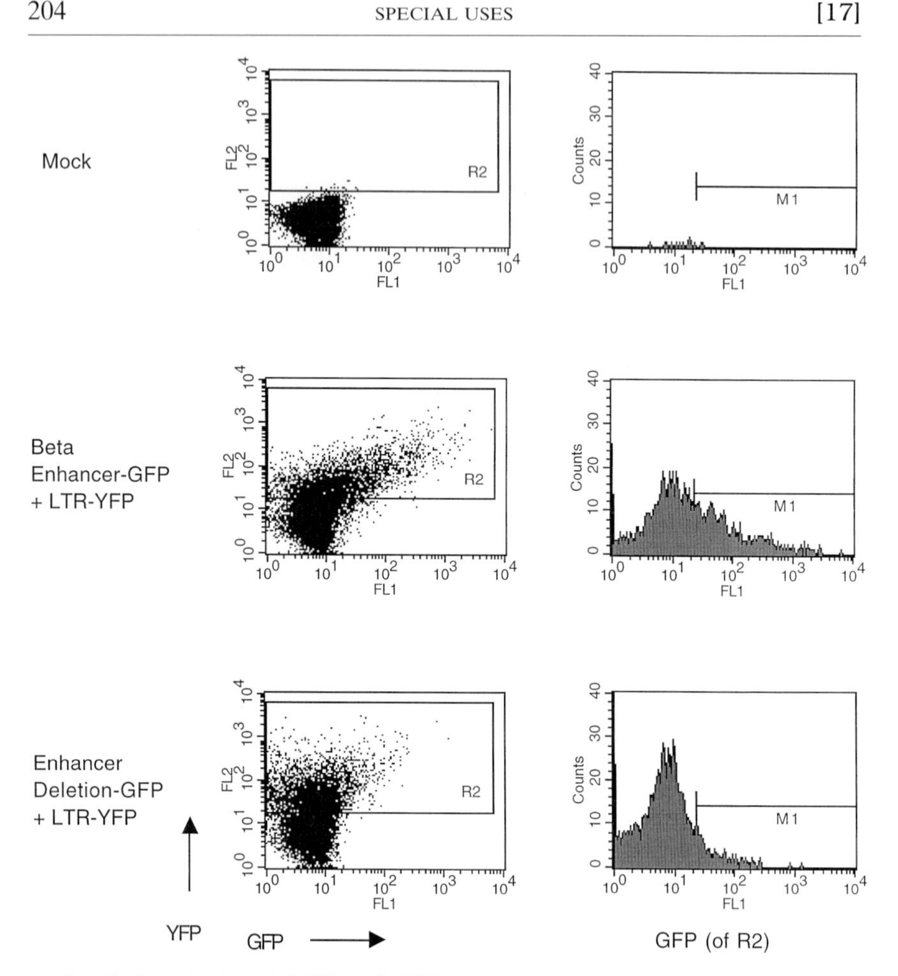

FIG. 1. Quantitation of EGFP and EYFP expression following transient transfection. ASL1-W cells were transfected via electroporation with 15 μg each of the indicated constructs. Region R2 (*left*) was set on the basis of the mock-transfected control. The single-parameter histograms (*right*) depict EGFP fluorescence of the cells within region R2 for each sample. Cells that fall within region M1 are considered EGFP positive and statistics from these cells were used to calculate the values shown in Table I.

also noteworthy that a vast difference is observed in the basal promoter activity between cell types.

From this experiment, it can be seen that transfection efficiency may vary from sample to sample with a given cell type (compare EYFP fluorescence units), although this variation may be typically small. However, should

TABLE I
COMPARISON OF T CELL RECEPTOR β-CHAIN ENHANCER ACTIVITY IN VARIOUS CELL TYPES

Transfection[a]	EYFP FU[b]	EYFP CF[c]	EGFP (of EYFP positive) FU	EGFP (of EYFP positive) Corrected FU[d]	Fold activation[e]
NIH 3T3 cells					
Enhancer(−)/EGFP and LTR/EYFP	647.6	1.0	98.43	—	—
Enhancer(+)/EGFP and LTR/EYFP	840.3	1.3	55.6	42.8	0.43
NS-1 cells					
Enhancer(−)/EGFP and LTR/EYFP	1,398	1.0	11,446	—	—
Enhncer(+)/EGFP and LTR/EYFP	1,206	0.86	7,786	9,054	0.79
ASL1-W cells					
Enhancer(−)/EGFP and LTR/EYFP	2,689	1.0	307.8	—	—
Enhancer(+)/EGFP and LTR/EYFP	1,934	0.72	4,200	5,833	19.0
ASL1-W cells (2X)[f]					
Enhancer(+)/EGFP and LTR/EYFP	4,708	1.75	9,929	5,673	18.4

[a] Each cell type was analyzed approximately 20 hr posttransfection with 15 μg of each indicated construct.

[b] FU, Fluorescence units; calculated for EGFP and EYFP as described in text.

[c] CF, Correction factor; represents the EYFP FU from enhancer(+) sample divided by the EYFP FU of the enhancer(−) transfection.

[d] Corrected FU equals EGFP FU of the enhancer(+) sample divided by the CF.

[e] Fold activation is the enhancer(+) EGFP corrected FU divided by the enhancer(−) EGFP FU.

[f] In this case, cells were transfected with 30 μg of each construct, but the CF and fold activation are based on the ASL1-W enhancer(−) sample using 15 μg.

the variation be marked, it is especially important to normalize the data for transfection efficiency. As an example, ASL1-W cells were cotransfected with twice the amount of TCR β-chain enhancer–EGFP construct (2×) and constitutive-EYFP construct (2×), to mimic a more efficient transfection. When these data are normalized against the reference construct at 1× concentration, the fold activation that is calculated is similar to the fold activation observed in the 1× TCR β-chain enhancer–EGFP sample (see Table I, ASL1-W 2× sample).

Discussion

The technique described here permits an accurate determination of reporter gene expression by correcting for variations in transfection efficiency from sample to sample and for the inherent variations in transfection efficiency seen among various cell lines. As with any such data normalization, care needs to be taken to assure that all of the assumptions of such normalization do, in fact, hold true for any given experimental condition. Of primary concern is to determine empirically that the response measured is proportional to the amount of DNA used for transfection. In other words, a "linear range" must be established for each fluorescent protein construct in each cell type.

When compared with more traditional methods of analyzing reporter expression, it seems likely that this system will not have the overall levels of sensitivity that are obtained with other reporters such as luciferase or chloramphenicol acetyltransferase. Nevertheless, FP reporters have several attractive advantages. First, the analysis can be limited exclusively to those cells in which transfection can be demonstrated by their expression of a control (EYFP) construct. Thus, the resulting data are not influenced by the large fraction of nontransfected cells in the entire population. Second, the analysis is rapid, with a total analysis time of a few minutes per sample, and sample processing and preparation time are minimal. Finally, because data from individual cells within a test population are collected with this technique, expression trends within the transfected population may be observed (such as discrete subfractions of this population that express different levels of FP); this would be difficult or impossible to detect using more conventional techniques. Thus, cotransfection with EGFP and EYFP reporter constructs should prove useful in the analysis of transcriptional activity in a wide variety of experimental systems.

Acknowledgments

The authors are grateful to Deborah Dempsey for flow cytometry assistance and to Dr. Steve Kain (Clontech) and Jim Grobholz (Omega Optical) for providing critical reagents. This work was supported by National Institutes of Health Grant R01 AI31567.

[18] Use of Coexpressed Enhanced Green Fluorescent Protein as a Marker for Identifying Transfected Cells

By YU FANG, CHIAO-CHIAN HUANG, STEVEN R. KAIN, and XIANQIANG LI

Introduction

Screening of stable mammalian cell clones that express a target gene is often a rate-limiting step in establishing such cell lines for functional studies of gene expression. Generally speaking, the target gene needs to be cotransfected with an antibiotic resistance gene that is used for selecting the transfected cells. Among the selected clones, only a few express the target gene owing to the low efficiency of plasmid vector-mediated transfection. Therefore, a large number of drug-resistant colonies are needed for screening in order to obtain such a positive clone. Reverse transcription-polymerase chain reaction (RT-PCR), Western blot analysis, and functional assays are often used to identify positive clones. Most of these screening procedures are tedious and time consuming.

Here we report a simplified screening procedure to identify clones that express the target gene by coexpression with enhanced green fluorescent protein (EGFP), using a bidirectional vector. The target gene and EGFP are simultaneously expressed by the bidirectional vector. The target gene expression can be easily identified by screening for EGFP expression. EGFP is a mutant of wild-type GFP with a 35-fold increase in fluorescence,[1,2] which is easily detected with a fluorescence microscope. We have successfully used the coexpressed EGFP as a marker to identify transfected clones that express a target gene, luciferase. We found that expression of each gene is well correlated. Therefore, it is possible to screen the positive clones expressing a gene of interest simply by identifying those cells that express EGFP.

Principle of Method

This method is based on the bidirectional vector expression system. The Tet expression system was developed by Gossen and Bujard.[3] In this system, the minimal region of the cytomegalovirus (CMV) promoter is combined with a tetracycline (Tc)-regulated element (TRE), and the pro-

[1] B. P. Cormack, R. H. Valdivia, and S. Falkow, *Gene* **173,** 33 (1996).
[2] T. T. Yang, L. Cheng, and S. R. Kain, *Nucleic Acids Res.* **24,** 4592 (1996).
[3] M. Gossen and H. Bujard, *Proc. Natl. Acad. Sci. U.S.A.* **89,** 5547 (1992).

0076-6879/99 $30.00

moter is activated by tetracycline-responsive transcriptional activator (tTA). Binding of tTA to the TRE is regulated by Tc. This system is further modified for simultaneous expression of two target genes.[4] With these bidirectional promoters, two genes are expressed in opposite orientation from a central promoter region containing a single TRE. Expression of both genes is therefore regulated by Tc. The system can be used for constitutive expression in the absence of Tc. Gene expression with this system needs to be carried out in tTA-pretransfected cells. Because tTA has been introduced in most commonly used cell lines, the application of the system is not limited.

In this chapter we describe the cloning of the EGFP gene on one side of the core promoter region and the use of EGFP as a marker to identify the expression of the target gene, luciferase, on the other side of the promoter core region. The fluorescence of EGFP can be directly visualized in living cells with a standard fluorescence microscope. Hence, a large number of colonies can be screened under the microscope and the green fluorescent clones can be easily identified. In addition, cells expressing EGFP can be selected by fluorescence-activated cell sorting (FACS), which can enrich the transfected cells.

Materials and Reagents

Expression of genes with the Tet system (Clontech, Palo Alto, CA) must be carried out in mammalian cells pretransfected with tTA. Most commonly used cell lines, such as Chinese hamster ovary (CHO-K1), have been transfected by tTA, and are commercially available (Clontech). We use CHO-K1 tet-off cells for expression in this study.

PCR amplification reagents, including reaction buffer, dNTPs, and *Pfu* DNA polymerase, are purchased from Stratagene (La Jolla, CA). DNA Engine is used for PCR amplification (PTC-200, Peltier thermal cycler; MJ Research, Watertown MA). All restriction enzymes and buffer are from New England BioLabs (Beverly, MA). T4 DNA ligase and DNA ligase buffer are from GIBCO-BRL (Gaithersburg, MD).

For cell culture, minimum essential medium (MEM; alpha modification), fetal bovine serum, and L-glutamine are from Sigma (St. Louis, MO). G418 sulfate and hygromycin B are from Calbiochem (La Jolla, CA).

General Methods

Construction of pBI-EGFP-Lu and pBI-EGFP-MCS

cDNA of EGFP is amplified by PCR with *Pfu* polymerase and cloned into the *Bam*HI and *Eco*RI sites of pBluescript SK vector (Stratagene)

[4] U. Baron, S. Freundlieb, M. Gossen, and H. Bujard, *Nucleic Acids Res.* **23,** 3605 (1995).

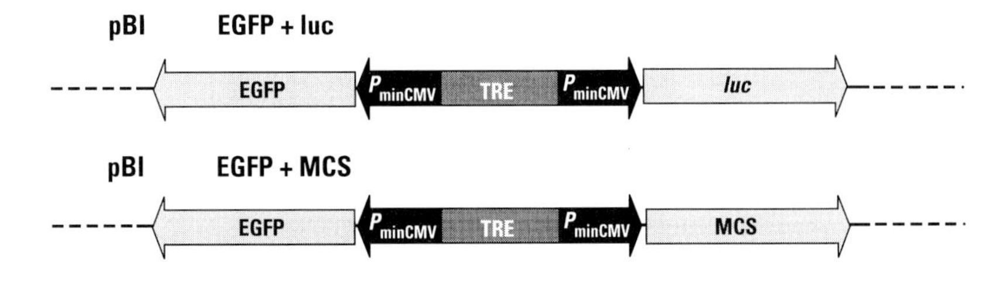

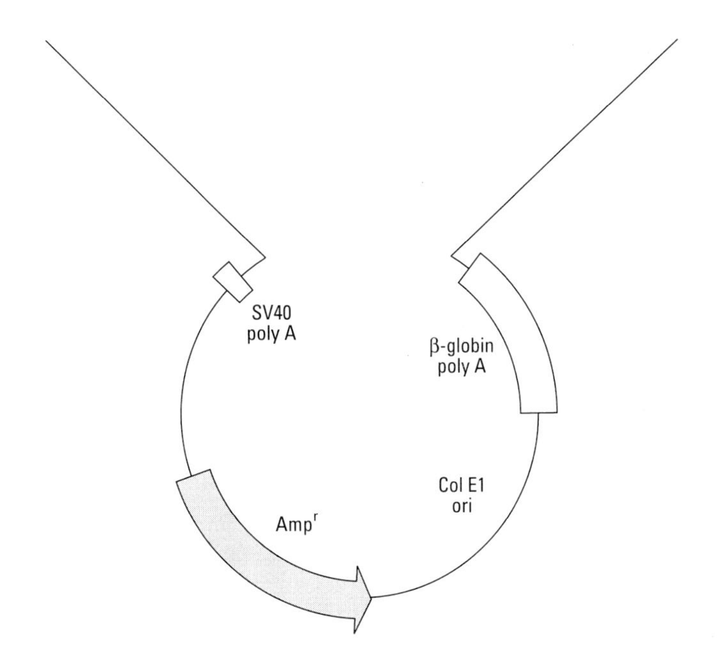

FIG. 1. Schematic maps of pBI-EGFP-Luc and pBI-EGFP-MCS plasmids. EGFP and luciferase genes were cloned into each side of the central promoter region of the pBI vector. In pBI-EGFP-MCS, one side of the promoter region is a multiple cloning site, containing *Mlu*I, *Nhe*I, and *Pvu*II. The central promoter region contains multiple copies of a Tet-responsive element (TRE) flanked by two minimal cytomegalovirus (CMV) promoters in opposite orientations.

with a 5′ primer with a *Bam*HI site sequence (GGGGATCCCCGCGGAT GGTGAGCAAGGGC) and a 3′ primer with an *Eco*RI site sequence (GGGAATTCTTACTTGTACAGCTCGTC). The insert is released by digestion with *Not*I and *Hin*dIII and then cloned into pBI-Luc bidirectional Tet expression vector at the *Not*I and *Hin*dIII sites (Clontech), to make pBI-EGFP-Luc (Fig. 1). The EGFP insert is further cloned into the *Not*I and *Sal*I sites of pBI bidirectional Tet expression vector (Clontech), to make pBI-EGFP-MCS. One side of the central promoter region within the vector is EGFP, and the other side is a multiple cloning site (MCS) that could be used for cloning of various target genes. Procedures for cloning are described as follows.

1. PCR amplification is performed in a 50-μl mixture: 5 μl of *Pfu* buffer, 1 μl each of 5′ and 3′ primer (50 p*M*), 1 μl of dNTPs (200 μM), 1 μl of *Pfu* polymerase (2 units), 10 ng of DNA template, and 40 μl of distilled water. The machine is set up as follows: 90° for 1 min, 55° for 1 min, and 72° for 2 min, for a total of 30 cycles.
2. Purified PCR product is used for restriction enzyme digestion: 2–7.6 μl of DNA (2 μg/ml), 2 μl of restriction enzyme buffer, 1 μl of each restriction enzyme, diluted with distilled water to a total volume of 20 μl. The reaction is carried out at 37° for 30 min. DNAs are loaded onto a 1% agarose gel and gel purified using a NucleoSpin extraction kit (Clontech).
3. Ligation is performed in a 10-μl mixture: 2 μl of T4 DNA ligase buffer, 3 μl each of DNA and vector, 1 μl of T4 ligase, and 1 μl of distilled water. The reaction is carried out at 16° overnight.
4. Transformation is performed in *Escherichia coli* DH5α competent cells (Clontech) and constructs are purified with Qiagen plasmid kits (Quigen, Chatsworth, CA)

Stable Transfection and Selection

CHO/tTA cells are cultured in a 10-cm dish in MEM supplemented with 10% FBS, 1% L-glutamine, and G418 (100 μg/ml) overnight at 37° in a CO_2 incubator. DNAs are transfected into the CHO/tTA cells with a liposome transfection reagent (Clonfectin; Clontech): 5 μg of pBI-EGFP-Luc and 1 μg of pTK-Hyg are mixed, and then incubated with 5 μg of Clonfectin in 0.5 ml of serum-free MEM at room temperature for 30 min. The mixture of DNAs and Clonfectin is diluted by adding 4.5 ml of serum-free MEM and overlaid on cells at 37° in a CO_2 incubator for 1.5 hr. The transfected cells are then rinsed once with phosphate-buffered saline (PBS) and cultured in MEM with 10% FBS, 1% L-glutamine, and G418 (100 μg/ml) overnight at 37° in a CO_2 incubator. Cells are split into two 10-cm dishes after 1 day and hygromycin B (250 μg/ml) is added for selection. After about 2 weeks of selection, the transfected cells form drug-resistant colonies. These drug-resistant colonies are screened under a fluorescent microscope, and green fluorescent colonies are marked on the bottom of the culture dish for isolation of single colonies (Fig. 2A). The individual green fluorescent clones are transferred and allowed to continue in culture (Fig. 2B).

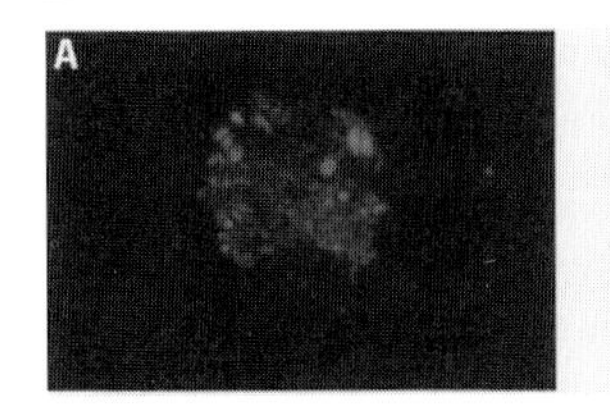

FIG. 2. Fluorescence microscopy of a green fluorescent colony and propagated cells. After transfected cells were selected in the presence of hygromycin (250 μg/ml), the drug-resistant colonies were examined under a fluorescent microscope, and the clones expressing EGFP were identified. One of these green fluorescent clones, CHO/tTA/EGFP/Luc, was recorded (A), as was its propagation (B).

Luciferase Assay

To determine whether the fluorescent clones in our experiment expressed the target gene, luciferase, we picked six green fluorescent colonies and one nonfluorescent colony as a control. One of these fluorescent clones, CHO/tTA/EGFP-Luc, is shown in Fig. 2. Each colony was propagated to about 10^6 cells, and cell lysate was prepared for luciferase assay. The luciferase activity of each fluorescent clone was determined with an assay kit provided by Promega (Madison, WI). Luciferase activity was detected in every fluorescent clone [clone numbers 2–7; luciferase (luc) activity range from 223.2 to 351.9 relative light units], but not in the nonfluorescent clone (clone number 1; luc activity is 0.009 (Fig. 3). The difference in luciferase activity between fluorescent and nonfluorescent clones is about 10^4-fold. Both EGFP and luciferase gene expression were regulated by Tc (data not

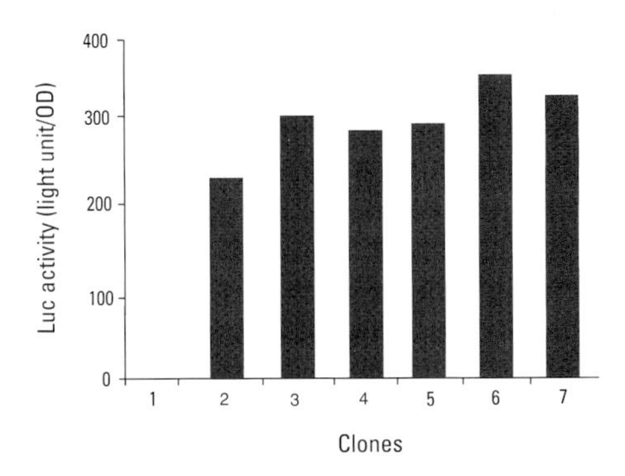

FIG. 3. Analysis of luciferase activity of green fluorescent colonies. Six green fluorescent colonies were identified under a fluorescence microscope and transferred to individual plates for propagation. The cell lysates were prepared and assayed for luciferase activity. One of the nonfluorescent colonies was used as a control (clone 1).

shown). These results suggest that the expression of EGFP and luciferase is simultaneous and well correlated in the system. Therefore, the coexpressed EGFP can be used as a marker to identify positive clones expressing the target gene.

Applications

First, coexpressed EGFP can be used as a universal indicator in the Tet expression system to show transfected colonies under a fluorescence microscope. This procedure allows the screening of a large number of drug-resistant clones for positive expression of the target genes. By using pBI-EGFP-MCS (described in Fig. 1), we cloned a rat purinergic receptor gene into the MCS, and obtained positive clones expressing the receptor gene by examination of the coexpressed EGFP. The expressed receptor was confirmed by Western blot and functional studies (our unpublished data, 1998).

Second, the expression of EGFP and the target gene is under the control of the TRE and CMV minipromoter. Therefore, their expression can be regulated by Tc. We found that gene expression of both cDNAs is tightly regulated by Tc. This is especially important when the genes under investigation express products that are toxic to cells, for example, apoptosis genes. Because EGFP can be monitored with microplate instruments such as FLIPR (Molecular Devices, Menlo Park, CA), individual clones with green fluorescence can be efficiently identified. This is especially useful in dealing with a large number of genes.

Concluding Remarks

The screening method described here is based on the coexpression of EGFP with a target gene by the pBI vector. In this system, two genes are simultaneously expressed and thus the expression of the target gene can be traced by EGFP. The simultaneous coexpression makes it possible to use EGFP to identify the clones expressing the target gene. Because the identification of coexpressed EGFP is straightforward, the screening procedure is much quicker than other screening methods for identifying positive clones that express the target gene. This method can achieve 100% screening efficiency in generating stable transfected cell lines. Therefore the screening method is highly effective in obtaining clones expressing the target gene.

[19] Jellyfish Green Fluorescent Protein: A Tool for Studying Ion Channels and Second-Messenger Signaling in Neurons

By L. A. C. BLAIR, K. K. BENCE, and J. MARSHALL

Introduction

A major aim of modern biology is to understand the regulation of cellular function. One of the most powerful tools with which to dissect intracellular interactions is the expression of specific genes encoding proteins of altered function. These altered proteins may be inactive for a variety of reasons. For example, they may be enzymatically inactive, unable to bind a normal target protein, or unable to achieve the proper subcellular localization through loss of myristoylation sites (e.g., "dominant-negatives"). Conversely, the altered proteins may be constitutively active, retaining the normal type of activity, but performing normal function(s) at a much higher level. However, whether the activity of a protein increases or decreases, its role in specific intracellular pathways and its importance to overall cellular function can be assessed by an array of techniques if it is first possible to transfect the cells of interest and identify unambiguously the transfectants.

Several problems and restrictions have until recently effectively limited the ability to express exogenous proteins in mammalian cell lines. Moreover, the absence of reliable methods to differentiate between transiently transfected and nontransfected cells led many researchers to establish stable cell lines carrying the gene of interest. Unfortunately, many of these genes encode proteins that are involved in survival and/or the regulation of long-term properties of the cells. The long-term expression of altered proteins or even overexpression of normal, wild-type proteins can dramatically alter basal cellular properties, making it difficult to establish a frame of reference, or can even confer lethality. Furthermore, stable transfection requires the maintenance of dividing cells in long-term culture, eliminating most primary cells and all nondividing cells.

Researchers have addressed these issues in several ways. The development of viral transfection systems has greatly expanded the types of cells that can be transfected and the introduction of inducible promoters is ameliorating the probelms associated with extended expression of potentially harmful genes. In addition, to identify successfully living transfectants, several groups have established techniques for transiently cotransfecting

0076-6879/99 $30.00

cells with a reporter gene that encodes a nontoxic marker, the jellyfish green fluorescent protein (GFP).[1–3] Importantly, this approach can be applied to cell lines, dividing primary cells, and postmitotic primary cells.[4]

GFP-based transfection techniques have proved particularly useful for the study of second-messenger pathways in neurons. Here, we focus on how to transfect neurons in primary culture, providing examples of several kinds of constructs (cotransfection with separate gene of interest and GFP cDNA constructs, transfection with a single construct containing the gene of interest fused to a GFP gene, and transfection with a single construct containing a separate gene of interest and GFP cassettes), and of several types of biological questions that can be resolved using this approach. We also address how to transfect neuronal ion channels into cell lines and some of their uses.

Primary Cultures of Cerebellar Granule Neurons

For most of our experiments on primary neurons we utilize cerebellar granule cell cultures. These cells were chosen for several reasons. First, it is possible to obtain large numbers of cerebellar granule neurons and to maintain them in relatively homogeneous cell culture; this enables the employment of biochemical as well as electrophysiological and immunocytochemical assays. In addition, because the cerebellum develops relatively late, cells are taken out for culture postnatally, obviating the necessity of operating to remove embryos. Specific to our work, our previous research using neuronal cell lines strongly suggested that receptor tyrosine kinases rapidly regulate specific ion channel activities[5]; it was known that granule neurons express both the receptor tyrosine kinases and the specific ion channel subtypes. Described here is a method largely based on Messer,[6] with the helpful additions from L. Nowak (personal communication, Cornell University, 1998). We include sufficient detail to enable researchers familiar with mammalian gross neuroanatomy to obtain a reasonable yield the first time. (On request, we will provide the complete four-page protocol.)

Prior to culturing, the culture surfaces, either tissue culture plastic or glass coverslips, are coated with a solution of high molecular weight

[1] D. C. Prasher, V. K. Eckenrode, W. W. Ward, F. G. Prendergast, and M. J. Cormier, *Gene* **111,** 229 (1992).

[2] J. Marshall, R. Molloy, G. W. J. Moss, J. R. Howe, and T. E. Hughes, *Neuron* **14,** 211 (1995).

[3] A. B. Cubitt, R. Heim, S. R. Adams, A. E. Boyd, L. A. Gross, and R. Y. Tsien, *Trends Biochem. Sci.* **20,** 448 (1995).

[4] L. A. C. Blair and J. Marshall, *Neuron* **19,** 421 (1997).

[5] R. H. Selinfreund and L. A. C. Blair, *Mol. Pharmacol.* **45,** 1215 (1994).

[6] A. Messer, *Brain Res.* **130,** 1 (1972).

poly-D-lysine [MW >150,000, dissolved in sterile double-distilled water (ddH_2O)]. Omission of pretreatment drastically reduces the yield; few granule neurons will attach to untreated tissue culture plastic and essentially none to untreated glass. Therefore, before starting the dissection, we coat culture dishes or glass coverslips with sterile-filtered, 0.1–1 mg/ml poly-D-lysine (~1 ml/35-mm tissue culture plate for at least 3–4 hr at room temperature, or overnight at 4°); the higher concentrations are used to coat glass surfaces. We then aspirate the polylysine solution, rinse with sterile ddH_2O, aspirate the ddH_2O, and dry the open plates under ultraviolet light for approximately 1 hr to ensure sterilization. Using this procedure, granule cells attach well (i.e., survive extensive washing), put out neurites within 6 hr of plating (often sooner), and survive and grow for at least 3 weeks.

For the dissection, cerebella are typically removed from 5-day-old rat pups (p5; tested range, p4–p10). The rats are decapitated, the skin over the top and sides of the skull is removed, and the back section of the skull is removed to reveal the cerebellum. Because the skin is a major source of bacterial infection, the head is repeatedly rinsed with 70% ethanol during all stages of the gross dissection. Note that the cerebellum is an outfolding of the myelencephalon; it floats on a stalk behind the colliculi and above the brainstem. Cerebella (usually 10–20) are collected and stored on ice in ice-cold Hanks' buffered saline (HBSS, pH 7.4, with penicillin–streptomycin) until all are obtained.

Cerebella are then washed in ice-cold Hanks' buffered saline and diced into small chunks (~1 mm^3). Tissue chunks are incubated in 10 ml of trypsin (0.5 mg/ml) in HBSS at 37° for 10 min (agitating every 2–3 min). Protease digestion is stopped by competition with serum. The chunks are gently pelleted (30 sec, lowest speed in a clinical centrifuge, ~600 rpm) and, after aspirating the trypsin–HBSS, resuspended in growth medium (Dulbecco's modified Eagle's medium [DMEM], 10% fetal bovine serum [FBS], 25 m*M* KCl, glucose [6 g/liter], 2 m*M* glutamine plus 10 U of penicillin and streptomycin [10 μg/ml]). The tissue pellet is then triturated using sterile, flame-polished, silanized pasteur pipettes until the solution is slightly cloudy; overtrituration destroys cells and reduces the overall yield. This solution is then briefly recentrifuged (lowest speed, ~30 sec) to pellet the larger chunks while keeping the single cells in solution in the supernatant. The supernatant containing the dissociated cells is plated into growth medium on the poly-D-lysine-coated culture surfaces. As shown in Fig. 1, fully dispersed and easily identifiable neurons are detected by 1 day in culture. Interestingly, though, granule neurons migrate in the dish, forming increasingly large clusters [compare Figs. 1 and 2A, left panel with Fig. 2A, right panel, where after 4 days in culture the neurons have already formed a

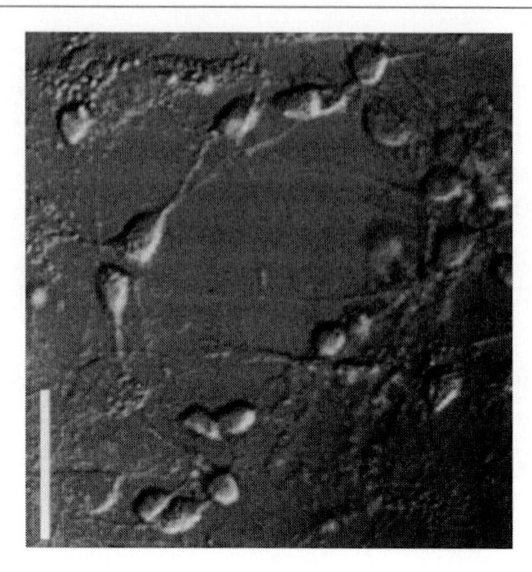

FIG. 1. Cerebellar granule neurons, 1 day *in vitro*, cultured from p5 rats. Scale bar: 10 μm.

massive group]. For many applications, such as kinase assays and enzymatic survival assays, isolated versus "lumped" cells makes no difference. However, for other types of assays, in particular, electrophysiological recordings, the success rate of the experiments depends on being able to utilize isolated cells; for these experiments, cells are plated at the lowest possible density and typically used before 7 days *in vitro*.

Granule neurons can be maintained for up to 3–4 weeks in a humidified 37° incubator (5% CO_2). Because nonneuronal support cells appear to condition the medium and aid in granule neuron survival, we often add, but only at the time of plating, basic fibroblast growth factor (bFGF, 20 ng/ml) to stimulate initially the support cells to help the neurons to get a strong start. Subsequently, however, cytosine arabinoside can be added to the culture medium (final concentration, 10 μM; starting on day 1–2 *in vitro* and maintained continuously thereafter) to eliminate the dividing cells and maintain a pure culture of granule neurons.

To maximize the yield, the trypsinization/trituration procedures are repeated several times on the remaining cell chunks, gradually reducing the volume of and exposure time to the trypsin solution and gradually switching to using smaller tip-bore, flame-polished, silanized pasteur pipettes. Each time the trituration is stopped as soon as the solution becomes cloudy, and those cells plated. Typically, approximately five rounds of trypsinization/trituration are performed. Importantly, the time spent triturating and the number of trypsinization/trituration rounds depend on more than the initial size of the tissue chunks; the age of the rat pup is also

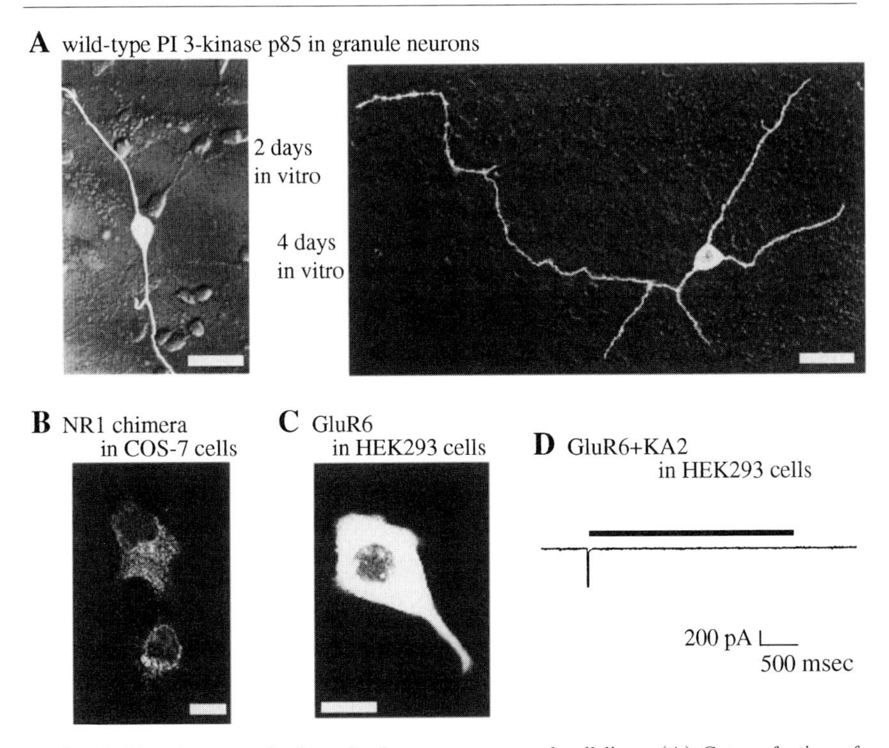

FIG. 2. Transient transfection of primary neurons and cell lines. (A) Cotransfection of cultured cerebellar granule neurons with cDNAs encoding a wild-type p85 subunit of PI 3-kinase and GFP. Overlays of fluorescence micrographs superimposed on the Nomarski images of the fields: *left*, 2 days in culture; *right*, 4 days in culture. Confocal microscopy; *z* sectioning was used to ensure full representation of the neurites, which, especially within clumps of neurons, traversed many optical sections. Cells were transfected by calcium phosphate precipitation. (B) Transfection of COS-7 cells with a single construct containing a single chimeric GFP-NR1 NMDA receptor subunit gene. Note that because the receptor subunit is directly tagged by the marker, fluorescence is detected only where NR1 is expressed. Confocal microscopy; a single optical section. The construct is shown in Fig. 4A. Transfection was via LipofectAMINE. (C) Transfection of HEK293 human embryonic kidney cells with a single construct containing, in separate cassettes, both the kainate receptor GluR6(Q) subunit and GFP genes. Although cells are again transfected with a single construct, the GFP is translated separately from the GluR6, resulting, as in (A), in diffuse, cytoplasmic expression of the marker. Confocal microscopy; one optical plane. The construct is shown in Fig. 4B. Transfection was by LipofectAMINE. (D) Coexpression of two kainate receptor subunits, GluR6(Q) and KA2, with GFP in HEK293 cells generates glutamate-induced currents that show rapid activation and rapid and complete desensitization. Glutamate, 100 μM (indicated by the bar), was applied by rapid superfusion. Cells were transfected by calcium phosphate precipitation. Scale bars: (A–C) 10 μm. [Data from E. P. Garcia, S. Mehta, L. A. C. Blair, D. G. Wells, J. Shang, T. Fukushima, J. R. Fallon, C. C. Garner, and J. Marshall, *Neuron* **21,** 727–739 (1999).]

significant. Cerebella from older pups, especially >p7, have considerably more extracellular matrix and require more trituration.

Green Fluorescent Protein and Gene of Interest Constructs

GFP, a 28-kDa autofluorescent protein that is naturally produced by the jellyfish *Aequorea victoria*, serves as the marker of successfully transfected cells. It is particularly useful because it requires no enzymatic reaction or cofactors to function and no special processing of the cells, meaning that it effectively serves as a vital dye to identify living cells. Moreover, because it is introduced into cells not as a protein, but as DNA, special injection procedures are not required. Cells are merely transfected with the GFP gene along with the gene of interest; the normal cellular machinery then transcribes and translates both messages. Detection is via standard fluorescein fluorescence optics.

A variety of GFP genes is now available. In addition to the wild-type gene, genetically engineered forms have been constructed to increase brightness and alter the emission spectrum of the chromophore to produce an autofluorescent protein emitting blue light[7,8] (BFP; Quantum Biotechnologies, Montreal, Quebec, Canada). Potentially, utilizing two proteins directly tagged with different chromophores could allow simultaneous tracing of the movements of two proteins of interest within cells. Currently, we employ SuperGlo (sg25) GFP (Quantum Biotechnologies),[9,10] which produces one of the brightest GFP proteins, detectable as early as 12 hr posttransfection in mammalian neurons and cell lines (our unpublished data, 1998). Early detection is especially important when testing overexpression of molecules expected to be harmful to the cells when produced over extended periods, such as dominant-negatives of proteins known to be involved in survival (see Applications to Biological Problems, below, and Fig. 3). Our earlier constructs utilized the pGreenLantern GFP (GIBCO-BRL, Gaithersburg, MD). Both of these GFP genes contain point mutations to increase fluorescence intensity and codon usage modifications (silent mutations) to stabilize the RNA and increase translation efficiency in mammalian cells.

GFP genes have been placed primarily into three kinds of constructs: (1) Alone in expression vectors such as pcDNA3 (Invitrogen, San Diego,

[7] R. Heim and R. Tsien, *Curr. Biol.* **6,** 178 (1996).

[8] R. M. Wachter, B. A. King, R. Heim, K. Kallio, R. Y. Tsien, S. G. Boxer, and S. J. Remington, *Biochemistry* **36,** 9759 (1997).

[9] G. J. Palm, A. Zdanov, G. A. Gaitanaris, R. Stauber, G. N. Pavlakis, and A. Wlodawer, *Nature Struct. Biol.* **4,** 361 (1997).

[10] R. H. Stauber, K. Horie, P. Carney, E. A. Hudson, N. I. Tarasova, G. A. Gaitanaris, and G. N. Pavlakis, *BioTechniques* **24,** 462 (1998).

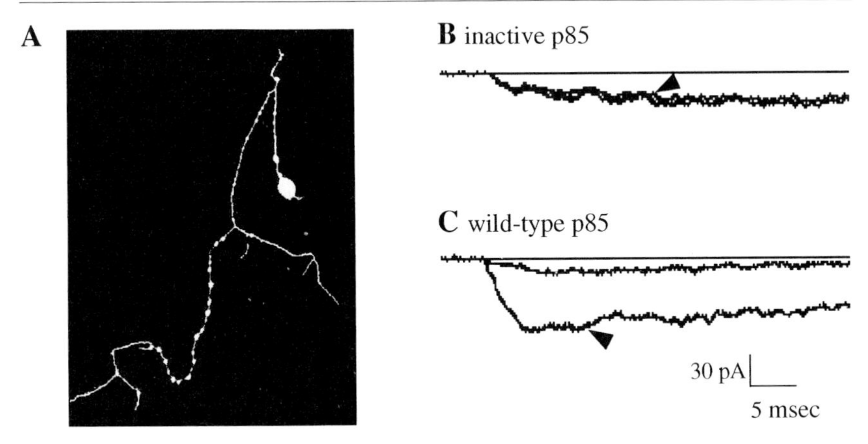

FIG. 3. Transfection of granule neurons with inactive PI 3-kinase subunits blocks IGF-I potentiation of calcium channel currents. (A) A cerebellar granule neuron cotransfected in culture by cDNAs encoding an inactive PI 3-kinase subunit [Δp85, gift of M. Kasuga; Δp85 is described in full in R. Dhand, K. Hara, I. Hiles, B. Bax, G. Panayotou, M. J. Fry, K. Yonezawa, M. Kasuga, and M. D. Waterfield, *EMBO J.* **13,** 511 (1994)] and GFP to identify the transfected cell. (B) IGF-I failed to potentiate calcium channel currents in a granule neuron transfected with Δp85. Currents recorded before and 10 and 60 sec after addition of IGF-I (100 ng/ml) superimpose. (B and C) Voltage-activated barium currents were evoked by depolarizing voltage pulses from −80 to 0 mV. Arrowheads indicate currents recorded after IGF-I addition. [Data from L. A. C. Blair and J. Marshall, *Neuron* **19,** 421 (1997), with permission.] (C) Calcium channel currents recorded 10 sec after IGF-I (20 ng/ml) addition increased manyfold in a wild-type p85- and GFP-transfected cell.

CA) under control of the cytomegalovirus (CMV) or simian virus 40 (SV40) promoter.[2,4] In this case, the GFP construct is cotransfected with a second construct carrying the gene of interest, and the GFP protein is expressed throughout the cytoplasm (Figs. 2A, 2C, and 3A); (2) a single construct that contains the gene of interest directly ligated to the GFP gene to create a fusion protein (Figs. 4A and 2B)[2] can also be made. Although the addition of the 28-kDa GFP to the protein of interest could potentially modify its function, this approach has the advantage that GFP is detected only where the protein of interest is expressed. It has proved particularly useful for examining the subcellular localization of proteins and the movements of proteins within cells[11–13]; (3) a number of groups have begun utilizing a

[11] J. Ellenberg, E. D. Siggia, J. E. Moreira, C. L. Smith, J. F. Presley, H. J. Worman, and J. Lippincott-Schwartz, *J. Cell Biol.* **138,** 1193 (1997).

[12] K. G. Wolter, Y. T. Hsu, C. L. Smith, A. Nechushtan, X. G. Xi, and R. J. Youle, *J. Cell Biol.* **139,** 1281 (1997).

[13] N. V. Burke, W. Han, D. Li, K. Takimoto, S. C. Watkins, and E. S. Levitan, *Neuron* **19,** 1095 (1997).

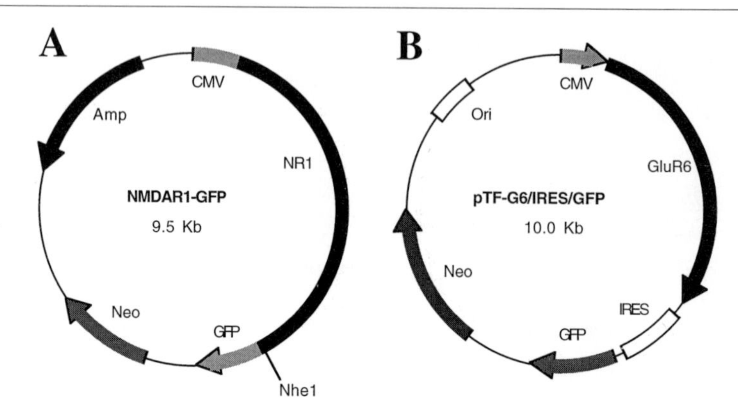

FIG. 4. Constructs used to express GFP-tagged proteins of interest and bicistronic transcripts encoding separately GFP and the protein of interest. (A) A construct encoding the NR1 NMDA receptor subunit directly tagged with GFP was made as described by J. Marshall, R. Molloy, G. W. J. Moss, J. R. Howe, and T. E. Hughes, *Neuron* **14,** 211 (1995). Briefly, to create the chimera, sequence encoding the last 13 amino acids of NR1 was replaced by sequence encoding alanine, followed by the GFP sequence encoding the full protein minus the first amino acid. In addition, to obtain brighter fluorescence, the wild-type GFP gene was replaced with that from pGreenLantern (GIBCO-BRL). Constructed by J. Edgerton. (B) A construct, pTF-G6/IRES/GFP, containing the coding sequences for the GluR6 kainate receptor (gift of J. Boulter and S. Heinemann) and the sg25 SuperGlo GFP (Quantum Biotechnologies), and driven by the CMV promotor-enhancer to ensure a high level of expression in mammalian cells, was made. It also contains the internal ribosome entry site (IRES) of the encephalomyocarditis virus, located between the GluR6- and GFP-coding sequences; this allows both the GFP and receptor proteins to be translated individually from a single bicistronic mRNA. Constructed by T. Fukushima.

third method. Here, a single construct that contains separate cassettes for the gene of interest and the GFP gene (Figs. 4B and 2C) is made. It contains a promoter (in our case, CMV), the gene of interest, an internal ribosome entry site (IRES), and the GFP gene. In this situation, a single transcript is formed, but the protein of interest and GFP are translated separately, leading to a diffuse distribution of the GFP marker in the cytosol. The advantage of this approach is that it allows correlating the level of expression of the protein of interest with the level of GFP fluorescence without affecting the function of the protein of interest. Although we are currently using a pBK-CMV (Stratagene, La Jolla, CA)-based expression vector (Fig. 4B; constructed by T. Fukushima, Brown University), this type of construct can also be put into viral vectors to increase transfection efficiency, as demonstrated by the use of an adenoviral expression system to transfect olfactory neurons *in situ* with odorant receptors.[14]

[14] H. Zhao, L. Ivic, J. M. Otaki, M. Hashimoto, K. Mikoshiba, and S. Firestein, *Science* **279,** 237 (1998).

Transient Transfection of Neurons by Calcium Phosphate Precipitation

To transfect cerebellar neurons, we use calcium phosphate precipitation procedures slightly modified from Chen and Okayama.[15] The yield of successful transfectants, as indicated by the presence of GFP fluorescence, is typically 5–10%. We have also tried lipid-mediated transfection techniques, which in our hands give a high efficiency in cell lines (see below), but a lower efficiency in primary neurons than calcium phosphate techniques.

Granule neurons are usually transfected 24–48 hr postplating. The efficiency of transfection sharply declines after 48 hr in culture owing, most likely, to secretion of extracellular matrix. Efficiency is also low when cells cultured less than 24 hr are transfected, possibly because the cells, still recovering from the dissociation procedures, are not yet strong enough to support the burden of overexpression of exogenous proteins. Typically, we transfect cells plated onto glass coverslips or directly into 35 × 10 mm dishes. To make the transfection cocktail, the following proportions are used per 35-mm dish: Up to 10 μg of DNA is diluted into 100 μl of 0.25 *M* $CaCl_2$ in a sterile tube; depending on the experiment, 1–10 μg of the gene of interest and 1–5 μg of GFP cDNA are used. Then, 100 μl of *N*,*N*-bis-(2-hydroxyethyl)-2-aminoethanesulfonic acid (BES)-buffered saline (50 m*M* BES, 280 m*M* NaCl, 1.5 m*M* Na_2HPO_4, pH 6.97; the pH is critical for maximizing the transfection efficiency) is added drop by drop to the DNA solution while gently bubbling the mixture with a sterile pasteur pipette. Starting solutions should be at room temperature and the resultant mixture should appear slightly cloudy. It is left at room temperature for 30 min and then added dropwise to the culture dish. Cells are exposed to the DNA solution for 6 hr in a 37°/3% CO_2 incubator. At the end of the 6 hr, the DNA-containing precipitate should appear very fine. If larger crystals appear, the efficiency of transfection will be significantly lower. Using solutions that have been warmed to room temperature will help the formation of a fine precipitate. After 6 hr, we replace the transfection medium with standard growth medium, and return the cells to a 5% CO_2 incubator. For most of the genes of interest that we have tested, expression is optimal 12–48 hr posttransfection.

LipofectAMINE Transfection in Cell Lines

We routinely transiently transfect cell lines using LipofectAMINE reagent (GIBCO-BRL). The efficiency varies with the line, but is high when using the kidney-derived COS-7 or HEK293 cells (typically 50–80%). It is,

[15] C. Chen and H. Okayama, *Mol. Cell. Biol.* **7,** 2745 (1987).

however, much lower (10–40%) when we transfect neuroblastoma lines, such as 401L or B104.

Cells are first plated onto poly-D-lysine-coated glass coverslips or tissue culture dishes and grown to 75% confluence in standard growth medium [DMEM, 5% FBS, 10 U of penicillin, and streptomycin (10 μg/ml)]. For each 35-mm dish, the following transfection cocktail is prepared in sterile 12 × 75 mm polystyrene tubes (do not use polypropylene tubes because lipids adhere to their surfaces): 1–2 μg of DNA is diluted into 100 μl of serum-free MEM and mixed with a solution of 5 μl of LipofectAMINE reagent in 100 μl of serum-free MEM. This mixture is left at room temperature for 15 min to allow DNA–liposome complexes to form; while complexes are forming, cells are rinsed once with serum-free MEM. For each transfection, 0.8 ml of Opti-MEM reduced-serum medium (GIBCO-BRL) is added to the DNA–liposome mixture, mixed gently to avoid breaking the relatively fragile DNA–liposome complexes, and gently pipetted over cells. Cells are then incubated in the presence of the DNA–liposome mixture at 37°/15% CO_2 for 4 hr; this mixture is then aspirated and replaced with posttransfection medium (DMEM, 10% FBS, with antibiotics). Peak expression of most proteins of interest is obtained 24–48 hr posttransfection.

Applications to Biological Problems

Cellular processes within the central nervous system have proved, for a variety of reasons, particularly difficult to study. In the past, nearly all work that attempted to examine second-messenger pathways and the regulation of ligand-gated and voltage-sensitive ion channels has utilized cell lines transfected either with regulatory receptors (G protein-coupled receptors or receptor tyrosine kinases, called RTKs) or ion channels or both. Studies of the movements of intracellular proteins have typically required "freeze-frame" immunocytochemistry using fixed tissue. The transfection of primary neurons with a nontoxic marker such as GFP provides a powerful approach to addressing these questions in the mammalian nervous system. Here, to provide specific examples of raw data, we focus on two of the applications that we have tested, the expression of ligand-gated ion channels in cell lines and the expression of signaling components that potentially could mediate the RTK regulation of voltage-sensitive ion channels.

Briefly, we had found that stimulation of the insulin-like growth factor I (IGF-I) RTK in cerebellar granule neurons rapidly induced a large increase in N and L calcium channel currents. However, the pathway mediating this response was unknown. To identify essential signaling components, we first used traditional pharmacological approaches, testing whether inhibitors of conceivable signaling components would inhibit the IGF-I/RTK regulation of the channels.[4] These experiments strongly implicated phosphatidyl-inositol 3-kinase (PI 3-kinase, EC 2.7.1.127) as a second messenger. Unfortunately, though, few pharmacological agents are as specific as researchers would usually prefer. We, therefore, adopted a molecular approach, transfecting the granule neurons in culture with cDNAs encoding an inactive, dominant-negative PI 3-kinase subunit (Δp85[16]) or the wild-type p85 subunit[4] (Fig. 3). Coexpression with GFP (pGreenLantern; GIBCO-BRL) was used to detect transfected neurons. As described above, neurons were cultured from p4–p5 rats. All cDNAs were subcloned into pcDNA3 expression vectors (Invitrogen) and were under the control of CMV promoters. Cells were transfected by the calcium phosphate precipitation techniques also described above. The ability of transfectants to respond to IGF-I was assessed by the permeabilized-patch variation of standard whole-cell recording techniques[17,18] and, to avoid the harmful long-term effects of overexpression of the inactive enzyme (PI 3-kinase is involved in granule neuron survival[19]), responsiveness was assayed as soon as possible after transfection.

We found that transient expression of the inactive, dominant-negative PI 3-kinase subunit fully blocked the rapid IGF-I potentiation of N and L channels (Fig. 3B). In addition, we found that overexpression of wild-type PI 3-kinase subunits not only permitted the IGF-I modulation to occur (Fig. 3C), but significantly increased IGF-I responsiveness over the level observed in untransfected and GFP-vector alone-transfected cells.[4] The results obtained using this approach provide strong evidence that PI 3-kinase is an essential and rate-limiting intracellular messenger in the signaling pathway used by the neuronal IGF-I/RTKs. The general biological interpretation of these experiments would be that the PI 3-kinase-dependent modulation, by controlling calcium channels, might also control an enormous variety of cellular calcium-dependent processes, including neurotransmitter release and IGF-I-dependent differentiation and/or survival.

The transfection approach differs from conventional pharmacological tests in important ways. First, by circumventing problems of pharmacological specificity, transfection with genes encoding inactive proteins of interest provides a fully independent means of assessing the importance of the normal protein. Moreover, the ability to transfect and detect wild-type

[16] R. Dhand, K. Hara, I. Hiles, B. Bax, G. Panayotou, M. J. Fry, K. Yonezawa, M. Kasuga, and M. D. Waterfield, *EMBO J.* **13,** 511 (1994).

[17] O. P. Hamill, A. Marty, E. Neher, B. Sakmann, and F. J. Sigworth, *Pflugers Arch.* **391,** 85 (1981).

[18] J. Rae, K. Cooper, P. Gates, and M. Watsky, *J. Neurosci. Methods* **37,** 15 (1991).

[19] H. Dudek, S. R. Datta, T. F. Franke, M. J. Birnbaum, R. Yao, G. M. Cooper, R. A. Segal, D. R. Kaplan, and M. E. Greenberg, *Science* **275,** 661 (1997).

and/or constitutively active proteins enormously expands the possible conclusions that can be drawn. For the experiments described above, the conclusion that PI 3-kinase is rate limiting would have been impossible if only pharmacological inhibitors had been used; it depends on being able to overexpress the wild-type protein in addition to the dominant-negative form (the equivalent of a pharmacological inhibitor). The expression of constitutively active proteins permits a separate set of conclusions. Taken in comparison with wild type and vector-alone transfections, analysis of cells transfected with a constitutively active enzyme allows a reasonable estimate of how a maximally driven signaling pathway will control the activity of downstream effectors.[20] In short, the expression of altered proteins of interest is a new and uniquely powerful tool for understanding second messenger-mediated systems.

We have, in addition, utilized GFP fluorescence to detect transfectants expressing specific glutamate receptor subunits (Figs. 2B–D and 4) and neuronal N and L calcium channel subunits in a variety of cell lines (K. Bence, unpublished data, Brown University, 1998). The purpose of the glutamate receptor experiments was to determine how coexpression of receptors with synapse-associated proteins (SAPs) regulates receptor clustering and ligand-induced responses.[21] Although GFP-tagged subunits can be used to detect receptors directly (Fig. 2B), the GFP marker was primarily used here to detect the expression of the gene of interest in cell lines (COS-7 and HEK293) for electrophysiological analysis (Fig. 2C and D). Glutamate responses, as regulated by expression of the different receptor subunit and SAP combinations, were determined. An example of the responses, assessed by standard techniques,[17,18] that are obtained when glutamate receptor subunits are expressed in the absence of SAP proteins is shown in Fig. 2D.

These few examples represent only a fraction of a multitude of possible applications. GFP, because it is a nontoxic and easily detectable marker of living transfected cells, is being applied to the study of dynamic processes within cells, including membrane and vesicular trafficking[13] and the movements/targeting of intracellular proteins.[11,12] Importantly, improvements in autofluorescent protein technologies are rapidly expanding their applications. In particular, the development of autofluorescent proteins with different chromophore emission spectra[7,8] (e.g., BFP; Quantum Biotechnologies) should provide a new and unique tool for studying dynamic protein interactions, the development of the brighter GFPs will enhance

[20] L. A. C. Blair, K. K. Bence, and J. Marshall, *Soc. Neurosci. Abstr.* **24** (1998).

[21] E. P. Garcia, S. Mehta, L. A. C. Blair, D. G. Wells, J. Shang, T. Fukushima, J. R. Fallon, C. C. Garner, and J. Marshall, *Neuron* **21,** 727–739 (1999).

their usefulness for dissecting signal transduction pathways via the expression of dominant-negative components, and the development of GFP- and gene of interest-containing viral vectors should dramatically enhance their usefulness for the study of that most recalcitrant of systems, central nervous function.

Acknowledgments

We thank Dr. John Wallenberg of Quantum Biotechnologies for information on the new GFP genes. We also thank Dr. Irwin Levitan for suggesting the IRES-containing construct, Teruyuki Fukushima for permission to discuss his pTF-G6/IRES/GFP construct, Jeremy Edgerton for making the construct for the NR1–GFP chimera and for assistance with the confocal microscopy, and Sunil Mehta for helpful advice. This work was supported by R29 NS 33914-02 (NIH) and SA047 (Council for Tobacco Research Scholar Award) to J.M.

[20] Expression of Green Fluorescent Protein and Inositol 1,4,5-Triphosphate Receptor in *Xenopus laevis* Oocytes

By ATSUSHI MIYAWAKI, JULIE M. MATHESON, LEE G. SAYERS, AKIRA MUTO, TAKAYUKI MICHIKAWA, TEIICHI FURUICHI, and KATSUHIKO MIKOSHIBA

Introduction

Green fluorescent protein is a sensitive and reliable fluorescent marker for examining levels and subcellular sites of expression of functional proteins in living cells. We report here efficient expression of green fluorescent protein in *Xenopus laevis* oocytes by direct microinjection of its plasmid DNA into the nucleus. Application of this method to studies on optical imaging of inositol 1,4,5-triphosphate-sensitive intracellular Ca^{2+} store sites and expression of inositol 1,4,5-trisphosphate receptor/Ca^{2+} release ion channels are discussed.

Inositol 1,4,5-triphosphate (IP_3) is a ubiquitous second messenger responsible for the release of Ca^{2+} from intracellular stores,[1] IP_3, following its release into the cytosol through phosphoinositide turnover stimulated with various extracellular signals such as growth factors, hormones and

[1] M. J. Berridge, *Nature* (*London*) **361,** 315 (1993).

0076-6879/99 $30.00

neurotransmitters, binds to an IP_3-gated intracellular Ca^{2+} channel, the IP_3 receptor (IP_3R; type 1, type 2 and type 3),[2] which in turn mediates the release of Ca^{2+} from the intracellular stores into the cytosol. IP_3-induced Ca^{2+} release is involved in spatio-temporal Ca^{2+} dynamics within a cell. The organization of intracellular Ca^{2+} stores is complex.[3] It is now generally accepted that the IP_3R resides on specialized regions, mostly the smooth endoplasmic reticulum (sER),[4–6] which constitutes IP_3-sensitive Ca^{2+} stores.[7] Intracellular Ca^{2+} dynamics is closely associated with organization of the ER network which can be altered depending on cell dynamics. Differences in subcellular distribution among IP_3R types as well as in organella as Ca^{2+} store present IP_3Rs need to be considered. Therefore, specific store sites and IP_3R types in certain cell-types and/or throughout cell stages have to be identified. In addition, little is known of mechanisms underlying intracellular targeting of the IP_3R to Ca^{2+} stores. We make use of green fluorescent protein (GFP), originally isolated from the jellyfish *Aequorea victoria,*[8] to monitor the expression level[9] and intracellular localization[10] of mouse type 1 IP_3R (mIP_3R1).[11] We expressed GFP and chimeric GFP-IP_3R fusion protein in *Xenopus laevis* oocytes. GFP fluorescence was observed, using a confocal laser scanning microscope, and relative expression levels of exogenous mIP_3R1 could be elucidated based on the GFP fluorescence intensity. Subcellular localization of GFP-mIP_3R1 can be monitored according to the GFP fluorescence pattern. It proved to be like that of the endogenous *Xenopus* IP_3R (xIP_3R).[12]

[2] T. Furuichi and K. Mikoshiba, *J. Neurochem.* **64,** 953 (1995).

[3] J. Meldolesi, L. Madeddu, and T. Pozzan, *Biochim. Biophys. Acta* **1055,** 130 (1990).

[4] H. Otsu, A. Yamamoto, N. Maeda, K. Mikoshiba, and Y. Tashiro, *Cell Struct. Funct.* **15,** 163 (1990).

[5] T. Satoh, C. A. Ross, A. Villa, S. Supattopone, T. Pozzan, S. H. Snyder, and J. Meldolesi, *J. Cell Biol.* **111,** 615 (1990).

[6] C. A. Ross, S. K. Danoff, M. J. Schell, S. H. Snyder, and A. Ullrich, *Proc. Natl. Acad. Sci. U.S.A.* **89,** 4265 (1992).

[7] T. Pozzan, R. Rizzuto, P. Volpe, and J. Meldolesi, *Physiol. Rev.* **74,** 595 (1994).

[8] D. C. Prasher, V. K. Eckenrode, W. W. Ward, F. G. Prendergast, and M. J. Cormier, *Gene* **111,** 229 (1992).

[9] J. M. Matheson, A. Miyawaki, A. Muto, T. Inoue, and K. Mikoshiba, *Biomed. Res.* **17,** 221 (1996).

[10] L. G. Sayers, A. Miyawaki, A. Muto, H. Takeshita, A. Yamamoto, T. Michikawa, T. Furuichi, and K. Mikoshiba, *Biochem. J.* **323,** 273 (1997).

[11] T. Furuichi, S. Yoshikawa, A. Miyawaki, K. Wada, N. Maeda, and K. Mikoshiba, *Nature (London)* **342,** 32 (1989).

[12] S. Kume, A. Muto, J. Aruga, T. Nakagawa, T. Michikawa, T. Furuichi, S. Nakade, H. Okano, and K. Mikoshiba, *Cell* **73,** 555 (1993).

Procedures

Green Fluorescent Protein Expression Plasmids

pMT3 (Genetics Institute, Cambridge, MA)[13,14] carries the adenovirus major late promoter and some sequence elements that increase mRNA stability and translatability,[13,14] and provides a system for expressing high levels of exogenous proteins in *Xenopus laevis* oocytes by direct nuclear injection.[14] We used this system to examine expression of exogenous GFP and IP_3R proteins in *Xenopus* oocytes.

pMT3-S65T: The coding region of GFP-S65T, a brighter form of the fluorophore by replacing chromophoretic Ser65 with Thr (S65T),[15] was amplified by PCR using as a template $pRSET_B$-S65T [GFP-S65T cloned in the *Bam*HI site of $pRSET_B$ (Invitrogen); provided by Drs. R. Heim, A. B. Cubbit and R. Y. Tsien, UCSD], and cloned into the *Not*I and *Sal*I sites of pMT3 as described.[9]

pMT3-S65T-IP_3R(ES): The coding region of GFP-S65T amplified by PCR using $pRSET_B$-S65T as a template and the IP_3R1(ES), the *Eco*RI fragment (named ES fragment; amino acids 2,216–2,749) of mouse type 1 IP_3R (mIP_3R1),[11] were fused in frame and cloned into the *Not*I and *Kpn*I site of pMT3.[10] The resultant pMT3-S65T-IP_3R(ES) contained 6 additional amino acids (Gly-Ser-Pro-Gly-Leu-Gln) which linked S65T to IP_3R(ES).

pMT3-IP_3R1: The *Sal*I fragment of the full length mIP_3R1 cDNA was cloned into the *Sal*I site of pMT3.[9]

Procurement and Microinjection of Xenopus laevis Oocytes

Adult female *Xenopus laevis* (purchased from Hamamatsu Seibutsu Company, Shizuoka, Japan) were subjected to hypothermia (by placing them on ice for approximately 20 min). Ovarian fragments were surgically removed, separated into small clumps, and washed in modified Barth's solution (MBS)[16] [88 m*M* NaCl, 1 m*M* KCl, 2.4 m*M* $NaHCO_3$, 10 m*M* HEPES (pH 7.5), 0.82 m*M* $MgSO_4$, 0.33 m*M* $Ca(NO_3)_2$, 0.41 m*M* $CaCl_2$; passed through a 0.22 μm Millex-GV filter (Millipore)]. Fully grown stage VI oocytes (1.2–1.3 mm diameter) were obtained by manual defolliculation

[13] R. J. Kaufman, M. V. Davies, V. K. Pathak, and J. W. B. Hershey, *Mol. Cell. Biol.* **9,** 946 (1989).

[14] A. G. Swick, M. Janicot, T. Cheneval-Kastelic, J. C. McLenithan, and M. D. Lane, *Proc. Natl. Acad. Sci. U.S.A.* **89,** 1812 (1992).

[15] R. Heim, A. B. Cubitt, and R. Y. Tsien, *Nature* (*London*) **373,** 663 (1995).

[16] A. Colman, "Transcription and Translation: A Practical Approach," pp. 271–302. IRL Press, Oxford, 1984.

using watchmakers forceps and a stereoscopic microscope, then were maintained in MBS at 18° until microinjection.

To prepare microinjection glass needles with gradual shanks, a borosilicate glass rod (G-1, Narishige, Japan) was pulled out with a micropipette puller (PN-3, Narishige) and the resultant tip was broken obliquely into a 6 μm outside diameter by gently bumping against a tip of microforge (MF-79, Narishige) or something else, e.g., tip of a forceps. Microinjection of individual oocytes was done using a stereoscopic microscope equipped with a double arm fiber optic system and a microinjection glass needle connected to a nitrogen air pressure (10 kgf/cm^2) microinjector (IM-300, Narishige) and a micromanipulator (M-152, Narishige). Before use of the glass needles, calibration of microinjection volumes was empirically estimated by the measurement of water drop sizes made on the tip of a needle, using a stereoscopic microscope with a micrometer while adjusting the pressure and time levels of the microinjector; drop volumes 1, 2, 5 and 10 nl, respectively, corresponded to diameters 124, 156, 212 and 267 μm. One to two microliters of plasmid DNA dissolved in injection buffer (88 m*M* NaCl, 1 m*M* KCl, 15 m*M* Tris-HCl, pH 7.5; filtrated through a 0.22 μm filter) placed onto Parafilm M (American National Can) was drawn into a needle [or backfilled with a Microloader (Eppendorf)]. Oocytes were immobilized by placing their animal hemisphere diagonally upward on a stainless-wire gauze (with approximately 1 mm mesh size, a little smaller than oocyte diameter) in a plastic petri dish containing MBS. Nuclei are generally present underneath the oocyte membrane of the animal pole. The glass needle set diagonal to the ground, that is, perpendicular to the oocyte pole, was microinjected into the nucleus of the oocyte by impaling the animal hemisphere (approximately 1–2 ng; e.g., 1–2 nl of 1 mg/ml DNA solution). All oocytes were maintained at 18° in MBS during all microinjection procedures and in MBS containing antibiotics (50 units/ml penicillin and 50 μg/ml streptomycin; filtrated through a 0.22 μm filter) after microinjection.

Fluorescence Measurements Using Confocal Microscopy

GFP fluorescent images after microinjection were monitored on either a Carl Zeiss laser scanning confocal microscope system (LSM 410) adapted to a Carl Zeiss inverted microscope Axiovert 135 TV (objective lens, ×10, ×20, ×40) using the Carl Zeiss LSM software package, or a Bio-Rad laser scanning confocal microscope system (MRC-500) adapted to an Olympus inverted microscope IMT-2 (objective lens, ×4, ×10) using CoMOS and TCSM software package (Bio-Rad). The focus of light was adjusted to the surface of the oocyte placed on a glass bottom plastic dish (35 mm diameter;

Mat Tek Corp., Ashland MA) and the confocal aperture was fully opened to maximize the collection of light. Conditions for FITC observation (light of excitation wavelength 488 nm and emission 510 nm, through a band path or long path filter) were utilized to visualize GFP fluorescence from microinjected oocytes. Oocytes positive in GFP fluorescence were selected for further experiments. Oocytes microinjected with vector pMT3 alone (vector control oocytes) were also monitored for fluorescence. All fluorescence measurements were carried out at room temperature.

GFP fluorescence from individual oocytes was monitored for one week after microinjection. Ten images were averaged by Kalman algorithm for each measurement, and relative fluorescence levels were determined by measuring the average value over the entire area of each oocyte, using the program NIH Image (NTIS, Springfield, VA).

Fixation of Oocytes for Confocal Imaging

Oocytes positive in fluorescence were fixed in 4% (w/v) paraformaldehyde in PBS for 2 hr at 4° and then washed in PBS. The oocytes were then manually sliced in half under a stereoscopic light microscope using a razor blade (thickness 0.1 mm) and along the plane perpendicular to the boundary between the animal and vegetal hemispheres of the oocyte. This enabled both hemispheres to be visualized using confocal microscopy. The sliced oocytes were then mounted on a *glass coverslip* and visualized for GFP fluorescence by confocal microscopy using a ×10, ×20 and ×40 objective lens.

Preparation of Xenopus Oocyte Microsomes

Microsomes of oocytes were prepared by a method modified from Parys *et al.*[17] Approximately 20 oocytes were homogenized in a glass-Teflon Potter homogenizer containing 4 vol of ice-cold buffer A [50 m*M* Tris-HCl (pH 7.25), 250 m*M* sucrose, 1 m*M* EGTA, 1 m*M* DTT, 0.1 m*M* PMSF, 10 μM leupeptin, and 1 μM pepstatin A]. The homogenate was centrifuged at 4500*g* for 15 min at 2° and the resulting supernatant was recovered. The pellet contained yolk and melanosomes.[17] The supernatant was centrifuged at 160,000*g* for 1 hr at 2° to sediment the microsomal fraction. The microsomal pellet was resuspended in ice-cold buffer B (20 m*M* Tris-HCl [pH 7.25], 300 m*M* sucrose, 1 m*M* EGTA, 1 m*M* DTT, 0.1 m*M* PMSF, 10 μM leupeptin, 1 μM pepstatin A), placed in liquid nitrogen and stored at −80°

[17] J. B. Parys, S. W. Sernett, S. DeLisle, P. M. Snyder, M. J. Welsh, and K. P. Campbell, *J. Biol. Chem.* **267,** 18776 (1992).

until use. Protein concentrations were measured using the Bio-Rad Protein Assay Kit according to the manufacturer's instructions.

Electrophoresis and Immunoblot Analysis

Microsomal and soluble fractions were analyzed by SDS–polyacrylamide gel electrophoresis (SDS–PAGE) (7.5%) according to the method of Laemmli[18] (approximately 10 μg protein per well). Following electrophoresis, proteins were transferred to a nitrocellulose membrane (Hybond ECL, Amersham) overnight at 2°, then the membranes were blocked for 1 hr at room temperature in 5% skimmed milk in 0.1% Tween 20/PBS. The nitrocellulose sheet was then washed in 0.1% Tween 20/PBS and the primary antibody, monoclonal antibody (mAb) 18A10 (raised against mouse cerebellar IP_3R1[19]), was applied to the sheet for 1 hr at room temperature. After washing the nitrocellulose membrane by gentle shaking in 0.1% Tween 20/PBS 3 times for 10 min each, the membrane was incubated with a secondary antibody (horseradish peroxidase conjugated with anti-rat IgG) for 1 hr at room temperature. The membrane was then washed 3 times (10 min each) in 0.1% Tween 20/PBS and bound the mAb18A10, which is specific for exogenous mIP_3R1, and at an even higher concentration very weakly cross-reacted with endogenous xIP_3R, and was visualized using the Enhanced Chemi-Luminescence system (ECL, Amersham).

Use of Green Fluorescent Protein for Evaluation of Relative Expression Levels of Exogenous Protein in *Xenopus* Oocytes

There was heterogeneity in the fluorescence intensity of oocytes microinjected with pMT3-S65T (Fig. 1). Fluorescence from GFP-S65T expressed was visible in some oocytes within 24 hr. By days 2 and 3 after microinjection, oocytes expressing GFP-S65T were easily identified as bright and dimly fluorescent ones. The fluorescence signal continued to increase over a 7-day period as GFP-S65T accumulated. Because GFP-S65T fluorescence is resistant to photobleaching, bright field and epifluorescence illumination can be used to identify positively expressing oocytes. Using conventional optics for FITC, GFP-S65T expression is observed as bright green fluorescence and non-fluorescence indicates a failure of expression or failure of injection into the nucleus. State of the art recombinant GFPs, such as enhanced or humanized GFPs, provided a stronger fluorescence in this amphibian expression system (data not shown).

When pMT3-S65T is co-injected with pMT3-IP_3R1, GFP-S65T fluorescence can be used not only to identify oocytes capable of transcription and

[18] U. K. Laemmli, *Nature (London)* **227,** 680 (1970).

[19] N. Maeda, M. Niinobe, and K. Mikoshiba, *EMBO J.* **9,** 61 (1990).

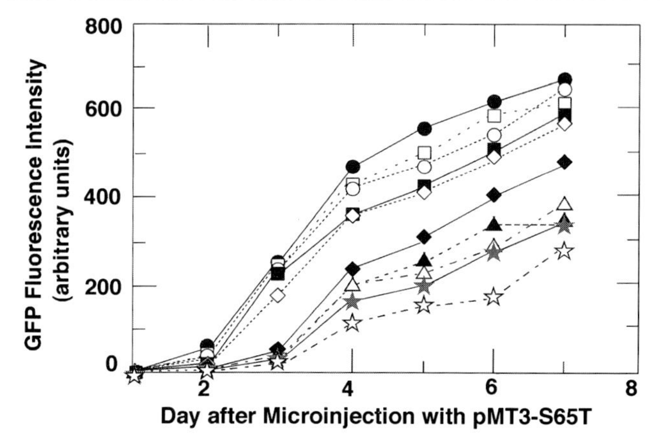

FIG. 1. Temporal changes in GFP-S65T fluorescence from 10 individual oocytes following pMT3-S65T microinjection. Each plot represents data from 10 individual oocytes.

translation of the IP_3R1 protein but also to estimate the relative expression level.[9] The correlation coefficient of fluorescence intensity from GFP-S65T and expression level of IP_3R1 protein estimated through Western blotting was 0.93 (n = 5) (Fig. 2). This high correlation makes it possible to evaluate relative levels of IP_3R expression in fluorescent oocytes. These results indicate that one can select oocytes highly expressing ion channels such as IP_3R1 or other molecules by co-injecting pMT3-S65T.

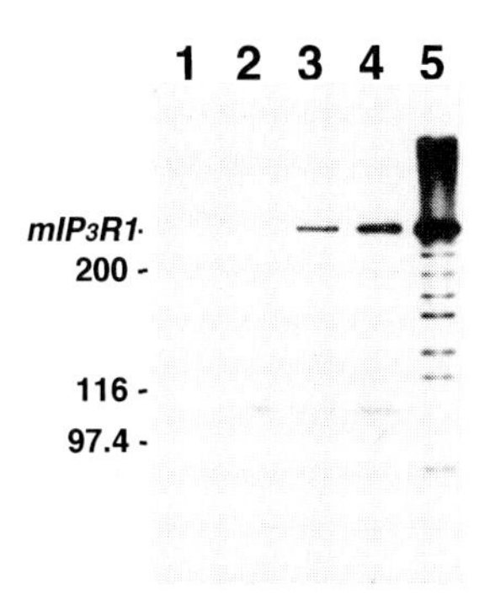

FIG. 2. Western blot analysis of expression levels of IP_3R1 in differently fluorescing oocytes, co-injected with pMT3-S65T and pMT3-IP_3R1. IP_3R1-specific monoclonal antibody 18A10, which barely cross-reacted with xIP_3R, was used. Lane 1, 5 μg of control microsome from uninjected oocytes; lanes 2–4, 5 μg of microsomes from oocytes prepared 3 days after co-injection of pMT3-S65T and pMT3-IP_3R1 exhibiting no fluorescence, weak fluorescence, and bright fluorescence, respectively; lane 5, 0.5 μg of mouse cerebellar microsomes as positive control. Molecular size markers (in kDa) are indicated on the left.

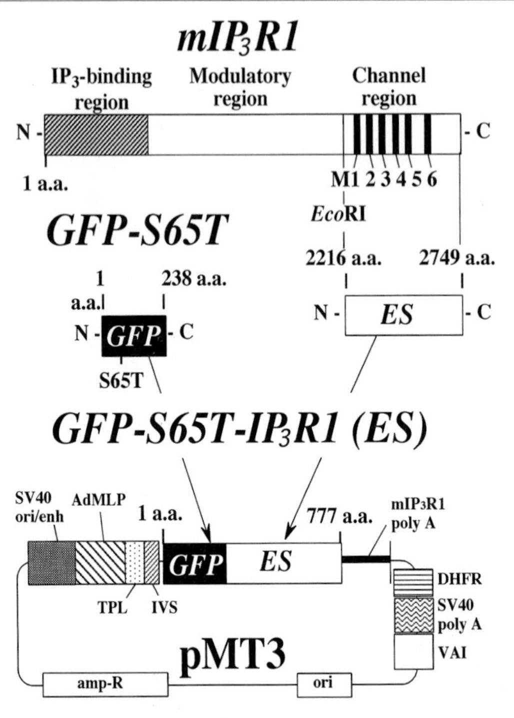

FIG. 3. Chimeric GFP-IP_3R1(ES) fusion protein. GFP-S65T was fused to the N-terminal end of a truncated mouse IP_3R1 protein (amino acids 2216–2749), IP_3R1(ES), which contained all six of the putative transmembrane-spanning domains (M1–6) and the following C-terminal tail, and cloned into the *Xenopus* oocyte expression vector pMT3, which is composed of the SV40 origin and enhancer element (SV40 ori/enh), the adenovirus major late promoter (AdMLP) containing the majority of the tripartite leader (TPL) present on adenovirus late mRNAs, a hybrid intervening sequence (IVS), a multicloning site, a DHFR-coding region, the SV40 early polyadenylation signal (SV40 poly A), and the adenovirus VAI gene.

Green Fluorescent Protein Localization of Intracellular Inositol 1,4,5-Triphosphate Receptor–Ca^{2+} Stores in *Xenopus* Oocytes

A chimeric fusion protein between GFP-S65T protein and truncated mouse IP_3R1 (mIP_3R1) mutant protein, the ES region (amino acids 2,216–2,749) containing the putative transmembrane channel domain composed of six membrane-spanning segments (M1–6) and the following C-terminal tail region, was expressed in *Xenopus* oocytes by nuclear microinjection with pMT3-S65T-IP_3R1(ES) (Fig. 3).[10] This system facilitated monitoring of the intracellular distribution of IP_3R1-Ca^{2+} pools, as GFP-S65T fluorescence within the cell. Following microinjection, the oocytes were incubated

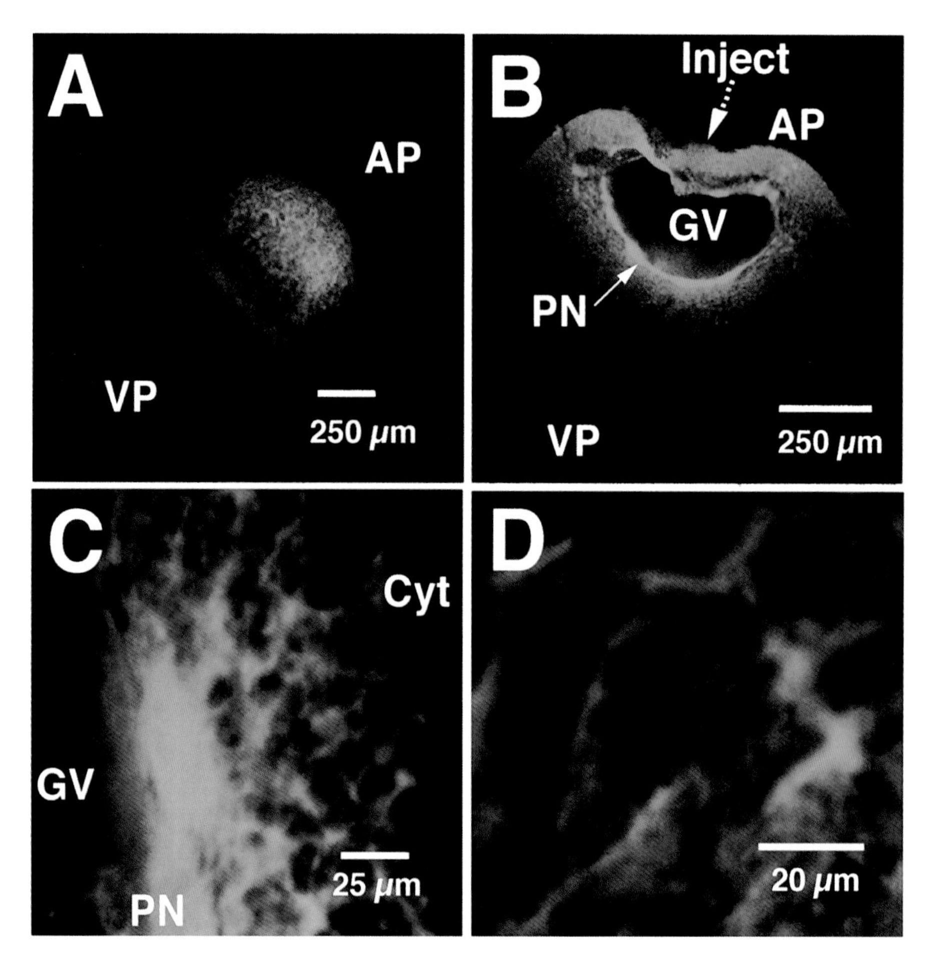

FIG. 4. Subcellular localization of GFP–IP_3R1(ES) in *Xenopus* oocytes as determined by fluorescence confocal microscopy. (A) Observation of GFP-S65T fluorescence in an intact, fully grown stage VI *Xenopus* oocyte, as viewed through a ×10 objective lens 3 days after microinjection with pMT3-S65T-IP_3R1(ES); enriched fluorescence under the animal pole (AP) of the oocyte is demonstrated, with little detectable fluorescence under the vegetal pole (VP). (B) Intracellular observation of GFP-S65T fluorescence from an oocyte microinjected with pMT3-S65T-IP_3R1(ES), viewed using a ×10 objective lens. Fluorescence is apparent from the AP (dashed arrow) into the nucleus, sliced in half after fixation in 4% paraformaldehyde. Intense areas of fluorescence can be observed in the perinuclear region (PN) around the germinal vesicle (GV) and in regions of the AP under the plasma membrane of the oocyte. Little fluorescence can be observed inside the VP. (C) Enlargement of a portion of the PN (×40 objective lens). Note the intense fluorescence observed in the PN surrounding the GV, which appears to exhibit continuity with a reticular network of fluorescence in the cytoplasm (Cyt) distant from the GV. (D) Enlargement of the reticular fluorescence pattern observed under the AP (×40 objective lens). The white arrow depicts regions of intense GFP-S65T fluorescence, possibly representing intracellular membranes that are the site of GFP-IP_3R1(ES) targeting (probably IP_3-sensitive Ca^{2+} pools).

for 3 days in MBS and observed under a confocal microscope. The expression of GFP-IP_3R1(ES) protein (calculated molecular weight 77,000) in oocytes microinjected with pMT3-S65T-IP_3R1(ES) positive in GFP-S65T fluorescence was confirmed by the presence of an intense band of approximately 80 kDa in Western blot analysis (data not shown). Using a confocal microscope, we detected the localization of GFP-S65T fluorescence on reticular structures under the animal pole (AP) of the oocyte microinjected with pMT3-S65T-IP_3R1(ES), with increased intensity in the perinuclear region (PN) and just beneath the plasma membrane (Fig. 4, see color insert).

The finding that expressed exogenous GFP-IP_3R1(ES) is selectively localized on the ER network, like endogenous *Xenopus* IP_3R1 (xIP_3R)[12,17] provides circumstantial evidence that GFP-IP_3R1(ES) and xIP_3R share similar targeting/retention mechanisms and are localized on the same Ca^{2+} store. Moreover, this study using the GFP-tagged IP_3R1(ES) paves the way for subsequent studies on not only the structure and function of mIP_3R1 necessary for ER targeting, but will facilitate studies on *in vivo* dynamics of IP_3R1-Ca^{2+} pools which apparently differ before and after fertilization.[12]

Acknowledgments

This study was supported by grants from the Ministry of Education, Science, Sports and Culture of Japan, the Ministry of Health and Welfare of Japan, and RIKEN (the Institute of Physical and Chemical Research).

[21] Expression of Green Fluorescent Protein in Transgenic Mice

By TATSUYUKI TAKADA, KENICHI YOSHIDA, KENJI NAKAMURA, KAZUKI NAKAO, GOZOH TSUJIMOTO, MOTOYA KATSUKI, and SUMIO SUGANO

Introduction

Developments in the making of genetically engineered mice offer new tools for studying the mechanism of gene expression *in vivo*.[1] The use of

[1] B. Hogan, R. Beddington, F. Costantini, and E. Lacy, "Manipulating the Mouse Embryo: A Laboratory Manual," 2nd Ed. Cold Spring Harbor Laboratory Press, Cold Spring Harbor, New York, 1994.

0076-6879/99 $30.00

reporter genes (luciferase, β-galactosidase, chloramphenicol acetyltransferase, alkaline phosphatase, alcohol dehydrogenase) has also contributed to monitoring gene expression and to analyzing promoter activity in mouse preimplantation embryos and transgenic mice.[2–11] Because these reporters are enzymes, it is necessary to load substrates inside the cell and to assay reporter activity. Green fluorescent protein (GFP), which is responsible for the bioluminescence of the jellyfish *Aequorea victoria*,[12,13] has been widely used as a new reporter in living cells.[14–17] GFP has a major advantage over the other reporters, because GFP is an autofluorescent protein and does not require any substrates or cofactors for its detection. This allows real-time observation of gene expression in living cells and even in organisms in a noninvasive manner. Purified GFP is a highly stable protein and has a quantum yield of 0.72–0.85,[16] suggesting GFP could be a suitable marker, especially for screening of living embryos and embryonic stem (ES) cells.

Application of GFP as a marker of gene integration is quite useful for the efficient production of transgenic animals. Transgenic animals have been made by microinjection of DNA solution (transgene) into the pronuclei of one-cell embryos, followed by transfer to pseudopregnant recipient animals.[1] However, the proportion of transgenic animals is small.[18] This low efficiency has been a serious problem, especially for the production of

[2] S. A. Lira, R. A. Kinloch, S. Mortillo, and R. M. Wassarman, *Proc. Natl. Acad. Sci. U.S.A.* **87,** 7215 (1990).

[3] M. Schickler, S. A. Lira, R. A. Kinloch, and P. M. Wassarman, *Mol. Cell. Biol.* **12,** 120 (1992).

[4] A. G. DiLella, D. A. Hope, H. Chen, M. Trumbauer, R. J. Schwartz, and R. G. Smith, *Nucleic Acids Res.* **16,** 4159 (1988).

[5] J.-Y. Nothias, M. Miranda, and M. L. DePamphilis, *EMBO J.* **15,** 5715 (1996).

[6] C. Bonnerot, D. Rocancourt, P. Briand, G. Grimber, and J. F. Nicolas, *Proc. Natl. Acad. Sci. U.S.A.* **84,** 6795 (1987).

[7] H. Le Mouellic, Y. Lallemand, and P. Brulet, *Proc. Natl. Acad. Sci. U.S.A.* **87,** 4712 (1990).

[8] A. L. Joyner, *BioEssays* **13,** 649 (1991).

[9] J. Sharpe, S. Nonchev, A. Gould, J. Whiting, and R. Krumlauf, *EMBO J.* **17,** 1788 (1998).

[10] K. Hanaoka, M. Hayasaka, T. Uetsuki, A. Fujisawa-Sehara, and Y. Nabeshima, *Differentiation* **48,** 183 (1991).

[11] L. L. Nielsen, and R. A. Pederson, *J. Exp. Zool.* **257,** 128 (1991).

[12] D. C. Prasher, V. K. Eckenrode, W. W. Ward, F. G. Prendergast, and M. J. Cormier, *Gene* **111,** 229 (1992).

[13] S. Inouye and F. I. Tsuji, *FEBS Lett.* **341,** 277 (1994).

[14] M. Chalfie, Y. Tu, G. Euskirchen, W. W. Ward, and D. C. Prasher, *Science* **263,** 802 (1994).

[15] S. Wang and T. Hazelrigg, *Nature* (*London*) **369,** 400 (1994).

[16] A. B. Cubitt, R. Heim, S. R. Adams, A. E. Boyd, L. A. Gross, and R. Y. Tsien, *Trends Biochem. Sci.* **20,** 448 (1995).

[17] T. Misteli and D. L. Spector, *Nature Biotech.* **15,** 961 (1997).

[18] R. L. Brinster, H. Y. Chen, M. E. Trumbauer, M. K. Yagle, and R. D. Palmiter, *Proc. Natl. Acad. Sci. U.S.A.* **82,** 4438 (1985).

transgenic livestock.[19] The gene integration event frequently happens at an early stage of development; thus, if we can select a transgenic embryo before transfer to the recipients, we would be able to produce transgenic animals selectively. Several attempts have been made to select transgenic embryos after microinjection, using luciferase or neomycin resistance.[20,21] None has worked successfully so far. Normal development of preimplantation embryos is enhanced by minimization of stress, such as that associated with biopsy or the addition of substrate in culture medium during *in vitro* culture. We used GFP as a selection marker for transgenic embryos and were able to produce transgenic mice with high efficiency. The noninvasive nature of GFP provides a unique opportunity to monitor gene integration by just observing it under excitation light.

Transgenic mice can be generated through ES cells. Transgenic mice generated by this method tend to have a low copy number of transgenes compared with those made using pronuclear injection. We also tested GFP for the selection of ES cells and succeeded in producing chimeras and transgenic mice through GFP-expressing ES cells. In this case, chimeras and transgenic mice were easily identified by monitoring GFP expression in a noninvasive manner. This simple procedure for detecting transgenic pups may greatly reduce the time and effort required to establish transgenic mouse lines.

This chapter describes a procedure for the efficient production of transgenic mice through zygotes and ES cells, using GFP as a marker of gene integration. We also describe the method used to detect expression of GFP in these transgenic mice.

Expression of Green Fluorescent Protein in Preimplantation Mouse Embryos

Superovulation

Mice are maintained under a constant light–dark cycle of 12–12 or 14–10 hr (8 AM–8 PM light, in our facility) at 22–25°. Fertilized one-cell embryos are obtained from female mice by superovulation. The mouse strain greatly influences the number of embryos obtained by superovulation, tolerance of microinjection, and the ensuing development *in vitro*.[1] Hybrid

[19] R. E. Hammer, V. G. Pursel, C. E. Rexroad, Jr., R. J. Wall, D. J. Bolt, K. M. Ebert, R. D. Palmiter, and R. L. Brinster, *Nature (London)* **315,** 680 (1985).

[20] E. M. Thompson, P. Adenot, F. L. Tsuji, and J.-P. Renard, *Proc. Natl. Acad. Sci. U.S.A.* **92,** 1317 (1995).

[21] N. Tada, M. Sato, K. Hayashi, K. Kasai, and S. Ogawa, *Transgenics* **1,** 535 (1995).

strains such as BDF1 (C57BL/6 × DBA/2) or B6C3F1 (C57BL/6 × C3H/He) respond to hormone treatment well (20–30 embryos per female), and a high percentage (>80%) of embryos develop to blastocyst stage in *in vitro* culture. Generally, F_1 or F_2 zygotes are easier and more efficient in generating transgenic mice than are the inbred strain. If an inbred strain is preferred, the FVB/N strain is recommended for transgenic mouse production, because FVB/N embryos have large and prominent pronuclei and develop to the blastocyst stage efficiently *in vitro.*[22] We use 5- to 10-week-old BDF1 females. They are intraperitoneally injected with 5 IU of pregnant mare's serum (PMS, 50 IU/ml in 0.9% NaCl) at 4–5 PM, followed 47–48 hr later by injection of 5 IU of human chorionic gonadotropin (hCG, 50 IU/ml in 0.9% NaCl). Aliquots of PMS and hCG solutions can be stored at −20° for several months. After hCG injection, the individual female is transferred into a cage with a BDF1 male for mating.

Collection of Mouse Embryos

The next morning, the female mice are examined for the presence of a vaginal plug and are sacrificed by cervical dislocation at 17–20 hr post-hCG injection. The mice are dissected and oviducts are collected in M2 medium (Sigma, St. Louis, MO) supplemented with hyaluronidase (300 μg/ml; Sigma) in a 35-mm culture dish. The swollen ampulla, which is located near the infundibulum (entrance of the oviduct), is identified with a dissecting stereomicroscope (×20–30 magnification). A clump of embryos can be seen through the wall of the swollen ampulla; both ends of the ampulla are held to the bottom of the culture dish with a pair of #5 watch-maker's forceps, and then its wall is torn with a 27- to 30-gauge needle attached to a disposable syringe. The clump of embryos surrounded by cumulus cells will easily flow out into the medium. Embryos are released from cumulus cells by incubation of the culture dish at 37° for several minutes. Do not expose cumulus-free embryos to hyaluronidase for a long time. Cumulus cell-free embryos are then collected using a mouth-controlled glass pipette (150- to 200-mm internal diameter) and are washed three times with M2 medium to remove hyaluronidase, cumulus cells, and debris. Transfer these embryos to microdrops of M16 medium (Sigma) under mineral oil and culture in a humidified incubator with 5% CO_2 and 95% air or 5% CO_2, 5% O_2, and 90% N_2 at 37° until microinjection.

[22] M. Taketo, A. C. Schroeder, L. E. Mobraaten, K. B. Gunning, G. Hanten, R. R. Fox, T. H. Roderick, C. L. Stewart, F. Lilly, C. T. Lansen, and P. A. Overbeek, *Proc. Natl. Acad. Sci. U.S.A.* **88,** 2065 (1991).

Green Fluorescent Protein Expression Vector

Wild-type GFP has a major excitation peak at 395 nm.[13] Exposure of living cells to such long-wave ultraviolet light may cause DNA damage to embryos. Furthermore, the excitation coefficient is not high, suggesting the intensity of fluorescence may not emit a signal strong enough to be detected. Amino acid mutation and codon optimization have been carried out extensively to change the excitation and emission wavelengths, brightness, and thermal stability.[23-28] We use the S65T mutant GFP gene, because its excitation wavelength is shifted to 489 nm and generates brighter fluorescence, four- to sixfold that of wild type.[23] We constructed a GFP expression vector that has the S65T mutant GFP (GFP-S65T) gene under the control of the cytomegalovirus immediate-early (CMV-IE) enhancer and the human elongation factor 1α promoter (CE promoter) needed for high-level expression in mammalian preimplantation embryos. The human elongation factor 1α promoter contains a part of the 5′ flanking region, complete exon 1, intron 1, and a part of exon 2 for efficient expression.[29]

Preparation of Injection Pipette

A glass capillary, 1 mm (o.d.) × 0.6 mm (i.d.) × 90 mm (GD-1; Narishige, Tokyo, Japan), is pulled with a micropipette puller (P-87; Sutter, San Rafael, CA). The procedure used for pulling is heat, 620; pull, 120; velocity, 70; time, 220; pressure, 300. The back end of the capillary is then heat polished and connected to a 10-ml disposable syringe by means of polyethylene tubing. The tip is dipped in 25% hydrofluoric acid (HF) for a short time, during which positive air pressure is provided from the syringe to make a tiny opening at the tip. The tip is dipped twice in water (filtered through a 0.22-μm pore size filter) to remove HF in the same way as HF dipping. This injection pipette may be bent at a position about 5 mm from the tip using a microforge at an angle (about 30°) to give horizontal movement to the bottom of the injection chamber.

Preparation of Holding Pipette

One of the pulled glass capillaries is cut at a position to give an outside diameter of approximately 80–100 nm, using a small glass bead melted on

[23] R. Heim, D. C. Prasher, and R. Y. Tsien, *Proc. Natl. Acad. Sci. U.S.A.* **91,** 12501 (1994).

[24] R. Heim, A. B. Cubitt, and R. Y. Tsien, *Nature (London)* **373,** 663 (1995).

[25] S. Delagrave, R. E. Hawtin, C. M. Silva, M. M. Yang, and D. C. Youvan, *Bio/Technology* **13,** 151 (1995).

[26] A. Crameri, E. A. Whitehorn, E. Tate, and W. P. C. Stemmer, *Nature Biotech.* **14,** 315 (1996).

[27] R. Heim and R. Y. Tsien, *Curr. Biol.* **6,** 178 (1996).

[28] B. P. Cormack, R. H. Valdivia, and S. Falkow, *Gene* **173,** 13 (1996).

[29] D. W. Kim, T. Uetsuki, Y. Kaziro, N. Yamaguchi, and S. Sugano, *Gene* **91,** 217 (1990).

the platinum filament of a microforge (MF-1; De Fonbrune, St. Louis, MO). The tip is then positioned close to the glass bead, and turns on the filament to polish the tip until the internal diameter becomes approximately 15–20 μm. The holding pipette is also bent 2 mm from the tip, using the microforge.

Injection Chamber

With Nomarski differential interference contrast (DIC) optics, we prepare an injection chamber using a Flexiperm (W. C. Heraeus GmbH, Hanau, Germany) and a glass slide. The inside partition of the Flexiperm is cut out before use, leaving the outer frame. The height of the Flexiperm can also be adjusted as desired by cutting. The siliconized glass slide is then wiped with 70% ethanol and attached to the modified Flexiperm, making certain the Flexiperm adheres firmly to the slide. This chamber provides a wider working area and clearer viewing than is provided by a depression slide. Micromanipulation is carried out in a microdrop (10 μl) of M2 medium on the glass slide of this chamber, which is covered with mineral oil. In the case of Hoffman optics, an inverted tissue culture lid is convenient.

Preparation of DNA Fragment for Injection

We normally use *Escherichia coli* DH5α for propagation of the plasmid. Supercoiled plasmid DNA purified by conventional cesium chloride (CsCl) dye density gradient[30] gives consistent results. Plasmid prepared by use of a commercial kit [plasmid kit (Qiagen Valencia, CA) or Wizard Plus DNA purification system (Promega, Madison, WI)] can also be used. The plasmid should be linearized by restriction enzyme digestion before injection, because the integration frequency of linear DNA is significantly higher than that of circular DNA.[18] It is recommended that the plasmid sequence flanking the transgene be minimized, because the presence of plasmid sequence sometimes may have a negative effect on transgene expression in transgenic mice.[31,32] The choice of restriction enzyme is important not only for the separation of the transgene insert from the vector but also for increasing the efficiency of transgene integration. The linear DNA molecule with blunt ends shows a lower integration frequency than does DNA with sticky ends.[18] After restriction enzyme digestion, the transgene fragment is separated

[30] J. Sambrook, E. F. Fritsch, and T. Maniatis, "Molecular Cloning: A Laboratory Manual," 2nd Ed. Cold Spring Harbor Laboratory Press, Cold Spring Harbor, New York, 1989.

[31] K. Chada, J. Magram, K. Raphael, G. Radice, E. Lacy, and F. Costantini, *Nature* (*London*) **314,** 377 (1985).

[32] T. M. Townes, J. B. Lingrel, H. Y. Chen, R. L. Brinster, and R. D. Palmiter, *EMBO J.* **4,** 1715 (1985).

from the vector backbone by agarose (SeaKem GTG; FMC, Rockland, ME) gel electrophoresis. The gel piece containing the transgene fragment is excised from the ethidium bromide-stained gel, and the DNA is either electroeluted followed by phenol–chloroform extraction and ethanol precipitation or purified using the glass milk procedure (Geneclean II; Bio 101, La Jolla, CA). DNA purity affects the efficiency of gene integration and embryo survival; contamination by agarose, organic solvents, and salts should be avoided.

Injection Buffer

The DNA fragments are dissolved in 10 m*M* Tris-HCl, pH 7.5, containing 0.1 m*M* EDTA and diluted to give a final concentration of 2–5 ng/μl with either the same buffer[18] or Dulbecco's phosphate-buffered saline [DPBs(–)] without magnesium and calcium, containing 0.1 m*M* EDTA. The DNA concentration is quantified using a spectrophotometer or by fluorescence emitted by ethidium bromide intercalation. The fluorescence method is particularly useful when DNA is prepared by the glass milk procedure, because only a small volume is required—as little as 1–5 ng of DNA can be detected.[30] After dilution, the DNA aliquot should be run on an agarose gel to check the size and integrity of the purified DNA fragment. Each aliquot (10 μl) can be stored at 4° or −20°. Before microinjection, an aliquot is centrifuged at 14,000*g* for 15 min at 4° and the upper two-thirds volume is carefully transferred to a new, clean microtube to eliminate particles that might clog the injection pipette. Approximately 0.5 μl of DNA solution is loaded into the back of the injection pipette, using a fine-drawn pasteur pipette or microloader tip (Eppendorf, Hamburg, Germany).

Microscope and Micromanipulator

An inverted microscope with Nomarski differential interference contrast is suitable for pronuclear microinjection. When using DIC optics, microinjection should be done with a glass injection chamber. Although DIC provides finer resolution, some investigators prefer the Hoffman module because microinjection can be done with a disposable plastic dish. We use an IX-70 (Olympus, Norwood, MA) with Hoffman optics, and micromanipulators MO-202 and MO-103 (Narishige) for injection and holding, respectively. Microinjectors IM-6, and IM-5B (Narishige) are used for injecting DNA solution and holding the embryo, respectively.

Microinjection

We perform microinjection under ×200 magnification. GFP transgene is injected into male pronuclei between 22 and 28 hr post-hCG injection.

To be sure of successful injection, it is important to observe obvious swelling of the pronuclei. We usually continue to inject DNA solution until the diameter of each pronucleus expands approximately 1.5- to 2-fold its orignal size. After injection, embryos are cultured in a microdrop of M16 medium under mineral oil in a humidified atmosphere of either 5% CO_2, 5% O_2, and 90% N_2 or 5% CO_2 and 95% air at 37° for further development. For culturing under the multigas condition, it is convenient to use a small multigas incubator (APMW-36; Astec, Fukuoka, Japan) or modular incubator (MIC101; Billups-Rothenberg, Del Mar, CA) purged with premixed gas.

Monitoring of Green Fluorescent Protein Expression

GFP expression is monitored at each developmental stage using either a confocal laser scanning microscope (GB-200; Olympus) with excitation at 488 nm and detection at 500–530 nm or a fluorescence microscope (IX-70; Olympus) with standard fluorescein isothiocyanate (FITC) filter set (BP470–490 and BA515). Embryos are transferred to a microdrop of M2 medium covered with mineral oil (Sigma) in a glass bottom culture dish (P35G; MatTek, Ashland, MA) for observation of GFP fluorescence by microscopy. In our experiment, GFP expression is first observed at the four-cell stage and continues to the blastocyst stage. We have noticed that in some GFP-S65T-expressing morulae expression becomes weak and sometimes even negative as they develop to the blastocyst stage.[33] However, this phenomenon does not seem to be the case when we use a brighter mutant (F64L + S65T) [enhanced GFP (EGFP; Clontech, Palo Alto, CA)], in which codon usage is optimized for human cells. We observe strong GFP expression at the blastocyst stage with this EGFP. Mosaic expression is also observed in some embryos. The pattern of expression and GFP-expressing cells are clearly identified using laser confocal microscopy (Fig. 1). For some unknown reason, embryos that develop abnormally or stop developing express GFP quite strongly.

Selective Production of Transgenic Mice

Normally, the population of transgenic mice is 10–30% of live-born mice. Transgenic mice carrying the GFP gene can be selectively produced by monitoring GFP expression at the expanded blastocyst stage and by transferring only GFP-expressing blastocysts to pseudopregnant females. After microinjection, embryos are cultured in microdrops of M16 medium

[33] T. Takada, K. Iida, T. Awaji, K. Itoh, R. Takahashi, A. Shibui, K. Yoshida, S. Sugano, and G. Tsujimoto, *Nature Biotech.* **15,** 458 (1997).

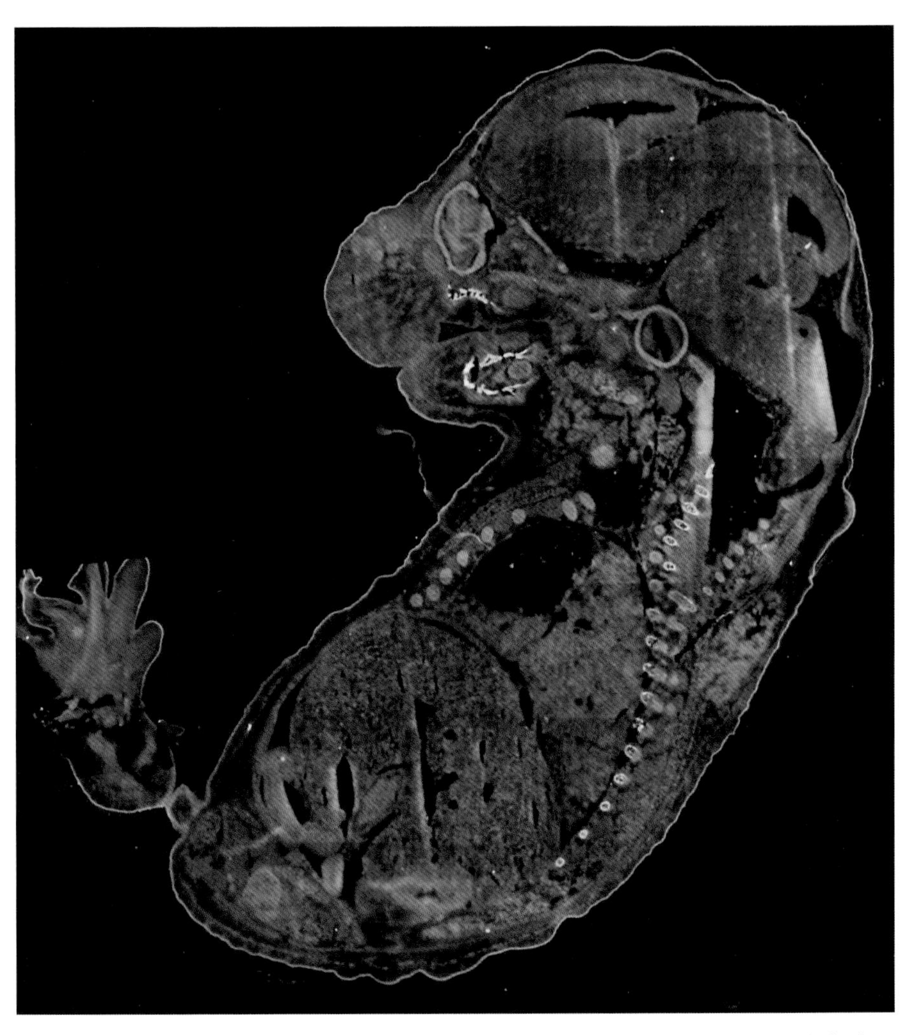

FIG. 2. Expression of GFP in a CE-EGFP transgenic mouse. Longitudinal whole-body frozen section was prepared from EGFP hemizygous F_1 embryo (18 days). GFP expression was observed using a fluorescence microscope with excitation filter (460–490 nm), dichroic mirror (505 nm), and absorption filter (510 nm). [Sectioned by M. Watanabe; photograph provided by courtesy of Olympus Optical Co., Ltd.]

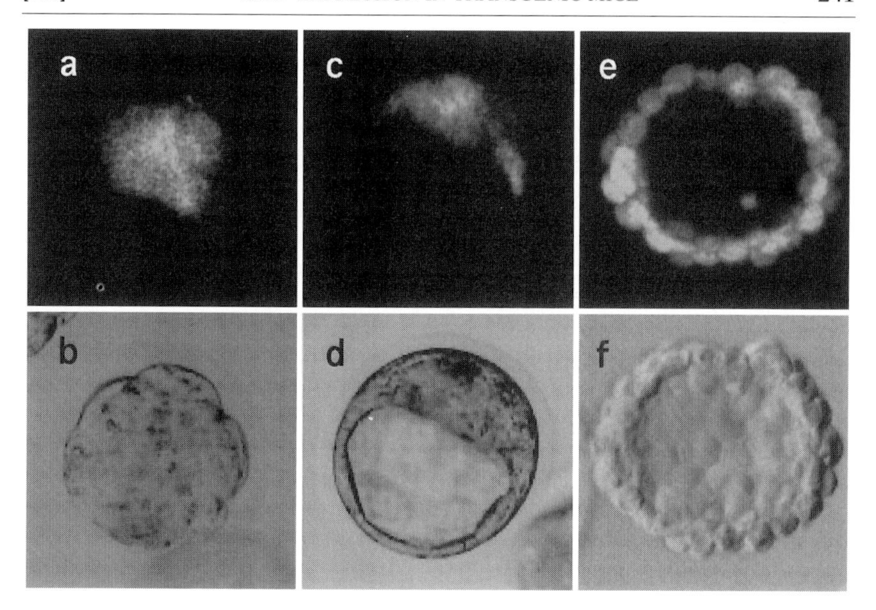

FIG. 1. Confocal image of GFP expression in preimplantation mouse embryos. GFP expression was monitored at morula (a and b), blastocyst (c and d), and hatched blastocyst (e and f) stages, using a confocal laser scanning microscope (GB-200; Olympus) with excitation at 488 nm and detection with a 500- to 530-nm bandpass filter. The confocal images were assembled in IPLab (Signal Analysis, Vienna, VA). (a, c, and e) Confocal flurorescent image; (b, d, and f) differential image. After microinjection of the GFP transgene, GFP expression was observed in preimplantation embryos. The pattern of GFP expression in internal cells of the morula (a), some part of the inner cell mass of the blastocyst (c), and most cells of the hatched blastocyst (e) was clearly visualized by confocal laser scanning microscopy. [Reprinted, with permission, from Takada *et al.*[33]]

in a humidified incubator of 5% CO_2, 5% O_2, and 90% N_2 at 37°. After 4 days of culture, embryos will develop to the expanded blastocyst stage. Several microdrops (3 μl) of M2 medium are prepared on a glass-bottom culture dish and covered with mineral oil; healthy expanded blastocysts are transferred to these microdrops, one embryo per microdrop. GFP expression of individual blastocysts is then monitored by fluorescence microscopy (IX-70; Olympus) under ×200 magnification with a standard FITC filter set (BP470–490 and BA515), and the GFP-expressing blastocysts are separated. We do not pick up embryos with ambiguous GFP expression. GFP-expressing embryos are transferred to day 3 pseudopregnant female mice (ICR); day 1 represents the day vaginal plugs appear. Pups will be born on day 20. If pups are not born on the due date, which happens in the case of small litter sizes, we perform a cesarean section in the afternoon of day 20. It is therefore recommended that foster mothers be prepared.

Usually, at the time of weaning, about 3 weeks of age, tails (5–10 mm) are cut and genomic DNA is prepared. Polymerase chain reaction (PCR) is then carried out using a pair of transgene-specific primers to screen for transgenic mice. In this experiment we analyzed transgenesis using fetuses 11 days after transfer, or newborn pups. The presence of the transgene is analyzed by PCR and Southern blot analysis using their genomic DNA. By this method, we were able to produce transgenic mice with high efficiency, i.e., 67%.[33] In practice, transgenic mice carrying the gene of interest can be produced by linking or coinjecting[34] this GFP gene with the transgene. Similar GFP expression is also observed in rat and bovine blastocysts[33] when the GFP gene is microinjected into the pronuclei of one-cell embryos, suggesting that this method can be directly applied for the efficient production of other transgenic animals.

Construction of Green Fluorescent Protein Transgenic Mice through Embryonic Stem Cells

Vector Construction

The GFP-S65T expression vector (Takada *et al.*[33]) is digested with *Hin*dIII and *Eco*RI. The resultant DNA fragment including the CE promoter, is separated from the rest of the vector by electrophoresis and recovered from an agarose gel. The gel-purified fragment is ligated into *Hin*dIII/*Eco*RI double-digested pEGFP-1 vector (Clontech). Plasmid DNA of the resultant expression vector pCE-EGFP-1 is prepared by CsCl equilibrium dye density gradient purification. To eliminate the pUC portion, purified pCE-EGFP-1 DNA is excised with *Hin*dIII and *Alw*NI. After determining that the digestion is complete, the DNA fragment, including EGFP cDNA under the control of the CE promoter and the G418 resistance gene, is separated from the pUC portion by electrophoresis. Ethidium bromide staining is performed and the linearized DNA fragment corresponding to the expected length is recovered as gel slices. We use gel electroelution for fragment isolation. Slices of gel are placed in a Biotrap BT1000 (Schleicher & Schuell, Keene, NH) containing 0.5× Tris–borate–EDTA (TBE) buffer. Using a microporous membrane set (BT1 and BT2; Schleicher & Schuell) DNA purification is performed. Direct current is passed between the electrodes at 120 V for 4 hr, and the polarity is then reversed at 200 V for about 20 sec to detach the DNA from the surface of the membrane. Electroelution is done at room temperature. We then proceed to phenol–chloroform and then chloroform extraction. After organic extrac-

[34] D. Methot, T. L. Reudelhuber, and D. W. Silversides, *Nucleic Acids Res.* **23,** 4551 (1995).

tion, the DNA is sodium acetate–ethanol precipitated and the pellet is then rinsed once with 70% ethanol. The pellet is allowed to dry for a few minutes, and is then dissolved in 10 m*M* Tris-HCl (pH 7.4)–1 m*M* EDTA (pH 8.0) at about 1 μg/μl. No dialysis of the DNA preparation is performed. A portion of the DNA is rechecked on a minigel before use to ensure that no degradation has taken place.

Electroporation

1. Culture ES cells up to 70–80% confluency on feeder cells (described below). In our experiments, the CCE ES cell line, established from the 129/SvEv strain, is used and cultured essentially as described.[35,36] We use antibiotic-free ES cell medium [Dulbecco's modified Eagle's medium (DMEM, No. 12800; GIBCO-BRL, Gaithersburg, MD), 20% fetal bovine serum, 0.1 m*M* 2-mercaptoethannol (Nakarai, Kyoto, Japan), nonessential amino acids (GIBCO-BRL), and leukemia inhibitory factor (LIF, 103 units/ml; GIBCO-BRL)]. STO cell lines derived from SIHM strains are used as the feeder layer. We use two 60-mm dishes for one experiment.
2. Within 2–3 hr of harvest, change the ES cell medium once.
3. After 2–3 hr, wash the dishes with 4 ml of phosphate-buffered saline without magnesium and calcium [PBS(–)].
4. Add 0.5 ml of trypsin–EDTA solution (0.25%/0.02%; Fujioka, Japan) and incubate at room temperature for 5 min.
5. Stop the trypsinization by adding 4.5 ml of ES cell medium and form a single-cell suspension by pipetting five times.
6. Replate the cells onto a 100-mm gelatinized dish [to gelatinize a dish, add 10 ml of 0.1% gelatin (Sigma) to the plate, then remove the gelatin solution by aspiration] and incubate for 12 min at 37° to allow feeder cells to attach to the plate.
7. Harvest the medium from the dish and collect the ES cells by centrifugation at 200*g* for 5 min at room temperature.
8. Aspirate the supernatant and resuspend the cells in 10 ml of PBS(–).
9. Repeat step 7.
10. Count the cells and adjust the cell concentration to 1×10^7 cells/ml.
11. Mix 1 ml of the cell suspension with 5 μg of linearized vector DNA.
12. Keep the cells at 4° for 5 min.

[35] E. Robertson, A. Bradley, M. Kuehn, and M. Evans, *Nature* (*London*) **323,** 445 (1986).
[36] A. Bradley, in "Teratocarcinomas and Embryonic Stem Cells: A Practical Approach" (E. J. Robertson, ed.), p. 113. IRL Press, Oxford, 1987.

13. Transfer the cells to an electroporation cuvette (flat pack chamber; BTX, San Diego, CA).

14. Adjust the conditions of an electroporator (e.g., Gene Pulser ECM 600; BTX) to 270 V, 500 μF.

15. Transfer the cuvette into the cuvette holder and deliver the pulse.

16. Remove the cuvette.

17. Electroporated ES cells are then resuspended in 40 ml of ES cell medium and immediately plated onto four 100-mm dishes (10 ml/dish).

18. Count the viable cells. One can expect about 30% viability.

19. Add 5 ml of the ES cell medium the next day.

20. On the second day after electroporation, refeed the cells with ES cell medium containing G418 (250 μg/ml; Sigma) for drug selection.

21. Change the medium every day for 8 days. At this point clearly visible G418-resistant colonies can be seen and are ready to be picked.

Identification of Green Fluorescent Protein-Expressing Embryonic Stem Cells

1. Have at hand previously prepared 24-well flat-bottomed (Sumiron, Tokyo, Japan) feeder plates and 24-well flat-bottomed gelatinized plates. Refeed the cells by adding 0.5 ml of ES cell medium to each well of the 24-well plates. Keep the plates at 37° in 5% CO_2 until step 7.

2. Prepare a 96-well flat-bottomed plate (Corning, Corning, NY) by adding 50 μl of trypsin–EDTA solution per well.

3. Aspirate the medium from the original 10-cm plate containing G418-resistant colonies of ES cells, and then add 10 ml of PBS(–) to cover the plate.

4. Place the plate on a dissecting microscope. Part or all of an ES colony is removed by applying gentle suction using a Pipetman micropipettor (Gilson, Villiers-le-Bel, France) and disposable sterile tips (1–20 μl, multiflex tip; Funakoshi, Tokyo, Japan). If the colony is to be expanded, the cells are transferred to the trypsin–EDTA solution in a well of the plate prepared in step 2.

5. After 48 colonies have been picked, keep the 96-well plate at room temperature for 10 min.

6. Add 50 μl of ES cell medium to each well of the 96-well plate and break up the cell clumps by gently pipetting up and down with a small tip (Gilson) on a Pipetman micropipettor (100–200 μl; Gilson) about 20–30 times.

7. Transfer 50 μl of the homogeneous suspension of each well to a

well in the 24-well feeder plate and to a well in the 24-well gelatinized plate refed in step 1. Change tips each time.

8. Incubate all of the plates at 37° in 5% CO_2.
9. Add 0.5 ml of ES cell medium the next day.
10. On the second day, change the ES cell medium.
11. When the wells are approaching 70–80% confluence, the 24-well plate containing feeder cells is frozen and stored at −80° as the master plate with frozen medium [fetal bovine serum containing dimethyl sulfoxide (DMSO) at 10%].
12. Cells growing on the gelatinized 24-well plates are trypsinized and replated onto 8-well Lab-Tek chamber slides (Nunc, Roskilde, Denmark) and onto new gelatinized 24-well plates without feeder cells. The Lab-Tek chambers are used to verify GFP expression; the 24-well plates are for PCR analysis.
13. Incubate the Lab-Tek chambers and 24-well plates at 37° in 5% CO_2.
14. When the cells in the wells of the Lab-Tek chambers are approaching confluence, the plastic portions of the Lab-Tek chambers are removed. The cells on the resultant slides are washed with PBS(–) several times and covered with a coverslip.
15. The expression of GFP is monitored by fluorescence microscopy, using FITC filter sets.
16. Once the desired clones have been identified, the appropriate wells of the master plate can be retrieved and expanded for blastocyst injection.

Generation of Chimeric Mice

In this experiment, 41 of 48 G418-resistant ES cell clones showed green fluorescence (about 85%). Some clones showed mosaic expression of GFP. Also, the level of expression varied among clones and even among the cells within the same clones. By semiquantitative PCR, most clones were found to have one to three copies of the transgene. We introduced GFP-expressing ES cells into C57BL/6J mouse embryos using the blastocyst injection procedure. In this system, chimeras will be recognized by the presence of agouti coat color. These chimeras will be apparent by day 10 after birth, when coat color becomes visible. For blastocyst injection, we selected GFP-expressing ES cell clones that expressed GFP uniformly and seemed to have only one copy of the transgene. Of four such clones (clones 11, 30, 33, and 36), three resulted in chimeric mice (clones 11, 30, and 36). In this case, we could discriminate chimeric mice also by monitoring GFP expression in the tail: when mice are about 3 weeks old, the tail or ear (approximately 2 mm is enough) is scissored out and monitored under a fluorescence microscope using excitation light. Only male chimeras are reserved for further bleeding.

Germ Line Transmission

The CCE ES cells used in this experiment are derived from the 129/SvEv mouse strain. The 129/SvEv strain is wild type at the agouti locus (*A/A*), which is dominant over the black, nonagouti C57BL/6J strain (*a/a*) from which blastocysts are obtained for injections. Male chimeras crossed with C57BL/6J females and germ line transmission are indicated by agouti offspring (*A/a*). Once germ line transmission is demonstrated, the male chimera is bred to a wild-type C57BL/6J female and successive back-crossing between GFP-expressing male offspring and wild-type C57BL/6J females is conducted up to 10–20 times, to produce a congenic background. Chimeras are also bred to 129/SvEv females to maintain an inbred background.

Chimeric mice generated from two clones (clones 11 and 36) had contributions of ES cells to their germ line and successfully transmitted the transgene to their offspring. By mating with C57BL/6 female mice, germ line chimeras sired agouti offspring in the resulting litters. One of two male chimeras derived from clone 11 and two of four male chimeras derived from clone 36 sired F_1 mice within three consecutive matings. The transgenic mice in both lines developed normally into adulthood and were fertile, indicating no deleterious effects of transgene expression. The clone 30 chimera was predominantly agouti by coat color but did not show germ line transmission.

The expression of GFP in F_1 pups ($\leq$3 days old) from each transgenic founder mouse was also examined by the naked eye under a UV illuminator. Transgenic pups showed weak green fluorescence and nontransgenic litters showed faint autofluorescence. Also, tails and ears of 3-week-old pups were monitored for GFP expression using fluorescence microscopy. An additional, tedious step such as DNA preparation for Southern blot or PCR analysis is not necessary. Unfortunately, clone 11 did not produce any littermates within three consecutive matings with 129/SvEv females. The low fecundity of 129/SvEv females makes it difficult to establish an inbred line.

Northern Blot

Expression of the transgene was also evaluated at the mRNA level by Northern blot analysis. Total RNA is isolated from various tissues of F_1 mice according to the method of Chomczynski and Sacchi.[37] Approximately

[37] P. Chomczynski and N. Sacchi, *Anal. Biochem.* **162,** 156 (1987).

20 μg of total RNA per lane is size fractionated on a 1% denaturing formaldehyde agarose gel and transferred onto a Hybond-N+ membrane (Amersham, Arlington Heights, IL). The *Not*I/*Sma*I-digested fragment (744 bp) of pEGFP-1 is labeled with [α-^{32}P]dCTP using a random primer DNA labeling kit (version 2.0; Takara, Kyoto, Japan). Hybridization is performed at 42° for 18 hr in a hybridization buffer containing 50% formamide, 5× SSC (1× SSC is 0.15 *M* NaCl plus 0.015 *M* sodium citrate), 5× Denhardt's solution, 0.1% sodium dodecyl sulfate (SDS), and denatured salmon sperm DNA (0.1 mg/ml). The membrane is then washed vigorously with 0.1× SSC and 0.1% SDS at 65° for 30 min, followed by autoradiography. All of the tissues we examined showed almost the same level of GFP mRNA expression (Fig. 2, see color insert).

Detection of Green Fluorescent Protein in Transgenic Mice

Fluorescence Microscopes and Their Setting

In fluorescence microscopy, the primary issue is the choice of filter sets. In general, conditions used for fluorescein isothiocyanate (FITC; excitation at 450–490 nm and emission at 520 nm) should give a good signal for the observation of GFP-S65T and EGFP. Some samples have a background autofluorescence. Most autofluorescence in mammalian cells is due to flavin coenzymes (FAD and FMN), which have an absorption/emission of 450/515 nm. These values are similar to those for GFP-S65T and EGFP (490/509 nm). To minimize autofluorescence, excitation near 490 nm is beneficial. A confocal laser scanning microscope with excitation at 488 nm and detection at 500–530 nm gives good results. To reduce the background, we also use an Olympus BHS-RFK (BH2-RFK) fluorescence microscope with a standard FITC filter set (BP490) and an EY475 accessory excitation filter, which reduces the wavelength to less than 475 nm. In the case of EGFP mice, GFP fluorescence can be directly photographed with the usual or charge-coupled device (CCD) camera equipped with an Olympus IF550 filter under a UV illuminator, or with blue light from a fluorescence microscope.

Detection of Green Fluorescent Protein in Fresh Tissues

In our experience, the strongest green fluorescence of GFP is observed with fresh tissues without fixation. As stated above, the expression of GFP in pups can be examined with the naked eye under a UV illuminator in the case of EGFP transgenic mice. Transgenic pups show weak green fluorescence and nontransgenic litters showed faint autofluorescence. Various organs can also be examined directly with a fluorescence microscope at low magnification. Autofluorescence is negligible, at least in the brain (cerebellum, hippocampus, and cerebral cortex), retina, heart, lung, liver, spleen, pancreas, kidney, skeletal muscle, lymph node, trachea, thyroid, small intestine, colon, and ovary.

Histological Analysis

For histological analysis, we use frozen sections followed by paraformaldehyde fixation. Freshly dissected tissues should be embedded in Tissue Tek (Miles, Elkhart, IN) and frozen on the surface of liquid nitrogen. The frozen tissues can be stored at $-80°$ without loss of GFP fluorescence for at least 6 months. We use a cryostat Leica CM1850 (Leica, Eaton, PA) to make thin sections (10 μm). Thin sections are air dried on nonfluorescent glass slides (Matunami Glass Industries, Tokyo, Japan) and are subjected to fixation or directly examined by fluorescence microscopy. When air-dried frozen sections are kept in dry, light-sealed boxes, they can be stored at room temperature for at least 1 month without loss of GFP fluorescence. Autofluorescence may be slightly increased on storage.

Fixation can be done by dipping the air-dried frozen sections into paraformaldehyde solution as follows.

1. Freshly prepare 4% paraformaldehyde solution in PBS (w/v) by dissolving paraformaldehyde powder (electron microscopy grade; Nakarai Tesque, Kyoto, Japan) in PBS. Heat the solution to 65–70° in order to dissolve the paraformaldehyde completely. Cool the solution to room temperature.
2. Dip the slides into the paraformaldehyde solution. Allow 10 min to fix.
3. Rinse the slides with several changes of distilled water to remove excess paraformaldehyde solution.
4. Air dry.

The sections can be mounted in glycerol and sealed with molten agarose or rubber cement or subjected directly to fluorescence microscopy.

Thin sections can also be fixed with organic solvents such as methanol and acetone without loss of GFP fluorescence. We fix air-dried frozen sections by dipping them into 100% methanol, 100% acetone, or methanol–acetone (1 : 1, v/v) for 10 min at 4°. After fixation, the slides are allowed to dry completely. The fixation with paraformaldehyde or organic solvents does not increase the background autofluorescence. In contrast, formaldehyde–PBS (3.7%, w/v, pH 7.4–7.6) fixation significantly increases the background. Tissues can be directly fixed with paraformaldehyde without doing a frozen section. In this case, tissues sliced about several millimeters thick are fixed in freshly made 4% paraformaldehyde for about 24 hr. Longer fixation will result in higher autofluorescence. These tissues will be subsequently processed for thin sections.

Choice of Green Fluorescent Protein Genes and Methods of Production of Green Fluorescent Protein Transgenic Mice

Several GFP transgenic mice have been produced.[33,38–40] As stated above, we produced transgenic mice carrying the GFP-S65T transgene or EGFP (Clontech) using the same promoter, which consists of the CMV-IE enhancer and the human elongation factor 1α promoter (CE promoter).[33] The GFP-S65T transgenic mice were generated by pronuclear microinjection. Although we used a promoter that is supposed to drive expression constitutively, visible GFP expression was observed only in the preimplantation embryo (four-cell blastocyst), and weakly in the testis of adult mice. It was not detected in fetuses, newborn pups, or other tissues of adult mice, albeit all transgenic mice had a transgene that was detected by PCR analysis. Interestingly, when a GFP-S65T transgenic hemizygous male is mated with a normal female, clear GFP expression is observed at the four-cell blastocyst stage in approximately half the embryos.

EGFP transgenic mice were generated through ES cells. In this case, ubiquitous GFP expression was observed in tissues examined, except in hair and in red blood cells, at various stages of development. Thus, we could efficiently select chimeras and transgenic mice through monitoring GFP fluorescence as a reporter. Okabe *et al.* also made GFP transgenic mice under the control of the chicken β-actin promoter, using either wild-type GFP or EGFP.[40] They used the pronuclear microinjection method for the generation of their transgenic mice. Although expression of wild-type GFP was mainly restricted to the muscle and pancreas, they observed ubiquitous GFP expression with EGFP except in hair and in red blood cells.

Our result together with those of Okabe *et al.*, suggest that the codon optimization seems to be responsible for the ubiquitous expression of GFP in transgenic mice. In our case, the low copy number of the transgene in ES cells could also be responsible for the difference. Thus, the proper choice of the promoter–GFP gene combination and the method of transgenic mouse generation is necessary for the production of GFP transgenic mice. For blastocyst selection, the CE promoter–GFP-S65T gene

[38] A. Chiocchetti, E. Tolosano, E. Hirsch, L. Silengo, and F. Altruda, *Biochim. Biophys. Acta* **1352,** 193 (1997).

[39] L. Zhuo, B. Sun, C.-L. Zhang, A. Fine, S.-Y. Chiu, and A. Messing, *Dev. Biol.* **187,** 36 (1997).

[40] M. Okabe, M. Ikawa, K. Komonami, T. Nakanishi, and Y. Nishimune, *FEBS Lett.* **407,** 313 (1997).

combination may be the correct choice, because the expression of GFP ceases in adult mice. For GFP expression in adult tissue, EGFP may be the first choice.

Acknowledgment

We thank Dr. Manabu Watanabe for excellent histological work.

[22] Green Fluorescent Protein as a Probe to Study Intracellular Solute Diffusion

By ALAN S. VERKMAN

The targeting of green fluorescent protein (GFP) to specific intracellular sites provides a powerful, noninvasive approach to measure solute diffusion in aqueous compartments and membranes of living cells. Photophysical studies have indicated that GFP has desirable properties for use as a probe of solute translational and rotational diffusion. GFP is fairly inert in cellular environments, and photobleaches with moderate efficiency and relatively little reversible photobleaching. GFP fluorescence is not depolarized by intramolecular segmental motions. Photobleaching experiments have demonstrated remarkably rapid translational diffusion of GFP in the aqueous phase of cytoplasm, the mitochondrial matrix, and the endoplasmic reticulum lumen. Much slower diffusion and transport of GFP-fusion proteins has been measured in cell plasma membranes and in membranes of organelles in the secretory pathway. The GFP labeling of intracellular sites and the use of time-resolved fluorescence methods thus provide a unique opportunity to define diffusive mechanisms in cells. In addition, the intrinsic pH sensitivity of GFP fluorescence has been exploited to measure pH at specific intracellular locations.

Introduction

The mobility of solutes in cellular aqueous and membrane compartments is a subject of long-standing interest in cell biology. Solute diffusion in the cytoplasm, nucleus, and aqueous lumen of organelles is important for a wide variety of cellular events such as nutrient and nucleotide transport, metabolism, signaling, and locomotion. The mobility of membrane-bound molecules is important in many cellular processes involving membrane protein function such as transport, signaling, binding, and recognition. The

0076-6879/99 $30.00

crowded nature of aqueous compartments and membrane phases poses a unique challenge to the measurement and analysis of solute diffusive mechanisms.

The tools to study solute mobility in cells include specific fluorescent labels and time-resolved fluorescence measurement methods. Until recently, exogenously introduced fluorescent probes were needed to label specific cellular sites. The introduction of GFP provides significant advantages over exogenous chemical fluorophores in terms of ease of use, site specificity, and noninvasive labeling with minimal cellular perturbation. GFP permits the labeling of many cellular sites that cannot be labeled by other methods. For measurements of translational diffusion, fluorescence recovery after photobleaching (FRAP) has been extremely useful and is the focus of the discussions to follow. Other potentially useful approaches for some applications include fluorescence correlation spectroscopy and single-particle tracking. For measurements of rotational diffusion, time-resolved fluorescence anisotropy is suitable for determination of rotational correlation times in the range of picoseconds to tens of nanoseconds. Other approaches for analysis of slower rotation include phosphorescence anisotropy decay and polarized photobleaching recovery.

This chapter describes the utility of GFP in studies of solute diffusion in living cells. The relevant photophysical properties of GFP are described, and examples of diffusive measurements in cellular compartments are provided. The potential pitfalls in such measurements are discussed, including the analysis of photobleaching data for solutes undergoing complex diffusion. Finally, the application of GFP as an intracellular pH sensor is mentioned.

Photobleaching and Polarization Properties of Green Fluorescent Protein

In photobleaching recovery, fluorophores in a defined volume are irreversibly bleached by a brief, intense laser beam; the subsequent kinetics of fluorescence recovery in the bleached region provides a quantitative measure of fluorophore translational diffusion. For use in photobleaching measurements, GFP should be brightly fluorescent when excited by available laser wavelengths and bleached irreversibly with moderate efficiency. Bleach efficiency should not be so high as to produce substantial bleaching by the probe beam used to monitor recovery. In addition, reversible photobleaching processes involving triplet state relaxation or other photochemical processes should be minimal, easily distinguished from irreversible photobleaching, or eliminated by maneuvers such as solution oxygenation. For analysis of GFP rotation by anisotropy decay and polarized photobleaching recovery, the intrinsic anisotropy of the GFP chromophore should be high, with minimal chromophore segmental motion. In addition, there should be little nonspecific binding of GFP to cellular components, which would slow diffusion and produce a heterogeneous population of diffusing species.

Of the available mutants, GFP mutant S65T was chosen for characterization because of its bright green fluorescence when excited with the 488-nm line of an argon laser.[1] Measurements of recombinant purified GFP-S65T in solution showed bright fluorescence (λ_{ex} 492 nm, λ_{em} 509 nm) with a lifetime of 2.9 nsec and averaged rotational correlation time (t_c) of 20 nsec.[2] The absence of a rapid rotational correlation time indicated that the chromophore is held rigidly in the GFP molecule, which is consistent with the reported GFP crystal structure.[3] The slow rotational correlation time of ~20 nsec is in general agreement with that predicted by the Stokes–Einstein relation for a GFP-sized spherical protein, recognizing that GFP is actually barrel shaped and undergoes anisotropic rotation. The absence of rapid depolarizations makes GFP a suitable rotational probe of microviscosity (see below).

Figure 1A shows spot photobleaching measurements of GFP-S65T in solutions containing various concentrations of glycerol. In response to a brief bleach pulse, GFP fluorescence promptly decreased and then recovered to initial levels by translational diffusion of unbleached GFP into the probed zone. As expected, the recovery rate was slowed with increasing glycerol concentrations, with a recovery half-time ($t_{1/2}$) that was proportional to solution viscosity. In saline, the recovery rate for fluorescein was approximately threefold greater than for GFP, giving a GFP diffusion coefficient of 8.7×10^{-7} cm^2/sec. The efficiency for irreversible photobleaching of GFP was several times lower than that of fluorescein. In air-saturated solutions of high viscosity, an additional faster recovery process was observed and ascribed to reversible photobleaching involving triplet state relaxation (Fig. 1B). The $t_{1/2}$ for reversible photobleaching was 1.5–5.5 msec (relative viscosity 5–250), independent of spot diameter, and not affected by O_2 or various triplet state or other quenchers [azide, triethylenediamine (DABCO), 2-mercaptoethanolamine, etc.]. The insensitivity of GFP to quenchers is consistent with the protected location of its triamino acid chromophore. The reader is referred to Refs. 4–6 for a more complete discussion about reversible photobleaching phenomena.

[1] R. Heim, A. B. Cubitt, and R. Y. Tsien, *Nature (London)* **373,** 663 (1995).
[2] R. Swaminathan, C. P. Hoang, and A. S. Verkman, *Biophys. J.* **72,** 1900 (1997).
[3] M. Ormö, A. B. Cubitt, K. Kallio, L. A. Gross, R. Y. Tsien, and S. J. Remington, *Science* **273,** 1392 (1996).
[4] N. Periasamy, S. Bicknese, and A. S. Verkman, *Photochem. Photobiol.* **63,** 265 (1996).
[5] L. Song, C. A. G. O. Varma, J. W. Verhoeven, and H. J. Tanke, *Biophys. J.* **70,** 2959 (1996).

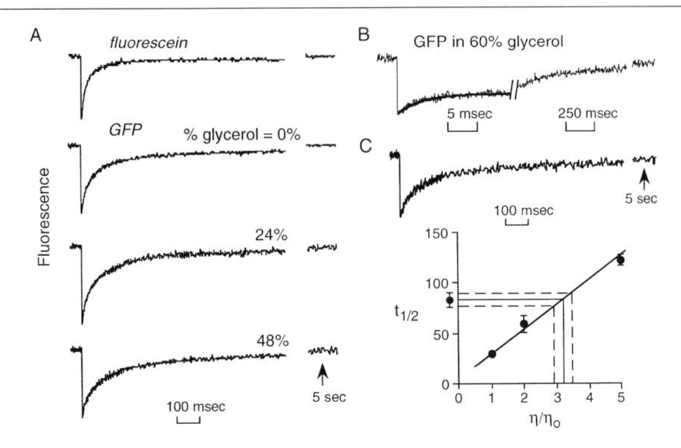

FIG. 1. Photobleaching of GFP-S65T. (A) Spot photobleaching of fluorescein (58 μM) in 0.1 N NaOH or GFP-S65T (66 μM) in PBS using 100-μsec bleach time and 5-μm spot diameter (×20 objective). Indicated percentages of glycerol were added to the PBS in GFP-S65T studies. (B) Reversible photobleaching of GFP-S65T. The solution contained 66 μM GFP-S65T in PBS containing 60% glycerol. Spot bleaching was done with a ×20 objective and 30-μsec bleach time. (C) *Top:* Spot photobleaching (×20 objective) of GFP in the cytoplasm of stably transfected CHO cells, using 50-μsec bleach time. *Bottom:* Determination of relative cytoplasmic viscosity. Recovery half-times ($t_{1/2}$) for GFP-S65T diffusion in PBS containing glycerol [as in (A)] plotted against relative solution viscosity. Data for GFP-S65T diffusion in cytoplasm are indicated.

It is pointed out that the initial rapid fluorescence recovery in viscous solutions does not represent GFP diffusion, and so if not identified, can result in a serious misinterpretation of the recovery data. It is recommended that experimental maneuvers be utilized to distinguish between irreversible and reversible bleaching.[2,4] Irreversible bleaching produces a recovery whose rate depends strongly on spot diameter whereas recovery from reversible photobleaching is independent of spot size. On the basis of the kinetics of triplet state relaxation, changing bleach beam intensity and time also gives predictable effects on reversible versus irreversible photobleaching efficiencies. The presence of reversible photobleaching does not preclude diffusion coefficient determination from analysis of the irreversible photobleaching component. The recovery from reversible photobleaching is generally much faster than that from irreversible photobleaching, or can be made so by increasing spot diameter. For chromophores other than GFP (such as fluorescein[4,5]), triplet state quenchers can be used to accelerate or eliminate the reversible photobleaching process. From these considerations, it is important to identify and account for reversible recovery processes in measurements of mobility and trafficking using GFP photobleaching.

[6] A. S. Stout and D. Axelrod, *Photochem. Photobiol.* **62,** 239 (1995).

Solute Diffusion in Cytoplasm

The physical structure of cell cytoplasm has been a topic of continued interest (reviewed in Refs. 7–9). The initial descriptions of cytoplasm were in terms of a viscous gelatinous mass without internal structure. However, the cytoplasm contains dissolved solutes and macromolecules in a complex array of microtubles, actin, and intermediate filaments organized into a lattice-like mesh. As more advanced biophysical techniques have been applied to study the physical state of cytoplasm, the notion is emerging that for diffusion of relatively small solutes, the cytoplasm is more like a bag of slightly viscous water than a complex gelatinous mass. One parameter describing cytoplasmic rheology is "fluid-phase viscosity," defined as the microviscosity sensed by a small solute in the absence of interactions with macromolecules and organelles. Measurements of the rotation of small fluorophores[10] and of intrinsically viscosity-sensitive fluorophores[11] indicated that the fluid-phase viscosity of cytoplasm is not much greater than that of water.

For transport of small solutes such as metabolites and nucleic acids, a more important parameter describing cytoplasmic viscosity is the translational diffusion coefficient. Kao *et al.*[12] measured the translational mobility of a small fluorescent probe in the cytoplasm of Swiss 3T3 fibroblasts utilizing spot fluorescence photobleaching. Diffusion of a fluorescein-sized solute in cytoplasm was about four times slower than in water. Three independently acting factors were identified that accounted quantitatively for the fourfold slowed diffusion: (1) slowed diffusion in fluid-phase cytoplasm, (2) probe binding to intracellular components, and (3) probe collisions with intracellular components. The latter factor, probe collisions, was determined to be the principal diffusive barrier that slowed the translational diffusion of small solutes. Similar results were obtained for small solute diffusion in membrane-adjacent cytoplasm using total internal reflection-FRAP.[13,14] In earlier studies,[15] spot photobleaching was used to measure the translational diffusion of larger solutes—microinjected, fluorescently

[7] A. B. Fulton, *Cell* **30,** 345 (1982).
[8] K. R. Porter, *J. Cell Biol.* **99,** 3s (1984).
[9] K. Luby-Phelps, *Curr. Opin. Cell Biol.* **6,** 3 (1994).
[10] K. Fushimi and A. S. Verkman, *J. Cell Biol.* **112,** 719 (1991).
[11] K. Luby-Phelps, S. Mujundar, R. Mujundar, L. Ernst, W. Galbraith, and A. Waggoner, *Biophys. J.* **65,** 236 (1993).
[12] H. P. Kao, J. R. Abney, and A. S. Verkman, *J. Cell Biol.* **120,** 175 (1993).
[13] S. Bicknese, N. Periasamy, S. B. Shohet, and A. S. Verkman, *Biophys. J.* **165,** 1272 (1993).
[14] R. Swaminathan, S. Bicknese, N. Periasamy, and A. S. Verkman, *Biophys. J.* **71,** 1140 (1996).
[15] K. Luby-Phelps, P. E. Castle, D. L. Taylor, and F. Lanni, *Proc. Natl. Acad. Sci. U.S.A.* **84,** 4910 (1987).

labeled dextrans and Ficolls. As dextran or Ficoll molecular size was increased, diffusion in cytoplasm progressively decreased relative to that in water, suggesting a cytoplasmic "sieving" mechanism that was proposed to involve the skeletal mesh. Data from our laboratory[16] also showed that very large solutes (>500 kDa) are impeded in their motion in cytoplasm; however, we were unable to demonstrate sieving of smaller solutes. It is noted that the measurements of dextran and Ficoll diffusion required direct cell microinjection, an invasive procedure with the potential for cell damage.

Cytoplasmically expressed GFP was used as a noninvasive probe to measure the rotational and translational diffusion of a protein-sized molecule.[2] Time-resolved microfluorimetry indicated a GFP rotational correlation time of 36 nsec, giving a relative viscosity (cytoplasm versus water) of 1.5. Photobleaching recovery of GFP in cytoplasm is shown in Fig. 1C (top). The fluorescence recovery was 82 ± 2% complete with a $t_{1/2}$ of 83 msec. Comparison of the $t_{1/2}$ in cytoplasm with the $t_{1/2}$ of GFP in aqueous solutions of known viscosities (Fig. 1C, bottom) gave a relative cytoplasmic viscosity of 3.2. This value agreed with the results obtained using microinjected fluorescein isothiocyanate (FITC)–dextrans and Ficolls.[2] GFP translational diffusion increased 4.7-fold as cells swelled from relative volume 0.5 to 2. The results in cytoplasm support the view that protein-sized molecules are mildly impeded in their motion, and that the primary barrier to GFP diffusion is collisional interactions between GFP and macromolecular solutes.

Diffusion in Intracellular Organelles

Solute diffusion within the aqueous lumen of intracellular organelles is involved in many processes, such as metabolism in mitochondria and protein processing and recognition in endoplasmic reticulum. The mitochondrial matrix is the aqueous compartment enclosed by the inner mitochondrial membrane and is a major site of metabolic processes. The matrix is a particularly interesting compartment because of its presumed high density of enzymes and other proteins. Theoretical considerations have suggested that the diffusion of metabolite- and enzyme-sized solutes might be severely restricted in the mitochondrial matrix.[17,18] It has been proposed that biochemical events might occur by a "metabolite channeling" mechanism, in which metabolites are passed from one enzyme to another in an organized complex without aqueous-phase diffusion.

The ability to target GFP to the mitochondrial matrix provided a unique

[16] O. Seksek, J. Biwersi, and A. S. Verkman, *J. Cell Biol.* **138,** 131 (1997).
[17] J. Ovadi, *J. Theor. Biol.* **152,** 1 (1991).
[18] G. R. Welch and J. S. Easterby, *Trends Biochem. Sci.* **19,** 193 (1994).

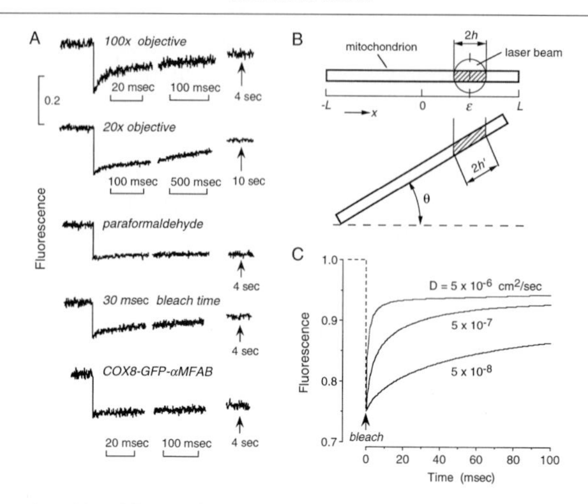

FIG. 2. Spot photobleaching of GFP in the mitochondrial matrix of CHO cells. (A) Photobleaching with 40-μsec bleach time, using $\times 100$ or $\times 20$ objective (top two curves). Photobleaching with $\times 100$ objective in cells after 6 hr of fixation in 4% paraformaldehyde (third curve). Photobleaching as in top curve with long (30 msec) bleach time (fourth curve). The fifth curve represents photobleaching (as in top curve) of a fusion protein of GFP with a matrix enzyme (α subunit of the mitochondrial β-fatty acid oxidation pathway). (B) Model of mitochondrion as a long thin cylinder with unobstructed lumen oriented at angle θ. See Ref. 19 for details. (C) Model predictions for fluorescence recovery at various D values for bleaching with $\times 100$ objective and short bleach time.

opportunity to test the hypothesis that solute diffusion is greatly slowed in the matrix. The diffusion of GFP-S65T was measured in the mitochondrial matrix of fibroblast, liver, skeletal muscle, and epithelial cell lines.[19] Spot photobleaching of GFP with a $\times 100$ objective (0.8-μm spot diameter) gave a half-time for fluorescence recovery of 15–19 msec with greater than 90% of the GFP mobile (Fig. 2A, top curve). As predicted for aqueous-phase diffusion in a confined compartment, fluorescence recovery was slowed or abolished by increased laser spot size (second curve), paraformaldehyde fixation (third curve), and increased bleach time (fourth curve). The fluorescence recovery data were analyzed using a mathematical model of matrix diffusion. As shown in Fig. 2B, mitochondria were modeled as long, continuously open cylinders with a specified orientational distribution. Fluorescence recoveries were computed from analytical solutions to the diffusion equation. Predicted recovery curves for different diffusion coefficients are shown in Fig. 2C. The fitted values for D were 2–3 $\times 10^{-7}$ cm^2/sec, only three- to fourfold less than that for GFP diffusion in water. Interestingly,

[19] A. Partikian, B. Ölveczky, R. Swaminathan, Y. Li, and A. S. Verkman, *J. Cell Biol.* **140,** 821 (1998).

little fluorescence recovery was found for bleaching of GFP in fusion with subunits of the fatty acid β-oxidation multienzyme complex (Fig. 2A, bottom curve) that are normally present in the matrix. Measurement of the rotation of free GFP by time-resolved anisotropy gave a rotational correlation time of 23 nsec, similar to that for GFP rotation in water. The rapid and unrestricted diffusion of GFP in the mitochondrial matrix suggested that because of the lack of diffusive barriers, metabolite channeling may not be required. It was proposed that the clustering of matrix enzymes in membrane-associated complexes might serve to establish a relatively uncrowded aqueous space in which solutes can freely diffuse.

A similar photobleaching study was carried out to measure solute diffusion in the endoplasmic reticulum (ER), the major compartment for the processing and quality control of newly synthesized proteins.[20] GFP was targeted to the ER lumen of CHO cells by transient transfection with cDNA encoding GFP (S65T/F64L mutant) with the C-terminus KDEL retention sequence and upstream preprolactin secretory sequence. Spot photobleaching using a $\times 100$ objective showed that nearly all GFP was mobile with a $t_{1/2}$ for fluorescence recovery of 22 msec. The $t_{1/2}$ increased to 45 msec with 30-msec bleach time ($\times 100$ objective), and to 143 msec using a $\times 40$ objective. Fluorescence recovery was abolished by fixation with paraformaldehyde. Utilizing a mathematical model for photobleaching that accounted for ER geometry, a GFP diffusion coefficient of $5\text{–}10 \times 10^{-8}$ cm^2/s was computed. The GFP rotational correlation time was measured to be 39 nsec, ~2-fold greater than that in water. The data suggested that GFP is mobile in the ER lumen and that its translational diffusion is mildly hindered. Measurements utilizing various expressed GFP-fusion protein should be useful in defining interactions between lumenal contents and the ER wall, such as the binding of misfolded proteins to molecular chaperones.

Quantitative Analysis of Solute Diffusion in Organelles

The ability to target GFP to specific cellular sites has permitted the measurement of diffusion in intracellular compartments. Because organelles often have a complex exterior geometry as well as internal barriers, new methods of analysis are required to quantitatively interpret photobleaching experiments. The mitochondrial matrix is seen by electron microscopy as a cylindrical compartment with internal barriers called cristae. The endoplasmic reticulum is generally thought of as an interconnected reticular network of cylinders or plates in three dimensions. From a practical experimental perspective, a strategy is needed to deduce intrinsic solute diffusion coefficients (for diffusion in the absence of boundaries or barriers) from

[20] M. Dayel, E. Hom, and A. S. Verkman, *Biophys. J.* in press (1999).

measured photobleaching recovery data. A more general issue is the elucidation of how organellar shape and barrier geometry influence particle diffusion *in vivo*. The analysis of solute diffusion in mitochondria described above utilized a model that assumed an open mitochondrial lumen without internal barriers. It is thus important to understand how cristae barriers affect the diffusional transport of metabolites along the mitochondrial axis.

A number of theoretical descriptions of particle diffusion in two-dimensional membranes have been reported for a variety of situations involving binding, crowding, and mobile and immobile obstacles.[21–24] We utilized Monte Carlo simulations in three dimensions to analyze effects of barrier properties on diffusion in the aqueous matrix of mitochondria and effects of reticular geometry on diffusion in endoplasmic reticulum, and to relate experimental photobleaching data to intrinsic diffusion coefficients.[25] Mitochondria were modeled as long closed cylinders containing fixed lumenal obstructions of variable number and size, and the ER was modeled as a network of interconnected cylinders of variable diameter and density. Particle trajectories were computed in each simulation for $>10^5$ particles, generally for $>10^5$ time steps. For mitochondria-like cylinders, significant slowing of diffusion required large or wide single obstacles, or multiple obstacles. In simulated spot photobleaching experiments, an $\sim$25% decrease in apparent diffusive transport rate (defined by the time to 75% fluorescence recovery) was found for a single thin transverse obstacle occluding 93% of lumen area, a single 53%-occluding obstacle of width 16 lattice points (8% of cylinder length), or 10 equally spaced 50%-occluding obstacles alternately occluding opposite halves of the cylinder lumen. Recovery curve shape with obstacles showed long tails, indicating anomalous diffusion. Related simulations indicated that significantly slowed particle diffusion can be produced by binding to barrier walls. For a reticulum-like network, particle diffusive transport was mildly reduced from that in unobstructed three-dimensional space. In simulated photobleaching experiments, apparent diffusive transport was decreased by 39–60% in reticular structures where 90–97% of space was occluded.

These computations provided an approach to analyze photobleaching data in terms of microscopic diffusive properties and supported the paradigm that organellar barriers must be quite severe to seriously impede solute diffusion. The observation that particles executing a random walk

[21] M. J. Saxton, *Biophys. J.* **66,** 394 (1994).

[22] M. J. Saxton, *Biophys. J.* **64,** 1053 (1993).

[23] B. Schram, J. F. Tocanne, and A. Lopez, *Eur. Biophys. J.* **23,** 337 (1994).

[24] M. J. Saxton, *Biophys. J.* **58,** 1303 (1990).

[25] B. P. Ölveczky and A. S. Verkman, *Biophys. J.* **74,** 2722 (1998).

have an uncanny ability to find sparsely distributed absorbing patches on the surface of a sphere was first recognized by Berg and Purcell.[26] In qualitative terms, the diffusing particle "feels" its way along an obstacle to locate a small opening, and then passes through the opening to escape on the opposite side. Further refinement of these simple notions will be needed as more data on intraorganellar solute transport are obtained.

Diffusion in Membranes

Green fluorescent protein has many potential applications to the study of lateral diffusion of proteins in membranes. Diffusion in membranes is remarkably slower than that in aqueous compartments and may be quite complex because of membrane crowding with proteins, distinct lipid domains, and membrane–cytoskeletal interactions (for review, see Ref. 27). As described for the analysis of solute diffusion in aqueous compartments, photobleaching recovery provides a quantitative method for analysis of GFP-fusion protein diffusion in membranes. Photobleaching measurements in membranes are technically easier than in aqueous compartments because of the slower recovery rates, giving improved detection signal-to-noise ratio. In addition, rapid reversible photobleaching is generally not a concern for measurement of slow lateral diffusion in membranes.

A few photobleaching studies of the diffusion of GFP-labeled membrane proteins have been reported. Barak *et al.*[28] found that a GFP-labeled β_2-adrenergic receptor diffused in cell plasma membranes freely with a diffusion coefficient of $4–12 \times 10^{-9}$ cm^2/sec. Cole *et al.*[29] found similar rates of diffusion ($3–5 \times 10^{-9}$ cm^2/sec) for GFP-labeled proteins in Golgi membranes with >90% of the proteins mobile. Ellenberg *et al.*[30] labeled the laminin B-receptor with GFP. The receptor was immobile when in the nuclear envelope but a subpopulation of the receptor in the endoplasmic reticulum diffused freely and rapidly (4×10^{-7} cm^2/sec). Our laboratory generated fusion proteins between GFP and members of the aquaporin water channel family.[31] Diffusion of several aquaporins was relatively fast,

[26] H. C. Berg and E. M. Purcell, *Biophys. J.* **20,** 193 (1977).

[27] P. F. F. Almeida and W. L. C. Vaz, *in* "Handbook of Biological Physics" (R. Lipowsky and W. Sackmann, eds.), p. 305, 1995.

[28] L. S. Barak, S. S. Ferguson, J. Zhang, C. Martenson, T. Meyer, and M. G. Caron, *Mol. Pharmacol.* **51,** 177 (1997).

[29] N. B. Cole, C. L. Smith, N. Sciaky, M. Terasaki, M. Edidin, and J. Lippincott-Schwartz, *Science* **273,** 797 (1996).

[30] J. Ellenberg, E. D. Siggia, J. E. Moreira, C. L. Smith, J. F. Presley, R. J. Worman, and J. Lippincott-Schwartz, *J. Cell Biol.* **138,** 1193 (1997).

[31] F. Umenishi and A. S. Verkman, *J. Am. Soc. Nephrol.* **9,** 27A (1998).

and in one case (AQP2) diffusion was regulated by cAMP. Photobleaching of GFP-labeled membrane proteins should be useful in studies of membrane protein associations, membrane organization (such as the identification of lipid domains), and cytoskeletal–membrane interactions.

Analysis of Photobleaching Data for Complex Diffusive Phenomena

The determination of diffusion mechanisms from photobleaching data requires quantitative analysis procedures. Photobleaching data in cell membrane and aqueous compartments have generally been analyzed in terms of a single diffusing component with or without an immobile fraction. For spot photobleaching in two dimensions (as in membranes), computational methods have been reported to estimate the fluorophore diffusion coefficient (D) from an ideal bleach spot profile and fluorescence recovery curve shape.[32–34] For analysis of photobleaching data in three dimensions, such as the aqueous phase of the cell cytoplasm, we introduced a calibration procedure in which the half-time ($t_{1/2}$) for fluorescence recovery in cells is compared with the $t_{1/2}$ measured in thin layers of fluorophores dissolved in artificial solutions of known viscosity.[12,16] These analytical and empirical methods are useful for determination of single, time-independent diffusion coefficients, but are not easily adapted to complex diffusive phenomena such as anomalous diffusion or diffusion of multiple species with different diffusion coefficients. The potential pitfalls in the assumption of simple diffusion and the significance of long-tail kinetics in diffusive phenomena have been recognized,[35] and papers have begun to consider how to interpret photobleaching data in systems with complex or anomalous diffusion.[25,36,37]

Solute diffusion is described as "simple" or "nonanomalous" in a homogeneous medium such as a liquid solvent, in which case solute transport is described adequately by a single diffusion coefficient. However, there are situations in which solute diffusion cannot be described in terms of a single diffusion coefficient, such as anomalous diffusion and diffusion of more than one species. The diffusion of a solute is defined to be anomalous if the mean squared displacement varies with time in a nonlinear manner. Here the diffusion coefficient is not constant, but time and/or space dependent. A number of different physical mechanisms giving anomalous diffu-

[32] D. Axelrod, D. E. Koppel, J. Schlessinger, E. Elson, and W. W. Webb, *Biophys. J.* **16,** 1055 (1976).
[33] G. Barisas and M. D. Leuther, *Biophys. J.* **10,** 221 (1979).
[34] E. J. J. Van Zoelen, G. J. Tertoolen, and S. W. de Laat, *Biophys. J.* **42,** 103 (1993).
[35] J. F. Nagle, *Biophys. J.* **63,** 366 (1992).
[36] T. J. Feder, I. Brust-Mascher, J. P. Slattery, B. Baird, and W. W. Webb, *Biophys. J.* **70,** 2767 (1996).
[37] F. P. Coelho, W. L. C. Vaz, and E. Melo, *Biophys. J.* **72,** 1501 (1997).

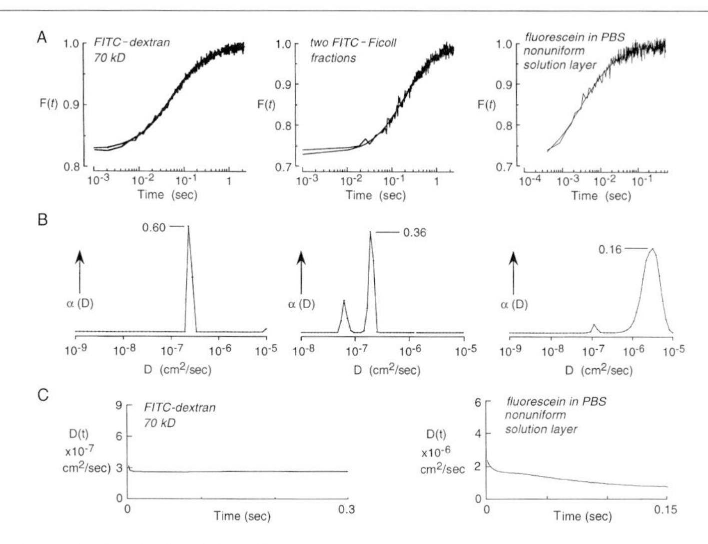

FIG. 3. Analysis of photobleaching recovery data in terms of continuous distributions of diffusion coefficients and time-dependent diffusion coefficients. (A) Fluorescence recovery data for size-fractionated FITC–dextran (70 kDa) in PBS (*left*), a mixture of two size fractions of FITC–Ficoll (*middle*), and a nonuniform solution of fluorescein in PBS (*right*). Experimental $F(t)$ data and fitted curve (smooth line) obtained by MEM analysis are shown. (B) Fitted distributions of diffusion coefficients $\alpha(D)$ by MEM analysis. (C) Time-dependent diffusion coefficient $D(t)$ determined from $F(t)$ for 70-kDa FITC–dextran (*left*) and fluorescein in a nonuniform solution layer (*right*).

sion have been described.[21–24,38–40] Another example of nonsimple diffusion is the presence of two or more diffusing species, each of which is described by a single diffusion coefficient. Without prior knowledge of the presence of multiple diffusing species, as might be the case in cellular environments where heterogeneous binding can occur, photobleaching data can be wrongly interpreted as anomalous diffusion.

We introduced the idea that photobleaching recovery data, $F(t)$, can be resolved in terms of a continuous distribution of diffusion coefficients, $\alpha(D)$.[41] A regression method to recover $\alpha(D)$ from $F(t)$ was developed on the basis of the maximum entropy method (MEM), a fitting procedure used previously to compute fluorescence lifetime distributions.[42] The MEM utilizes a "basis" recovery curve for simple diffusion obtained using a reference sample such as a thin uniform film of fluorescein in saline. Figure

[38] J. P. Bouchaud and A. Georges, *in* "Disorder and Mixing" (E. Guyon, J. P. Nadal, and Y. Pomeau, eds.), p. 19. Kluwer Academic Publishers, Dordrecht, The Netherlands, 1988.

[39] J. M. Drake and J. Klafter, *Phys. Today* **44,** 46 (1990).

[40] R. Artuso, *Phys. Rep.* **290,** 37 (1997).

[41] N. Periasamy and A. S. Verkman, *Biophys. J.* **75,** 557 (1998).

[42] J. B. Brochon, *Methods Enzymol.* **240,** 262 (1994).

3A shows $F(t)$ (on a log time scale) for a single diffusing species (70-kDa FITC–dextran; Fig. 3A, left), a mixture of two species with different diffusion coefficients (two FITC–Ficoll side fractions; Fig. 3A, middle), and simulated anomalous diffusion in which a nonuniform layer of fluorescein in saline was created (Fig. 3A, right). The fitted smooth curves through the data were obtained by MEM regression for the $\alpha(D)$ distribution shown in Fig. 3B. A unimodal narrow distribution was obtained for the FITC–dextran and a bimodal distribution was obtained for the FITC–Ficoll mixture. The sample simulating anomalous diffusion produced a broad distribution. Analysis of various theoretical cases of anomalous diffusion defined characteristic $\alpha(D)$ patterns including broad and asymmetric distributions.

An independent method to identify anomalous diffusive processes from $F(t)$ data was developed in which an apparent time-dependent diffusion coefficient, $D(t)$, was computed directly from $F(t)$ and the reference recovery curve.[41] Figure 3C (left) shows a constant $D(t)$ for simple diffusion of FITC–dextran, indicating that the intrinsic curve shapes for the reference recovery curve and that for recovery of FITC–dextran are identical up to a time scaling factor. [$D(t)$ was not generated for the case of two diffusing species each undergoing simple diffusion.] For the simulated anomalous subdiffusion sample, $D(t)$ decreased with time as predicted. The results indicate that determination of $\alpha(D)$ and $D(t)$ from photobleaching data provides a systematic approach to identify and quantify simple and anomalous diffusive phenomena. Further work using artificial and biological systems will be required to assess the utility of the various available methods to analyze photobleaching data.

Green Fluorescent Protein as a pH Sensor

In addition to its use as a probe for solute diffusion, GFP has considerable potential as a noninvasive sensor of biologically important intracellular functions. Energy transfer was exploited in the design of a protease sensor in which blue and green fluorescent GFP mutants were linked by a spacer containing a trypsin cleavage site.[43] A GFP-based fluorescence sensor of calcium utilizes energy transfer between blue and green fluorescent GFP mutants linked by a calmodulin-binding sequence.[44,45] Innovative GFP-fusion constructs will likely be useful for measurements of other parameters, such as the potential sensitivity conferred by fusion of GFP and a potassium channel domain.[46]

[43] R. Heim and R. Y. Tsien, *Curr. Biol.* **6,** 176 (1996).
[44] V. A. Romoser, P. M. Hinkle, and A. Persechini, *J. Biol. Chem.* **272,** 13270 (1997).
[45] A. Miyawaki, J. Lloopis, R. Heim, J. M. McCaffery, J. A. Adams, M. Ikura, and R. Y. Tsien, *Nature* (*London*) **388,** 882 (1997).
[46] M. S. Siegel and E. Y. Isacoff, *Neuron* **19,** 735 (1977).

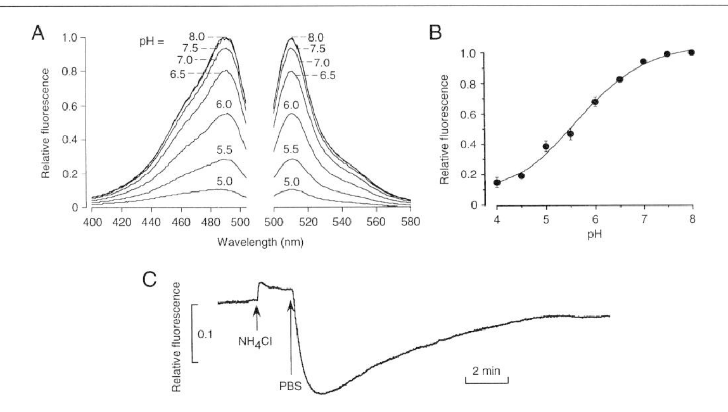

FIG. 4. GFP as a pH sensor. (A) Fluorescence excitation and emission spectra of GFP-S65T (~20 μg/ml) in PBS at the indicated pH. (B) Titration of GFP fluorescence versus pH in LLC-PK1 cells expressing GFP-F64L/S65T. Cells were initially perfused with PBS and then with PBS containing ionophores at the indicated pH. Averaged fluorescence (SE, n = 4) with fit to a single-site titration model. (C) Time course of cytoplasmic GFP-F64L/S65T fluorescence in response to replacement of 30 mM NaCl by NH_4Cl and subsequent return to NaCl.

On the basis of early studies of native GFP, in which it was noted that the absorbance or fluorescence of GFP was sensitive to pH,[47] we examined the possibility that GFP could serve as a noninvasive intracellular pH indicator.[48] It was found that the fluorescence of GFP is sensitive to pH *in vitro* and *in vivo,* and that its pH sensitivity could be modified by point mutations. pH titrations of purified recombinant GFP mutants indicated >10-fold reversible changes in absorbance and fluorescence, with pK_a values of 6.0 (GFP-F64L/S65T), 5.9 (S65T), 6.1 (Y66H), and 4.8 (T203I), and with apparent Hill coefficients of 0.7 for Y66H and ~1 for the other proteins. The pH titration for GFP-S65T is shown in Fig. 4A. For GFP-S65T in aqueous solution in the pH range 5–8, the fluorescence spectral shape, lifetime (2.8 nsec), and circular dichroism (CD) spectra were pH independent, and fluorescence responded reversibly to a pH change in <1 msec. At lower pH, the fluorescence response was slowed and not completely reversed. These findings suggest that GFP pH sensitivity involves simple protonation events at pH >5, but both protonation and conformational changes at lower pH.

GFP was evaluated as an intracellular pH indicator using transfected cells in which GFP-F64L/S65T was targeted to cytoplasm, mitochondria, Golgi, and endoplasmic reticulum. Calibration procedures were developed to determine the pH dependence of intracellular GFP fluorescence utilizing

[47] S. H. Bokman and W. W. Ward, *Biochem. Biophys. Res. Commun.* **101,** 1372 (1981).
[48] M. Kneen, J. Farinas, Y. Li, and A. S. Verkman, *Biophys. J.* **74,** 1591 (1998).

ionophore combinations (nigericin and CCCP) or digitonin. The pH sensitivity of GFP-F64L/S65T measured in cytoplasm (Fig. 4B) and organelles was similar to that of purified GFP-F64L/S65T in saline. NH_4Cl pulse experiments indicated that intracellular GFP fluorescence responds rapidly to a pH change (Fig. 4C). Applications of intracellular GFP were demonstrated including cytoplasmic and organellar pH measurement, pH regulation, and response of mitochondrial pH to protonophores. The results established the application of GFP as a targetable, noninvasive indicator of intracellular pH.

Acknowledgments

I thank Drs. J. Farinas and N. Periasamy for critical reading of this manuscript. This work was supported by NIH Grants DK43840, DK35124, HL51854, HL60288, DK16095, and HL59198, and by Grant R613 from the National Cystic Fibrosis Foundation.

[23] Application of Green Fluorescent Protein–Protein A Fusion Protein to Western Blotting

By Takashi Aoki, Katherine S. Koch, Hyam L. Leffert, and Hiroyuki Watabe

Introduction

Green fluorescent protein (GFP), isolated from the jellyfish *Aequorea victoria,* noncatalytically produces an intense and stable greenish fluorescence.[1] *Aequorea* GFP maximally absorbs blue light at 395 nm and emits green light with a peak at 509 nm. GFP is a protein of 238 amino acids with a molecular mass of 27–30 kDa as determined by sodium dodecyl sulfate–polyacrylamide gel electrophoresis (SDS–PAGE).[2–4] The hexapeptide segment, beginning at residue 64, functions as a fluorescent chromophore formed on cyclization of the residues Ser · dehydro-Tyr · Gly within the hexapeptide, by posttranslational modification.[5] GFP has a unique structure[6,7] and interesting physical properties (e.g., high stability to denaturing

[1] H. Morise, O. Shimomura, F. H. Johnson, and J. Winant, *Biochemistry* **13,** 2656 (1974).
[2] F. G. Prendergast and K. G. Mann, *Biochemistry* **17,** 3448 (1978).
[3] O. Shimomura, *FEBS Lett.* **104,** 220 (1979).
[4] D. C. Prasher, V. K. Eckenrode, W. W. Ward, F. G. Prendergast, and M. J. Cormier, *Gene* **111,** 229 (1992).
[5] C. W. Cody, D. C. Prasher, W. M. Westler, F. G. Prendergast, and W. W. Ward, *Biochemistry* **32,** 1212 (1993).
[6] M. Orm, A. B. Cubitt, K. Kallio, L. A. Gross, R. Y. Tsien, and S. J. Remington, *Science* **273,** 1392 (1996).
[7] F. Yang, L. G. Moss, and G. N. Phillips, Jr., *Nature Biotechnol.* **14,** 1246 (1996).

0076-6879/99 $30.00

reagents or proteases).[8] GFP fluorescence occurs without cofactors[9] and this property allows GFP fluorescence in nonnative organisms in which GFP is expressed. Although GFP is relatively large to serve as a fusion tag, GFP-tagged proteins retain their original functions in many cases.[10,11] Therefore, GFP has been used as a reporter for gene expression, as a tracer of cell lineage,[12] and as a fusion tag to investigate protein localization[13] and secretion systems *in vivo*.[14–16]

The reports of GFP fusions to protein A[17] and to streptavidin,[18] and of a simple immunoassay system using a GFP tag,[19] also indicate a wide range of *in vitro* applications. In this chapter we review the construction of a protein A–GFP fusion (PA–GFP) and its use as a labeled antibody-specific ligand in immunoblotting. Immunoblotting requires a labeled antibody or antibody-specific ligand (such as protein A) and a system specifically for detection of the label. Labeling reagents frequently used have been enzymes such as peroxidase, alkaline phosphatase, and β-galactosidase; gold particles, radioisotopes, and fluorochromes have also been used. Now the fluorescence of protein A tagged with GFP may allow simple and rapid detection of blotted proteins.

Construction and Expression of Protein A–Green Fluorescent Protein

Bacterial Strains and Plasmids

Escherichia coli HB101 and N4830-1[20] are used as host strains for amplication of pRIT2T,[21] a protein A gene fusion vector (Pharmacia Biotech, Uppsala, Sweden). The λ right promoter in pRIT2T is induced by shifting the growth of N4830-1 from 30 to 42° for 90 min; and HB101 is used for constitutive expression of the promoter. Plasmid TU#65,[9] containing

[8] S. H. Bokman and W. W. Ward, *Biochem. Biophys. Res. Commun.* **101,** 1372 (1981).
[9] M. Chalfie, Y. Tu, G. Euskirchen, W. W. Ward, and D. C. Prasher, *Science* **263,** 802 (1994).
[10] H.-H. Gerdes and C. Kaether, *FEBS Lett.* **389,** 44 (1996).
[11] T. Misteli and D. L. Spector, *Nature Biotechnol.* **15,** 961 (1997).
[12] D. Tannahill, S. Bray, and W. A. Harris, *Dev. Biol.* **168,** 694 (1995).
[13] S. Wang and T. Hazelrigg, *Nature (London)* **369,** 400 (1994).
[14] C. Kaether and H.-H. Gerdes, *FEBS Lett.* **369,** 267 (1995).
[15] R. Y. Hampton, A. Koning, R. Wright, and J. Rine, *Proc. Natl. Acad. Sci. U.S.A.* **93,** 828 (1996).
[16] A. E. Pouli, H. J. Kennedy, J. G. Schofield, and G. A. Rutter, *Biochem. J.* **331,** 669 (1998).
[17] T. Aoki, Y. Takahashi, K. S. Koch, H. L. Leffert, and H. Watabe, *FEBS Lett.* **384,** 193 (1996).
[18] C. Oker-Blom, A. Orellana, and K. Keinanen, *FEBS Lett.* **389,** 238 (1996).
[19] T. Aoki, M. Kaneta, H. Onagi, Y. Takahashi, K. S. Koch, H. L. Leffert, and H. Watabe, *J. Immunoassay* **18,** 321 (1997).
[20] M. Zabeau and K. K. Stanley, *EMBO J.* **1,** 1217 (1982).
[21] B. Nilsson, L. Abrahmsén, and M. Uhlén, *EMBO J.* **4,** 1075 (1985).

Aequorea GFP cDNA, is provided by M. Chalfie (Columbia University, New York, NY). *Escherichia coli* JM109 (pHTK503), a strain producing human neuron-specific enolase (NSE),[22] is used for blotting studies.

Buffer Solutions

TEN buffer: 20 m*M* Tris-HCl, pH 8.0; 10 m*M* EDTA; 30 m*M* NaCl
Elution buffer: 20 m*M* Tris-HCl, pH 8.0; 10 m*M* EDTA; 30 m*M* NaCl; 9 *M* urea
TBS (Tris-buffered saline): 50 m*M* Tris-HCl, pH 7.4; 200 m*M* NaCl
Protein sample buffer (3×) for SDS–PAGE: 50% (v/v) glycerol; 20% 2-mercaptoethanol; 10% SDS; 300 m*M* Tris-HCl, pH 6.8; bromphenol blue (0.5 mg/ml)

Plasmid Construction and Expression

A modified GFP cDNA is prepared from plasmid TU#65 by standard polymerase chain reaction (PCR) using the following primers: 5′-CCCGAATTCCATGAGTAAAGGAGAAGAACTTTTCAC-3′ and 5′-AAAGGATCCCTATTATTTGTATAGTTCATCCA-3′ (*Eco*RI and *Bam*HI sites are underlined). The fragment is cloned into pRIT2T by standard procedures. Enzymes are purchased from GIBCO-BRL (Gaithersburg, MD) or from Nippon Gene (Tokyo, Japan). DNA is transformed into *E. coli* HB101 or N4830-1 with $CaCl_2$,[23] and transformants are selected on LB agar plates containing ampicillin (50 μg/ml). Recombinant plasmids are isolated as described by Birnboim and Doly[24] with slight modifications, and analyzed by agarose gel electrophoresis. DNA is sequenced using a fluorescence imaging analyzer FMBIO-100 (Takara Shuzo, Tokyo, Japan).

The amplified fragment (0.75 kb), containing the preceding *Eco*RI and *Bam*HI sites at each end, is digested with these enzymes and cloned into pRIT2T. The resulting plasmid, pPAGFP127, contains a presumed protein A–GFP cDNA encoding a protein of about 60 kDa, which would be composed of 500 amino acids and the initiating methionine residue downstream from the λ right promoter (Fig. 1a). Cell extracts prepared from *E. coli* HB101 strains harboring pRIT2T or pPAGFP127 are analyzed by SDS–PAGE (Fig. 1b). Proteins are separated on 0.1% SDS–14% polyacrylamide gels according to Laemmli.[25] SDS–PAGE molecular mass standard kits are purchased from Bio-Rad Laboratories (Hercules, CA). The separated proteins are stained with 0.25% Coomassie Brilliant Blue R-250 (CBB). PA–GFP is not detected as a major band by CBB staining (Fig. 1b, lane

[22] T. Aoki, M. Kimura, M. Kaneta, H. Kazama, J. Morikawa, and H. Watabe, *Tumor Biol.* **14,** 261 (1993).
[23] M. Mandel and A. Higa, *J. Mol. Biol.* **53,** 159 (1970).
[24] H. C. Birnboim and J. Doly, *Nucleic Acids Res.* **7,** 1513 (1979).
[25] U. K. Laemmli, *Nature* (*London*) **227,** 680 (1970).

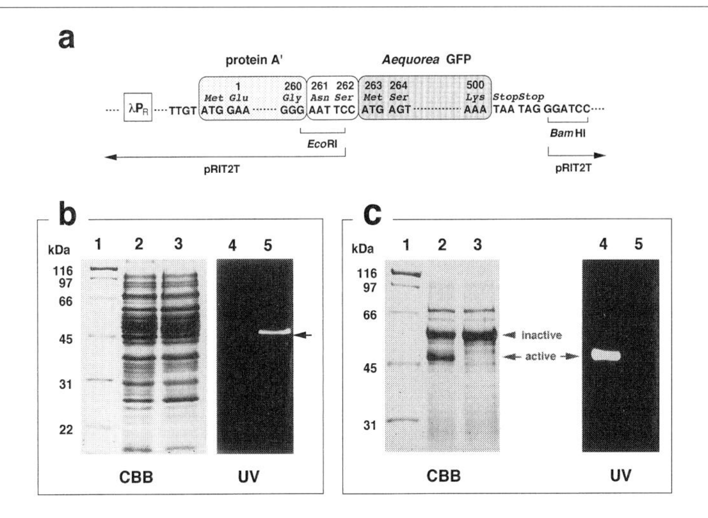

FIG. 1. Construction of the recombinant plasmid pPAGFP127 and SDS–PAGE of PA–GFP expressed in *E. coli.* (a) A 0.75-kb *Aequorea* GFP cDNA was modified and amplified by PCR, digested with *Eco*RI and *Bam*HI, and cloned into pRIT2T. Transcription of the fusion protein (PA–GFP) was controlled by the λ right promoter in pRIT2T. (b) *Escherichia coli* HB101 strains harboring pRIT2T or pPAGFP127 were grown sequentially on LB agar plates at 37° (1 day) and 25° (3 days). Cells collected from two plates (about 0.3 g) were suspended in 1 ml of TEN buffer and sonicated. Supernatant (1 μl) obtained after centrifugation was resolved by SDS–PAGE. Cell extracts, prepared from strains harboring pRIT2T (lanes 2 and 4) or pPAGFP127 (lanes 3 and 5), were stained with CBB (lanes 2 and 3) or irradiated by UV light at 365 nm (lanes 4 and 5). Molecular mass markers are shown in lane 1. Fluorescent PA–GFP is indicated by the arrow. (c) PA–GFP was purified as described in text. After dialysis, the fraction containing PA–GFP (5.2 ml) was resolved by SDS–PAGE. Purified PA–GFP (20 μl) was loaded directly onto the gel (lanes 2 and 4) or after being heated at 95° for 5 min (lanes 3 and 5). PA–GFP was stained with CBB (lanes 2 and 3) or irradiated by UV light at 365 nm (lanes 4 and 5). Fluorescent and nonfluorescent PA–GFP bands are indicated as "active" (arrow) and "inactive" (arrowhead), respectively. Molecular mass markers are in lane 1. [(a) through (c) are taken from Ref. 17 with permission.]

3). However, a strong greenish fluorescent band of 47 kDa is visualized by UV irradiation in the sample from the strain harboring pPAGFP127 (Fig. 1b, lane 5). This molecular mass differs from the theoretical value of 60 kDa for PA–GFP.

The λ right promoter is commonly induced by shifting the growth temperature from 30 to 42° for 90 min in *E. coli* N4830-1, which produces a temperature-sensitive repressor.[20] However, fluorescent PA–GFP is not expressed by this standard induction procedure. Instead, fluorescent PA–GFP is constitutively and most efficiently expressed in HB101, at or below 25°, on agar plates rather than in liquid medium. The fluorescence intensity increases gradually, reaching a maximum at 3 to 5 days at low temperature under oxygen-rich conditions on plates, rather than by the faster expression at higher temperatures. Also, fluorescence is greatest in newly transformed cells, decreasing with replating. Inouye and Tsuji have reported that the fluorescence intensity of intact *Aequorea* GFP expressed in *E. coli* increases in harvested cells stored overnight at room temperature.[26] Lim *et al.* have also reported the thermosensitive formation of fluorescent GFP in *Saccharomyces cerevisiae.*[27] Codon-usage modifications of GFP cDNAs have created new forms of GFP that fluoresce at high temperature.[28] Use of such "thermoresistant GFP" could require alteration of the growth conditions described herein.

Purification, Electrophoretic and Hydropathy Analysis of Protein A–Green Fluorescent Protein

To clarify molecular properties of fluorescent PA–GFP from *E. coli* cells, PA–GFP from cell extracts of HB101 harboring pPAGFP127 was isolated by passage through a column containing rabbit IgG coupled to Sepharose 6B.

Purification Procedure

The HB101 strain harboring pPAGFP127 is grown on LB agar plates (8.5 cm in diameter) at 37°. One day after plating, the incubation temperature is shifted to 25°. Cells are collected from 19 plates after a 3-day incubation of 25°. Approximately 3.4 g wet weight of cells is suspended in 9.5 ml of TEN buffer, and disrupted by sonic oscillation. Two milliliters of the supernatant, obtained by centrifugation at 27,000 *g*, is diluted 1 : 10 with TEN buffer and used as the "PA–GFP working solution" for the blotting studies. The remaining supernatant is applied to a column (1.5 × 10 cm) containing rabbit IgG immobilized on Sepharose 6B. The column is washed with TEN buffer and the bound material recovers with the elution buffer (TEN buffer containing 9 *M* urea). Fractions emitting green fluorescence under a UV light at 365 nm (Manaslu-Light; Manaslu Chemical Industry, Tokyo, Japan) are pooled and dialyzed against TEN buffer.

Electrophoretic Analysis

As shown in Fig. 1c, purified PA–GFP migrated as major CBB bands at 47 and 57 kDa. However, only the 47-kDa band was fluorescent. When the sample was heated to 95°, the lower band was no longer detected by CBB stain or by fluorescence, while only a nonfluorescent band was detected at the 57-kDa position, corresponding to the upper band in lane 2 (Fig. 1c).

[26] S. Inouye and F. I. Tsuji, *FEBS Lett.* **351,** 211 (1994).

[27] C. R. Lim, Y. Kimata, M. Oka, K. Nomaguchi, and K. Kohno, *J. Biochem.* (*Tokyo*) **118,** 13 (1995).

[28] Y. Kimata, M. Iwai, C. R. Lim, and K. Kohno, *Biochem. Biophys. Res. Commun.* **232,** 69 (1997).

The theoretical molecular mass of PA–GFP is ~60 kDa. Thus, the upper band in lane 2 (Fig. 1c) may be a nonfluorescent inactive PA–GFP; and the lower band a fluorescent active PA–GFP, although its migration distance does not correspond to its theoretically predicted molecular mass. It is unclear if the inactive PA–GFP is an artifact of purification or a natural GFP metabolite.

Intact *Aequorea* GFP expressed in *E. coli* migrated at ~40 kDa as a greenish fluorescent band on SDS–PAGE.[26] This mobility pattern also did not correspond to the theoretical molecular mass of 28 kDa; on the other hand, heat-denatured nonfluorescent GFP migrated as predicted by theoretical mass. Therefore, it would appear that when active GFP molecules were inactivated, the mobility on SDS–PAGE shifted from the upper (40 kDa) to the lower position (28 kDa). In contrast, our results indicate that the mobility of protein A–GFP fusion molecules shifted from the lower (47 kDa) to the upper position (57 kDa) on denaturation. The effect of the fusion on mobility in SDS–polyacrylamide gels is not clearly understood. Significant solubility differences between GFP and PA–GFP molecules are not predicted from their mean hydrophobicity values (0.53 and −0.86, respectively)[29]; furthermore, the hydrophobicity profiles of their active hexapeptide centers (F·S·dehydro-Y·G·V·Q) are identical (data not shown). However, from these results, we can speculate as follows: (1) GFP has a high stability to SDS[8] because of its special structure,[6,7] and this stability influences its mobility on SDS–PAGE; (2) PA–GFP is not denatured completely by SDS alone; and (3) non- or partially denatured PA–GFP migrates at a position that differs from that of PA–GFP completely denatured by SDS and heating, because the two molecules differ in conformation and in amount of bound SDS.

Application of Protein A–Green Fluorescent Protein to Immunoblotting

Blotting and Detection Procedure

Proteins separated on 0.1% SDS–14% polyacrylamide gels are electroblotted onto nitrocellulose membranes. Electroblotting is performed using Trans-Blot SD (Bio-Rad) and Towbin buffer at a constant voltage of 20 V for 90 min. Proteins are also dot-blotted using the Easy-Titer enzyme-linked immunoflow assay (ELIFA) system (Pierce Chemical, Rockford, IL). The blotted proteins are visualized by fluorescence detection of PA–GFP or by conventional color detection of horseradish peroxidase (HRP)-labeled second antibody, as outlined in Fig. 2.

Fluorescence Detection Using Protein A–Green Fluorescent Protein. After blocking with TBS containing 3% bovine serum albumin (BSA) for

[29] J. Kyte and R. F. Doolittle, *J. Mol. Biol.* **157,** 105 (1982).

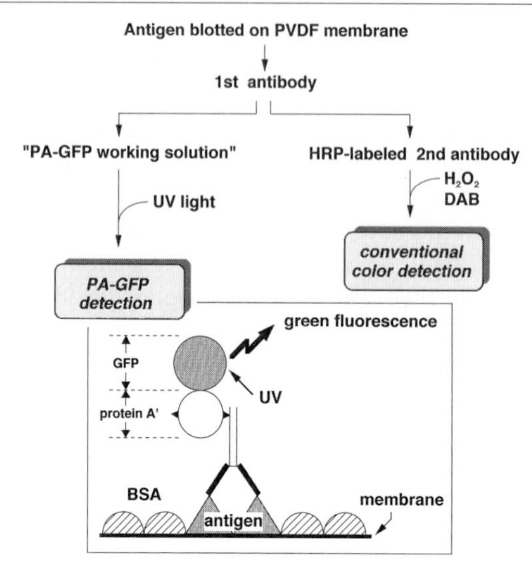

FIG. 2. Detection of blotted antigen using PA–GFP. PA–GFP bound to the antibody was irradiated by UV light and the green fluorescence emission was recorded on Polapan 3200B (Polaroid). Details are given in text.

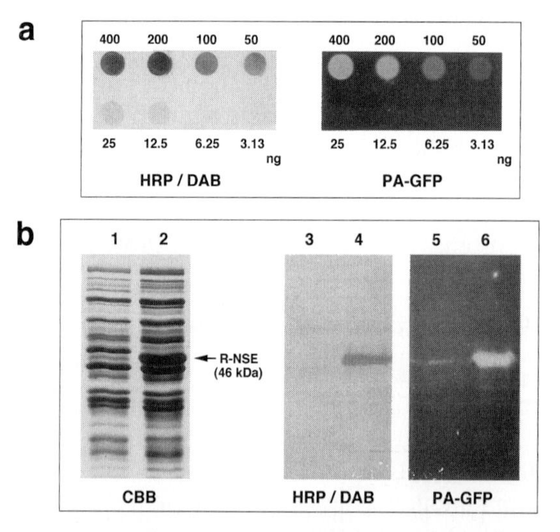

FIG. 3. Comparison between antigen detection methods employing PA–GFP and conventional HRP–DAB detection reagents. (a) Purified R-NSE[22] (3.13–400 ng) was blotted onto a nitrocellulose membrane as 4-mm spots. (b) *Escherichia coli* extracts prepared from R-NSE-nonproducing strain harboring pKK223-3 (lanes 1, 3, and 5), or from an R-NSE-producing strain harboring pHTK503 (lanes 2, 4, and 6), were resolved by SDS–PAGE and electroblotted as described.[22] The separated proteins were stained with CBB (lanes 1 and 2), and the R-NSE band was detected with HRP–DAB (lanes 3 and 4) or PA–GFP (lanes 5 and 6) reagents as described in text. [Taken from Ref. 17 with permission.]

1 hr at 25°, and incubation with rabbit anti-human NSE (Eiken Chemical, Tokyo, Japan) for 1 hr at 37°, the bound antibodies are visualized by incubation in PA–GFP working solution for 20 min at 37°. The greenish fluorescent bands or spots are detected under a UV light at 365 nm, and recorded on Polapan 3200B (Polaroid; Eastman Kodak, Rochester, NY) using a green filter.

Conventional Color Detection. After blocking and incubation with anti-NSE, membranes are incubated in TBS containing second antibody (HRP-labeled anti-rabbit IgG, 2 μg/ml) and 5% BSA for 1 hr at 37°. HRP is visualized by incubation with TBS containing 0.02% 3,3′-diaminobenzidine tetrahydrochloride (DAB; Sigma Chemical, St. Louis, MO) and 0.01% H_2O_2 for 10 min as described by Hawkes *et al.*[30]

Detection of Human Neuron-Specific Enolase

Dot and Western blots were incubated with PA–GFP working solution. The sensitivity and specificity were compared with those of the standard HRP–DAB system. Recombinant human NSE (R-NSE)[22] was used as a model antigen, and the dot blots are shown in Fig. 3a. The HRP–DAB system was slightly more sensitive than the PA–GFP system, but the sensitivity of PA–GFP detection still may be sufficient, and it provides rapid and easy screening. The cell extract prepared from the R-NSE-producing strain[22] was also examined by Western blotting. As shown in Fig. 3b, the specificity of PA–GFP detection was identical to that of the HRP–DAB system.

PA–GFP detection is performed with commonly used laboratory equipment. Although special (and expensive) equipment is not required, the computer-controlled fluorescence scanning system facilitates rapid and precise quantitation. In addition, thermoresistant GFP[28] and mutants[31–33] with red-shifted excitation spectra and with increased fluorescence signals may lead to even more sensitive PA–GFP detection systems.

Additional features of PA–GFP detection are as follows: (1) chemicals for detection are not required, therefore the procedure is simple, rapid, and inexpensive; (2) reactions are controlled easily by monitoring the fluorescent band under UV light; (3) fluorescence is stable to long UV irradiation; (4) fluorescence and binding activities do not decrease when PA–GFP is stored (for at least 1 month at −20°); and (5) the purification of PA–GFP is not required for blotting, as the crude cell extract of the strain harboring

[30] R. Hawkes, E. Niday, and J. Gordon, *Anal. Biochem.* **119,** 142 (1982).

[31] A. B. Cubitt, R. Heim, S. R. Adams, A. E. Boyd, L. A. Gross, and R. Y. Tsien, *Trends Biochem. Sci.* **20,** 448 (1995).

[32] R. Heim, A. B. Cubitt, and R. Y. Tsien, *Nature (London)* **373,** 663 (1995).

[33] A. Crameri, E. A. Whitehorn, E. Tate, and W. P. C. Stemmer, *Nature Biotechnol.* **14,** 315 (1996).

pPAGFP127 can be utilized directly. In conclusion, PA–GFP is a useful tool for blotting studies; and the sensitivity and specificity of PA–GFP detection may be sufficient for rapid and easy screening, particularly for monitoring the expression of recombinant proteins such as NSE.[22]

Acknowledgments

This work was supported in part by Grant-in-Aid for High Technology Research Program from the Ministry of Education, Sciences, Sports and Culture of Japan.

[24] Green Fluorescent Protein as a Reporter for Promoter Analysis of Testis-Specific Genes in Transgenic Mice

By P. PRABHAKARA REDDI, MARKO KALLIO, and JOHN C. HERR

Introduction

Spermatogenesis, a complex cell differentiation process culminating in the production of spermatozoa, is accompanied by the expression of a number of genes encoding novel proteins that make up the unique architecture of the sperm. Although a number of testis-specific genes have been identified and characterized,[1] information on the mechanisms of transcriptional regulation in the testis is limited. This, in large part, can be attributed to the lack of a convenient cell culture system that allows spermatogenesis to proceed *in vitro*.[2] Promoter analysis of testis-specific genes, therefore, necessarily has involved generation of transgenic mice. Using transgenic technology, the functional domains in the promoters of a few testis-specific genes have been identified.[3–8]

[1] D. J. Wolgemuth and F. Watrin, *Mammalian Genome* **1,** 283 (1991).
[2] M. J. Wolkowicz, S. A. Coonrod, P. P. Reddi, J. L. Millan, M. C. Hofmann, and J. C. Herr, *Biol. Reprod.* **55,** 923 (1996).
[3] J. J. Peschon, R. R. Behringer, R. L. Brinster, and R. D. Palmiter, *Proc. Natl. Acad. Sci. U.S.A.* **84,** 5316 (1987).
[4] T. A. Stewart, N. B. Hecht, P. G. Hollingshead, P. A. Johnson, J. A. Leong, and S. L. Pitts, *Mol. Cell. Biol.* **8,** 1748 (1988).
[5] M. M. Gebara and J. R. McCarrey, *Mol. Cell. Biol.* **12,** 1422 (1992).
[6] T. Howard, R. Balogh, P. Overbeek, and K. E. Bernstein, *Mol. Cell. Biol.* **13,** 18 (1993).
[7] Z. Galcheva-Gargova, J. P. Tokeson, L. K. Karagyosov, K. M. Ebert, and D. L. Kilpatrick, *Mol. Endocrinol.* **7,** 979 (1993).
[8] N. L. Shaper, A. Harduin-Lepers, and J. H. Shaper, *J. Biol. Chem.* **269,** 25165 (1994).

0076-6879/99 $30.00

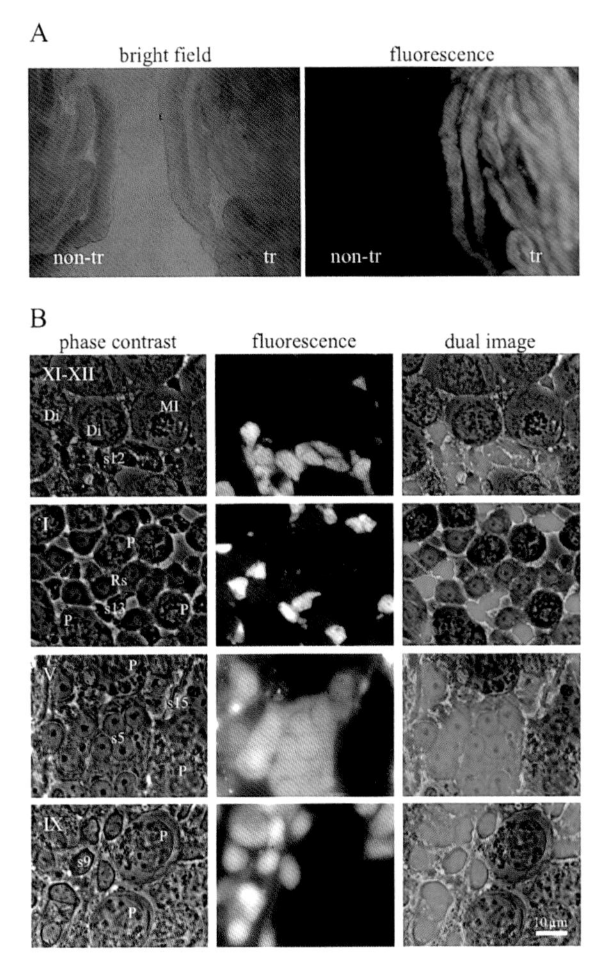

FIG. 2. Expression of GFP in the testis of SP-10-GFP transgenic mice. (A) Bright-field and fluorescence views of the seminiferous tubules from nontransgenic (non-tr) and transgenic (tr) mice. Under the control of SP-10 promoter fragment, GFP was produced in the seminiferous tubules of the transgenic mice (tr). (B) Identification of stage- and germ cell type-specific transgene expression. Photomicrographs of stages XI–XII, I, V, and IX of the seminiferous epithelium cycle, are arranged vertically; the phase-contrast, fluorescence, and dual (phase and fluorescence) images of the same fields are placed horizontally. At stages XI–XII of the cycle the diplotene spermatocytes (Di), meiotically dividing spermatocytes (MI), and step 12 spermatids (s12), are identified (phase contrast). The lack of GFP signal in the spermatocytes (fluorescence) indicates that the reporter gene is not transcribed during meiosis. The haploid step 12 spermatids, on the other hand, are GFP positive (fluorescence). At stage I, round spermatids (Rs), pachytene spermatocytes (P), and step 13 spermatids (s13) can be seen (phase contrast). The onset of GFP expression is first detected as dim fluorescence over the round spermatids (Rs) at stage I (fluorescence and dual images). The intensity of the GFP signal has increased in the step 5 spermatids (s5) at stage V, reflecting increased transcriptional activity and accumulation of the protein. However, even after the transcription has ceased, the bright fluorescent signal persisted in later steps of spermatid development (s9 at stage IX, and s12, s13, s15 at stages XI–XII, I, V, respectively), owing to the long half-life of GFP. In summary, GFP production under SP-10 promoter control first begins in the early round spermatids (Rs).

Identification of the stage of spermatogenesis, and the germ cell type in which the transgene transcription has been initiated, is critical to the analysis of testis-specific promoters. The choice of the reporter gene and the assays required to detect reporter gene activity play an important role in determining the spatial and temporal gene expression pattern imparted by the promoter fragment. Green fluorescent protein (GFP), derived from the jellyfish *Aequorea victoria,*[9] is particularly useful as a reporter molecule in the testis. Because GFP emits bright green fluorescence when excited by blue light, live spermatogenic cells expressing GFP under the control of a testis-specific promoter can be identified by fluorescence microscopy. When coupled with a transillumination-assisted microdissection technique,[10] which permits the isolation of stages of spermatogenesis from freshly dissected seminiferous tubules, the developmental stage- and cell type-specific onset of promoter activity can be accurately determined. Visualization of GFP does not require incubation with exogenous substrates or cofactors. This contrasts with other reporter genes previously expressed in the testis, such as the bacterial β-galactosidase or the mammalian growth hormone genes, which require fixation of the tissues, sectioning, and cytochemical or immunological staining for detection.[3,4,6] Using the GFP reporter system, we have dissected the promoter of a testis-specific gene, SP-10, in transgenic mice and identified the -408 to $+28$ base pair (bp) promoter region to be sufficient for round spermatid-specific gene expression.[11–15] In this chapter, we describe experimental procedures for the use of GFP as a reporter for promoter analysis of testis-specific genes, and include examples of GFP expression in the testes of transgenic mice.

Strategy for Generation of Transgenic Mice to Analyze Testis-Specific Promoters

Green Fluorescent Protein Reporter Gene Constructs

As a first step in promoter analysis, the transcriptional start point (+1 site) of the testis-specific gene under study must be determined using a

[9] D. C. Prasher, V. K. Eckenrode, W. W. Ward, F. G. Prendergast, and M. J. Cormier, *Gene* **111,** 229 (1992).

[10] M. Parvinen and T. Vanha-Perttula, *Anat. Rec.* **174,** 435 (1972).

[11] J. C. Herr, R. M. Wright, E. John, J. Foster, T. Kays, and C. J. Flickinger, *Biol. Reprod.* **42,** 377 (1990).

[12] B. E. Kurth, K. Klotz, C. J. Flickinger, and J. C. Herr, *Biol. Reprod.* **44,** 814 (1991).

[13] B. E. Kurth, R. M. Wright, C. J. Flickinger, and J. C. Herr, *Anat. Rec.* **236,** 619 (1993).

[14] P. P. Reddi, S. Naaby-Hansen, I. Aguolnik, J.-Y. Tsai, L. M. Silver, C. J. Flickinger, and J. C. Herr, *Biol. Reprod.* **53,** 873 (1995).

[15] P. P. Reddi, C. J. Flickinger, and J. C. Herr, in preparation (1999).

primer extension assay and/or RNase protection assay, following standard methods.[16] The DNA sequence upstream of the +1 site is generally considered as the promoter region responsible for gene regulation. To begin testing for promoter activity, a −3 kb to +20 bp promoter region (where +1 is the transcriptional start point) may be cloned upstream of the reporter gene coding region. Two or more additional reporter gene constructs may also be made that contain a 5′ truncated version of the promoter to aid in identification of the region responsible for testis-specific expression. The choice of the type of GFP cDNA to be used as the reporter is made on the basis of the application. For promoter analysis in transgenic mice, a GFP that has been optimized for expression in mammalian cells is preferred. The promoterless cloning vector pEGFP1 (Clontech, Palo Alto, CA) encodes a red-shifted variant of the wild-type GFP that has been optimized for brighter fluorescence and higher expression in mammalian cells.[17,18] Moreover, because the maximal excitation peak of this enhanced GFP (EGFP) is at 488 nm, commonly used filter sets such as fluorescein isothiocyanate (FITC) optics that illuminate at 450–500 nm can be used to visualize GFP fluorescence (see [32] in this volume[18a]). pEGFP1 proved to be useful as a reporter vector for promoter analysis in transgenic mice (Okabe *et al.*[19] and our unpublished results[15]).

There are two ways in which the putative promoter fragments of the gene of interest can be prepared (usually from a parent phage clone containing 8–10 kb of genomic DNA for cloning into the multiple cloning site of a GFP reporter vector. In the first method, restriction endonucleases are used to excise the promoter fragments to be inserted into the reporter vector. For example, if *Eco*RI and *Bam*HI sites were present at −3 kb and +20 bp positions of the promoter fragment, then the −3 kb to +20 bp fragment can be generated by digestion with *Eco*RI and *Bam*HI. However, the feasibility of this method depends on the availability of proper restriction endonuclease sites in the promoter fragment. In the second, and more commonly used method, the required promoter fragment is amplified by the polymerase chain reaction (PCR)[20] using oligonucleotide primers bearing the appropriate sites for restriction endonuclease cleavage. In this method, the sequence necessary for restriction cleavage is included at the

[16] J. Sambrook, E. F. Fritsch, and T. Maniatis, "Molecular Cloning: A Laboratory Manual." Cold Spring Harbor Press, Cold Spring Harbor, New York, 1989.

[17] B. P. Cormack, R. Valdivia, and S. Falkow, *Gene* **173,** 33 (1996).

[18] J. Haas, E.-C. Park, and B. Seed, *Curr. Biol.* **6,** 315 (1996).

[18a] G. J. Paler and A. Wlodawer, *Methods Enzymol.* **302,** [32], 1999 (this volume).

[19] M. Okabe, M. Ikawa, K. Kominami, T. Nakanishi, and Y. Nishimune, *FEBS Lett.* **407,** 313 (1997).

[20] R. K. Saiki, D. H. Gelfand, S. Stoffel, S. J. Scharf, R. Higuchi, G. T. Horn, K. B. Mullis, and H. A. Erlich, *Science* **239,** 487 (1988).

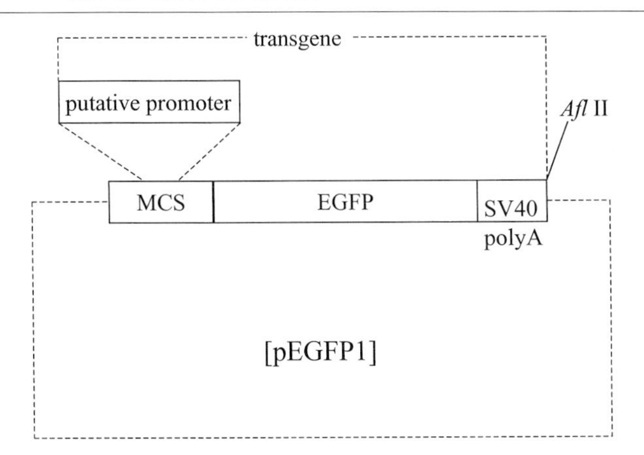

FIG. 1. Schematic of the GFP reporter gene construct used for promoter analysis of testis-specific genes in transgenic mice. The putative promoter fragment is inserted into the multiple cloning site (MCS) of the promoterless vector pEGFP1 (Clontech) upstream of the coding sequence for EGFP. EGFP encodes a red-shifted variant of the wild-type GFP that has been optimized for brighter fluorescence as well as expression in mammalian systems. The SV40 polyadenylation signals, located at the 3′ region, direct the processing of EGFP mRNA. The transgene fragment to be excised for microinjection into the fertilized eggs is indicated.

5′ end of the forward and reverse primers that flank the promoter fragment to be amplified. After PCR amplification, the appropriate ends are generated by restriction digestion of the PCR product. The promoter fragments, generated by either method, are then ligated into the multiple cloning site of the reporter vector following standard cloning procedures.[16] (*Note:* It is recommended that the DNA sequence of the PCR-generated promoter fragments in the constructs be verified prior to generation of transgenic mice.) The resulting reporter gene construct will contain the putative promoter fragment located upstream of the GFP cDNA (Fig. 1).

Purification of Transgene for Microinjection

For microinjection of fertilized eggs, a linear DNA fragment (the transgene; Fig. 1) containing the promoter, the reporter gene, and the polyadenylation signals is excised from the reporter gene construct. The transgene is gel purified by the electroelution method; any traces of impurities are further removed by passage through an Elutip D column (Schleicher & Schuell, Dassel, Germany).

Reagents and Materials

TAE buffer (1×: 0.04 *M* Tris-acetate, 0.001 *M* EDTA, pH 8.0)
High-salt buffer (1.0 *M* NaCl, 20 m*M* Tris-HCl, 1.0 m*M* EDTA, pH 7.5)

Low-salt buffer (0.2 *M* NaCl, 20 m*M* Tris-HCl, 1.0 m*M* EDTA, pH 7.5)
Injection buffer (10 m*M* Tris, 0.1 m*M* EDTA, pH 7.5, prepared with Milli-Q quality water)
Horizontal gel electrophoresis system, dialysis tubing, Elutip D columns, 5-ml syringes

Protocol

1. Digest 50–100 μg of the reporter gene construct using appropriate restriction endonucleases to release the transgene fragment (see example in Fig. 1).

2. Resolve the restriction endonuclease-cleaved products in a 1% (w/v) agarose gel containing ethidium bromide (0.5 μg/ml) and TAE buffer at 5–6 V/cm. Observe the gel on a UV transilluminator (use a long-wavelength UV lamp to reduce nicking of DNA); locate the transgene band by size and carefully excise the gel piece containing the required band.

3. Add the gel slice and 1 ml of 0.5× TAE buffer to a dialysis bag (which has been boiled in 1 m*M* EDTA, pH 8.0, for 10 min[16]), and fasten the ends.

4. Submerge the dialysis bag containing the gel piece in a horizontal gel electrophoresis chamber containing 0.5× TAE buffer. Electrophorese at 5–6 V/cm for 45 min. Reverse the current flow in the electrophoresis chamber for 1 min before stopping the run (this is to release any DNA that may be attached to the wall of the dialysis tube).

5. Collect the TAE buffer from the dialysis bag into a fresh eppendorf tube. This contains the electroeluted DNA. The gel piece may be observed on the UV transilluminator to ascertain that the electroelution of the DNA is complete. The electroeluted DNA sample is further purified by passage through Elutip D columns.

6. Use a 5-ml syringe to apply solutions to the Elutip D column, avoiding reverse flow. Prewash the matrix of the column with 1–2 ml of high-salt buffer, followed by a wash with 5 ml of low-salt buffer.

7. Slowly load the solution containing the electroeluted DNA. Wash the column with 2–3 ml of low-salt buffer. Elute the DNA in 0.4 ml of high-salt buffer. Add 2 vol of cold 95% ethanol to precipitate the DNA.

8. Collect the DNA by centrifugation in a microcentrifuge at 14,000 *g* for 10 min; carefully remove the alcohol without disrupting the DNA pellet. Wash the pellet at least twice with 70% (v/v) ethanol, and dry the pellet. (The washing and drying steps are important, as residual salt and ethanol are lethal to the developing embryos.) Resuspend the DNA in injection buffer.

9. Determine the concentration of the purified transgene DNA fragment by measuring the A_{260} value ($A_{260} = 1$ for 50 μg of DNA per milliliter), using a spectrophotometer. DNA prepared in this manner is suitable for microinjection into fertilized mouse eggs.

Generation of Transgenic Mice

The purified transgene fragment is microinjected into the male pronuclei of fertilized eggs obtained from B6 CBA females by standard methods.[21] Tail biopsies are collected from 3-week-old pups and genomic DNA isolated.[21] Founder transgenic mice are identified by PCR analysis of genomic DNA, using GFP primers (5′ primer, 5′ ATGGTGAGCAAGGGCGAG GAGCTG; and 3′ primer, CTTGTACAGCTCGTCCATGCCGAG) that are complementary to the GFP cDNA (a 716-bp region between bp 97 through 813 in the pEGFP1 vector, GenBank accession number U55761) in the reporter gene construct. The PCR is carried out in a volume of 100 μl containing 1 μg of genomic DNA, in 1× reaction buffer supplemented with 0.2 mM dNTPs, 2 mM $MgCl_2$, 100 picomoles of each of primer, and 2.5 units of *Taq* polymerase (Promega, Madison, WI). Each of the 30 PCR cycles consists of denaturation at 94° for 1 min, annealing at 54° for 1 min, and extension at 72° for 1 min. The founder mice may be identified by the presence of the 716-bp GFP PCR product. The founder mice are then mated with C57BL partners to generate transgenic F_1 lines of mice.

Analysis of Transgenic Mice Expressing Green Fluorescent Protein

Tissue Specificity of Green Fluorescent Protein Expression

Sexually mature F_1 transgenic males are analyzed to determine whether the putative promoter conferred the expected spatial and temporal expression pattern to the GFP reporter. Because expression of GFP results in fluorescent labeling of living cells (Okabe *et al.*[19] and our unpublished results[15]), a rapid screening for tissues in which the transgene is expressed can be performed using a fluorescence microscope equipped with FITC optics.

Method

1. Sacrifice 6 to 8-week-old transgenic male mice by CO_2 inhalation, or any other procedure approved by the institution.
2. Remove testis, liver, heart, kidney, lung, spleen, muscle, intestine, and brain into separate petri dishes containing sterile phosphate-buffered saline (PBS) at 37°.

[21] B. Hogan, "Manipulating the Mouse Embryo: A Laboratory Manual." Cold Spring Harbor Laboratory Press, Cold Spring Harbor, New York, 1986.

3. Cut a small piece (1–2 mm) from each tissue, place on a glass slide in 30 μl of PBS, and mince finely with forceps. Gently apply a coverslip on top and seal the edges using paraffin oil to prevent evaporation.

4. Observe the slides at ×10–40, under fluorescence optics. The cells from the tissues in which GFP gene is expressed will fluoresce bright green.

5. Observe tissue samples from a nontransgenic littermate processed in the same way. This serves as a negative control for autofluorescence.

The SP–10–GFP mice, in which GFP is placed under the control of the SP-10 promoter,[15] were analyzed following this procedure. Bright green fluorescence was observed only in the testicular cells. All somatic tissues obtained from the SP-10–GFP mouse as well as all tissues from the control mouse were negative for GFP, indicating that the SP-10 promoter fragment used in this study directed testis-specific gene expression. The seminiferous tubules from these mice appeared bright green in color whereas those from the control mice did not fluoresce (Fig. 2A, see color insert). Thus the GFP reporter construct allowed rapid visualization of signal in living tissues, without the need for fixing, sectioning, and cytochemical or immunological staining of the tissue. The next important question in the promoter analysis addresses the spatiotemporal pattern of transgene expression during spermatogenesis.

Green Fluorescent Protein Detection in Isolated Stages of Mouse Spermatogenesis

The development and differentiation of spermatozoa occur inside seminiferous tubules in three main phases. First, *mitotic proliferation of spermatogonia* produces mature spermatogonia (type B) that divide to form preleptotene spermatocytes. This initiates the second phase, *meiosis.* After meiotic DNA synthesis is completed, the spermatocytes initiate a long pachytene phase during which many meiosis-specific events occur, including pairing of the homologous chromosomes (synapsis) and formation of the synaptonemal complex. Two successive cell divisions at the end of meiosis give rise to haploid spermatids, which subsequently undergo many morphological changes to form mature spermatozoa during the differentiation phase, called *spermiogenesis.*

In the seminiferous epithelium, different germ cell populations are regularly associated with each other at specific phases of development (Fig. 3). These cell associations, also called the stages of the seminiferous epithelium, follow each other in a wavelike fashion along the seminiferous epithelium. One stage progresses to the next in a cyclic fashion at regular time intervals, without any longitudinal movement along the axis of the tubule, creating

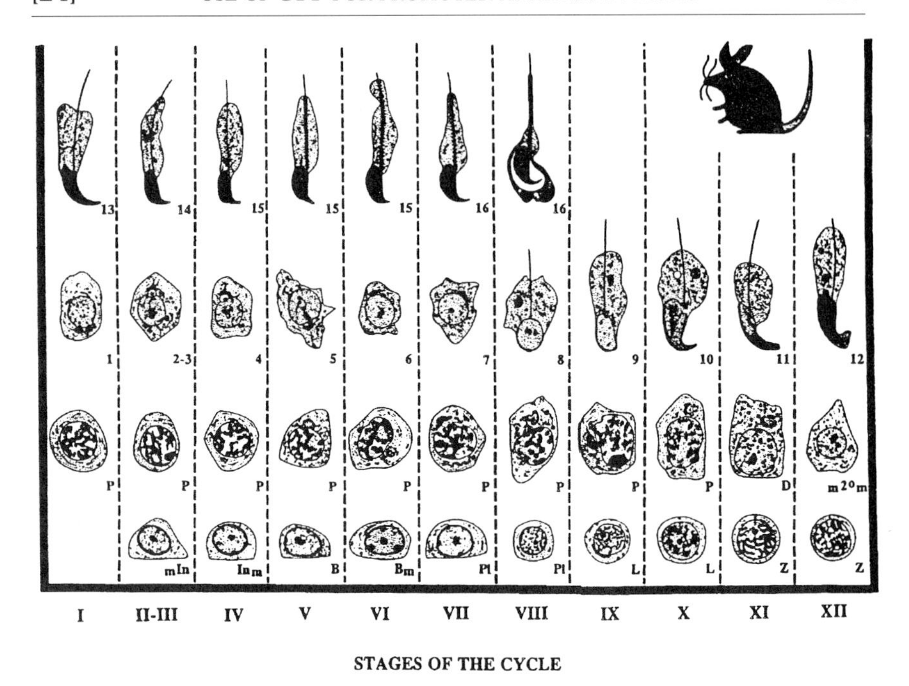

FIG. 3. Stages of the seminiferous epithelium cycle in the mouse. [Reproduced by permission of Cache River Press from L. D. Russel, R. A. Ettlin, A. P. S. Hikim, and E. D. Clegg, "Histological and Histopathological Evaluation of the Testis" (1990).]

what has been termed the cycle of the seminiferous epithelium.[22] In the mouse, the cycle of the seminiferous epithelium is divided into 12 stages (Fig. 3) that are classically identified according to the morphology of the developing acrosomes and the nucleus of early spermatids.[23]

Transillumination-Assisted Microdissection Method for Identification and Isolation of Stages of Mouse Spermatogenesis

In the early 1970s, a transillumination-assisted microdissection method enabling fast and accurate identification and isolation of mouse and rat stages was developed and later improved.[10,24] This technique has been successfully applied to a variety of morphological, biochemical, and toxico-

[22] C. P. Leblond and Y. Clermont, *Ann. N.Y. Acad. Sci.* **55,** 548 (1952).

[23] E. F. Oakberg, *Am. J. Anat.* **99,** 507 (1956).

[24] M. Parvinen, J. Toppari, and J. Lahdetie, *in* "Methods in Toxicology" (R. E. Chapin and J. J. Heindel, eds.), Vol. 3, Part A, p. 142. Academic Press, New York, 1993.

logical studies on mouse and rat spermatogenesis.[25–30] The method is based on the specific light absorption patterns of the mouse and rat seminiferous tubules caused by the cyclic changes in the chromatin condensation of haploid germ cells and in the localization of elongating spermatids within the seminiferous epithelium.[31] In brief, an isolated segment of mouse seminiferous tubule may contain four main light absorption zones (Fig. 4A). The "pale" zone, consisting of stages IX–XI, allows passage of most of the light. The "weak spot" zone, containing stages XII–III, is more light absorbent because of the formation of bundles of elongating spermatids at steps 12–14. As the formation of spermatid bundles increases and the association of step 15 spermatids with the Sertoli cells deepens, the "weak spot" zone turns into a "dark spot" zone (stages IV–V). The "dark" zone (stages VI–VIII) is characterized by the release of the bundle arrangement of the step 15 spermatids and localization of the step 16 spermatids at the lumenal edge of the seminiferous epithelium. When spermiation occurs at the end of stage VIII of the seminiferous cycle the dark zone stops and the pale zone reappears.

Method. A number of basic tools are required for transillumination-assisted microdissection: a stereomicroscope with transmitted light and magnification ranging from ×1 to ×10, normal scissors, two fine forceps, and small iridectomy scissors. The procedure is shown schematically in Fig. 4B.

1. Decapsulate a testis and transfer it to a petri dish containing testis isolation medium [TIM: 105 m*M* NaCl, 45 m*M* KCl, 1.2 m*M* $MgSO_4$, 6.0 m*M* Na_2HPO_4, 0.7 m*M* glucose (pH 7.2), 0.05% phenol red, 1.2 m*M* $CaCl_2$] that has been prewarmed at 37°. Place the petri dish under a stereomicroscope with transmitted light.
2. Gently pull the tubules apart with fine forceps, using low magnification (×1–3), until the transillumination pattern of the parts of the individual tubules can be observed.
3. Using small iridectomy scissors and fine forceps, separate a long segment of the tubule from the rest of the testis. Preferentially isolate a whole wave of seminiferous epithelium running from the start of the pale

[25] M. Parvinen and N. B. Hecht, *Histochemistry* **71,** 567 (1981).
[26] T.-L. Penttilä, L. Yuan, P. Mali, C. Höög, and M. Parvinen, *Biol. Reprod.* **53,** 499 (1995).
[27] M. Kallio, T. Mustalahti, T. Yen, and J. Lähdetie, *Dev. Biol.* **195,** 29 (1998).
[28] W. Yan, A. West, J. Toppari, and J. Lähdetie, *Mol. Cell. Endocrinol.* **132,** 137 (1997).
[29] T.-L. Penttilä, H. Hakovirta, P. Mali, W. W. Wright, and M. Parvinen, *Mol. Cell. Endocrinol.* **113,** 175 (1995).
[30] T. Sjöblom and J. Lähdetie, *Oncogene* **12,** 2499 (1996).
[31] M. Parvinen and A. Ruokonen, *J. Androl.* **3,** 211 (1982).

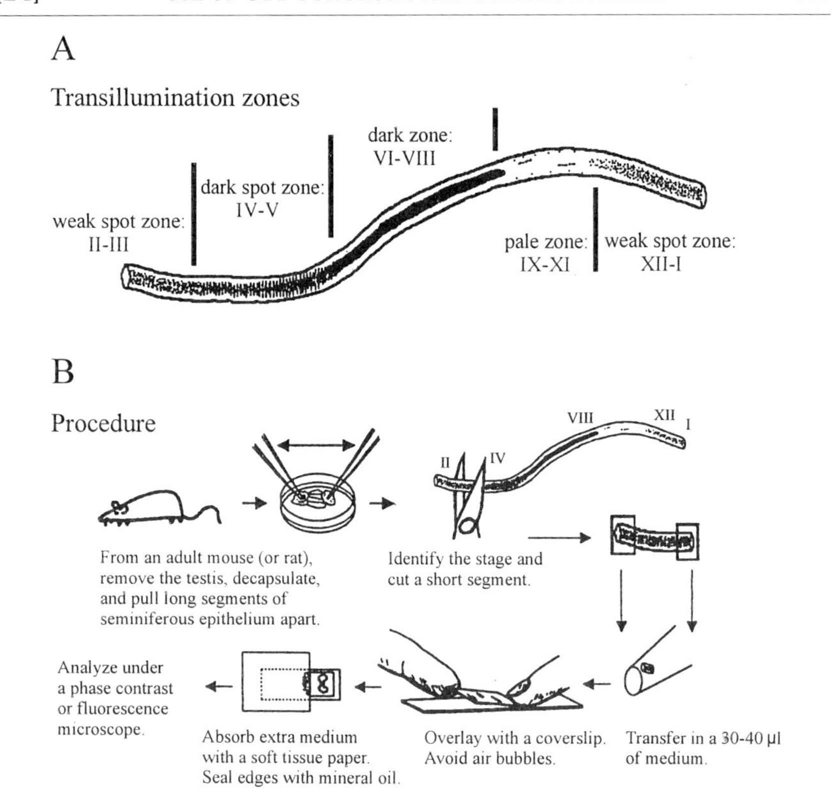

FIG. 4. (A) Drawing of the transillumination pattern of mouse seminiferous tubule. A section of the tubule containing one whole wave of spermatogenesis showing the different light absorption zones is depicted here. (B) Diagrammatic representation of the transillumination-assisted microdissection procedure. [Adapted and modified from Ref. 24, with permission from Academic Press.]

light absorption zone to the end of the dark zone (approximately a 2-cm segment). A clear difference in tubular transillumination can be found at stage VIII, where the dark zone abruptly ceases.

4. Transfer this long segment to a clear part of the petri dish and identify different light absorption zones using a higher magnification ($\times 3$–10).

5. When a desired zone or stage of the cycle has been localized, cut a short (~2-mm) tubule segment from this region and transfer the segment in an aliquot of 30–40 μl of TIM or tissue culture medium [Hams F12–Dulbecco's modified Eagle's medium (DMEM), 1 : 1] onto a clean objective slide and cover with a 22 $\times$ 22 mm coverslip.

6. The weight of the coverslip pushes the cells out of the tubule segment and a slightly flattened monolayer of cells is formed. This process can be

aided by absorbing extra medium from beneath the coverslip, using a soft tissue paper. Different germ cell populations flow out from a tubule segment in an order dictated by the localization of the germ cell types within a particular segment of the seminiferous epithelium. The first cells to come out are those close to the lumen of the seminiferous tubule. These include postmeiotic round spermatids and elongating spermatids. These are followed by meiotic cells such as pachytene spermatocytes and, last, by different types of spermatogonia and the Sertoli cells. (*Note:* The key is to spread the cells as a monolayer. Avoid absorbing too much medium from under the coverslip, as this may cause the cells to disrupt.)

7. For live cell analysis, seal the edges of the coverslip with mineral oil to prevent gas exchange and drying of the sample. Sealing also prevents cell movement under the coverslip. The sample can now be analyzed with a phase-contrast and/or fluorescence microscope.

Stage- and Cell Type-Specific Expression of Green Fluorescent Protein under Control of SP-10 Promoter

Stages of the seminiferous epithelium from 6- to 8-week-old SP-10–GFP mice as well as nontransgenic littermates were identified and isolated using the transillumination-assisted microdissection technique. A slide containing a segment from the desired stage of the cycle was placed under an inverted Nikon (Melville, NY) Diaphot-TMD microscope equipped with a GFP filter and an electronically controlled Lambda 10-2 shutter (Sutter Instrument Co., Novato, CA). The cells were analyzed using a Nikon Planapochromat 60× (NA 1.4) objective and the images of GFP–SP-10-expressing cells and control cells were captured using a cooled CCD SenSys camera (Photometrics, Tucson, AZ) and processed with Metamorph image analysis software (Universal Imaging, West Chester, PA). Images of monolayered germ cells from each stage of the cycle were captured and analyzed. To limit the fading of the GFP signal, exposure times were kept as low as possible (0.2 to 0.8 sec).

Photomicrographs representing stages XI–XII, I, V, and IX from SP-10–GFP mice are shown in Fig. 2B. The stages were identified on the basis of the morphology of the developing acrosomes and the nuclei of the spermatids. At stage I, two steps of spermatids may be identified; early postmeiotic step 1, and elongated step 13 spermatids. The step 1 spermatids have round nuclei with prominent nucleoli, whereas the step 13 spermatids have prominent hook-shaped nuclei (Stage I, s13 in Fig. 2B). The step 5 spermatids at stage V are characterized by the developing acrosomic system spreading over the nuclear envelope. At stage IX, the only haploid cells are step 9 spermatids with elongated nuclei. Stage XII is characterized by the presence of meiotic divisions. Identification of fluorescent germ cells at various stages of spermatogenesis indicated that the expression of GFP under the control of the SP-10 promoter starts only after meiotic divisions. No GFP signal was observed during meiosis: preleptotene, leptotene, zygotene, pachytene, and meiotically dividing spermatocytes from SP-10–GFP transgenic mice did not fluoresce green. The GFP signal was first observed in early postmeiotic step 1 spermatids, in which a dim green fluorescence was present diffusely throughout the cells (Fig. 2B, Rs). This is consistent with the expression pattern of the endogenous SP-10 gene. We have previously observed[13] that mRNA and protein of SP-10 appear postmeiotically, in the early round spermatids. The GFP signal, which was more intense in the later steps of spermatid development, persisted in all postmeiotic germ cell types from step 1 to step 16 spermatids, probably owing to the long half-life of the protein. Analysis of the stages of nontransgenic littermates (control mice) showed no fluorescent cells at any stages of the cycle (Fig. 5).

Summary

The application of green fluorescent protein (GFP) as a reporter in transgenic mice is particularly useful for the study of testis-specific gene promoters. A major advantage is that GFP can be detected directly in freshly isolated spermatogenic cells without the need for cytochemical or immunohistochemical reactions. When combined with the transillumination-assisted microdissection technique, expression of GFP in the testis permits precise identification of the germ cell type, and the stage of the seminiferous epithelium cycle at which the putative promoter is active. We have demonstrated the utility of this approach using a SP-10-GFP transgenic mouse model in which GFP is expressed in the haploid male germ cells.

Future Applications

Because the spermatogenic cells of the SP-10-GFP transgenic mice carry a fluorescent marker for postmeiotic gene expression and terminal differentiation (acrosome biogenesis) of male germ cells, these mice may be useful for several applications. First, the development of immortalized spermatogenic cells capable of differentiating into spermatids has yet to be achieved. Transformed spermatogonial or spermatocyte cell lines derived from SP-10-GFP mice may be conveniently monitored in culture for gene expression representative of acrosome biogenesis and hence entry into the haploid phase. Second, the SP-10-GFP mice may be ideal as a donor source for spermatogonial stem cell transplantation experiments, as it will be easier

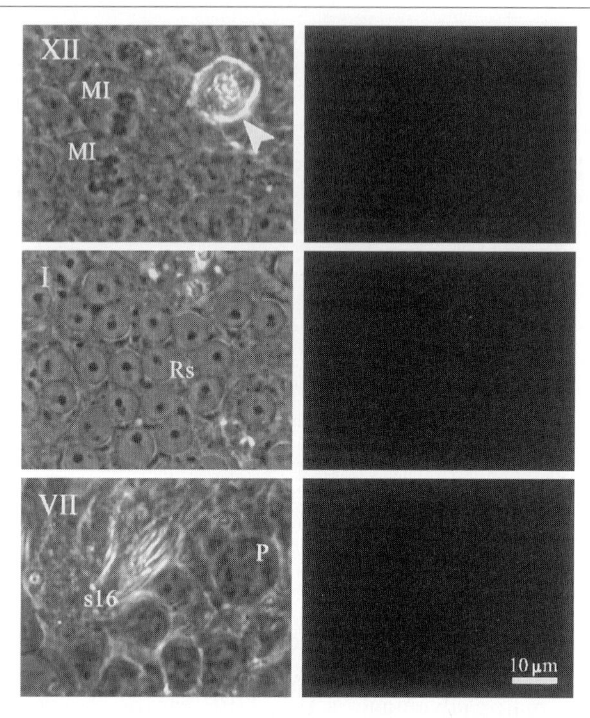

FIG. 5. Nontransgenic littermates of the SP-10–GFP mice do not express GFP in the seminiferous epithelium. *Left:* Phase-contrast micrographs from stages XII, V, and VII. *Right:* Fluorescence image of the same field. All cell types in the seminiferous epithelium, including the haploid spermatids, are GFP negative. MI, First meiotic division; Rs, round spermatids; P, pachytene spermatocytes; s16, step 16 spermatids. The arrowhead in the stage XII micrograph denotes a degenerating spermatocyte in first meiotic division, arrested at metaphase.

to address the fate of the GFP-producing donor germ cells within the recipient testis. Third, the SP-10-GFP mice may provide *in vivo* and *in vitro* experimental models to test new methods of male contraception that involve targeting drugs to cause arrest in spermatogenesis. For example, primary cultures of SP-10-GFP germ cells may be used for high throughput screening of combinatorial libraries to search for compounds that could cause meiotic arrest. Finally, it may be possible to develop assay systems, using SP-10-GFP germ cells, to identify transcriptional antagonists which prevent DNA–protein interactions between testis-specific transcription factors and the SP-10 gene-promoter sequences.

Acknowledgments

This work was supported by NIH U54HD29099, and by a grant from the Andrew W. Mellon Foundation. M. K. was supported by the Academy of Finland.

[25] Green Fluorescent Protein in the Measurement of Bacteria–Host Interactions

By LUIZ E. BERMUDEZ, FELIX J. SANGARI, and AMY PARKER

Introduction

Light is produced by the bioluminescent jellyfish *Aequorea victoria*. Green fluorescent protein (GFP) is a 238-amino acid, of 27 kDa, that emits green light when excited by blue light, the source of which in *A. victoria* is a calcium-activated photoprotein aequorin. The fluorophore is formed by an autocatalytic cyclization of three amino acid residues within GFP.[1] The ability to generate fluorescence *in situ* by expressing the gene for GFP has opened up enormous possibilities for the study of host–pathogen interaction by continuously monitoring for gene expression, for direct localization of bacteria within cells, and for protein tracking in living cells. Traditionally, bacterial pathogens have been detected indirectly by using fluorescent antibodies. More recently, investigators have directly coated microorganisms with fluorochromes to study real-time interactions with live host cells. These approaches, however, can be of limited value for following bacterial pathogens within live host cells. For example, division dilutes the coated fluorescence signal during long-time infection, which does not happen in the case of GFP expression. Because GFP is a stable molecule, it is feasible for use in reporter systems for gene expression.

In this chapter we describe some of the techniques that can be employed to enhance our understanding of host–bacteria interaction.

Properties of Green Fluorescent Protein

Green fluorescent protein absorbs light with a major excitation peak of 395 nm and a secondary peak at 470 nm, and fluoresces with a maximum emission of 510 nm. This occurs in the absence of cofactors, is oxygen dependent, and occurs gradually after transduction.[2] The presence of GFP can be monitored using standard fluorescein isothiocyanate (FITC) excitation–emission filter sets for the minor absorption peak at 470 nm.

GFP has several unique features that are extremely useful in the study of host–pathogen interactions: (1) GFP is a small cytoplasmic protein with

[1] M. Chalfie, Y. Tu, G. Euskirchen, W. W. Ward, and D. C. Prasher, *Science* **263,** 802 (1994).
[2] M. Ormö, A. B. Cubitt, K. Kallio, L. A. Gross, R. Y. Tsien, and S. J. Remington, *Science* **273,** 1392 (1996).

0076-6879/99 $30.00

TABLE I
PROPERTIES OF SEVERAL GREEN FLUORESCENT PROTEIN MUTANTS

Protein	Excitation (nm)	Emission (nm)	Main features (intensity relative to wild type)	Ref.
wGFP	395 (470)	510	Wild type, green fluorescence	1
GFPmut1[a]	488	507	Red-shifted excitation maximum (35×)	4
GFPmut2	481	507	Red-shifted excitation maximum (19×)	4
GFPmut3	501	511	Red-shifted excitation maximum (21×)	4
GFPuv	360–400	509	Optimally excited by UV light	5
sg12	398	509	(9–12×)	6
sg11	471	508	(19–38×)	6
sg25	474	509	(50–100×)	6
d2EGFP	488	507	Destabilized mutant, human codon optimized (35×)	3
BFP	383	447	Blue mutant	7
sg42	386	450	Blue mutant (27×)	6
sg49	386	450	Blue mutant (37×)	6
sg50	386	450	Blue mutant (63×)	6
EBFP	380	440	Blue mutant, human codon optimized (2–3×)	3, 7
EYFP	513	527	Yellow mutant, human codon optimized	2, 3

[a] EGFP is GFPmut1, human codon optimized.

low toxicity, therefore its presence should have minimal effects on the expression of other genes, including determinants of pathogenicity; (2) GFP can be continuously synthesized, which minimizes the effect of bacterial replication on fluorescence dilution; and (3) GFP is easily imaged and quantitated, and this can be done with live cells and bacteria over a period of time.

The wild-type GFP has proved to be a malleable protein that can be easily reengineered by mutagenesis to improve some of its properties and shift the excitation and emission wavelengths, creating different colors and potentially new applications (Table I[2–7]). A number of GFP mutants have been developed to expand its applications. These mutants generally show an increase in fluorescence intensity as well as a shift in the spectral properties. The increase in fluorescence depends on several factors, such as increased protein expression, more efficient protein folding, increased absorption at the frequency used to excite the protein, or faster chromophore

[3] *CLONTECHniques* **XIII**(2), 16 (1998).
[4] B. P. Cormack, R. Valdivia, and S. Falkow, *Gene* **173,** 33 (1996).
[5] A. Crameri, E. A. Whitehorn, E. Tate, and W. P. C. Stemmer, *Nature Biotechnol.* **14,** 315 (1996).
[6] R. H. Stauber, K. Horie, P. Carney, E. A. Hudson, N. I. Tarasova, G. A. Gaitanaris, and G. N. Pavlakis, *BioTechniques* **24,** 462 (1998).
[7] R. Heim, A. B. Cubitt, and R. Y. Tsien, *Nature* (*London*) **373,** 663 (1995).

formation. Some of these variants have also been codon optimized to increase the expression in certain hosts. Other mutations produce a higher amount of soluble protein and, thus, a higher fluorescence, although the fluorescence per soluble unit of protein remains unchanged. In addition, some mutations result in a shift in the maximal excitation. And, finally, the expression of some of the GFP mutants is significantly faster than that of wild-type GFP, 8 min versus 1–2 hr after induction, allowing a more sensitive measure of gene expression. At the same time, other mutations are responsible for a shift in the emission maxima, producing different colors; an example is T66H, which results in a shift in emission maxima to 448 nm. One mutant shows great potential for gene expression experiments. It is a destabilized version of enhanced GFP (d2EGFP)[3] with the same excitation and emission properties as EGFP, but with a half-life of only 2 hr. The protein carries a short sequence that targets the protein for degradation.[3] It is ideally suited for accurate analysis of cis-acting elements in studies of gene regulation, and may facilitate generation of stable cell lines without excess buildup of GFP. Furthermore, it should allow study of rapidly degraded proteins, because unlike EGFP, d2EGFP should not artificially stabilize the protein to which it is fused.

Applications of Green Fluoroscent Protein Expression in Host–Pathogen Interactions

Optimizing Mutants Expressing Green Fluorescent Protein

A number of vectors have been constructed in which GFP is under the control of a strong promoter. The *lacZ* promoter has been successfully used for both *Salmonella*[8] and *Yersinia,*[8] while the heat shock protein hsp60 and L5 promoters have been used for mycobacteria.[9,10] Other promoters have been used to express GFP in *Bacillus*[11] and *Candida albicans.*[12] Once the transformants are created using published techniques, the brightest clone must be selected. This can be accomplished by using a fluorimeter, by fluorescence microscopy or, more efficiently, by fluorescence-activated cell sorting (FACS).

In studying the relationship between bacteria and its host, it is useful to be able to visualize the location of the bacteria, as well as the response

[8] R. H. Valdivia, A. E. Hromockyz, D. Monack, L. Ramakrishnan, and S. Falkow, *Gene* **173,** 472 (1996).

[9] S. Dhandayuthapani, L. E. Via, C. A. Thomas, P. M. Horowitz, D. Deretic, and V. Deretic, *Mol. Microbiol.* **17,** 901 (1995).

[10] A. Parker and L. E. Bermudez, *Microb. Pathogenesis* **22,** 193 (1997).

[11] C. D. Webb, A. Decatur, A. Teleman, and R. Losick, *J. Bacteriol.* **177,** 5906 (1995).

[12] M. Brenton and S. Falkow, *Microbiology* **143,** 2161 (1977).

of the bacillus to changing external conditions. GFP provides the means of constitutively labeling a living bacteria without the use of antibodies or other substrates, and a system for visualizing gene expression as a quantifiable marker. We present three protocols representing different experimental perspectives that can help illuminate the bacteria–host interaction. The protocols presented are general, as there are many different bacteria and hosts that could be used; however, the references cited give specifics for certain infection models.

Construction of Plasmids. The construction of plasmids will depend on the microorganism used and the purpose of the system. Detailed instructions on commonly used molecular techniques are not included in this protocol but have been referenced.[13] In the case of *Mycobacterium avium,* we constructed the plasmid pWES4 by ligation of *gfp* downstream of the bacillus Calmette–Guérin 60-kDa heat shock protein (hsp60) promoter.[10] We also constructed a different vector with the L5 promoter upstream of *gfp* in an hsp60-deleted pMV261. In this case, we cloned the *gfpmut2* mutant of *gfp,*[4] which resulted in a significant increase in the brightness of the bacteria. This system has two advantages compared with the previously described system: first, as previously said, GFPmut2 is significantly brighter than GFP and, second, the L5 promoter is a stronger promoter in mycobacteria than is the hsp60 promoter (Fig. 1).

Methods for Green Fluorescent Protein Detection. The most common means of measuring levels of GFP are by fluorescence microscopy, FACS, or by using a fluorometer. Depending on the goal of the experiment, one or the other may be more appropriate. The following outlines some of the advantages and shortcomings of these methods.

Microscopy Studies. Recombinant *M. avium* expressing GFP as well as infected macrophages can be visualized using a fluorescence microscope, The macrophage nuclei can be stained with Hoechst 33342 (0.5 μg/ml) nuclear dye for 5 min at room temperature or incubated for 5 min at room temperature with a 0.1-μg/ml concentration of propidium iodide (ideal for confocal microscopy). Slides are excited at 488 nm and visualized with GFP or FITC filter.

Advantages: One of the main advantages of microscopy is the ability to specifically localize either an expressed protein or bacteria relative to other structures. For example, GFP-expressing bacteria could be dynamically followed in an unfixed preparation over time.

Disadvantages: Screening or quantifying pools of fluorescent bacteria in this manner is tedious; using a method that can directly measure the fluorescent emissions is more practical.

[13] F. M. Ausubel, "Current Protocols in Molecular Biology." John Wiley & Sons, Boston, 1987.

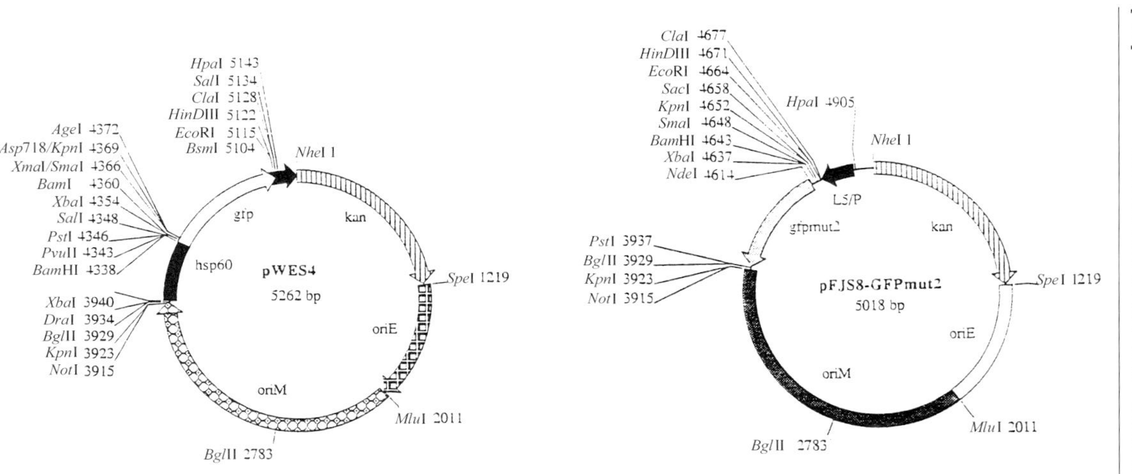

FIG. 1. GFP expression vectors. Symbols: *oriC*, origin of replication in *E. coli; oriM*, origin of replication in mycobacteria; kan, kanamycin resistance-encoding gene; L5, mycobacteriophage L5 promoter; hsp60, heat shock protein 60-encoding gene. Plasmids were electroporated into *M. avium* strain 104 according to a published protocol.[10] Briefly, *M. avium* was grown in 7H9 broth as described to a density of 0.5 ($A_{600\ nm}$). Bacteria were washed three times with 10% glycorol and resuspended in a fraction of the initial volume. Approximately 1 μg of either pWES4 or FJS8-GFPmut2 and washed cells were placed in an electroporation cuvette and subjected to a pulse of 1.25 kV, 25 μF, and 1000 Ω. Warm 7H9-OADC broth was then added to the cells, followed by incubation at 37°. Transformants were selected on 7H10-OADC agar plates supplemented with kanamycin (200 μg/ml).

Fluorometric Dectection. Recombinant *M. avium* expressing GFP and untransformed *M. avium* are harvested from exponentially growing cultures. The cells are washed twice with phosphate-buffered saline (PBS) and resuspended in PBS to a final concentration of 1×10^8 bacilli per milliliter. With our construct, GFP expression can be detected with 10^6 bacteria with an excitation wavelength of 355 nm and emission at 538 nm (a standard setting on our fluorimeter). The results of the experiment should be expressed in fluorescence intensity units at 538 nm (for example, in a 96-well format).

Advantages: The technique is effective for screening individual clones or pools of clones quantitatively for GFP expression.

Disadvantages: It provides no spatial information and manual isolation is required to sort clones from pools.

Flow Cytometry. Flow cytometry and FACS are performed using standard procedures. Illumination is done with a 200-mW, 488-nm argon ion laser and emission light is detected through a 530/30-nm bandpass filter. Data should be collected at 50,000 individual events per sample.

Advantages: Flow cytometry allows clones to be obtained that express or do not express GFP under the conditions of interest. The technique is an excellent choice for screening expression libraries, and can be used to sort expression of specific GFP mutants.[14]

Disadvantages: It is expensive and not always readily available to all researchers. For information about the location of labeled proteins or bacteria, it is necessary to follow up with fluorescence microscopy analysis.

Cell Culture Assays. Techniques to culture macrophages as well as epithelial cells are described in the following sections.

Macrophage Infection. Cells used can be human monocyte-derived macrophages (purified as described previously).[15] Monocytes or macrophages should be seeded at 5×10^5 to 1×10^6 cells per well. Cell monolayers can be infected (dependent on the goal of the experiment) with a ratio of from 0.1 bacterium per cell to 100 bacteria per cell. If more bacteria are used, the monolayer will remain viable for fewer days. After infection, which can take from 1 to 4 hr, the extracellular bacteria should be removed by successive washings. If the bacteria bind to the plastic, use of an antibiotic that kills extracellular bacteria but does not enter cells is necessary. Ingestion of the microorganisms is followed by fixation of cells at appropriate times thereafter. For immunofluorescence staining in macrophages, we use standard fixation techniques employing paraformaldehyde or methanol. Autofluorescence can be a significant problem with these cells and may be

[14] L. Lybarger, D. Dempsey, G. H. Patterson, D. W. Piston, S. R. Kain, and R. Chervenak, *Cytometry* **31,** 147 (1998).

[15] L. E. Bermudez, A. Parker, and J. Goodman, *Infect. Immun.* **65,** 1916 (1997).

less prominent with methanol fixation. We routinely use paraformaldehyde (3%) fixation for 90 min. Staining of intracellular markers [such as early phagosome, late phagosome, or lysosomal markers (Rab-5 Lamp-1, Rab-7 Lamp-2)] to investigate trafficking can be done using a different fluorochrome such as Texas red.

Visualization of the bacteria can be improved by staining the macrophage nuclei with propidium iodide (0.1 μg/ml), which will show as orange through an FITC filter.

The choice of cell line can be critical for establishing invasiveness by a particular pathogen. The physiological state of that cell line can also have a major influence on the outcome of invasion and gene expression. The variables are related to the presence and distribution of receptors necessary for bacterial invasion.

Epithelial Cell Infection. A single-cell suspension of the desired epithelial cells is prepared by trypsinization. It is important that a uniform suspension be prepared because epithelial cell clumps result in an increased background. The monolayers should be used when 90 to 100% confluent. Details can be obtained in Ref. 11. The number of cells constituting such a confluent monolayer is dependent on the cell line being used and can range from 7×10^4 to 5×10^5.

Infection should be achieved by the same methods described for macrophage monolayers. The growth of bacterial inoculum can be important for an efficient invasion. In general, mid-log-phase bacteria (agitated or without agitation, depending on the microorganism) can be harvested by centrifugation and then resuspended to 1×10^6 or 5×10^6 per ml in tissue culture medium or balanced salt solution. The standard multiplicity of infection (MOI) is 10 but can be varied to optimize invasion. When adding bacteria to the monolayer, however, it is important to shake the slide, briefly and gently, to ensure uniform distribution of bacteria. After the period of infection (minutes to 4 hr, usually), an antibiotic (gentamicin or amikacin, 100 μg/ml) should be added for 1 to 2 hr to kill extracellular bacteria. After the treatment with antibiotic, the supernatant should be removed and the monolayer fixed. The process of fixation is the same as described for macrophages.

Background fluorescence is a problem with these cells, and staining of the nucleus may improve visualization.

Bacterial Localization

In Vitro

GFP-expressing bacteria can be used in an assay for colocalization with mammalian cell antigens, using confocal or fluorescence microscopy. Invasion should be carried out according to the protocol described above.

Fixation of the material with 2% paraformaldehyde will preserve expressed GFP. The use of filters and fluorochromes (Texas red, rhodamine, and FITC) allows for observation of mammalian cell antigens related to the bacteria. For the study of cytoskeleton rearrangement following entry of the cell by the bacteria, the use of a bacterium expressing GFP can facilitate the visualization of the organism (when compared with phase-contrast microscopy) but is not ideal. In this particular case, it is preferable to use specific antibodies that can bind to extracellular bacteria in a nonpermeabilized monolayer and to intracellular bacteria and extracellular bacteria when incubated with a permeabilized monolayer. GFP-expressing bacteria are more useful as a system for the study of intracellular trafficking as described above, or for determination of the pH of an intracellular compartment occupied by the bacteria. In this specific case, bacteria can be easily detected, and the pH of the vacuole can be determined, by the use of fluorochromes such as dextran-NERF (manufactured by Molecular Probes, Eugene, OR).

In Vivo

The visualization of GFP-expressing bacteria in infected animal tissues represents potentially another use for GFP in the study of pathogenesis. For example, one can use GFP-producing bacteria as probes to determine precise host cell interactions in infected animals. It will make it possible to identify subsets of infected mammalian cells. The potential use of GFP as a biological tracer will allow for the following of colonization, persistence, and transmission of pathogenic bacteria in live animals. It will also allow for the identification of subsets of infected mammalian cells in complex environments.

To detect GFP-expressing bacteria in tissues, animals, after being sacrificed, should have the organ tissue removed, minced in small pieces and embedded in O.C.T. frozen on solid CO_2, and cryosectioned in 5- to 10-μm sections for histopathology and fluorescence microscopy analysis. Fluorescent organisms are readily visualized by fluorescence microscopy in frozen tissue sections. To confirm the findings of fluorescence microscopy, tissue should be stained for the bacterium.

As an alternative, tissue fixation done with 3% paraformaldehyde will also conserve the expressed GFP.

Bacterial Gene Expression

GFP can be used for the detection of gene expression in both extracellular and intracellular bacteria. It has been demonstrated to be a sensitive reporter of bacterial gene expression in *Salmonella*[8] and *Mycobacterium*[9] (Fig. 2).

BACTERIAL LOCALIZATION PLASMID CONSTRUCTION

Materials

Transformable strain of bacteria of interest: We use *M. avium* 104, a transformable strain of *M. avium* (the majority of *M. avium* strains cannot be transformed)

Selectable plasmid vector with multicloning site downstream of a functionally strong promoter (can be either a replicative or integrative plasmid)

Plasmid containing a copy of the *gfp* gene (or a mutant) with the appropriate restriction sites flanking it (or, alternatively, PCR primers designed to amplify the *gfp* gene containing restriction sites at their 5′ ends)

Procedure

By simply cloning *gfp* from another plasmid into a vector that contains the appropriate selectable marker, origin of replication, and promoter for the bacteria with which one works, an easily visible, constitutively expressing strain can be created. This strain should be visible in macrophages, epithelial cells, and tissues from infected hosts. Transformants expressing GFP should be subjected to a screening process with the goal of obtaining a bright clone. It can be achieved using any of the methods of detection described in text; however, FACS and fluorimetry are the most efficient

If a copy of *gfp* with the appropriate restriction sites is not available, and the PCR route is chosen instead, a proofreading polymerase such as Pfu polymerase (PFU) should be used to reduce the risk of misincorporated bases in the *gfp* product

If the bacterium under research has a low efficiency of transformation and a vector is available that can be shuttled into *E. coli* first, create the construct in *E. coli* before transforming the purified supercoiled plasmid into the bacteria of interest

In addition to localizing the GFP-expressing bacteria, GFP expression can also be used as a method of bacterial quantification by monitoring the increase in brilliance with the amount of GFP produced by growing bacteria

After the creation and selection of the desired mutant, virulence-associated properties such as the ability to invade cells, and intracellular replication, should be examined in comparison with the nontransformed parent strain. The expression of foreign protein in bacteria occasionally results in total or partial loss of virulence

FIG. 2. General plasmid construction for constitutive GFP expression.

Determination of Extracellular Gene Expression

Bacteria regulate gene expression according to the surrounding environment. In the host, bacteria are exposed to changing challenge conditions and the ability to adapt rapidly to an existing condition is crucial for bacterial survival.

Therefore, exposure of bacteria to laboratory environments that mimic

BACTERIAL GENE EXPRESSION PLASMID CONSTRUCTION

Materials

Transformable strain of bacteria of interest
Selectable plasmid vector with a multicloning site without a functional promoter (can be either a replicative or integrative plasmid)
Plasmid containing a copy of the *gfp* gene (or a mutant) with the appropriate restriction sites flanking it (or, alternatively, PCR primers designed to amplify the *gfp* gene containing restriction sites at their 5′ ends)
The 5′ regulatory region of the gene of interest or a library of small fragments for expression screening

Procedure

The approach here is virtually the same as with the bacterial localization protocol in Fig. 2, with a few significant differences. The primary difference is that one will be creating a *gfp* vector that is promoterless and therefore replicates in the bacterium without expressing *gfp*
It is useful to clone the promoterless *gfp* downstream of a multicloning site to ease the cloning of specific gene regulatory regions or a library
Once the promoterless *gfp* construct has been created it is ready to have a promoter of interest cloned upstream and *gfp* expression monitored under conditions of interest (e.g., altered pH or O_2 tension, in the presence of a host cell, *in vivo*)
If the goal of the gene expression experiment is to identify genes of interest under certain conditions then it is appropriate to clone a library of 500 two-kilobase genomic DNA fragments upstream on the promoterless *gfp* for screening

FIG. 3. General plasmid construction for gene expression assays.

conditions found *in vivo* has the potential of triggering the expression of genes that are expressed during the life inside of the host.

Bacteria containing a construct in which *gfp* is under the control of a promoter library should be washed and then incubated under the desired conditions for 2 and 4 hr. In our experience, 2 hr is sufficient to observe protein synthesis in mycobacteria. After the period of incubation, the culture should be analyzed by FACS or fluorescence microscopy. If FACS is chosen, sorting parameters need to be set to discriminate on the basis of fluorescence intensity. If a fluorometer is used to determine gene expression, in our system a minimum of 10^5 bacteria is needed for detection (Fig. 3).

Determination of Intracellular Gene Expression

The recommended enrichment for GFP-expressing bacteria was originally described by Valdivia and Falkow.[16] Tissue culture wells seeded with

[16] R. H. Valdivia and S. Falkow, *Mol. Microbiol.* **22,** 367 (1996).

5×10^5 cells are infected with 5×10^6 bacteria, and centrifuged over the cell layer (2000*g* for 5 min). After the infection (approximately 30 min), the monolayers are washed with fresh medium and incubated for 3–4 hr. Monolayers are then washed three times with PBS, placed on ice, and lysed with 0.1 ml of Triton X-100 in PBS for 5 min. Samples should be kept on ice until analyzed. The kinetics of GFP expression by intracellular bacteria are determined by infecting macrophages for 0.5, 1, 2, 4, and 6 hr.

The assay may also be carried out in a fluorometer. In this case, pools (usually of 20 clones) of bacteria carrying constructs with GFP under the control of a promoter library are used to infect macrophage monolayers established in 96-well plates. The limit of detection of our assay is approximately 10^5 bacteria; therefore, we use an inoculum of 10^7 bacteria. Extracellular bacteria should be removed by extensive washing. The plate should be read at each time point and positive pools (strong fluorescence) should be plated onto 7H10, individualized, and then screened individually for GFP expression.

Green Fluorescent Protein-Tagged Proteins

Techniques to measure the precise location of proteins within mammalian cells have become potentially important for the investigation of host–pathogen interactions, as, for example, to determine the correct cellular localization of involved proteins.

The mammalian cell needs to be transfected with a plasmid, or mRNA, as described.[17] RNA needs to be electroporated into adherent cells using a small-volume electroporation device. The vector used for transcription should consist of a 5′ untranslated region of the desired protein, a Kozac consensus sequence, a *gfp* insert, and a 3′ untranslated region of the desired protein.

In some cases, the efficiency of the transcription is increased by polyadenylation of the RNA (polyR polymerase). Efficient translation usually requires 1 μl of mRNA at a concentration of 1 μg/μl.

Cells can be observed by fluorescence microscopy or confocal microscopy. This approach, for example, allows localization of intracellular proteins such as Rab-5, Rab-7, cathepsin D, Lamp-1, and Lamp-2, which have their trafficking altered during mycobacterial infection of macrophages.

[17] H. Yokoe and T. Meyer, *Nature Biotechnol.* **14,** 1252 (1996).

[26] Flow Cytometric Analysis of Transgene Expression in Higher Plants: Green Fluorescent Protein

By DAVID W. GALBRAITH, LEONARD A. HERZENBERG, AND MICHAEL T. ANDERSON

Introduction

The green fluorescent protein (GFP) of *Aequorea victoria* provides a unique means to study living cells, because transgenic expression of the coding sequence is sufficient to produce fluorescence within the cells of many different pro- and eukaryotic species.[1] The availability of GFP as a transgenic marker has spurred interest in methodologies and instruments that can accurately measure cellular fluorescence emission, of which flow cytometry is one. Flow cytometric instrumentation[2,3] is designed to examine the optical properties of suspensions of single cells or cellular homogenates as they are constrained to pass rapidly in a fluid stream through focused illumination. The individual cells or subcellular particles scatter light and, if they contain fluorochromes, also absorb light that is subsequently reemitted in the form of fluorescence. These signals are detected using sensitive photomultiplier tubes (PMTs) and photodiodes screened by appropriate wavelength-selective filters. After conversion to voltage-versus-time pulse waveforms, the intensity values are digitized and stored for analysis or further processing; the values are stored either on a cell-by-cell basis ("list-mode"), or in the form of frequency distributions. Flow cytometric data are then presented in the form of multidimensional population distributions, which are graphically represented as histograms. This method of data accumulation and display, because it preserves differences between cells, is different from that commonly encountered in biochemical determinations, which provide only population means. Flow cytometry therefore allows identification of subpopulations of cells having different optical characteristics; these typically are observed as individual clusters within two-dimensional representations. Flow cytometry is also inherently quantitative, in that the intensities of the individual light scatter and fluorescent signals are accurately recorded. Either linear or logarithmic scales can be employed, the latter being preferred when signals of large dynamic ranges are encountered.

[1] P. M. Conn (ed.), *Methods Enzymol.* **302,** 99 (1999).

[2] M. R. Melamed, T. Lindstrom, and M. L. Mendelsohn, "Flow Cytometry and Cell Sorting." John Wiley & Sons, New York, 1990.

[3] H. M. Shapiro, "Practical Flow Cytometry," 3rd ed. John Wiley & Sons, New York, 1995.

0076-6879/99 $30.00

Some of the optical characteristics of cells, from which flow cytometric measurements derive, reflect the physical properties of the cells, for example, the intensities of light scattered at angles close to the axis of illumination (forward-angle light scatter, FALS) or orthogonally to the axes of illumination and fluid flow (90° light scatter). Fluorescent signals produced from cells reflect either the presence of endogenous autofluorescence or result from addition of fluorochromes (synthetic chemicals or natural products). Some of these have intrinsic specificity, such as the various fluorochromes that tightly bind to DNA or intercalate nucleic acids. Others have no intrinsic specificity, but through being covalently linked to other molecules, such as antibodies, can be provided with particular binding specificities. On the basis of the commercial availability of many different types of fluorescent molecules, flow cytometric methods and reagents have been devised to provide a sensitive measure of a large variety of cellular parameters.

Flow cytometry finds particular power when it is combined with cell sorting. Sorting is done following laser interrogation and analysis of the flow stream, which technically occurs either in air or within an enclosed cuvette. In either case, the stream subsequently emerges through a precisely engineered flow orifice, having a typical diameter in the range of 40–400 μm. The technique of cell sorting relies on the principle that cylindrical fluid streams in air are unstable and, as a consequence of minimizing surface free energy, decay into droplets. Imposition of a periodic oscillation on the fluid stream at the point of emergence, using a piezoelectric or electromechanical drive, constrains the stream to break into droplets at a precisely defined point below the flow tip, which therefore corresponds to a precise point in time after a given object has been analyzed. An electrical voltage is applied to the stream at the precise time that the desired object is entering the "last-attached" droplet. This leaves a residual charge on the newly formed droplet surface, and the droplet can then be collected following electrostatic deflection by passage through a fixed high-voltage field. Flow sorting is implemented through computerized sort decisions, based on regions selected on the population distributions that were previously collected. Switching of the sort charge can be done rapidly, and this means that high rates of sorting can be achieved, and large numbers of purified cells obtained for further characterization.[2,3]

Considerable effort has been devoted to the development of flow cytometry and cell sorting for the analysis of the fluorescence derived from transgenic expression of GFP.[4,5] This chapter describes transient and

[4] S. Welsh and S. A. Kay, *Curr. Opin. Biotechnol.* **8,** 617 (1997).

[5] D. W. Galbraith, M. Anderson, and L. A. Herzenberg, *Methods Cell Biol.* **58,** 315 (1998).

transgenic expression of GFP in plants, targeting of GFP to the nucleus, and the use of flow cytometry to analyze this expression in protoplasts and in isolated nuclei.

Materials. Enzymes for modification of DNA molecules are available from a number of sources. Macerase and cellulysin are obtained from Calbiochem (La Jolla, CA), and MS (Murashige and Skoog) medium from GIBCO (Grand Island, NY). Other chemicals are available from the Sigma Chemical Company (St. Louis, MO). Sterile plastic centrifuge tubes (Falcon 2098; Becton Dickinson Labware, Lincoln Park, NJ) and culture dishes (Falcon 3047) are used in the preparation and manipulation of protoplasts. Disposable microcentrifuge tubes, pipette tips, and 12 × 75 mm test tubes are available from general scientific supply companies.

Methods

Recombinant DNA Molecules

Routine manipulation of DNA follows the general methods of Sambrook *et al.*[6] The design and construction of the plasmids employed in this work have already been described: 35SCPPDK-GFP comprises[7] a hybrid promoter formed between the cauliflower mosaic virus (CAMV) 35S enhancer and the basal promoter and 5′ nontranslated region of the maize C4PPDK gene. This hybrid promoter was used to drive expression of the wild-type GFP coding sequence.[8] pRJG2 comprises[9] the wild-type GFP-coding sequence under the transcriptional regulation of a doubled CAMV 35S promoter and the polyadenylation signal of the same gene.[10] In pRJG8, the CAMV 35S promoter is placed upstream from a modified version (mGFP[11]) of the GFP coding sequence, in which the codon usage is altered to prevent improper mRNA processing. The GFP-coding sequence is followed by the transcriptional terminator from nopaline synthetase. In pRJG23, the doubled CAMV 35S promoter and the polyadenylation sequence regulate expression of a chimeric protein,[12] containing the minimal

[6] J. Sambrook, E. F. Fritsch, and T. Maniatis, "Molecular Cloning: A Laboratory Manual," 2nd Ed. Cold Spring Harbor Laboratory Press, Cold Spring Harbor, New York, 1989.

[7] J. Sheen, S. Hwang, Y. Niwa, H. Kobayashi, and D. W. Galbraith, *Plant J.* **8,** 777 (1995).

[8] M. Chalfie, Y. Tu, G. Euskirchen, W. W. Ward, and D. C. Prasher, *Science* **263,** 802 (1994).

[9] D. W. Galbraith, J. Sheen, G. M. Lambert, and R. J. Grebenok, *Methods Cell Biol.* **50,** 3 (1995).

[10] F. Guerineau, S. Woodston, L. Brooks, and P. Mullineaux, *Nucleic Acids Res.* **16,** 11380 (1988).

[11] J. Haseloff, K. R. Siemering, D. C. Prasher, and S. Hodge, *Proc. Natl. Acad. Sci. U.S.A.* **94,** 2122 (1997).

[12] R. J. Grebenok, E. A. Pierson, G. M. Lambert, F.-C. Gong, C. L. Afonso, R. Haldeman-Cahill, J. C. Carrington, and D. W. Galbraith, *Plant J.* **11,** 573 (1997).

nuclear targeting sequence from the C2 protein[13] of *Nicotiana tabacum*, sGFP,[14] and β-glucuronidase (GUS[15]). pRJG37 was produced[16] by transferring the expression cassette of pRJG23 into *Kpn*I-cut pBin19.[17]

Plant Growth and Maintenance

Maize hybrid lines FR9^{cms} × FR37 or FR992 × FR637 are from Illinois Foundation Seed (Champaign, IL). Seeds are germinated (25 seeds per pot, 6 inches in diameter) and plants grown either in sterile potting soil (Hyponex, Marysville, IL) or a 2:1 vermiculite:peatlite mix (Grace Sierra, Milpitas, CA) within an environmental chamber (Revco or Conviron) in darkness at 25°. Etiolated leaves are harvested 11–12 days after germination. Plantlets are watered, but no additions of fertilizer are made during growth.

Axenic *Nicotiana tabacum* cv. Xanthi plants are individually grown from surface-sterilized seed in Magenta boxes on 50 ml agar-solidified MS medium containing 3% sucrose. Plants are maintained within standard growth chambers (Revco) at 22°, under constant illumination. Plantlets are subcultured by excising the apical meristem with one or two lateral nodes and subtending leaves, which is then transferred onto fresh medium.

Preparation of Transgenic Plants and of Nuclei

Transgenic tobacco plants are produced by *Agrobacterium*-mediated transformation, using routine methods.[18] For preparation of homogenates, tobacco leaves (200–500 mg) are excised and transferred into a plastic petri dish (60-mm diameter), placed on a prechilled (4°) ceramic tile. Ice-cold chopping buffer (3 ml) is added, and the tissues chopped for 1.5 min using a single-edged razor blade. The chopping buffer (CB) comprises 45 m*M* $MgCl_2$, 30 m*M* sodium citrate, 20 m*M* morpholinepropanesulfonic acid (MOPS), and 0.1% (w/v) Triton X-100, pH 7.0. Prior to analysis, the homogenates are filtered through 15-μm pore size nylon mesh (Tetko, Briarcliff Manor, New York).

[13] C. L. Afonso and D. W. Galbraith, *In Vitro Cell Dev. Biol. (Plants)* **30,** 44 (1994).
[14] W. L. Chiu, Y. Niwa, W. Zeng, T. Hirano, H. Kobayashi, and J. Sheen, *Curr. Biol.* **6,** 325 (1996).
[15] R. A. Jefferson, T. A. Kavanagh, and M. W. Bevan, *EMBO J.* **6,** 3901 (1987).
[16] R. J. Grebenok, G. M. Lambert, and D. W. Galbraith, *Plant J.* **12,** 685 (1997).
[17] M. Bevan, *Nucleic Acids Res.* **12,** 8711 (1984).
[18] G. Rogers, R. B. Horsch, and R. T. Fraley, *Methods Enzymol.* **118,** 627 (1986).

Protoplast Preparation

Maize protoplasts are prepared using the methods described by Sheen *et al.*[7] Second leaves from 11- to 12-day-old maize plants are excised, and stacked (20–25 high) in a sterile plastic 9-cm-diameter petri dish. The leaves are trimmed to remove 4 cm from the top. The remainder is cut into 0.5-mm strips using single-edged razor blades sterilized with 95% ethanol. The strips are placed in a 250-ml sterile side-arm filtration flask containing 20 ml of filter-sterilized enzyme solution: 1% cellulase RS, 0.25% macerozyme R10, 0.6 *M* mannitol, 10 m*M* 2-(*N*-morpholino)ethanesulfonic acid (MES), 1 m*M* $CaCl_2$, 5 m*M* 2-mercaptoethanol, and 0.1% bovine serum albumin (BSA). Vacuum is then applied for 15 min with constant swirling. The flask is placed on a rotary shaker at 40 rpm for 2 hr, and then at 100 rpm for 10 min. The suspension is filtered through a funnel covered with 60-μm pore size nylon mesh into a 50-ml sterile plastic centrifuge tube (Falcon 2098), and centrifuged at 100 *g* for 5 min. The pelleted protoplasts are resuspended in electroporation buffer comprising 0.6 *M* mannitol, 20 m*M* KCl, 4 m*M* MES, and 5 m*M* EGTA, pH 5.7. The centrifugation step is repeated and the protoplasts resuspended to a concentration of $1–2 \times 10^6$/ml.

Tobacco protoplasts are prepared from wild-type and transgenic plants according to the following procedure; all steps are done under standard sterile conditions. Fully expanded leaves are excised (~500 mg) and transferred into a sterile plastic petri dish containing 20 ml of filter-sterilized [Millipore (Bedford, MA) GSWP 047] digestion medium (0.1% driselase, 0.1% macerase, and 0.1% cellulysin, dissolved in 0.5 *M* mannitol, 10 m*M* $CaCl_2$, and 3 m*M* MES, pH 5.7). Incubation is continued overnight (18–24 hr) at 22° in darkness. The protoplast suspension is filtered through two layers of sterile cheesecloth into a 50-ml disposable plastic centrifuge tube, and is centrifuged at 50 *g* for 10 min. The pelleted protoplasts are gently resuspended in 20 ml of 25% (w/v) sucrose dissolved in modified T0 medium.[19] The suspension is overlaid with 5 ml of W5 medium,[20] and centrifuged at 50 *g* for 10 min. Viable protoplasts accumulate as a band at the interface, and are removed using a pasteur pipette. The protoplasts are diluted by addition of 2 vol of W5 medium, collected by centrifugation at 50 *g* for 10 min, and resuspended to a final concentration of 10^6/ml in W5 medium; protoplast numbers are determined by hemocytometry. Samples are filtered through 50-μm pore size nylon mesh immediately prior to flow analysis.

[19] K. R. Harkins, R. A. Jefferson, T. A. Kavanagh, M. W. Bevan, and D. W. Galbraith, *Proc. Natl. Acad. Sci. U.S.A.* **87,** 816 (1990).

[20] I. Negrutiu, R. Shillito, I. Potrykus, G. Biasini, and F. Sala, *Plant Mol. Biol.* **15,** 363 (1987).

Protoplast Transfection

Protoplast transfection is done either by using polyethylene glycol (PEG), or by electroporation. For PEG-mediated transfection, tobacco leaf protoplasts are resuspended in a sterile 15-ml disposable centrifuge tube in 600 μl of W5 medium at a concentration of $\sim 1.6 \times 10^6$ protoplasts/ml. Plasmid DNA and carrier (calf thymus) DNA are added to final concentrations of 20 and 50 mg/ml, respectively. The side of the tube is tapped to ensure the protoplasts are thoroughly resuspended, then 1.5 vol of PEG solution is added with gentle stirring using a 1-ml disposable pipette tip. The PEG solution comprises 40% (w/v) PEG 4000, 0.4 *M* mannitol, and 0.1 *M* $Ca(NO_3)_2 \cdot 4H_2O$, pH 7.0. Incubation is done at room temperature for 25 min, then 8 vol of culture medium [medium NTTO modified[19] to contain 0.18 *M* glucose and 0.15 *M* mannitol, and supplemented with ampicillin (75 mg/ml)] is added. Incubation is then continued overnight (18–24 hr) in darkness at room temperature.

For transfection of maize protoplasts by electroporation,[9] the protoplast pellet is resuspended in electroporation buffer comprising 0.6 *M* mannitol, 20 m*M* KCl, 4 m*M* MES, and 5 m*M* EGTA, pH 5.7. The protoplasts are collected by centrifugation at 100 *g* for 10 min, are resuspended (1.5×10^5) in 0.3 ml of electroporation buffer, and are transferred into a plastic electroporation cuvette (gap width, 4 mm; BTX, San Diego, CA). Plasmid DNA is added [25–50 μg in 30–60 μl of Tris–EDTA (TE)]. Electroporation is done using three pulses, each of 10-msec duration, a field strength of 400–500 V/cm, and a 200-mF capacitance setting. The protoplasts are transferred into plastic six-well culture plates (Falcon 3046), kept on ice for 10 min after electroporation, and diluted by addition of electroporation solution (0.7 ml/well). The culture plates are covered with foil, and incubated in darkness at room temperature overnight.

Setting up Flow Cytometers

Coulter Elite. The Coulter (Hialeah, FL) Elite flow cytometer is operated using a single 20-mW air-cooled argon laser (488-nm emission), a forward-angle light scatter (FALS) detector, and four photomultiplier tubes (PMTs). The standard optical path assigns 90° light scatter to PMT1, 505 to 545-nm fluorescence (green) to PMT2, 555- to 595-nm fluorescence (orange) to PMT3, and 670- to 680-nm (red) fluorescence to PMT4. The laser beam intercepts the flow stream within a 250-μm square quartz cuvette, and emerges through a 100-μm diameter orifice.

The sheath pressure is adjusted to 8.0 lb/in^2 and biparametric histograms of log 90° light scatter versus log green fluorescence, and of log green fluorescence versus log red fluorescence (chlorophyll) are collected at a sample flow rate of 70 particles/sec, with the FALS discriminator being set to 50, and all other discriminators being switched off. PMT voltages and amplification settings should be empirically adjusted, but typical values are as follows: FALS, 260/10; PMT1, 350/7.5; PMT2, 850/10; PMT4, 950/5.

Cytomation MoFlo. The Cytomation (Fort Collins, CO) MoFlo flow cytometer is equipped with a single water-cooled argon laser providing ultraviolet (UV) and 488-nm excitation, an FALS detector, and 5 PMTs. The optical path is organized such that 90° light scatter is assigned to PMT1, 505- to 545-nm fluorescence (green) to PMT2, 555- to 595-nm fluorescence (orange) to PMT3, and 670- to 680-nm (red) fluorescence to PMT4. The laser beam intercepts the flow stream in air after it emerges from a 70- or 100-μm-diameter orifice.

The laser is adjusted to produce 488-nm light at a power output of 180 mW. The sheath pressure is adjusted to 15 lb/in^2, and biparametric histograms collected of log 90° light scatter versus log green fluorescence, and log green versus log red fluorescence, to a total event count of 10^5, at a sample flow rate of 100–200/sec. Typical photomultiplier voltages and amplification settings are as follows: PMT1, 400/1.0; PMT2, 520/1.0; PMT3, 430/1.0; PMT4, 600/1.0. The system is triggered on 90° light scatter, with the threshold being adjusted to eliminate detection of debris.

For protoplasts, the sheath fluid comprises 154 m*M* NaCl, 125 m*M* $CaCl_2$, 5 m*M* KCl, and 5 m*M* glucose, pH 5.7; it is filtered through a 0.22-μm pore size filter prior to use. For nuclei, the sheath fluid comprises CB lacking Triton. The cytometers are aligned using calibration fluorospheres (DNA Check; Coulter Electronics) diluted by addition of 10 vol of deionized water. Adjustment of the optics is continued until population coefficients of variation (CVs) for the various fluorescence and light scatter signals are minimized; these are typically $<2\%$ for fluorescence, and slightly higher (2–3%) for light scatter.

Analyzing Protoplasts and Subcellular Organelles

For analysis of protoplasts, biparametric histograms of log 90° light scatter versus log green fluorescence, and log red fluorescence (chlorophyll) versus log green fluorescence, are collected to a total count of $0.1–1 \times 10^5$. The sample flow rate was 100–200/sec. For analysis of subcellular homogenates,[16] biparametric histograms are similarly collected, with the exception that it is recommended that triggering be done on fluorescence signals rather than light scattering. Homogenates contain a large majority of nonfluorescent, light-scattering particles, and triggering on light scatter can obscure the contributions made by the minor populations of fluorescent organelles.

Sorting

Data from experiments involving sorting are not presented, because they are beyond the scope of this chapter. However, for completeness, the experimental setup required for sorting is provided below.

Coulter Elite. Using the 100-μm flow tip, sheath and sample pressures are set to 12 and 11.5 lb/in^2, respectively, and the piezoelectric drive to about 17 kHz. For large-particle sorting (>10 μm diameter), the sheath and sample pressures are lowered to 8.0 and 7.5 lb/in^2, respectively. Conditions that produce a stable sorted stream can then be found within the range of 12–17 kHz at a 70% drive amplitude. Lower drive frequencies in general improve the efficiency of recovery of larger particles.[21,22] Sort optimizing involves the following: the transducer is allowed to warm up for 60 min at 70% amplitude. The droplet deflection assembly is moved upward to a point as close as possible to the flow cell tip without blocking the laser beam. The video field is adjusted so that both the ground plate of the deflection assembly (to the left edge of the screen) and the laser beam intercept (to the right) can be observed using only the "pan" function. The drive frequency and amplitude are adjusted to provide the shortest possible droplet breakoff point. In sort test mode, the delay setting is adjusted in 0.1-drop increments ("phase" adjustment) to give a sorted stream that is stable. The deflection plate assembly is then lowered so that two or three free droplets are visible above the ground plane. The last-attached droplet should have a well-rounded profile on the left-hand (non-attached) side, and should be obviously connected to the flow stream on the right-hand side. If smaller satellite droplets are observed, it should be established whether these are "fast" or "slow," i.e., whether they merge with the major droplet ahead of or behind the satellite. The amplitudes and frequencies of the piezoelectric transducer are then adjusted until only "fast" satellites are produced. The cursor is next moved to superimpose the second undulation to the right of the last-attached droplet. A point is marked between the first two free drops above the ground plane. The number of droplets separating the two cursors is counted and entered into the delay-calculation program. Histograms are acquired, using the particles of interest, and appropriate sort windows are defined. The delay setting is now empirically optimized through doing a "sort matrix analysis," which

[21] K. R. Harkins and D. W. Galbraith, *Physiol. Plantarum* **60,** 43 (1984).

[22] D. W. Galbraith and G. M. Lambert, *in* "Flow Cytometric Applications in Cell Culture" (M. Al-Rubeai and A. N. Emery, eds.), p. 311. Marcel Dekker, New York, 1995.

aims to achieve a maximal (near 100%) sort efficiency. This involves batch sorting of groups of 25 particles onto standard glass microscope slides, counting the numbers actually recovered under the microscope as a function of the sort delay setting. Between each batch, the sort delay setting is adjusted in one-step increments to span a range of ± five around the calculated delay setting. For one-drop sorts, the delay is adjusted in 0.1-drop increments. In the case of large, fragile particles, such as protoplasts, sorting efficiencies must be optimized using indestructible particles of similar sizes, such as pollen.[22] The deflection plate high-voltage and phase settings are adjusted to obtain side streams lacking droplet "fanning." Histograms are accumulated for the particles of interest, the appropriate sort windows are defined, and sorting is enabled. For larger tips, droplet formation usually occurs below the field of view of the video camera, and the cursor-based computation cannot be done. In this case, the sort delay has to be determined through a sort matrix.

If protoplasts are to be cultured after sorting, sterile sorting procedures are required. Sheath and rinse tanks are cleaned with detergent, then filled with 70% ethanol. Ethanol is back-flushed through the sample uptake opening, then a sample tube containing 70% ethanol is run through the system for 15 min. These steps are repeated with sterile water, before filling the system with sterile sheath fluid. Samples are sorted into sterile 15-ml centrifuge or microcentrifuge tubes. In standard, three-droplet sort mode, we routinely obtain 100% sort efficiencies using a transducer drive frequency of 16.7 kHz and a sort delay of 15. Three-droplet sorting provides the greatest margin of error for recovery of the desired particles, but it also increases the probability that unwanted particles will be included within the sorted droplets. The unwanted particles can be eliminated by enabling the anticoincidence circuit, as long as they trigger data acquisition, although this may result in unacceptably low recoveries. If the unwanted particles produce signals lower in value than the threshold set on the active discriminator, they will not trigger data acquisition. Under these circumstances the only available means to eliminate contamination is to perform one-droplet sorts. Because the error margin for sorting the desired particle is reduced in one-droplet mode, optimizing sort recoveries requires careful adjustment of the phase setting.

Cytomation MoFlo. For sorting of protoplasts or nuclei using the 100-μm tip, we employ sheath and sample pressures of 15 and 14.3 lb/in^2, a drive frequency of 21.5 kHz, and a drive amplitude of 60 V. These settings produce symmetrical droplets having six to eight fast satellites visible below the last attached droplet. In this system, the deflection assembly is fixed in position so no adjustment is required. Using the test pattern function, the charge phase and deflection amplitude values are adjusted to produce stable side streams with minimal fanning. An estimate of the delay is first determined from the stroboscopic camera image of the flow stream, by counting the number of undulations observed between the last attached droplet and the flow cell tip. This is then used as a starting point for performing a sort matrix, using the methods described in the previous section. It is particularly critical, when employing the one-drop sorting mode, to find via the sort matrix the nearest one-sixteenth drop setting that gives 100% recovery.

For sterile sorting, the sheath tank is filled with 70% ethanol and the instrument is allowed to run for 15 min, thereafter being rinsed with sterile water before filling the sheath tank with sterile sheath fluid. Samples are sorted into sterile 12 × 75 mm tubes.

Results

Flow Cytometric Detection of Green Fluorescent Protein in Transfected Protoplasts

Figure 1 illustrates the flow cytometric detection of accumulation of cytoplasmic GFP following transfection of maize leaf protoplasts with 35SCPPDK-GFP. Biparametric analysis of red versus green fluorescence emission intensity indicates that control (nontransfected) protoplasts characteristically occupy a discrete cluster having a low level of green autofluorescence and a somewhat higher level of red autofluorescence (Fig. 1A). This is a consequence of the presence of endogenous pigments within these etiolated cells. In contrast, protoplasts examined under the microscope 10–18 hr after transfection with the GFP constructs exhibit green fluorescence,[12] and flow cytometric analysis of these protoplasts reveals a second cluster, having a greatly increased level of green fluorescence but unaltered red fluorescence (Fig. 1B). This is consistent with expression of GFP within these protoplasts. The proportion of protoplasts expressing GFP is variable between experiments. In the illustrated case, approximately 14% express GFP, and 86% do not. However, the lower cluster includes nonviable protoplasts, and this means that the overall transfection percentages are almost certainly higher than 14%.

Flow cytometric analysis of surface granularity (90° light scatter) can also be combined with measurement of green fluorescence emission for detection of GFP expression in protoplasts. Again, nontransfected protoplasts occupy a single cluster (Fig. 1C), whereas protoplasts analyzed 18 hr posttransfection are distributed between two clusters (Fig. 1D), the upper of which (in this experiment, 34% of the protoplasts) represents the GFP-expressing cells. Statistical analysis can be automatically done on the clusters to provide mean, median, and mode values. The mean fluorescence emission

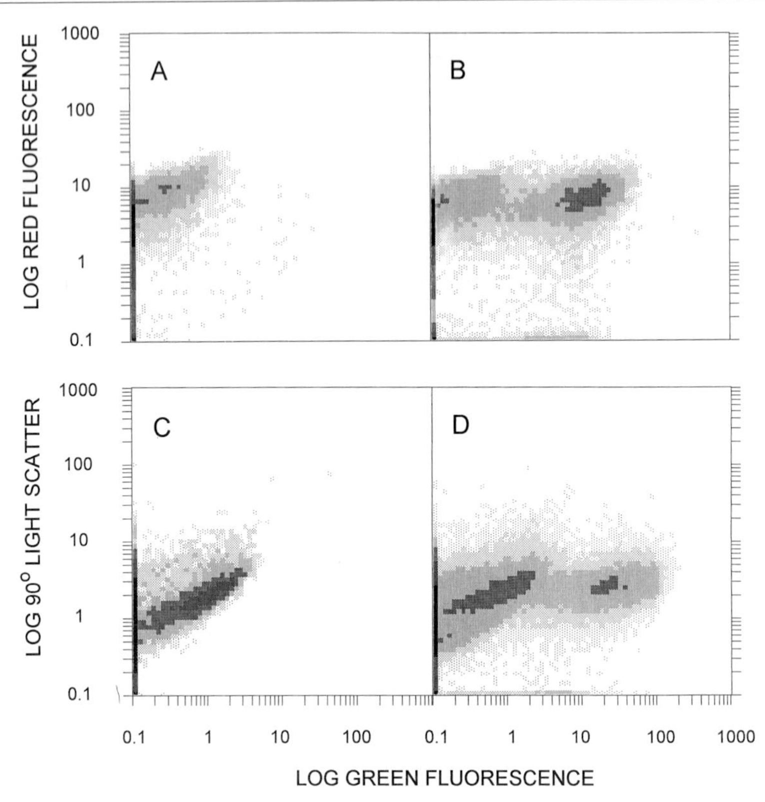

FIG. 1. Transient expression of GFP in maize leaf protoplasts. (A) Biparametric flow analysis of control protoplasts, based on green and red fluorescence emission. (B) Biparametric flow analysis, based on green and red fluorescence emission, of transfected maize protoplasts. (C) Biparametric flow analysis of control protoplasts, based on green fluorescence and 90° light scatter. (D) Biparametric flow analysis of transfected protoplasts, based on green fluorescence and 90° light scatter (from Ref. 9, with permission).

intensity from individual GFP-expressing protoplasts in these cases is about 80-fold that of the nontransfected controls.

Similar results are obtained using pRJG2, pRJG8, and 35SCPPDK-GFP, and using tobacco leaf protoplasts, although the proportion of tobacco protoplasts that express GFP is generally lower than for maize. Particularly important to successful transfection is the use of plants growing under optimal conditions. Such plants produce high yields of protoplasts, which are intact and viable, and can tolerate transfection without excessive cell rupture. As a general rule, if initial protoplast yields are poor, transfection efficiencies will be drastically reduced.

Flow Cytometric Detection of Green Fluorescent Protein in Protoplasts Prepared from Transgenic Plants

Protoplasts from the leaves of dark-grown maize seedlings contain etioplasts, rather than chloroplasts, and therefore appear as translucent spheroids approximately 25 μm in diameter, within which is the nucleus, a large vacuole, and cytoplasmic strands containing numerous granules.[7] In contrast, protoplasts from tobacco leaves are about 35 μm in diameter, and contain approximately 70 mature chloroplasts.[21] Biparametric analysis of tobacco protoplasts based on 90° side scatter versus log red fluorescence identifies two clusters (Fig. 2A and C). The cluster of higher fluorescence comprises intact protoplasts. The cluster of lower red fluorescence comprises free chloroplasts that are derived from broken protoplasts that, owing to the large numbers of chloroplasts within individual cells, inevitably contaminate protoplast populations, even those purified by sucrose gradient flotation.[21]

Tobacco plants transformed with pRJG37 accumulate green fluorescence within the nuclei.[16] No differences are observed in biparametric analyses of 90° side scatter versus red fluorescence between the control and transgenic populations (Fig. 2A and C). Biparametric analysis of 90° light scatter versus log green fluorescence of control tobacco protoplasts also reveals two clusters (Fig. 2D). However, for the transgenic population (Fig. 2B), an additional cluster is observed, having an approximately fourfold increase in green fluorescence. It can be assumed that all tobacco protoplasts produce low levels of endogenous green fluorescence due to contributions from chloroplast and vacuolar pigments having emission spectra overlapping that of GFP. This modest overall increase in green fluorescence may also reflect in part the small size of the nuclear compartment within which the GFP molecules are accumulated. Interestingly, there is evidence that the transgenic population contains some protoplasts that express little or no GFP (in Fig. 2B, some protoplasts remain in a cluster occupying the same position as that of the nontransgenic protoplasts in Fig. 2D).

Flow Cytometric Detection of Green Fluorescent Protein in Nuclei Isolated from Transgenic Plants

The problem of high endogenous fluorescent backgrounds provided by protoplasts can be alleviated by analysis of subcellular homogenates. When leaves from transgenic tobacco plants transformed with pRJG37 are homogenized by chopping, intact nuclei are released. Biparametric flow analysis of 90° side scatter versus log green fluorescence identifies a discrete cluster of green-fluorescent nuclei that are not present in homogenates prepared

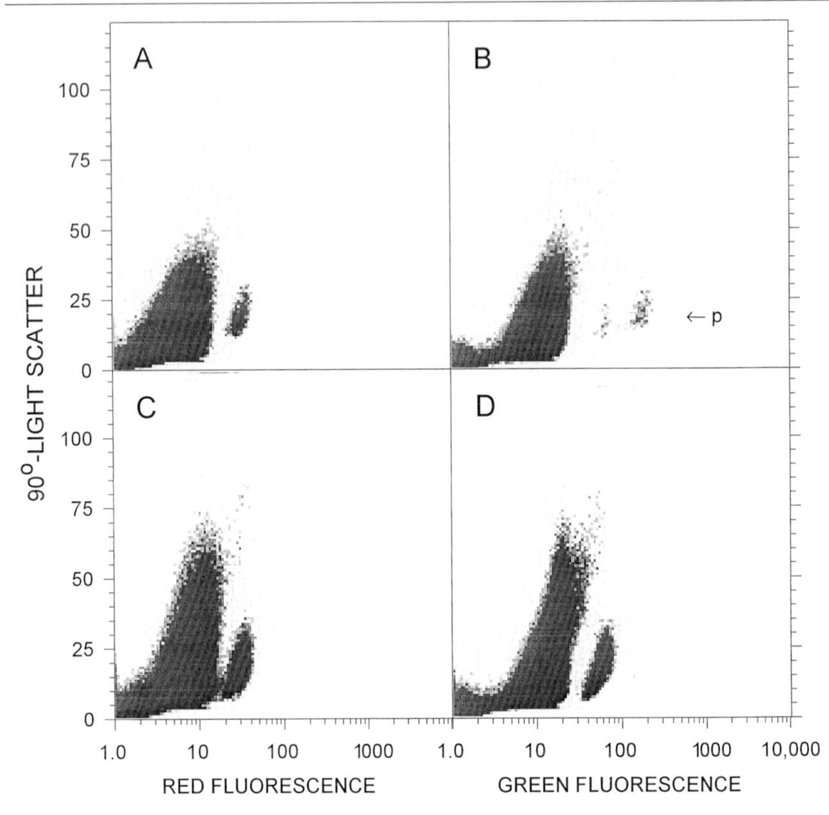

FIG. 2. Flow cytometric analysis *in vivo* of GFP targeting to nuclei. Protoplasts from leaves of transgenic plants (A and B), or nontransgenic controls (C and D), were subjected to biparametric analysis of 90° light scatter either versus red fluorescence (A and C), or versus green fluorescence (B and D). No differences are observed between the transgenic and nontransgenic populations in terms of red fluorescence. In the transgenic population an additional cluster of protoplasts (p) is visible; this cluster has an enhanced level of green fluorescence that is absent from the nontransgenic control.

from the nontransgenic controls (Fig. 3A and B). A similar analysis can be done employing log green fluorescence versus log orange fluorescence (Fig. 3C and D). In this case, the nuclei appear as a cluster oriented diagonally. This is due to spillover of the long-wavelength portion of the GFP fluorescence into the orange fluorescence detector (555–595 nm). In comparison with the situation observed in analysis of intact protoplasts, the cluster of nuclei is completely separated from the other cluster corresponding to subcellular organelles and miscellaneous debris, indicating the endog-

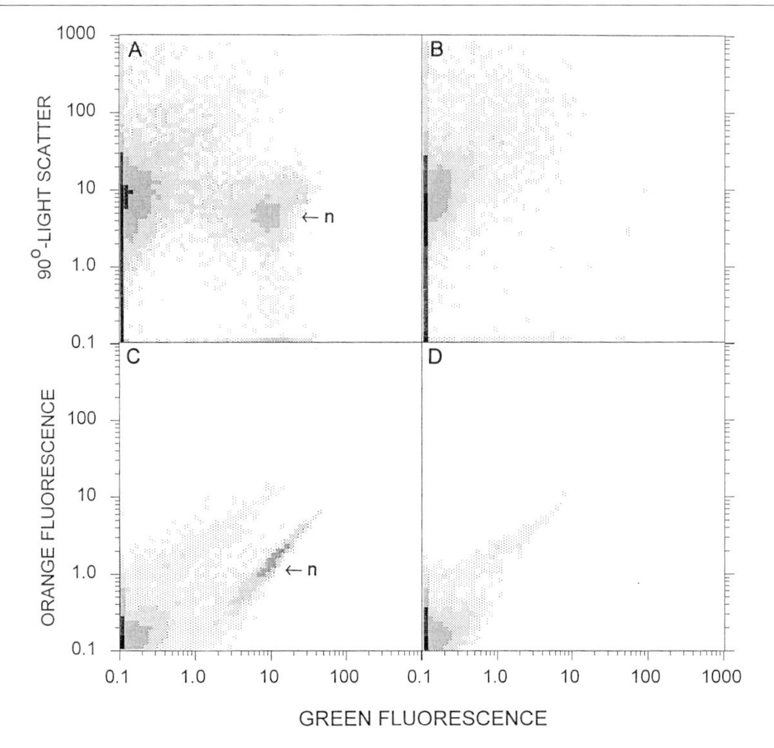

FIG. 3. Flow cytometric analysis *in vitro* of GFP targeting to nuclei. Homogenates from leaves of transgenic plants (A and C), or nontransgenic controls (B and D), were subjected to biparametric analysis of 90° light scatter versus green fluorescence (A and B), or of orange versus green fluorescence (C and D). The regions corresponding to nuclei (n) containing GFP are absent from the controls (from Ref. 16, with permission).

enous levels of non-GFP green fluorescence within cellular homogenates are low.

Examination of different transgenic plants from independent transformations reveals variation in the level of GFP fluorescence produced by the nuclei, which can be quantitatively analyzed through reference to fluorescent beads as internal standards.[16] The degree of nuclear GFP accumulation is significantly correlated ($r^2 = 0.81$) with GUS activity measured in total cellular homogenates. GUS activity is also topologically restricted to the nuclei as indicated by histochemical staining.[16] The expression cassette contained in pRJG23 and pRJG37 therefore represents an example of a truly bifunctional reporter molecule.

Discussion

Flow cytometry provides a powerful means for the accurate analysis of the amounts of fluorescence contained within populations of cells and within subcellular organelles.[2,3] For flow cytometry to be employed for the analysis of transgenic expression of marker proteins, it is necessary to link detection of a fluorescent signal to the level of expression of the transgene. Traditionally, this has been done through use of cell surface markers, which are subsequently detected via immunofluorescence labeling.[2,3] The emergence of GFP as a marker transgene has considerably expanded the range of experiments that can be done using flow cytometry. The unique attributes of GFP, which provide particular advantages over other markers, are primarily that the production of fluorescence requires only the translation of the primary sequence of the protein, and again that observations can be made *in vivo,* and do not require fixation or other cellular perturbations. A further, practical feature of GFP is that it can be efficiently excited using the argon lasers with which most flow cytometers are routinely equipped.[5] Finally, in the specific case of higher plants, the characteristic fluorescence emission of GFP is, for many tissues and plant species, spectrally distinct from other autofluorescent pigments found within the cells, most notably the pigments of the photosynthetic apparatus.

Our results demonstrate that flow cytometry can be readily employed to monitor the expression of GFP within higher plants, both within intact protoplasts and, following targeting, within subcellular organelles.[7,9,12,16] In the flow cytometric analysis of plants, the major problems first concern reducing native organs and tissues into suspensions of cells or subcellular organelles that can pass through the restricted diameter of the flow cell tip. This problem is avoided either through enzymatic production of protoplasts, which are spherical, wall-less cells, or through mechanical tissue disruption to produce homogenates containing suspensions of intact organelles. The second problem is the production of a GFP-specific fluorescent signal that can be readily detected over the background of cellular fluorescence. Endogenous autofluorescence is largely associated with three cellular structures, the cell wall, the vacuole, and the plastids; other objects within the cell, for example, nuclei, mitochondria, and miscellaneous vesicles, are largely translucent and nonfluorescent. The fluorescent contribution of cell walls is irrelevant in the analysis of both protoplasts and subcellular homogenates. For homogenates, tissue disruption converts the large vacuoles typical of plant cells into a myriad of tiny vesicles, the individual fluorescent contributions of which turn out to be negligible. For subcellular homogenates, the problem then reduces to providing a GFP molecule that is efficiently targeted to a location that is structurally distinctive, and that can be preserved in an intact state following homogenization. Targeting of GFP to the nucleus satisfies these criteria, and it then becomes a question of distinguishing green fluorescent nuclei from other organelles having endogenous autofluorescence that overlaps the emission spectrum of GFP. For protoplasts, the problem is more complex, because a single protoplast contains a fluorescent background comprising the autofluorescent contribution of its vacuole plus the summed contributions of all its plastids. It should be noted that, for some tissues (tobacco roots), the presence of green intracellular autofluorescence significantly impedes examination of nontargeted (i.e., cytoplasmic) GFP expression in transgenic plants.[16] In this regard, targeting of GFP to the nucleus provides an unambiguous means to detect transgene expression over the background autofluorescence, because this background is diffusely dispersed throughout the cell, and never occupies punctate regions as seen for accumulation of GFP within the nucleus.

The means that we have employed to achieve GFP expression involves one of two possible experimental procedures: transfection of protoplasts resulting in transient expression, or production of transgenic plants. We have shown that both approaches result in GFP expression that can readily be observed using epifluorescence and confocal microscopy, and, through including appropriate modifications to the GFP primary sequence, allows unambiguous observation of targeting to the nucleus. Two modifications are needed. The first is to include an effective nuclear localization signal (NLS) translationally fused to the GFP coding sequence. We have employed a short NLS taken from an orphan tobacco transcriptional factor (the C2 protein[13]), which appears functional in a wide variety of different tissues, and in at least two plant species. The second modification involves increasing the size of the GFP molecule. Native GFP approximates 26 kDa in size, which is smaller than the exclusion limit of nuclear pores. For this reason, GFP appears free to diffuse into and out of the nucleus,[12] and is not retained within the nucleoplasm even by an effective NLS. We increased the size of GFP through fusing it to the complete coding region of GUS; the chimeric protein has a molecular size in excess of 100 kDa. Correspondingly, attachment of the C2 NLS efficiently targets this chimeric molecule into the nucleus. We have also reported the construction of GFP fusions with NLS-containing polypeptide sequences taken from the tobacco etch virus.[12] These effectively target GFP to the nucleus, and a common feature is an increase in molecular mass, respectively, to 48 and 76 kDa. Reports relating to the interaction of the molecular size of GFP with the process of nuclear targeting include analysis of the process of signal-mediated nuclear export.[23] In this case, a fusion protein (~68 kDa) was synthesized comprising a GFP

[23] K. Stade, C. S. Ford, C. Guthrie, and K. Weis, *Cell* **90,** 1041 (1997).

dimer coupled to the simian virus 40 (SV40) T antigen NLS and to the leucine-rich nuclear export signal (NES) of the inhibitor of the cAMP-dependent protein kinase. The increased size of this reporter prevented passive diffusion of the molecule across the nuclear pore, and thereby permitted genetic analysis of the mechanisms of NLS- and NES-mediated nucleocytoplasmic shuttling.

The presence of GUS and GFP within a contiguous polypeptide provides a bifunctional reporter. As far as we are aware, this is the first study of bifunctional reporter molecules in higher plants. Because the two different functions can be separately assayed, this means that an examination is possible of the interactions of these domains. Similar values were observed for the specific activities of GUS, c2-NLS-GUS, and c2-NLS-GFP-GUS in transgenic tobacco plants, and in transfected protoplasts.[12,15] The different transgenic plants produced in these experiments, derived from separate transformation events, displayed different levels of GFP accumulation. We observed a strongly positive correlation between the fluorescence intensity of the leaf nuclei and the total GUS activity measured in leaf extracts, implying that all of the bifunctional reporter is located in the nuclei. This also suggests that gene expression, rather than protein targeting, limits the levels of fluorescence that were observed. The residual variance in the regression analysis might represent interference between the two protein domains during folding, because GUS forms homo-tetramers,[15] whereas GFP has the ability to dimerize under various conditions,[24–26] which results in fluorescence quenching.

For the protoplasts isolated from transgenic plants, in contrast to the situation observed for transfected protoplasts, the total levels of cellular green fluorescence measured through flow cytometry were only modestly increased over those found in the nontransgenic controls. This is in part due to the presence of endogenous autofluorescence, which overlaps the emission spectrum of GFP. In part, the low contribution may also reflect the possibility that the nucleus provides a limited storage capacity for GFP. It is also possible that transgenic expression of GFP might be, at some level, toxic to the cell. A persistent level of comment on the Fluorescent Proteins Newsgroup on Bionet (*www.bio.net*), concerns the potential toxic effects of cytoplasmic accumulation of GFP. This contrasts with published reports from various groups of high levels of cytoplasmic GFP expression in

[24] A. B. Cubitt, R. Heim, S. R. Adams, A. E. Boyd, L. A. Gross, and R. Y. Tsien, *Trends Biol. Sci.* **20,** 448 (1995).

[25] M. Ormö, A. B. Cubitt, K. Kallio, L. A. Gross, R. Y. Tsien, and S. J. Remington, *Science* **273,** 1392 (1996).

[26] F. Yang, L. G. Moss, and G. N. Phillips, Jr., *Nature Biotechnol.* **14,** 1246 (1996).

transgenic plants.[27,28] In common with most studies in transgenic organisms, qualitative rather than quantitative methods are typically employed for estimating the amounts of GFP fluorescence. This can be fluorescence microscopy, although there is considerable interest in the possibility of employing hand-held illumination sources for large-scale screening of transformants in the greenhouse or in the field. From the results that have appeared, it is impossible to determine whether specific limits exist as to the levels of GFP that can accumulate within individual plant cells without encountering toxicity. It is self-evident that the expression of any protein beyond specific limits will, by definition, prove toxic to any cell. The situation for GFP is further complicated by photochemical effects, which are almost certainly a function of the intensity of the incident illumination.

Our results indicate that transgenic nuclear accumulation of GFP can be achieved at levels sufficient to provide an easily measured signal that can be recorded using epifluorescence or confocal microscopy, or flow cytometry. Under these conditions, the plants appear normal, and progress through their life cycles in a normal manner. It should be noted that the use of different promoters, coding sequences, targeting signals, and terminators prevents determination of the reasons for the different levels of expression exhibited between the transfected protoplasts and transgenic plants. This is further complicated by the known occurrence of position effects in transgenic plants, as well as between-experiment differences in transfection efficiencies. In our experiments, the highest levels of expression were typically found in transfected protoplasts. Variation was associated with transfection efficiencies, but not with the type of plasmid that was employed. From this, we conclude that, at least in tobacco, the levels of expression of cytoplasmic GFP that were achieved are not toxic over the short term, and are not seriously compromised by improper mRNA processing. The systematic use of flow cytometry to compare overall levels of GFP expression between different constructions, target locations, and tissue types, should be instructive. It should also allow an unambiguous determination as to whether transgenic GFP expression beyond a certain concentration limit becomes toxic to the cell, and whether targeting to various subcellular compartments ameliorates or augments this toxicity.[11]

Future developments in the flow cytometric analysis of GFP expression in plants will take advantage of an increasing number of spectral variants of this molecule. The wild-type GFP fluorophore has a bimodal excitation

[27] C. Reichel, J. Mathur, P. Eckes, K. Langenkemper, C. Koncz, J. Schell, B. Reiss, and C. Maas, *Proc. Natl. Acad. Sci. U.S.A.* **93,** 5888 (1996).

[28] S. Z. Pang, D. L. DeBoer, Y. Wan, G. Ye, J. G. Layton, M. K. Neher, C. L. Armstrong, J. E. Fry, M. A. W. Hinchee, and M. E. Fromm, *Plant Physiol.* **112,** 893 (1996).

spectrum, having peaks of absorption at 395 and 475 nm. Its emission spectrum is unimodal, with a maximum at about 509 nm. Amino acid substitutions either in the fluorophore or in the surrounding sequences are found to alter the emission and excitation spectra, the photostability, and the quantum yield of GFP. Spectral variants include those characterized by a unimodal absorption spectrum in which one of the two wild-type absorption peaks is eliminated, but without an alteration in the emission spectrum. One class of these variants has an absorption maximum at 395 nm. These variants[29,30] share a threonine-to-isoleucine substitution at amino acid 203, and for flow cytometric applications can be conveniently excited using the Innova 300 krypton laser.[31] Members of a second, large class of spectral variants exhibit unimodal excitation spectra having a maximum near 475 nm.[32–34] The excitation maxima of these variants are typically at a slightly longer wavelength than the 475-nm peak of wild-type GFP. This means that they are more efficiently excited by the 488-nm argon laser line, and therefore have an increased brightness over wild type. The extinction coefficients and quantum yields of both of these classes of variants are at least as great as wild-type GFP and comparable to that of fluorescein. As a general rule of thumb, in flow cytometry it has been found that at least 500 fluorescein molecules must be associated with a specific particle or cell to produce a fluorescent signal that is above the background derived from cellular autofluorescence and instrument noise. In this respect, optimized[14] S65T GFP (a member of the second class of variants) produces roughly equivalent amounts of fluorescence as fluorescein. GFP variants also have been described that absorb in the UV and emit in the blue regions of the spectrum,[29] and that absorb in the green and emit in the yellow.[25] In these cases, modifications to codon usage to increase overall levels of expression are also incorporated; such variants are typically termed "enhanced." Although the codon usage is human, we have found that such enhanced GFPs provide excellent signals in higher plant cells. Biparametric analysis of simultaneous expression of two GFP variants has been described[31,35] in

[29] R. Heim, D. C. Prasher, and R. Y. Tsien, *Proc. Natl. Acad. Sci. U.S.A.* **91,** 12501 (1994).

[30] T. Ehrig, D. J. O'Kane, and F. G. Prendergast, *FEBS Lett.* **367,** 163 (1995).

[31] M. Anderson, I. M. Tjioe, M. C. Lorincz, D. R. Parks, L. A. Herzenberg, G. P. Nolan, and L. A. Herzenberg, *Proc. Natl. Acad. Sci. U.S.A.* **93,** 8508 (1996).

[32] S. Delagrave, R. E. Hawtin, C. M. Silva, M. M. Yang, and D. C. Youvan, *Bio/Technology* **13,** 151 (1995).

[33] B. P. Cormack, R. H. Valdivia, and S. Falkow, *Gene* **173,** 33 (1996).

[34] R. Heim, A. B. Cubitt, and R. Y. Tsien, *Nature (London)* **373,** 663 (1995).

[35] J. Ropp, C. Donahue, D. Wolfgang-Kimball, J. Hooley, J. Chin, R. Hoffmann, R. Cuthbertson, and K. Bauer, *Cytometry* **24,** 284 (1996).

animal cell systems, and it will be of interest to determine whether these techniques can be applied to higher plants.

In terms of instrumentation, the illumination that can be applied to the cells is only limited by the financial resources available to the flow cytometric facility. Base model flow cytometers are equipped solely with an air-cooled argon laser, emitting light of 488 nm. This evidently restricts analysis of GFP molecules to those absorbing at that wavelength. Upgrades to these instruments provide additional lasers with illumination either in the UV or at various visible wavelengths from 456 to 799 nm. Various numbers of lasers can be arranged to independently intersect the flow stream. The unofficial record of numbers of independent flow cytometric parameters simultaneously made on a single population of cells is nine, and required three-laser excitation.[36] The availability of multiple different excitation wavelengths considerably expands the numbers of different GFP variants that might simultaneously be examined using flow cytometry, although the present costs of such equipment will inevitably be restrictive. Further developments in laser technologies, particularly for diode lasers, are certainly likely to reduce future costs, and increase the different numbers of lines available for use in flow cytometry. An alternative strategy is to search for novel, spectrally distinct variants that can be simultaneously excited using 488-nm light. An example of this is a report describing dual-color flow cytometric analysis of expression in murine cells of enhanced GFP (i.e., $E_{max} = 490$ nm; codon optimized) and enhanced yellow fluorescent protein ($E_{max} = 513$ nm), using argon laser excitation at 488 nm.[37] Finally, it is most likely that additional novel fluorochromatic polypeptides will be recovered from oceanic species other than *Aequorea,* and these may well prove useful in flow cytometric analyses.

Acknowledgments

The author thanks Georgina Lambert for expert technical assistance in the flow cytometric analyses. This work was supported by grants from the National Science Foundation, and the United States Departments of Agriculture, Energy, and the Army.

[36] N. Baumgarth, D. R. Parks, M. Bigos, R. T. Stovel, M. A. Anderson, R. A. Gerstein, U. C. Klug, G. C. Jager, O. C. Herman, L. A. Herzenberg, and L. A. Herzenberg, *Cytometry Suppl.* **9,** 38 (1998).

[37] L. Lybarger, D. Dempsey, G. H. Patterson, D. W. Piston, S. R. Kain, and R. Chervenak, *Cytometry* **31,** 147 (1998).

[27] Continual Green Fluorescent Protein Monitoring of Promoter Activity in Plants

By PETER E. URWIN, SIMON G. MØLLER, JENNIFER K. BLUMSOM, and HOWARD J. ATKINSON

Introduction

Both luciferase[1] and β-glucuronidase (GUS)[2,3] have been widely used as reporters of gene expression in plants. A principal advantage of luciferase is that it is not stable in plants and so it can provide real-time monitoring. It shares with GUS the disadvantage of requiring uptake of exogenous substrates. Luciferin is an expensive substrate and must be applied repeatedly for continual monitoring. Other limitations of this approach can include the need for highly sensitive detection systems such as single-photon counting to detect lower levels of light emission[4] and the limited subcellular localization that can be achieved. The standard GUS-based procedure is sensitive, uses an inexpensive substrate, and requires only a basic bright-field microscope to observe reporter activity. It has the considerable disadvantage of requiring destructive sampling of plants and of not providing quantitative information on promoter activity. The exogenously applied substrates Imagene Red and Green (Molecular Probes, Cambridge, UK) raise the cost of the assay but they do avoid the need for destructive sampling. The product is not lethal to plant cells and can be visualized by standard fluorescence.

Green fluorescent protein (GFP)[5] offers features that overcome many of the disadvantages of both GUS and luciferase-based reporter systems. It provides a reporter gene system *in planta* that avoids destructive sampling and it does not require substrate addition.[6] It can be observed using a hand-held, fluorescent light source in a limited range of applications but it is usually visualized using a standard fluorescence microscope. The potential of the system can be enhanced using charge-coupled device (CCD)-based

[1] D. W. Ow, K. V. Wood, M. Deluca, J. R. de Wet, D. R. Helinski, and S. H. Howell, *Science* **234,** 866 (1986).
[2] R. A. Jefferson, *Plant Mol. Biol. Rep.* **5,** 387 (1987).
[3] R. A. Jefferson, T. A. Kavanagh, and M. W. Bevan, *EMBO J.* **6,** 3901 (1987).
[4] A. J. Miller, S. R. Short, N. H. Chua, and S. A. Kay, *Plant Cell* **4,** 1075 (1992).
[5] D. C. Prasher, V. K. Eckenrode, W. W. Ward, F. G. Prendegast, and M. J. Cormier, *Gene* **111,** 229 (1992).
[6] J. Sheen, S. Hwang, Y. Niwa, H. Kobayashi, and D. W. Galbraith, *Plant J.* **8,** 777 (1995).

0076-6879/99 $30.00

systems (see Fluorescence Microscopy of GFP Emissions). GFP produces a bright green fluorescence in prokaryotic or eukaryotic systems when cells expressing it are illuminated with the correct excitation wavelength. The fluorescence is independent of cell type or location and the protein requires the presence of oxygen[7] but no cofactors to emit light. Several variants of the original wild-type GFP have been engineered.[8] Variants are now commercially available with both altered emission wavelength and intensity (e.g., Clontech, Basingstoke, UK), which will further increase the applications of GFP as a reporter *in planta.* Here we provide examples of the potential of GFP for continual, nondestructive monitoring of promoter activity *in planta* using mGFP [kindly provided by J. Haseloff, Medical Research Council (MRC), Cambridge, UK].

Green plant tissue has a naturally high background fluorescence, primarily owing to the presence of chlorophyll. Sheen *et al.*[6] demonstrated a considerable change in the visualized color of maize mesophyll protoplasts, according to the ratio of GFP and chloroplast emissions, with a spectrum of coloration from green through yellow and orange to red according to their relative intensities. Plautz *et al.*[9] describe a similar phenomenon following GFP expression in cells of soybean cultured in suspension. Consequently, GFP is not readily used in green tissue of appreciable thickness owing to its quenching by chlorophyll and the intense light emission by this fluorophore. In contrast, roots exhibit much lower levels of autofluorescence, which makes it much easier to detect and quantify levels of GFP expression in such tissue. Therefore much of the work in this chapter focuses on continual monitoring of GFP in roots. However, the advent of GFP variants with improved levels of fluorescence and color shifts in their emissions[10] may ensure many of the approaches described here will soon be adaptable to aerial tissue.

Green Fluorescent Protein as a Plant Transformation Marker

Green fluorescent protein has been visualized in green tissue under particular circumstances. Baulcombe *et al.*[11] inoculated plants with potato virus X harboring GFP and photographed regions with high viral particle densities within infected plants. This required a hand-held 100-W UV light source, a green or yellow secondary filter (Wratten 58 or 8, respectively,

[7] R. Heim, D. C. Prasher, and R. Y. Tsien, *Proc. Natl. Acad. Sci. U.S.A.* **91,** 12501 (1994).

[8] M. Staci, S. A. Mabon, and C. N. Stewart, *BioTechniques* **23,** 912 (1997).

[9] J. D. Plautz, R. N. Day, G. M. Daily, S. B. Welsh, J. C. Hall, S. Halpain, and S. Kay, *Gene* **173,** 83 (1996).

[10] C. N. Stewart, Jr., *Nature Biotech.* **14,** 682 (1996).

[11] D. C. Baulcombe, S. Chapman, and S. Santz Cruz, *Plant J.* **7,** 1045 (1995).

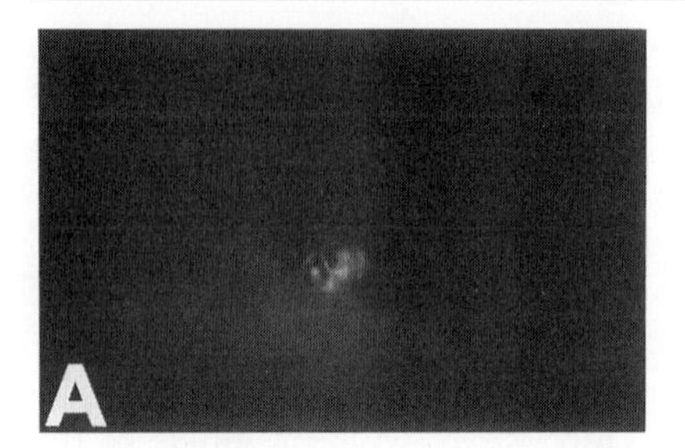

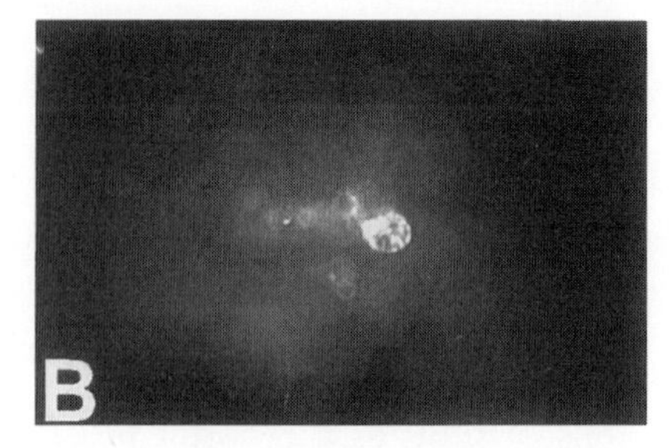

FIG. 1. Expression of GFP in callus of *Arabidopsis thaliana* C24 at (A) 7 days and (B) 14 days. Cells in the plane of focus are seen as brighter. Images were viewed with an Olympus (model BH2) microscope, using an excitation wavelength of 395–415 nm (violet light) and a cutoff emission filter at 495 nm. Images were collected with an Olympus 35-mm OM4Ti, with Kodak Ektachrome color reversal film 160T with exposures of between 10 and 15 sec.

Kodak, Manchester, UK), a 35-mm camera, and reversal (transparency) film [400 ASA; Kodak (Rochester, NY) Ektachrome Panther] plus exposure times of up to 70 sec. Kohler *et al.*[12] used a hand-held 365-nm UV light source to excite GFP in green tobacco callus and photographed the emissions from GFP with a 35-mm camera. Nagatani *et al.*[13] also monitored the expression of GFP, which in this case was driven by the ubiquitin promoter *ubi1*. Detection was in rice calli following heat induction of the promoter by excitation with near-UV light; images were collected with a CCD camera.

GUS has often been used as a marker for transient expression studies or in cotransformation to determine transformation efficiencies.[14] GFP has clear potential for such use. We have used GFP to select positive transformants from putatively transformed *Arabidopsis* growing on a selection medium. A cauliflower mosaic virus (CaMV35S) and GFP construct was electroporated into *Agrobacterium tumefaciens* and subsequently introduced into *Arabidopsis thaliana* ecotype C24.[15] This procedure involves the redifferentiation of a plant from callus tissue. GFP fluorescence was monitored at this early stage to establish those calli that were transformed (Fig. 1). This can be achieved using a fluorescence microscope. Owing to the nondestructive nature of the analysis no tissue must be sacrificed and the selected calli can be allowed to differentiate into mature plants. GFP could be used either in tandem with another transgene or as part of a construct to be translated as a fusion protein with other transgene products.

[12] R. H. Kohler, W. R. Zipfel, W. W. Webb, and M. R. Hansen, *Plant J.* **11,** 613 (1997).

[13] N. Nagatani, S. Takumi, M. Tomiyama, T. Shimada, and T. Tamiya, *Plant Sci.* **124,** 49 (1997).

[14] P. Christou, T. L. Ford, and M. Kofron, *Bio/Technology* **9,** 957 (1991).

[15] P. E. Urwin, S. G. Møller, C. J. Lilley, H. J. Atkinson, and M. J McPherson, *Mol. Plant Microbe Interact.* **10,** 394 (1997).

This would provide a simple and rapid basis for screening putative transgenics and for monitoring subsequent events such as gene silencing. Plantlets can be selected, using this system, by examining them under a low-power objective of a fluorescence microscope. The use of a simple hand-held system may also be of value in monitoring whole plants to determine the potential spread of genes from transgenic crops to wild relatives.[8]

Fluorescence Microscopy of Green Fluorescent Protein Emissions

The basic requirements for GFP detection *in planta* are normally present in most fluorescence microscopes. Essentially, the conditions required to visualize fluorescein isothiocyanate (FITC) will often suffice to observe mGFP. We have used a standard microscope (model BH2; Olympus, Norwood, MA) equipped with a 100-W mercury lamp and an epiillumination system with a dichromic mirror involving excitation with violet light at 395–415 nm and a secondary filter passing >495 nm. We have also used an infinity-corrected microscope (model DMRB; Leica, Eaton, PA) with an FITC filter set (450 to 490-nm bandpass excitation filter and a secondary filter passing >515 nm). Confocal laser scanning microscopy is also often employed to study the expression of GFP. Such microscopes are supplied by several manufacturers [e.g., Bio-Rad (Hercules, CA), Molecular Dynamics (Sunnyvale, CA), Zeiss (Thornwood, NY), and Leica]. A laser able to excite GFP is used and images are viewed through an emission filter. Such systems have the advantage of focusing in one specific plane and in allowing the capture of "optical sections," which can later be reconfigured using suitable software (see Fig. 4).

GFP has a characteristic luminous lime green coloration and so can usually be discriminated by eye from the coloration of autofluorescence such as that associated with plant cell walls. Panchromatic film is sensitive to green light and so emissions can usually be photographed if they are at an intensity that can be observed. Most fluorescence microscopes accommodate 35-mm cameras and those capable of taking automatic exposures for at least 30 sec are of particular value. Exposure compensation may be required particularly in the absence of spot metering if the GFP output is highly localized in a generally dark field of view. We normally use a reversal (transparency) film because it provides a primary image of high color fidelity. Film speed influences image definition but films of c100 ASA provide acceptable images for all but extreme enlargements. Color negative film of c1000 ASA or more is useful when GFP emission is not bright to the eye. Panchromatic black-and-white films are also sensitive to green light but have the disadvantage in some applications of not discriminating between autofluorescence and emissions from GFP. This film type can provide excellent image quality or high film speed as required. CCD cameras with sensitivity similar to that of the human eye are widely available and have the added value of providing immediate confirmation that the image has been captured. Many such cameras are available. We have routinely used such cameras with one of the many image analysis hardware and software systems available (Leica Quantimet 500 and 500C). We use both CCD color (CF15 MCC; Kappa, Optivision, Leeds, UK) and monochrome (DC10; Sony, Optivision, Leeds, UK) cameras to replace 35-mm cameras on our fluorescence microscopes. CCD cameras with a higher sensitivity to light than the human eye are required for some applications and are considered in the next section.

Image Capture and Analysis for Continual, Quantitative Study of Green Fluorescent Protein

Microscope and specialized interference filter manufacturers (Chroma Technology, Brattleboro, VT) can supply filter sets with narrow bandpass excitation filters to enhance detection of a given GFP variant relative to any other fluorescence emitted by the specimen. Similarly, the emission bandpass can also be matched to that of the GFP variant being used. Narrow filter bandpasses limit the observable color range and the total light emissions. CCD cameras with a sensitivity greater than the human eye may be required to detect such fluorescence. When this occurs there is a need to establish systems that assure measurements are due to GFP and not other sources of fluorescence.

An excitation monochromator provides a particularly effective means of limiting GFP excitation wavelengths that has an inherent ability to support reliable detection below the sensitivity of the eye. We have used a monochromator driven by software (TILL Photonics and Openlab, respectively, both supplied by Improvision, Coventry, UK). The monochromator is capable of providing wavelengths between 260 and 680 nm with a bandpass of ±5 nm, and is motor driven. It allows excitation of the sample by any two wavelengths in succession within 4 msec. We use a Zeiss microscope (Axiovert 135TV) with two camera ports to which a chilled, 3CCD color camera and dual-mode cooled black-and-white camera are connected (both from Hamamatsu Photonics Systems, Bridgewater, NJ). Dedicated software (Openlab) automates the process of repeated image collection over a long period of time. It provides a simple means of assembling a macro program by dragging and dropping icons (a specific example is described in Derepression of GFP Expression, below). This software also has the ability to use digital techniques to determine and remove out-of-focus input from microscope images, a process known as deconvolution. This imaging system captures optical sections using a standard fluorescence microscope and emulates the outputs of a confocal microscope at a lower cost.

Observing Green Fluorescent Protein Expression in Root Systems

Observing one region of a root system over an extended period can be difficult when plants grow in soil or in tissue culture without restriction. The direction of root growth in the plane of focus can be limiting for many studies. One approach involves use of observation chambers. Plastic spacers (1.5 mm thick) are placed around the sides and bottom edge of a glass plate and a fourth spacer is placed 5 cm from the top edge. Medium is poured onto the plate between the spacers and allowed to set before the top spacer is removed. Next, a small sterile plant (e.g., an ~2-week-old *Arabidopsis*) is placed on the plate with just its roots on the medium and all the aerial tissue in the top portion of the plate. A further set of three spacers (1.5 mm thick) is added on top of the first set, followed by a second glass plate. The sides and bottom are sealed with waterproof tape and the top edge with micropore tape. The lower portion of the cassette is covered with aluminum foil to prevent exposure to light. The plates are held together with small "fold-back" clips. The plate is then placed in a plant growth chamber between measurements, which can extend for several weeks. Roots can also be studied continually over a short period of time on microscope slides. Agarose medium is placed on a sterile microscope slide, and seedlings are layered onto this and covered with dialysis membrane. This system has been used to study the early events of symbiosis between *Rhizobium meliloti* and alfalfa.[16]

Green Fluorescent Protein and Reductions in Promoter Activity

The GUS reporter system can be used to analyze declines in promoter activity in plant cells but repeated destructive sampling is required.[17] GFP only offers a basis for continual measurements in such declines if the changes under study are not rapid. As yet there are no measurements of the half-life of GFP in plant cells from which to determine the extent of its stability; this limits its use for monitoring a decline in promoter activity. Sheen *et al.*[6] concluded that the GFP chromophore, in plants, is as stable as the reporter chloramphenicol acetyltransferase (CAT), which persists for up

[16] D. J. Gage, T. Bobo, and S. R. Long, *J. Bacteriol.* **178,** 7159 (1996).
[17] C. Gatz, C. Frohberg, and R. Wendenburg, *Plant J.* **2,** 397 (1992).

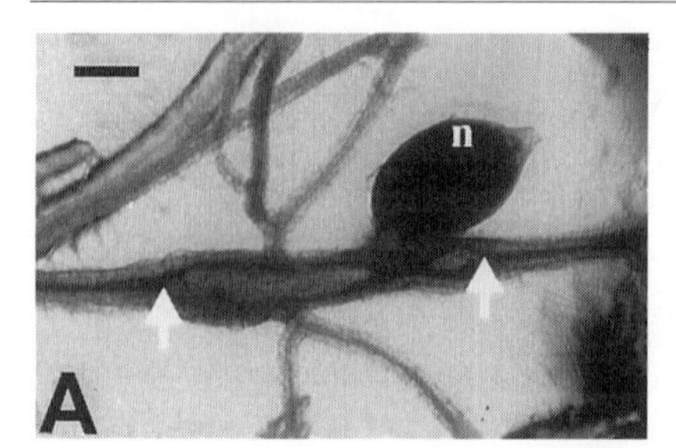

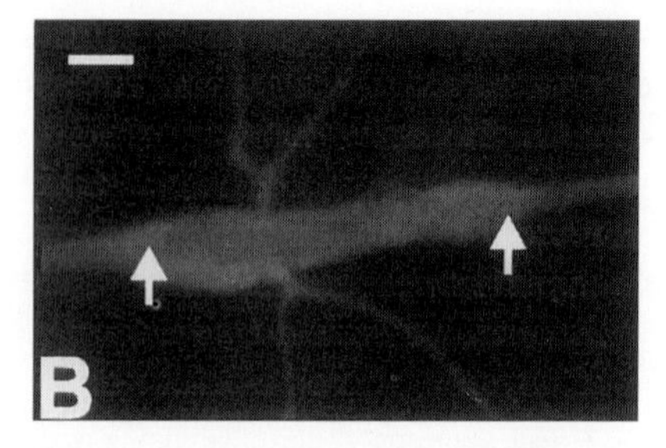

FIG. 2. Downregulation of the CaMV35S promoter by *Heterodera schachtii,* monitored by GFP expression in *Arabidopsis thaliana* C24. Image capture as for Fig. 1. Exposure times of ~7 sec. (A) Viewed by bright-field microscopy; (B) viewed by fluorescence microscopy. n, Nematode; arrows indicate the extent of the feeding site. Scale bar: 100 μm.

to 50 hr in cultures of mammalian cells.[18] Clontech has developed a destabilized variant of their enhanced GFP (EGFP), termed d2EGFP. This variant contains a 39-amino acid sequence from the C terminus of mouse ornithine decarboxylase (MODC). This sequence contains a PEST sequence that targets the protein for degradation, resulting in the protein having a half-life of 2 hr. The modification does not alter the fluorescence properties of the EGFP chromophore.[19] The work has currently been tested only in mammalian systems and awaits analysis in plants.

The stability of older versions of GFP is unlikely to be a problem if the time course of interest has a duration of days rather than hours. A progressive decline in CaMV35S promoter activity has been measured in roots over 5 weeks.[15] Using a growth system as described above we observed the effect of the nematode parasite *Heterodera schachtii* on CaMV35S promoter activity in roots. This nematode modifies root cells to form a syncytium and the female nematodes feed from this structure continually for weeks. A decline in GFP in the syncytium occurred 10 days after root infection by the parasite. This was obvious as a reduction in intensity of GFP emission in the syncytium relative to cells around it (Fig. 2). It could also be visualized as a change in coloration of emissions from roots exposed to low-light levels. Under such conditions, syncytia produce chloroplasts and as a consequence chlorophyll emission occurs under the fluorescence microscope. The syncytium, but not surrounding cells, changes color progressively from green to yellow or orange as GFP levels decline, just as occurs in protoplasts expressing the reporter.[6] The decline in GFP emission for the syncytium of *H. schachtii* contrasted with the upregulation of the

[18] J. F. Thompson, L. S. Hayes, and D. B. Lloyd, *Gene* **103,** 171 (1993).
[19] Clontech, *Clontechniques* **XIII,** 16 (1998).

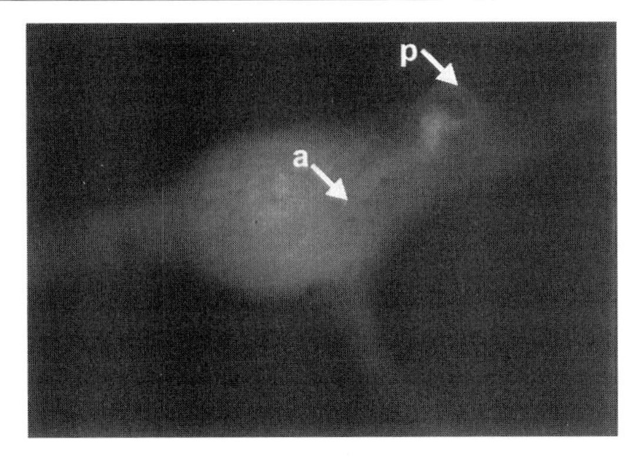

FIG. 3. Ingestion of GFP by *Meloidogyne incognita.* A "stream" of GFP can be seen in the animal, showing the extent of the digestive system. The anterior (a) and posterior (p) of the animal are marked. Image capture as for Fig. 2.

same CaMV35S promoter by another nematode pathogen, *Meloidogyne incognita.*[15] This nematode causes the formation of multinucleate plant cells from which it feeds. These structures showed increased GFP fluorescence relative to other plant cells around them.[15] Using some of the methods described in this chapter, the ingestion of GFP by the plant pathogen *M. incognita* can be detected (Fig. 3). This ingestion can be confirmed by Western blot.[15] This has led to valuable insights into the feeding habits of this nematode, work that is receiving further attention in our laboratory.

The previous examples demonstrate two uses for GFP. The first describes expression of GFP by the plant to study, for example, promoter activity in the plant. The second uses expression of GFP by the plant but subsequently analyzes GFP in the pathogen. A third possibility is the analysis of GFP expressed by the plant pathogen. The group of Reinhold-Hurek[20] has studied the association of the bacterium, *Azoarcus* sp. BH72, and rice roots. They used the expression of GFP in the bacteria to establish nondestructively the site of infection within the plant root. After infection of rice with the bacteria, roots were incubated for 1–2 hr at 4°. Micrographs were recorded on Kodak Ektachrome 64 film or captured with a Hamamatsu Photonic Systems 3-chip RGB color camera, mounted on a Zeiss Axioplan 2 microscope. A bandpass excitation filter (BP450–490) was used in conjunction with a barrier emission filter (LP520). Initially, laser scanning confocal microscopy was used but provided disappointing images. This

[20] T. Egener, T. Hurek, and B. Reinhold-Hurek, *Mol. Plant Microbe Interact.* **11,** 71 (1998).

FIG. 4. Endophytic expression of nitrogenase genes of *Azoarcus* sp. BH72 inside a rice root, using a transcriptional *nifH–gfp* fusion. *Right:* Original nonenhanced image. *Left:* After deconvolution using the Openlab Confocal Imaging module. Image capture is described in text. The bright white areas indicate the extent of GFP fluorescence and therefore the localization of the bacteria, although some of the less intense white areas represent autofluorescence. The two can be distinguished in color, GFP being green white autofluorescence of plant tissue is apparent as an orange/red color. [Image kindly supplied by B. Reinhold-Hurek via Improvision.]

work used the Openlab Confocal Imaging module to remove haze from *z* sequences of the image (Fig. 4).

Derepression of Green Fluorescent Protein Expression by Tetracycline *in Planta*

A CaMV35S derivative, which is tightly repressed by the Tn*10*-encoded tetracycline repressor in the plasmid Bin-HYG-TX, has been described.[17] The construct has three copies of the *Tet* operator in the CaMV35S promoter fragment, one upstream and two downstream of the TATA box. This triple operator arrangement bound the *Tet* repressor and effectively repressed activity of the CaMV35S promoter. As a consequence it blocked expression of the downstream GUS reporter. This repression can be removed by adding tetracycline, which competes successfully with the *Tet* repressor, resulting in GUS reporter activity.[17] Our expectation was that the time course for this derepression should be more rapid than when changes in gene regulation induce promoter activity. Therefore we have used this system to determine the time lapse between first GFP translation and its detection for a highly active promoter such as CaMV35S.

We replaced GUS with GFP in the plasmid Bin-HYG-TX kindly provided by G. Gatz (Albrecht-v.-Haller-Institut für Pflanzenwissenschaften, Gotten, Germany). The modified construct was incorporated into *Agrobacterium rhizogenes* by triparental mating. Transgenic tobacco seeds expressing the Tet repressor (also donated by G. Gatz) were germinated on 1% water agar plates in the dark and the base of each hypocotyl was removed. Hypocotyls were shaken for 30 min in a culture of *A. rhizogenes* grown for 16 hr in YMB (2 g of $MgSO_4$, 0.5 g of K_2HPO_4, 0.1 g of NaCl, 10 g of mannitol, 0.4 g of yeast extract, made to 1 liter, pH 7.0). Hypocotyls were blotted between sterile filter paper, then put onto 0.5× MS30 plates (Murashige and Skoog basal medium containing sucrose at 30 g $liter^{-1}$) and left in the light for 2 days. The plates were then moved to the dark and the roots were allowed to grow. When they were about 10 mm long they were moved onto selective plates [0.5× MS30 containing hygromycin (5 μg ml^{-1}) and cefatoxime (250 μg ml^{-1})]. Having established resistance to hygromycin, hairy roots were maintained on 0.5× MS30 plates.

Hairy roots growing on an agar plate were focused under an inverted microscope (Axiovert; Zeiss). The lid of the agar plate was removed and 10 μl of tetracycline solution (1, 10, 100, or 1000 μg ml^{-1}) was added topically at the root tip. Sterility could have been maintained if prolonged observation or reuse was anticipated. The monochromator (see above) was set to 397 nm prior to capturing an image and then reset to 680 nm immediately after this process was completed. A high-sensitivity black-and-white camera (C5810; Hamamatsu) was used to collect images.

An initial measurement was made immediately prior to the addition of tetracycline and subtracted from later values for all roots including those not treated with tetracycline. Repeated measurements were collected from the same sample over a period of time. The collection of successive images can be facilitated using the Openlab automation feature, which stacks collected images in the computer memory (Fig. 5).

A single region of interest was defined in the root and the mean gray values measured for this area in all the images of the stack. A macro to carry out this process can also be automated within the software. A number of stacks were collected and measured from different samples. The measurements were transferred to a spreadsheet (Excel 5; Microsoft, Redlands, WA) for analysis. We also chose to export data sets to a statistical package

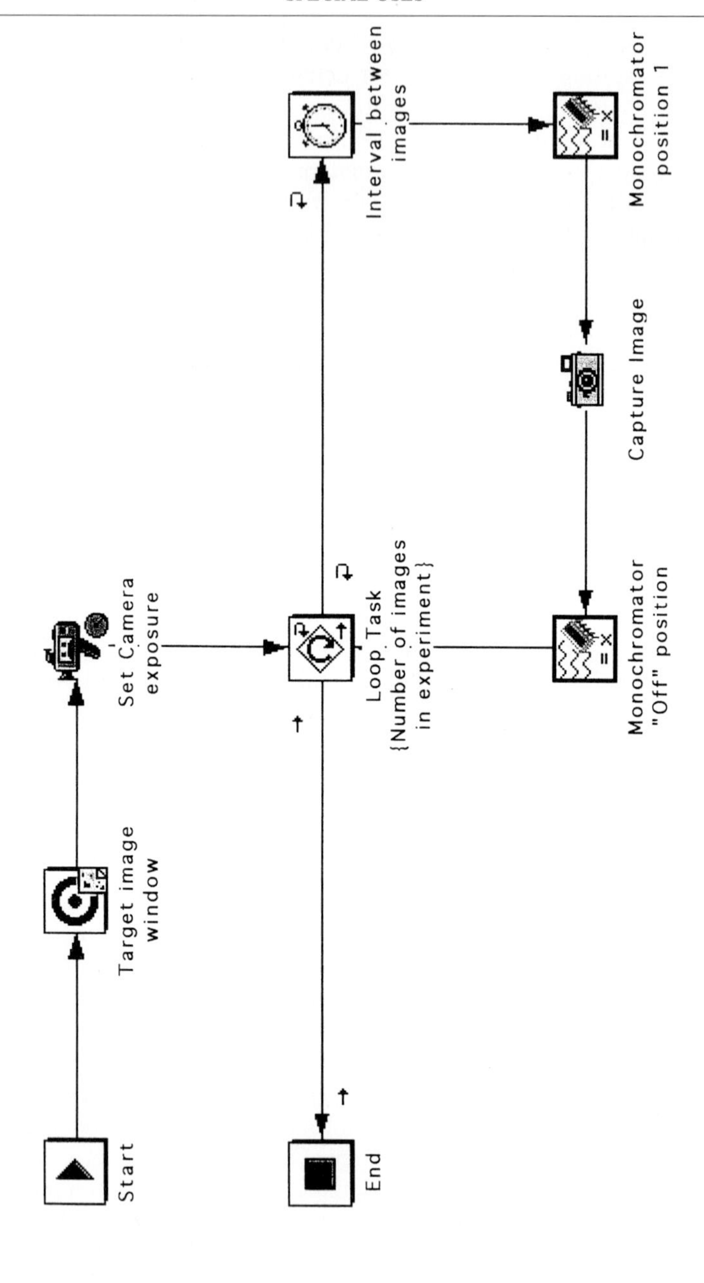
Start
Target image window
Set Camera exposure
End
Loop Task {Number of images in experiment}
Interval between images
Monochromator "Off" position
Capture Image
Monochromator position 1

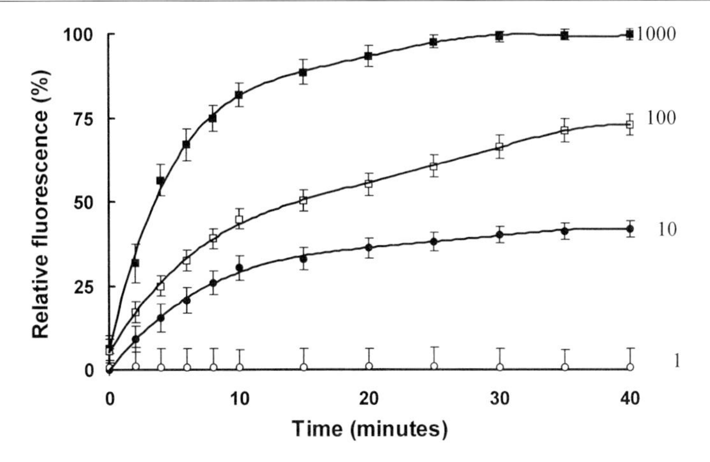

FIG. 6. The loss of repression of the CaMV35S triple Tet operator as monitored by the GFP reporter. Derepression was carried out at tetracycline concentration of (■) 1000, (□) 100, (●) 10, and (○) 1 μg ml^{-1}.

(Statistical Package for the Social Sciences; SPSS, Chicago, IL) resident on the hard drive of our computer. This package was used to carry out regression and one-way ANOVA.[21]

The data collected are summarized in Fig. 6. The concentration of tetracycline added topically to roots influences the maximum recorded light emission from GFP. The lowest concentration of tetracycline used (1 μg ml^{-1}) failed to induce a level of GFP expression that could be detected by the CCD camera. Gatz *et al.*[17] indicate that at 1 μg ml^{-1} of chlorotetracycline, derepression of the promoter as measured by GUS reporter activity occurred after 4 days in leaves, with a maximum GUS transcript occurring

[21] G. W. Snedecor and W. G. Cochran, "Statistical Methods." Iowa State University Press, Ames, Iowa, 1989.

FIG. 5. A representation of Openlab automation, used to collect multiple images of a single sample. The automation proceeds at Start; Target Image Window instructs the software where to save the images as they are collected; Set Camera Exposure allows specific settings to be used and saved with the automation; Loop Task designates the number of images that will be collected; Interval Between Images designates the duration of the delay between the capture of images; Monochromator Position 1 designates the excitation wavelength used when the image is collected, Capture Image activates the camera; Monochromator "Off" Position moves the wavelength away from that used to collect the image (important if photobleaching is considered a problem). The automation returns to the loop a further *n* times before exiting at End.

at 2 days. The authors suggest the delay in the production of the gene product may be typical of the GUS reporter system, or it may be caused by factors associated with the triple operator elements of the promoter. We expected roots to take up tetracycline rapidly when it was applied topically to them at an extreme concentration of 1000 μg ml^{-1}. Our results established that under such conditions repression is lost rapidly and 50% maximum GFP expression occurred in 4 min. This demonstrates that GFP detection in roots does not involve a considerable lag after transcription of mRNA. With high-sensitivity detection, GFP can clearly monitor an increase in promoter activity without prolonged delay. More work is required to determine if GUS shares this ability.

GFP does have a number of advantages as a real-time reporter for plants. It does not require addition of an exogenous substrate. It is not phytotoxic in many applications and destructive sampling of the plant is not required. It is also available with a range of excitation and emission wavelengths, offering the possibility of dual-reporter studies. In many applications, it can be visualized using a fluorescence microscope without additional specialized equipment. CCD cameras allow quantification of its emissions. Its detection at levels below the sensitivity of the eye is also possible with specialized CCD cameras. Such sensitive detection also allows narrow bandpass filters to be used to enhance the specificity of excitation and emission wavelengths. The delay between transcription of the GFP message and translation of the gene product is short in root cells of hairy tomato roots. The principal disadvantages of the approach are the equipment costs for sensitive detection and the stability of GFP in plants. This ensures it also has limited value as a real-time reporter of a reduction in promoter activity. This second disadvantage may be overcome soon if dEGFP proves to be of value *in planta.*

Acknowledgments

This work was supported by funding from the Scottish Office of Agriculture, Environment and Food Department, and BBSRC (LINK and JREI grant schemes). S. G. Møller is the recipient of a Norwegian Research Council stipend, NFR-107499/410.

[28] Retroviral Expression of Green Fluorescent Protein

By ILYA A. MAZO, JOHN P. LEVY, REBECCA R. MULDOON, CHUCK J. LINK, JR., and STEVEN R. KAIN

Introduction

Retroviral vectors are powerful tools for gene transfer *in vitro* and *in vivo*. They have a number of convenient features that explain their multiple experimental and clinical applications. Most importantly, retrovirus-mediated gene transfer leads to stable proviral integration in the target cell genome, allowing for the transmission of foreign genetic material to the progeny of infected cells.[1] Furthermore, retroviral vectors can be modified by deleting all viral protein-coding sequences (*gag, pol, env*) without a significant decrease in the efficiency of infection.[2,3] Helper-free virus stocks can be obtained for safe gene transfer by using retrovirus packaging cell lines, which supply all of the viral proteins required for vector transmission.[4–7] In addition, defective recombinant retroviruses allow for efficient gene transfer into a broad range of mammalian and avian cells derived from different species and tissues.[1,2]

Retroviral vectors are especially useful in achieving stable and efficient transduction of a gene or genes into cells that are not easily transfected, such as primary cells of various types and cells *in vivo*.[8,9] Of special importance is also the broad application of retroviruses as vectors in gene therapy approaches.[10] In fact, the majority of gene therapy clinical trials currently underway utilize retroviral vectors. Success of complicated experimental

[1] J. M. Coffin, S. H. Hughes, and H. E. Varmus (eds.), "Retroviruses." Cold Spring Harbor Laboratory Press, Cold Spring Harbor, New York, 1997.

[2] F. M. Ausubel, R. Brent, R. E. Kingston, D. D. Moore, J. G. Seidman, J. A. Smith, and K. Struhl (eds.), "Current Protocols in Molecular Biology," Supplement 36, Section III. John Wiley & Sons, New York, 1996.

[3] A. D. Miller and G. J. Rosman, *BioTechniques* **7,** 980 (1989).

[4] J. P. Morgenstern and H. Land, *Nucleic Acids Res.* **18,** 3587 (1990).

[5] R. Mann, R. C. Mulligan, and D. Baltimore, *Cell* **33,** 153 (1983).

[6] A. D. Miller and F. Chen, *J. Virol.* **70,** 5564 (1996).

[7] W. S. Pear, G. P. Nolan, M. L. Scott, and D. Baltimore, *Proc. Natl. Acad. Sci. U.S.A.* **90,** 8392 (1993).

[8] L. Cheng, C. Du, D. Murray, X. Tong, Y. A. Zhang, B. P. Chen, and R. G. Hawley, *Gene Ther.* **4**(10), 1013 (1997).

[9] M. F. A. Bierhuizen, Y. Westerman, T. P. Visser, A. W. Wognum, and G. Wagemaker, *Biochem. Biophys. Res. Commun.* **234,** 371 (1997).

[10] C. L. Patricia (ed.), "Somatic Gene Therapy." CRC Press, Boca Raton, Florida, 1995.

0076-6879/99 $30.00

and clinical approaches that involve retroviral vectors depends greatly on optimal vector design, careful tuning of experimental protocols, and an adequate reporter/marker gene system. An ideal marker would allow for a fast, easy, and noninvasive detection and would enable a researcher to follow the distribution of viruses and the fate of infected cells *in vivo* and *in vitro*.

It is not surprising that the discovery of green fluorescent protein (GFP)[11] and its variants[12] had an immediate impact on the field of retrovectorology. Detection of GFP and its variants can be performed in living samples, and is amenable to real-time analysis of molecular events. When expressed in either eukaryotic or prokaryotic cells and illuminated by blue or ultraviolet (UV) light, GFP emits a bright green fluorescent signal that is easily detected by fluorescence microscopy, flow cytometry, or other fluorescence imaging techniques. Light-stimulated GFP fluorescence is species independent and does not require additional cofactors, substrates, or gene products. Therefore, GFP is superior to other markers used for the selection of viable cells such as antibiotic resistance (e.g., *neo, hyg*), which requires extensive culture in selective medium, or cell surface markers, which require staining with specific antibodies.

Green Fluorescent Protein Variants Used for Mammalian Expression

A number of variants of wild-type (wt) GFP have been developed that show improved expression in mammalian cells.[12,13] Most of them contain one or more amino acid substitutions in the chromophore region of the protein, resulting in the shift of the major fluorescence excitation peak toward the red, from 395 nm in wt GFP to 488–490 nm. The major excitation peak of the red-shifted variants encompasses the excitation wavelength of commonly used fluorescence filter sets, so the resulting signal is much brighter relative to wt GFP. Similarly, the argon ion laser used in most flow cytometry instruments and confocal scanning laser microscopes emits at 488 nm, so excitation of the red-shifted GFP variants is much more efficient than excitation of wt GFP.

It is interesting to note that retrovirus-mediated gene transfer was employed by several groups to compare the fluorescence intensities of different GFP mutants in a variety of cell lines.[9,14] Transduced populations can be

[11] D. C. Prasher, V. K. Eckenrode, W. W. Ward, F. G. Prendergast, and M. J. Cormier, *Gene* **111,** 229 (1992).

[12] S. R. Kain, *in* "Biotechnology International" [Review] (T. H. Connor and C. F. Fox, eds.), p. 79. Universal Medical Press, Inc., San Francisco, California, 1997.

[13] R. Heim and R. Y. Tsien, *Curr. Biol.* **6,** 178 (1996).

[14] R. R. Muldoon, J. P. Levy, S. R. Kain, P. A. Kitts, and C. J. Link, Jr., *BioTechniques* **22,** 162 (1997).

TABLE I
PROPERTIES OF ENHANCED GREEN FLUORESCENT PROTEIN VARIANTS[a]

Variant	Mutations	Excitation maximum (nm)[b]	Emission maximum (nm)[b]	E_m	QY
Wild type	None	395 (470)	509	21,000 (7,150)	0.77
EGFP	F64L, S65T	488	507	~55,000	~0.7
EBFP	F64L, S65T, Y66H, Y145F	380	440	31,000	~0.2
EYFP	T203Y, S65G, V68L, S72A	513 (488)	527	84,000	0.63
ECFP	F64L, S65T, Y66W, N146I, M153T, V163A	433 (453)	475 (501)	26,000	0.4

[a] The E_m (extinction coefficient) reflects the efficiency of light absorption; QY (quantum yield) shows what percentage of absorbed energy is released in the form of fluorescence.
[b] Values in parentheses represent secondary peaks.

easily obtained that contain one copy of GFP per cell and are therefore accurate for such comparison. Retroviral transduction eliminates the need to select for stable clones or use cotransfection markers as a control for the copy number. The measurement of the fluorescence profile of infected cells by fluorescence-activated cell sorting (FACS) immediately provides statistically significant information about relative intensities of analyzed GFP mutants.

The two red-shifted mutants most commonly used for retroviral protocols (and mammalian expression in general) are humanized S65T and EGFP (enhanced GFP).[14–17] EGFP (Table I) contains the same serine-to-threonine substitution at position 65 in the chromophore as S65T, plus a phenylalanine-to-leucine mutation at position 64. Humanized variants also have silent base changes to contain codons preferentially found in highly expressed human proteins. The "humanized" backbone used in EGFP contributes to efficient translation of this variant in mammalian cells and subsequently bright fluorescent signals. On the basis of spectral analysis of equal amounts of soluble protein, EGFP fluoresces approximately 35-fold more

[15] P. B. Robbins, X. J. Yu, D. M. Skelton, K. A. Pepper, R. M. Wasserman, L. Zhu, and D. B. Kohn, *J. Virol.* **17**(12), 9466 (1997).
[16] J. P. Levy, R. R. Muldoon, S. Zolotukhin, and C. J. Link, Jr., *Nature Biotechnol.* **14**, 610 (1996).
[17] D. Klein, S. Indraccolo, K. von Rombs, A. Amadori, B. Salmons, and W. H. Gunzburg, *Gene Ther.* **4**, 1256 (1997).

intensely than wt GFP when excited at 488 nm,[18] owing to an increase in its extinction coefficient (Table I). When expressed in cells EGFP can be detected at concentrations as low as 30 nM[19] or ~4000 molecules per cell. Other advantages of EGFP include improved solubility, faster chromophore oxidation to form the fluorescent form of protein, and reduced rates of photobleaching.

More recently, GFP variants have been developed that differ from wt GFP in emission as well as excitation spectra. Such variants can be used in double-labeling experiments or when cellular green autofluorescence is a concern. Table I summarizes features of three humanized variants capable of producing distinct colors.

The blue emission mutant EBFP has two more mutations in addition to those present in EGFP. EBFP has a major shift in both the excitation and emission maxima, with the point mutation at position 66 (tyrosine to histidine) being primarily responsible for a cobalt blue signal produced by this mutant.[20] Another variant, termed "cyan" (ECFP), has spectral characteristics intermediate between those of EGFP and EBFP. This mutant can be excited by the same frequency as EGFP (although less efficiently) and is an ideal candidate for double-labeling experiments.

Finally, EYFP shows dramatic red shifts in both the excitation and emission maxima at 513 and 527 nm, respectively. This variant produces green–yellow fluorescence discernible from wt GFP or EGFP. This phenotype is due to a threonine-to-tyrosine mutation at position 203, which was shown to be critically associated with Tyr-66 in the GFP chromophore.[21] The valine-to-leucine and serine-to-alanine mutations (at positions 68 and 72, respectively) improve the folding of this variant at 37°, but are not significantly responsible for the shift in excitation or emission maxima.[18] This variant has extinction coefficient (E_m) and QY values similar to those for S65T, and therefore yields a bright fluorescent signal.

In summary, "humanized" GFP variants are available that produce distinct colors and are well suited for use in retroviral protocols as well as other expression applications in mammalian cells.

Retroviral Vector Design

Most of the currently existing vector/packaging line systems are based on the mammalian C-type Moloney murine leukemia virus (Mo-MuLV).[3–5]

[18] B. P. Cormack, R. Valdivia, and S. Falkow, *Gene* **173,** 33 (1996).
[19] D. Piston, personal communication (1998).
[20] R. Heim, D. C. Prasher, and R. Y. Tsien, *Proc. Natl. Acad. Sci. U.S.A.* **91,** 12501 (1994).
[21] M. Ormo, A. B. Cubitt, and K. Kallio, *Science* **273,** 1392 (1996).

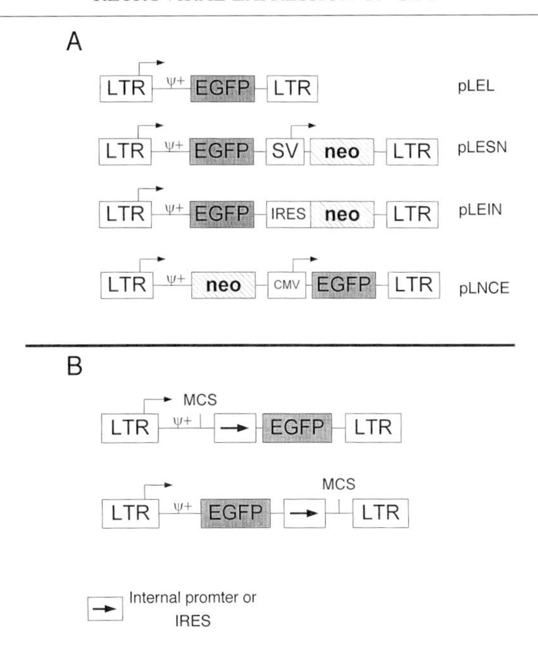

FIG. 1. Scheme of various retroviral vectors expressing EGFP. (A) Vectors in which GFP is expressed as a surrogate for the gene of interest; (B) vectors in which GFP is used as a selectable marker. SV, Simian virus 40; IRES, internal ribosomal entry site; CMV, cytomegalovirus; MCS, multiple cloning site; ψ^+, extended packaging signal.

Modern retroviral vectors generally retain only the minimal cis-acting sequences necessary for genomic transcription, packaging, reverse transcription, and integration. The functions of viral proteins Gag, Pol, and Env are supplied by packaging cell lines that express these polypeptides from constructs lacking the packaging signal.

The choice of a vector for GFP expression is dictated by such parameters as titer, stability of expression, and the strength and specificity of the promoter. Figure 1 outlines the design of a number of vectors used to express EGFP. (Although protocols and vector designs in this section are limited to EGFP, they can be easily generalized to other GFP variants as well.) The vectors are arranged by titer in descending order (Fig. 1A), with an overall difference of an order of magnitude. The highest titer can be produced by the simple vector pLEL. This vector also shows the highest expression of EGFP of all three vectors that have EGFP driven by the 5′ long terminal repeat (LTR) (Fig. 2). The lack of a selection marker, which is an obvious disadvantage of this vector, can be circumvented by creating

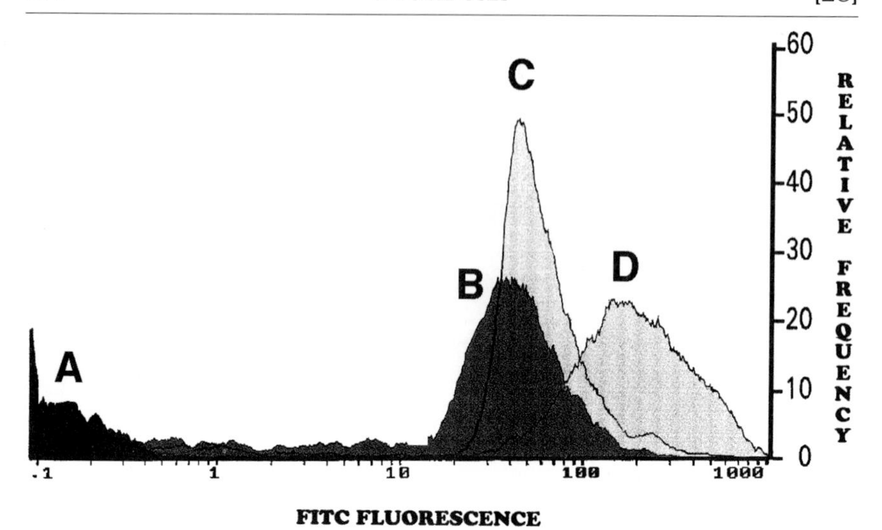

FIG. 2. EGFP expression by A375 cells infected with different retroviral vectors. Fluorescence intensities are shown for (A) noninfected cells, (B) cells infected with pLEL, (C) cells infected with pLNCE, and (D) cells infected with pLESN.

EGFP fusions with a gene encoding a selectable marker [e.g., *hyg*[22] or thymidine kinase (*tk*)[23]] or a gene of interest (e.g., *p53*[24]).

The classic two-promoter vectors such as pLESN and pLNCE offer the versatility of different selection markers and promoter elements that can be used to express EGFP (Fig. 2). The use of a strong internal promoter (pLNCE) can be critical for expression of GFP mutants that have higher detection limits (e.g., EBFP), and in cell types that do not support expression from the MuLV LTR. In addition, an internal promoter can be substituted by tissue-specific or inducible promoters.

On the negative side, promoter interference effects intrinsic to two-promoter vectors decrease the expression from the 5′ LTR and subsequently the viral titer.[25] For the same reason, *neo* selection of infected cells provides a growth advantage to the cells that have acquired proviruses with a rearranged or deleted second promoter.[26] The problem is further aggravated by the fact that high levels of EGFP expression can be cytotoxic.

[22] L. Lybarger, D. Dempsey, K. J. Franek, and R. Chervenak, *Cytometry* **25,** 211 (1996).
[23] S. Loimas, J. Wahlfors, and J. Janne, *BioTechniques* **24,** 614 (1998).
[24] P. S. Norris and M. Haas, *Oncogene* **15**(18), 2241 (1997).
[25] M. Emerman and H. M. Temin, *Cell* **39,** 459 (1984).
[26] B. Schott, E. S. Iraj, and I. B. Roninson, *Somatic Cell Mol. Genet.* **22**(4), 291 (1996).

Accordingly, vectors such as pLNCE were reported to be unstable in both producer and infected cells[27] and generally have lower titers.[17]

On the contrary, we found bicistronic vectors such as pLEIN (Fig. 1A) to be stable and to produce titers more than 2×10^6 from pools of virus-producing cells (PT67) after several months in culture. Good correlation between expression of *neo*, which can be selected for by increasing concentrations of G418, and EGFP was also observed.

In addition to the vectors discused above, in which GFP is expressed as a surrogate for a clinically important gene or a gene to be studied, vectors in which GFP is used as a selectable marker are becoming increasingly popular (Fig. 1B). The gene of interest in that case can be transcribed from a second promoter[28] or as part of a bicistronic message driven by the 5′ LTR.

Methods

Virus Production and Transduction of Mammalian Cell Lines in Culture

Retroviral vectors are manipulated in the form of plasmid DNA and can be converted into the form of infectious retroviral particles by transfection (or transduction) of packaging cell lines. The retrovirus-packaging cells provide all of the viral proteins necessary for encapsidation of vector RNA into virions and for subsequent infection, reverse transcription, and integration of the vector into the genomic DNA of infected cells.

Classic packaging cell lines are based on NIH 3T3 cells that contain constructs expressing viral proteins (Gag, Pol, and Env).[4–6] Stable virus-producing populations can be easily obtained from such cell lines that yield viral titers of 10^5–10^6 colony-forming units (CFU)/ml. Individual high-titer clones (titers up to 10^7 CFU/ml) can then be isolated. More recently, a number of 293-based packaging cell lines have been developed.[7] Such cell lines can be efficiently transfected by either calcium phosphate or liposome-mediated techniques, and are ideally suited for producing high-titer viral stocks (up to 10^7 CFU/ml) in a transient fashion. However, morphological properties of 293 cells make it almost impossible to obtain stable virus-producing populations or clones from these cells.

In this section we recommend a protocol that combines the advantages

[27] Y. Hanazono, J. M. Yu, C. E. Dunbar, and R. V. Emmons, *Hum. Gene Ther.* **8**(11), 1313 (1997).

[28] L. Alexander, H. Lee, M. Rosenzweig, J. U. Jung, and R. C. Desrosiers, *BioTechniques* **23,** 64 (1997).

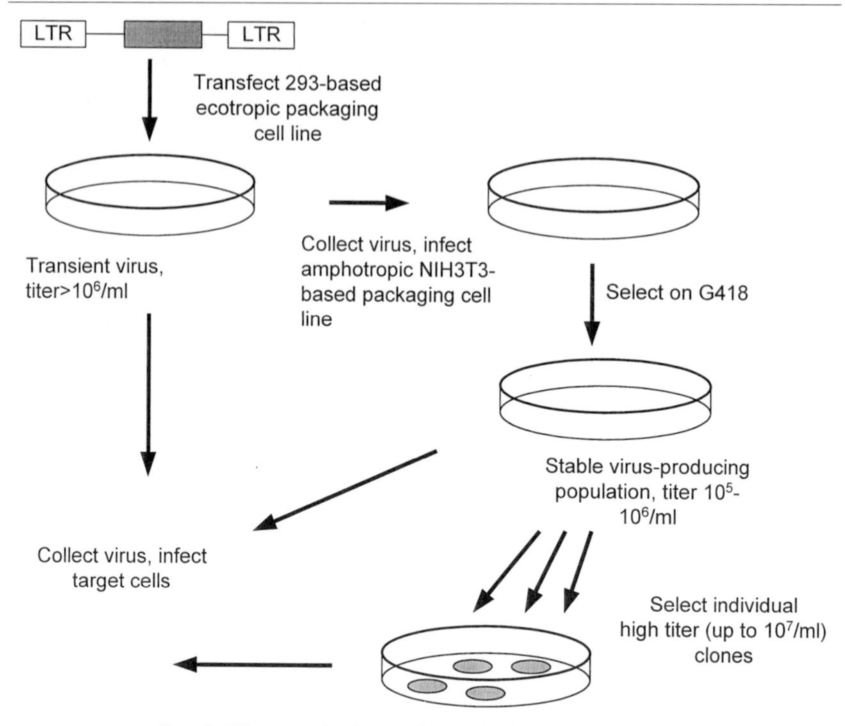

FIG. 3. Virus production and transduction of cells in culture.

of different cell lines (Fig. 3). First, vector DNA is transfected into a 293-based cell line. Transient virus is then collected and used to infect NIH 3T3-based packaging cells. Infected cells are subsequently subjected to antibiotic selection or FACS (for GFP-expressing vectors) to obtain stable virus-producing populations. This step can often be eliminated, as the titer of transient virus produced by 293-based cell lines is usually high enough to result in infection efficiency close to 100%. Additional reasons for this procedure rather than direct isolation of vector-transformed populations are that transduced cells are more stable, have more consistent copy number (one or two per cell), and generally produce higher titer virus.

It is important to note that the two packaging cell lines used in this procedure should express envelopes of different tropism. The overexpression of *env* in a packaging cell line leads to binding and blocking of a cognate receptor. As a result, packaging cells cannot be infected with a virus of the same tropism. It is usually desirable to have cells producing stable virus of amphotropic or other broad host range. In that case an ecotropic 293-based packaging cell line can be used (e.g., Bosc23,

Phoenix-E, or EcoPack 293). For more information on available packaging cell lines with different tropism, please refer to Ref. 2.

For some vectors (e.g., pLNCE) it is difficult to obtain stable virus-producing lines. Transient virus should be used directly to infect target cells. Protocols describing the production of transient virus stocks in 293-based packaging cell lines[7] and the use of episomal retroviral platforms[29] have been described elsewhere.

Transfection of Retroviral Vectors

1. Plate the 293-derived packaging cells at a density of 4–6 $\times$ 10^6 cells per 100-mm plate 12–24 hr before transfection (5–6 $\times$ 10^5 cells per 100-mm plate for NIH 3T3-derived cell lines). One to 2 hr before transfection, replace the medium with fresh medium.
2. Each 100-mm plate should be transfected with 10–15 μg of plasmid DNA [CsCl- or Qiagen (Chatsworth, CA) purified], using the desired method. Standard calcium phosphate or liposome-mediated procedures work well.
3. Twelve to 18 hr posttransfection, aspirate the medium, wash the cells once with phosphate-buffered saline (PBS), and add 5 ml of complete medium.
4. Incubate the culture for an additional 12–48 hr. The viral titer reaches a maximum ~48 hr after transfection and is generally at least 30% of maximum between 24 and 72 hr after transfection.

Because the viral titer is proportional to the efficiency of transfection, it is important to estimate the percentage of transfected cells. For GFP-expressing vectors, that can be done simply by counting green cells by inverted-phase fluorescence microscope. Note, however, that GFP must undergo a number of posttranslational modifications to become active. It takes generally 48–72 hr before GFP expression reaches its maximum.

Selection of Stable Virus-Producing Cell Lines

To obtain stable virus-producing cell lines, the NIH 3T3-derived packaging cells are infected (see the next section) or transfected with retroviral plasmids and plated in selection medium 2–3 days after transfection. Most retroviral constructs carry the neomycin gene as a selectable marker. For G418 selection, cells are cultured in the presence of G418 (0.5 mg/ml) for 1 week. Vectors carrying other selectable markers such as genes encoding puromycin, bleomycin, or hygromycin can be used to obtain stable virus-

[29] F. Grignani, T. Kinsella, A. Mencarelli, M. Valtieri, D. Riganelli, F. Grignani, L. Lanfrancone, C. Peschle, G. P. Nolan, and P. G. Pelicci, *Cancer Res.* **58**(1), 14 (1998).

producing cell populations as well. The selected cell populations usually produce titers of 10^5–10^6 recombinant virus particles per milliliter. This amount is suitable for most purposes. Cells can be frozen at that stage and provide a practically limitless supply of a retroviral vector.

For some experiments, higher titer clones may be required. In this case, after selection, pick individual clones (use cloning cylinders, cloning disks, or limiting dilution) for propagation. The viral titer is determined as described below. Usually, screening of 20–50 clones is required to guarantee isolation of a clone of acceptably high titer. Although it is appealing to use GFP expression in virus-producing cells as a predictor of viral titer, we found no correlation between level of GFP expression and the virus titer produced.

Infection of Target Cells

1. Plate the target cells (e.g., NIH 3T3) 12–18 hr before infection at a cell density of 3–5 $\times$ 10^5 per 100-mm plate. If the infected cells are to be used for a biological assay, make sure that the control cells are treated with an insert-free virus under identical conditions.

Note: Retrovial transduction (specifically, nuclear entry of the viral integration complex) can occur only in actively dividing cells.[30]

2. For infection, collect medium from the packaging cells, filter through a 0.45-μm pore size filter, add Polybrene to the medium to a final concentration of 4 mg/ml, and add to the target cells.

Note: The filter used should be cellulose acetate or polysulfonic (low protein binding), but not nitrocellulose. Nitrocellulose binds proteins present in the membrane of retrovirus and destroys the virus.

Until you have determined the titer of the supernatant, use as much virus-containing medium as possible for the infection. Remaining fluid can be frozen at $-80°$ but the titer will decrease ~2- to 10-fold per each freeze–thaw cycle.

3. Replace the medium after 24 hr of incubation.

4. A small subpopulation of the infected cells may be subjected to antibiotic (such as G418) treatment to determine the efficiency of infection. The infected cells should be used for experiments or for selection as soon as possible but not earlier than 48 hr after the last infection.

Half-maximal infection occurs after 5–6 hr of exposure of cells to virus; maximal infection occurs after ~24 hr of exposure.[31] The actual reverse transcription and integration take place within 24–36 hr of infection, de-

[30] D. G. Miller, M. A. Adam, and A. D. Miller, *Mol. Cell. Biol.* **10**(8), 4239 (1990).

[31] J. R. Morgan, J. M. LeDoux, R. G. Snow, R. G. Tompkins, and M. L. Yarmush, *J. Virol.* **69**(11), 6994 (1995).

pending on cell growth kinetics. Expression can start to be observed at 24 hr, and reaches a maximum at ~48 hr.

Growth of some target cells is strongly affected by medium conditioned by the packaging cells. A number of precautions can be taken to avoid an adverse effect induced by the packaging cell-derived supernatants.

- Dilute virus-containing medium at least twofold with fresh medium.
- Expose target cells to the virus for 4–6 hr and then replace with fresh medium. Alternatively, infections can be done sequentially, 12 hr apart. This further increases the efficiency of infection.

Determining Viral Titer

This is a standard protocol applicable to the most adherent cell lines and any vectors that have selection markers. The viral titer produced by transiently transfected or stable virus-producing packaging cell lines is determined as follows.

1. Collect virus-containing medium from packaging cells.
2. Add Polybrene to a final concentration of 4 mg/ml and filter the medium through a 0.45-μm pore size filter.
3. Prepare serial dilutions (six 10-fold serial dilutions are usually prepared). To dilute virus, use fresh medium containing Polybrene (4 mg/ml).
4. Infect target cells (NIH 3T3, plated 1 day before in six-well plates, 5×10^4–1×10^5 cells per well, in 4 ml of medium) by adding virus-containing medium to the wells.
5. Forty-eight hours later, subject to G418 selection (0.5 mg/ml) for 1 week (other selection agents can also be used).
6. The titer of virus corresponds to the number of colonies multiplied by the dilution factor. For example, the presence of four colonies in the 10^5 dilution would represent a viral titer of 4×10^5 CFU/ml.

Use of Green Fluorescent Protein as a Marker to Estimate Transduction Efficiency and Viral Titer

The virus titration methods described above are not easily applicable to suspension cell lines or primary cells. Besides, they are rather time consuming. One of the advantages of GFP-containing vectors is the ability to estimate viral titer and efficiency of infection by using cell sorting with GFP as a marker.

Assessment of Green Fluorescent Protein Transduction Efficiency. Transduction efficiency of GFP-expressing retroviral vectors can be quantitated by FACS analysis. Target cells are transduced in six-well dishes and trypsinized 48 hr later. Cells are washed with Hanks' balanced salt solution (HBSS) and resuspended in HBSS at a concentration of 1×10^5 cells/ml. Cytometry of cells has previously been demonstrated using an EPICS Profile II Analyzer (Coulter, Miami, FL) with an excitation source of 488 nm.[14] Cells were analyzed using a 525-nm bandpass filter set (Coulter). All FACS analyses use the FL1 emission channel to monitor green fluorescence [normally a fluorescein isothiocyanate (FITC) monitor]. Nontransduced cells are used as a negative standard for the assay.

A histogram will be produced, with the *x* axis representing the logorithmic axis of fluorescence and the *y* axis representing the cell count. The negative standard will form a bell-shaped curve starting on the *x* axis at the intersection with the *y* axis. Pure populations of GFP cells will shift the bell curve away from the *y* axis, depending on the level of GFP expression and intensity (Fig. 2). Mixed cell populations containing GFP will show two populations and or a broad bell-shaped curve with a shift in the bell away from the *y* axis. Settings on the FACS analyzer will quantitate the portion of cells in the negative region and the positive region, allowing one to obtain the transduction efficiency by using the positive percent value.

Titer Assay Based on Percentage of Green Fluorescent Protein-Positive Cells. Titer of GFP-containing retroviruses can be calculated if the transduction efficiency and the exact number of cells subjected to virus are known.

1. Plate target cells (e.g., NIH 3T3) 12–16 hr before infection.
2. Immediately before infection use one plate to collect and count the cells (*N*). A known number of NIH 3T3 cells is transduced with different dilutions (1×–10,000×) of retroviral supernatant.
3. Two days later, collect the cells that were exposed to virus and determine the percentage of infected cells by FACS as described above.
4. The virus titer is estimated as the percentage of infected cells multiplied by the number of cells *N*. For higher accuracy, transduction efficiencies are plotted versus different dilutions of virus. The titer is calculated from dilutions corresponding to the linear slope of the curve.[32]

Concluding Remarks

The applications of retroviral vectors expressing GFP are many. GFP is a convenient tool with which to optimize retroviral gene transfer protocols, specifically for *in vivo* studies,[15] as well as retroviral vector design. For example, in addition to the vectors discussed above, EGFP was cloned into a number of tetracycline (Tet)-inducible retroviral vectors.[33] This allowed

[32] A. Limon, J. Briones, T. Puig, M. Carmona, O. Fornas, J. A. Cancelas, M. Nadal, J. Garcia, F. Rueda, and J. Barquinero, *Blood* **90**(9), 3316 (1997).

[33] S. M. Cunningham and I. A. Mazo, unpublished data (1998).

for quick comparison of such parameters as titer, background of expression, and induction level.

On the other hand, utilization of retroviral vectors will be beneficial for many areas of research that involve GFP: promoter studies, protein–protein interactions, and protein localization. For example, GFP is extensively used as a protein fusion partner to track subcellular localization of proteins. In yeast, an approach to comprehensive analysis of protein localization was taken that utilized a gene library in which genomic sequences of *Schizosaccharomyces pombe* were fused to GFP.[34] Similarly, retroviral vectors may be used for construction of complex cDNA–GFP fusion libraries. On transduction of mammalian cells, such libraries would permit high-scale identification of genes on the basis of their localization.

[34] K. E. Sawin and P. Nurse, *Proc. Natl. Acad. Sci. U.S.A.* **94,** 15146 (1997).

[29] Confocal Imaging of Ca^{2+}, pH, Electrical Potential, and Membrane Permeability in Single Living Cells

By JOHN J. LEMASTERS, DONNA R. TROLLINGER, TING QIAN, WAYNE E. CASCIO, and HISAYUKI OHATA

Introduction

Responses of single cells to stimuli are often heterogeneous. Bulk measurements by conventional biochemical and physiological techniques may fail to represent accurately the magnitude and time course of individual cell changes. Spatial heterogeneity of responses within single cells also occurs. For this reason, microscopic techniques with good three-dimensional resolution are needed to study individual cells as they respond to imposed stimuli and stresses. Increasingly, confocal microscopy of parameter-specific fluorophores is permitting direct observation of single-cell physiology with high spatial and temporal resolution.

Formation of Optical Sections by Confocal Microscopy

Pinhole Principle

The lateral resolving power of conventional optical microscopy approaches 0.2 μm, but axial resolution is much less because the effective depth of field for all but the smallest objects is 2–3 μm even at highest

0076-6879/99 $30.00

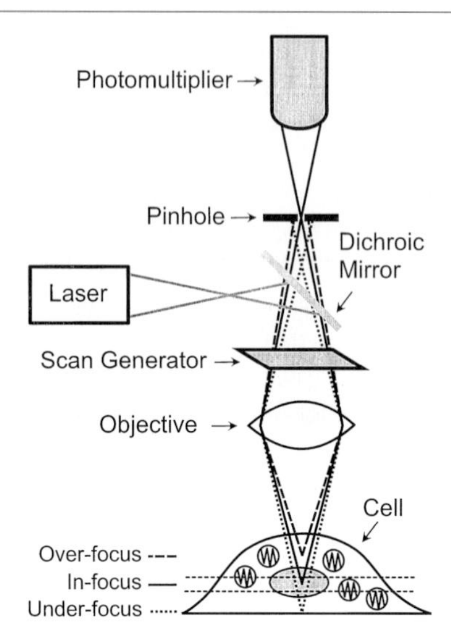

FIG. 1. Principles of laser scanning confocal microscopy. Laser light reflects off a dichroic mirror and is focused to a small spot by the microscope objective, which is scanned across the specimen by the scan generator. Light fluoresced by the specimen passes through the dichroic mirror and is focused by the microscope objective lens on a pinhole aperture. The pinhole aperture transmits in-focus light but blocks out-of-focus light. Thus, the photomultiplier detects only light arising from the in-focus specimen plane.

magnification. Thus, superimposition within this thick plane of focus obscures structures that might otherwise be resolved. In addition, in specimens more than a few microns thick, light arising from out-of-focus planes is projected on the in-focus image. Out-of-focus light decreases image contrast and limits applications that require quantitative analysis. This effect of out-of-focus light is especially severe in fluorescence microscopy, where imaging of thick, densely stained specimens is almost impossible by wide-field microscopy.

The technique of laser scanning confocal microscopy acts to eliminate out-of-focus light and produces remarkably detailed images whose depth of field is less than 1 μm. The principle of confocal microscopy, originally described by Minsky,[1] is quite simple (Fig. 1). Excitation light, most often from a laser, is focused by the microscope objective to a small spot within the specimen. At the cross-over point, spot diameter is diffraction limited, or about 0.2 μm for a high numerical aperture (NA) objective lens. Light reflected or fluoresced by the specimen is collected by the objective lenses, separated from the excitation light using a partially reflecting prism or

[1] M. Minsky, U.S. Patent 3,013,467 (1961).

dichroic mirror, and focused onto a small pinhole. This light, which originates from the in-focus cross-over point of the illuminating light beam, then passes through the pinhole to a light detector, typically a photomultiplier. However, light from above and below the focal plane strikes the edges of the pinhole and is not transmitted (Fig. 1). This selective transmission by the pinhole of in-focus light and rejection of out-of-focus light forms the basis for creating thin optical sections through thick specimens. Pinhole diameter determines the z-axis resolution of the microscope. As the pinhole is made smaller, the thickness of the confocal slice decreases until the system becomes diffraction limited, after which further decreases of pinhole diameter produce no additional improvement of z-axis resolution. Theoretical z-axis resolution is[2]

$$z_{\mathrm{min}} = 2\lambda\eta/(\mathrm{NA})^2 \tag{1}$$

where z_{min} is diffraction-limited z-axis resolution, λ is the wavelength of emitted light, η is the index of refraction of the object medium, and NA is the numerical aperture of the objective lens. Assuming that λ is 0.54 μm (fluorescein emission), η is 1.33 (water), and NA is 1.4, then z_{min} is about 0.7 μm.

Scanning

To create two-dimensional images, the beam must be translated across the specimen, either by moving the specimen underneath a stationary beam or, more commonly, by scanning the beam in x and y directions using vibrating mirrors in the light path.[3] The scan generator, which moves the beam across the specimen, also "descans" the returning light so that it can be focused on the stationary pinhole. The output of the photomultiplier is then stored in computer memory and displayed on a video monitor as the confocal image. In theory, any light source can be used to generate the scanning beam, but in practice only lasers provide powerful enough point illumination. The laser, scan generator, microscope objective, and pinhole together form the essential elements of the laser scanning confocal microscope (Fig. 1).

Precautions against Photodamage and Photobleaching

Confocal microscopy requires the use of brighter excitation light than conventional wide-field fluorescence microscopy for the simple reason that

[2] S. Inoué, *in* "Handbook of Biological Confocal Microscopy" (J. B. Pawley, ed.), 2nd Ed., p. 1. Plenum Press, New York, 1995.

[3] J. G. White, W. B. Amos, and M. Fordham, *J. Cell Biol.* **105,** 41 (1987).

confocal microscopy images light from a thin optical slice rather than from the entire specimen itself. In addition, the more complicated optics of confocal scan generators invariably involve some additional light loss compared with conventional fluorescence microscopy. Finally, the quantum efficiency for photon detection of photomultipliers is less than that of the best cooled charge-coupled devices (CCDs) used in conventional fluorescence microscopy, although the temporal response of photomultipliers remains far superior. As a consequence, specimens viewed by confocal microscopy are more vulnerable to photobleaching and photodamage, especially during serial imaging of living cells. The following strategies are helpful to minimize photobleaching and photodamage.

1. Light collection should be optimized by using a high-NA objective lens. For oil immersion, an NA of 1.3–1.4 should be used. For water immersion lenses, which are needed to image deeper than about 10 μm into aqueous samples, an NA greater than 1.0 is desirable. Empirical measurement of light transmission by individual lenses may be necessary, because light throughput also depends on the design of the lens and the glasses used in its construction. In particular, objectives made from ordinary glass transmit ultraviolet light poorly.

2. Laser power should be attenuated to the maximum extent possible using neutral density filters. Photomultiplier voltage (sensitivity) must be increased to compensate for the weaker fluorescent signal. Because lasers are so powerful, excellent image quality can often be obtained using an attenuation of 300- to 3000-fold. However, too much attenuation will increase spatial signal to noise and produce a snowy (noisy) image, much like reception of a distant television signal. This is because too few photons are measured from each picture element (pixel).

3. Because a larger pinhole transmits more light, the pinhole may be opened to increase image brightness and decrease spatial signal to noise without increasing laser illumination. Most laser scanning confocal microscopes incorporate adjustable pinhole apertures for this purpose. For a particular application, the resulting small loss of axial resolution may be acceptable. However, for maximum axial resolution, the pinhole diameter should be set to slightly less than the size of the diffraction-limited Airy disk projected on the pinhole. Airy disk diameter at the pinhole increases as the magnification factor of the objective lens increases but decreases as NA increases. Thus, the pinhole diameter needs to be readjusted when the objective lens is changed. Many commercial systems use software to calculate the Airy disk and adjust the pinhole diameter automatically.

4. Casual viewing, which might cause unnecessary photobleaching, should be avoided. A typical beginner's mistake is to view a sample at high

laser excitation. Initially, excellent images are obtained, but photobleaching soon makes collection of any image almost impossible. Instead, survey images should be collected at the lowest possible illumination. When an area of interest is found, laser intensity may then be increased to achieve the desired image brightness and spatial signal to noise.

5. The zoom feature of laser scanning confocal microscopes permits magnification to be varied continuously over a 5- to 10-fold range. Zooming simply limits the x and y excursion of the beam across the specimen. As zoom magnification increases, the same laser energy concentrates into a smaller volume of the sample, causing accelerated photobleaching and photodamage. Thus, zoom magnification should be minimized to the maximum extent that is consistent with the resolution requirements of the experiment. However, pixel size must remain less than half the diameter of the smallest object to be discerned.

6. The choice of fluorophores is also critically important. Stable fluorophores can be serially imaged with little phototoxicity. These include many of the rhodamine derivatives. Other fluorophores, such as fluorescein and acridine orange, rapidly bleach or generate toxic photoproducts. A simple strategy to reduce photobleaching is to use higher fluorophore concentration, which permits greater attenuation of the excitation light for a given intensity of fluorescent signal.

Preparation of Cells for Microscopy

Cells must be viewed through glass coverslips for virtually all types of high-resolution light microscopy. For most objective lenses, cells should be attached to #1$\frac{1}{2}$ coverslips, a designation that means that the glass is 170 μm thick. Generally, cultured cells grow poorly on glass. To improve both adherence and growth on glass, coverslips must be treated with materials such as type I collagen or laminin. Coverslips are first rinsed in ethanol for sterilization and dried. Subsequently, 100 μl (1–2 drops) of collagen (1 mg/ml in 0.1% acetic acid) or laminin (0.1 mg/ml in Tris-buffered saline) is spread across the glass surface. After air drying overnight, the coverslips are placed inside petri dishes for plating with cells in the usual fashion. For viewing in the microscope, a chamber is needed into which the coverslip is placed. These chambers can be built in a machine shop or purchased commercially (FCS2 chamber; Bioptechs, Butler, PA). An open chamber requires use of an inverted microscope, a configuration that permits easy access to the cell medium. Closed chambers usually permit continuous perfusion with buffers of interest and can be adapted to both upright and inverted microscopes. For studying living cells, adequate temperature regulation is important, which heaters built into the coverslip chamber or a stream of tempered air can provide.

Multicolor Fluorescence

Commercially available lasers for confocal microscopy emit narrow lines of light at fixed wavelengths, a fact that dictates the types of fluorophores that can be used. The most common lasers and their wavelengths of emission are as follows: argon, 488 and 514 nm; argon–krypton, 488, 568, and 647 nm; helium–cadmium, 442 nm; helium–neon, 543 nm; and UV–argon, 351–364 nm. The argon–krypton laser with well-separated blue, yellow, and red emissions is perhaps the most versatile single laser for biological applications. Using a multiline laser, multiple spectrally distinct fluorophores can be excited simultaneously. Fluorescence emitted from multiple fluorophores can be separated by wavelength and directed to different photomultipliers for simultaneous detection.

Imaging Electrical Potential with Cationic Fluorophores

In living cells under normal conditions, the plasma membrane and mitochondrial inner membrane maintain a negative-inside electrical potential difference ($\Delta\Psi$). As a consequence lipophilic cations, such as rhodamine 123, tetramethylrhodamine methylester (TMRM), and tetramethylrhodamine ethylester (TMRE), accumulate in the cytosol and the mitochondrial matrix in response to these membrane potentials.[4–6] At equilibrium, accumulation of permeant monovalent cations is related to $\Delta\Psi$ by the Nernst equation:

$$\Delta\Psi = -60 \log F_{in}/F_{out} \tag{2}$$

where F_{in} and F_{out} represent the concentration of monovalent cation inside and outside the membrane, respectively. By quantifying the intracellular distribution of cationic fluorophores using confocal microscopy, a map of intracellular electrical potential can be generated using Eq. (2).

For most cells, plasma membrane $\Delta\Psi$ is -30 to -90 mV and mitochondrial $\Delta\Psi$ is -120 to -180 mV. These $\Delta\Psi$ values are additive. Thus, mitochondria are as much as 270 mV more negative than the extracellular space. From Eq. (2), this potential difference corresponds to a cation concentration

[4] L. V. Johnson, M. L. Walsh, B. J. Bockus, and L. B. Chen, *J. Cell Biol.* **88,** 526 (1981).

[5] R. K. Emaus, R. Grunwald, and J. J. Lemasters, *Biochim. Biophys. Acta* **850,** 436 (1986).

[6] B. Ehrenberg, V. Montana, M.-D. Wei, J. P. Wuskell, and L. M. Loew, *Biophys. J.* **53,** 785 (1988).

ratio of more than 10,000 to 1 inside mitochondria relative to the cell exterior. With a conventional memory of 256 gray levels per pixel (8 bits), the large gradients of fluorescence corresponding to this concentration ratio cannot be stored using a linear scale. Either 16-bit memory or a nonlinear logarithmic (gamma) scale must be used instead.[7] Gamma scales, long used in scanning electron microscopy, condense the input signal into the available 256 gray levels of pixel memory.

To measure electrical potential, cells are first loaded with a potential-indicating fluorophore (50–500 n*M*), such as TMRM. A loading time of 15 min is usually adequate. The loading buffer is then replaced with experimental medium containing a small amount of fluorophore (20–150 n*M*) to maintain equilibrium distribution of the fluorophore. Equation (2) assumes ideal electrophoretic distribution of the fluorophore, but some probes do not behave ideally. Rhodamine 123, for example, binds nonspecifically to the mitochondrial matrix.[5] Moreover, the fluorescence of rhodamine 123 and other fluorophores becomes quenched as accumulation into mitochondria occurs. This quenching is concentration dependent and may be decreased by using smaller loading concentrations. Fluorophores can reach millimolar concentrations inside mitochondria, which may cause metabolic inhibition. For example, rhodamine 123 at such high concentrations strongly inhibits the oligomycin-sensitive mitochondrial F_1F_0-ATPase that catalyzes ATP synthesis during oxidative phosphorylation.[5] DiOC(6), a cationic fluorophore frequently used in flow cytometry, is an even more potent inhibitor of mitochondrial respiration.[8] TMRM and TMRE seem to lack many of these undesirable characteristics,[6,9] but low fluorophore concentrations (<500 n*M*) should always be employed to minimize quenching and metabolic effects.

In cells loaded with a permeable cationic fluorophore, confocal images are collected. The stored images are then processed to produce a map of intracellular electrical potential. Average fluorescence intensity in the extracellular space is divided into intracellular fluorescence on a pixel-by-pixel basis to determine F_{in}/F_{out} of Eq. (2). Equation (2) is then applied to each pixel to calculate the electrical potential difference between each point inside the cell and the extracellular space. Finally, the calculations are displayed as a pseudocolor map that shows the intracellular distribution of electrical potential.[7]

[7] E. Chacon, J. M. Reece, A.-L. Nieminen, G. Zahrebelski, B. Herman, and J. J. Lemasters, *Biophys. J.* **66,** 942 (1994).

[8] H. Rottenberg and S. Wu, *Biochem. Biophys. Res. Commun.* **240,** 68 (1997).

[9] D. L. Farkas, M.-D. Wei, P. Febbroriello, J. H. Carson, and L. M. Loew, *Biophys. J.* **56,** 1053 (1989).

Figure 2 (see color insert) illustrates intracellular electrical potential measured by this procedure for a cultured hepatocyte. In the cytosol and nucleus, pseudocoloring reveals an electrical potential of between −20 and −40 mV. Because the electrical potential of the extracellular medium is zero, plasma membrane $\Delta\Psi$ is about −30 mV, as expected for hepatocytes. Throughout the cytoplasm are also distributed electronegative mitochondria. These mitochondria show an apparent heterogeneity in their electrical potential. Heterogeneity is due, at least in part, to the fact that not all mitochondria extend completely through the confocal section. Consequently, fluorophore uptake for many mitochondria is underestimated. In Fig. 2, maximum mitochondrial potential calculated by Eq. (2) was −160 mV. The difference between this value and the cytosolic potential, −130 mV, represents a minimum estimate of mitochondrial $\Delta\Psi$.

Principle of Ratio-Imaging

A wide variety of fluorophores is available to measure ions and other parameters within individual living cells. Typically these fluorophores are loaded as their membrane-permeant ester derivatives. Inside cells, esterases release and trap the parameter-sensitive free acid form of the fluorophores. The fluorescence of these fluorophores depends not only on the parameter measured but also on the amount of fluorophore present in the light path. With some fluorophores, a ratioing procedure may be used to correct for variations in fluorophore concentration provided that parameter binding causes a shift of the fluorescence excitation or emission spectrum. Images are acquired at two different excitation or emission wavelengths, one that increases as the parameter of interest increases and one that decreases or stays the same. Background images of cell-free regions must also be collected at both wavelengths at the same microscope settings. These backgrounds are then substracted on a pixel-by-pixel basis from the experimental images. After background subtraction, the image at one wavelength is divided by the image at the second wavelength to form a ratio image. Such ratioing removes the contribution of fluorophore concentration to signal strength. The ratios are then compared with a standard curve and converted to units signifying parameter concentration. These values are often displayed in the image using a pseudocolor scale. The advantage of ratio imaging is the elimination of variations in signal due to differences in regional fluorophore concentration, dye leakage over time, photobleaching, and accessible volume.

For laser scanning confocal microscopy, ratioing of emission wavelengths must usually be employed. Lasers provide only one or a few narrow lines of monochromatic light, which severely limits the selection of excitation wavelengths. A good ratiometric fluorophore for confocal microscopy is carboxyseminaphthorhodafluor 1 (SNARF 1), a pH indicator. When SNARF 1 is excited with the 568-nm line of an argon–krypton laser, fluorescence emission at 640 nm increases as pH increases but emission at 585 nm remains unchanged.[7] Thus, the ratio of the two wavelengths is proportional to pH. For *in situ* calibration, SNARF-1-loaded cells are incubated with 10 μM nigericin and 5 μM valinomycin in modified Krebs–Ringer buffer, in which NaCl and KCl are replaced by their corresponding gluconate salts to minimize swelling.[7] SNARF 1 readily enters both the cytosolic and mitochondrial compartments after ester loading, and confocal ratio imaging of SNARF-1 demonstrates marked heterogeneity of pH within cells (Fig. 3, see color insert). Cytosolic and nuclear regions have a pH near 7.2, whereas mitochondrial pH is close to 8. Thus, a ΔpH of 0.8 pH units exists across the mitochondrial membrane, as predicted by P. Mitchell's chemiosmotic hypothesis (Mitchell, 1966).[10] Because the protonmotive force (Δp) equals $\Delta\Psi - 59\ \Delta\text{pH}$, Δp in intact living hepatocytes can be estimated to be approximately -180 mV.

Compartmentation of Ester-Loaded Fluorophores

Warm versus Cold Loading

Many fluorophores useful for confocal microscopy are multivalent organic anions. These multiple charged molecules are impermeant to membranes. To load these molecules into cells, the carboxyl groups of these acids must be neutralized by forming acetate or acetoxymethyl esters, which are membrane permeable. When cells are incubated with these derivatives, endogenous esterases regenerate the free acid form of the fluorophores, which becomes trapped inside the cell. Esterases are found within several organelles, including the cytosol, mitochondria and lysosomes. Thus after ester loading, fluorophores can be heterogeneously distributed in several different compartments.

For many fluorophores, the temperature of loading strongly affects intracellular distribution.[11] This is illustrated for calcein, a pentacarboxylic acid fluorophore whose fluorescence is independent of physiological changes of intracellular ions. When primary cultured hepatocytes or cardiac myocytes are incubated with the acetoxymethyl ester (AM) of calcein at

[10] P. Mitchell, *Biol. Rev.* **41,** 445 (1966).

[11] A.-L. Nieminen, A. K. Saylor, S. A. Tesfai, B. Herman, and J. J. Lemasters, *Biochem. J.* **307,** 99 (1995).

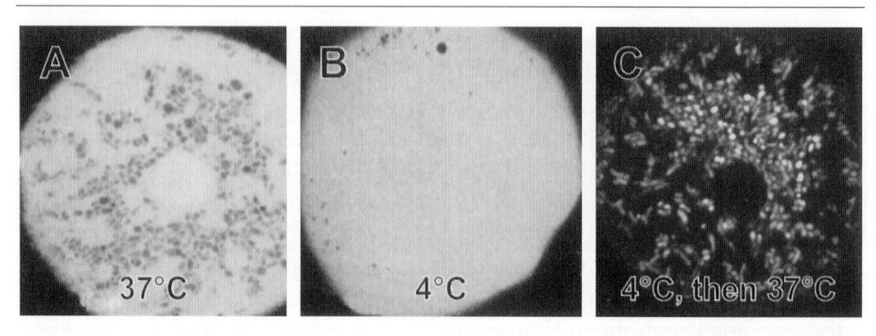

FIG. 4. Temperature dependence of calcein loading in hepatocytes. One-day cultured rat hepatocytes were loaded with 1 μM calcein AM for 15 min at 37° (A), with 1 μM calcein AM for 2 hr at 4° (B), or with 5 μM calcein AM on ice for 2 hr followed by incubation at 37° for 12 hr (C). Green fluorescence was imaged by laser scanning confocal microscopy. Note dark round voids of fluorescence in (A), representing mitochondria. These voids are absent in (B). In (C), mitochondria themselves are labeled with calcein with little fluorescence in the cytosol. [After Nieminen *et al.*[11] and T. Qian and J. J. Lemasters, unpublished, 1997.]

37°, calcein fluorescence enters the cytosol and nucleus (Fig. 4A). Under these loading conditions, calcein is excluded from mitochondria, leaving small round voids in the green fluorescence of calcein. These voids correspond exactly to the punctate red fluorescence of TMRM-labeled mitochondria (Fig. 5). By contrast, after ester loading at 4°, calcein enters both the cytosol and the mitochondria (Fig. 4B). Presumably during warm loading, cytosolic esterases are so active that calcein AM is hydrolyzed before it has a chance to enter mitochondria. At 4°, however, enzymatic activity is inhibited, allowing movement of unhydrolyzed ester into the mitochondria where mitochondrial esterases can cleave the ester and trap the free acid. The temperature dependence of loading is both probe and cell type specific. For example, the Ca^{2+}-indicating fluorophore Fluo 3 loads similarly to calcein, whereas Indo 1, another Ca^{2+} indicator, and SNARF 1 load into mitochondria even at 37°.[7,12]

Cold Loading followed by Warm Incubation

Fluorophores loaded into the cytosol gradually leak across the plasma membrane during warm incubation through an organic anion carrier.[13] However, fluorophores trapped in mitochondria and other organelles are lost much more slowly. Thus, when cold ester loading is followed by several hours of warm incubation, fluorophore loading is almost exclusively mito-

[12] H. Ohata, E. Chacon, S. A. Tesfai, I. S. Harper, B. Herman, and J. J. Lemasters, *J. Bioenerget. Biomembr.* **30,** 207 (1998).
[13] E. D. Wieder, H. Hang, and M. H. Fox, *Cytometry* **14,** 916 (1993).

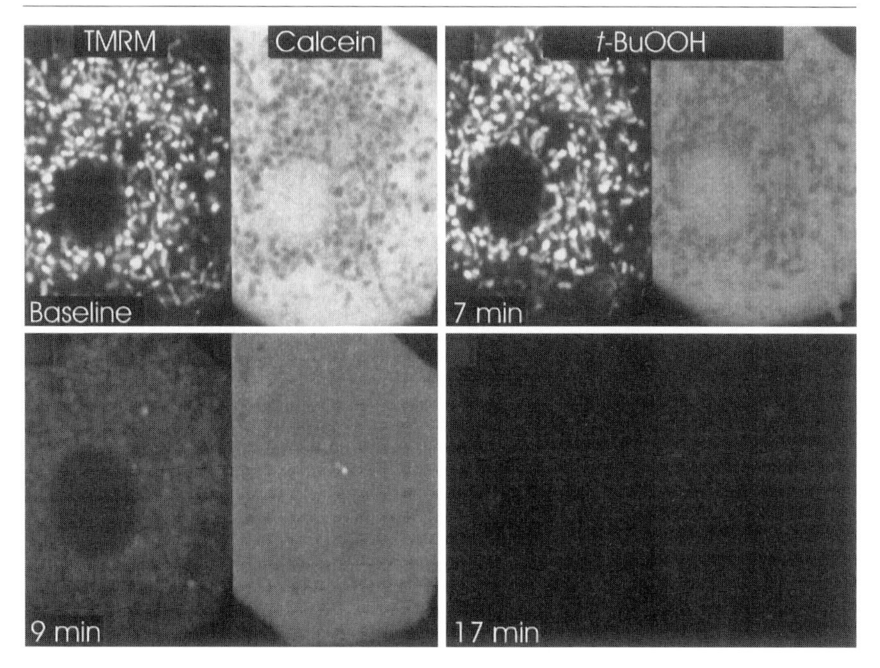

FIG. 5. Onset of the mitochondrial permeability transition in hepatocytes induced by *tert*-butylhydroperoxide (*t*-BuOOH). A cultured rat hepatocyte was coloaded with TMRM and calcein. Pairs of red TMRM and green calcein fluorescence images were collected by laser scanning confocal microscopy. In the baseline images, note that TMRM-labeled mitochondria correspond to dark voids in the calcein fluorescence. Within 9 min of adding 100 μM *t*-BuOOH to produce oxidative stress, calcein redistributed from the cytosol into the mitochondria and TMRM fluorescence was lost from the mitochondria, events signifying the onset of the MPT. These changes were followed by loss of cell viability, documented by loss of cytosolic calcein fluorescence after 17 min. [After Nieminen *et al.*[11]]

chondrial (Fig. 4C).[14] Thus, by manipulating the temperature of loading and the duration of the subsequent incubation, ester-loaded fluorophores can be directed to the cytosol, the cytosol plus mitochondria, or just the mitochondria.[11,12,14]

Membrane Permeability

Directed compartmental loading of calcein allows direct observation in cells of changes of permeability of membranes. For example, at the onset

[14] D. R. Trollinger, W. E. Cascio, and J. J. Lemasters, *Biochem. Biophys. Res. Commun.* **236,** 738 (1997).

of cell death after toxic and hypoxic injury, trapped cytosolic calcein is lost almost instantaneously (see Fig. 5).[15] Similarly, when calcein is in the extracellular space, the fluorophore enters the cell interior at onset of cell death. Other extracellular fluorophores enter also, such as propidium iodide, which binds to DNA with an enhancement of fluorescence.

After warm ester loading of calcein, calcein redistribution from the cytosol into the mitochondria indicates onset of the mitochondrial permeability transition (MPT). The MPT is caused by opening of a high-conductance pore in the mitochondrial inner membrane with a molecular weight cutoff of about 1500.[16] Onset of the MPT leads to mitochondrial depolarization and uncoupling of oxidative phosphorylation. At onset of the MPT in hepatocytes coloaded with TMRM and calcein, mitochondria release TMRM, which indicates depolarization, and fill with calcein, which indicates permeability of the mitochondria to small solutes (Fig. 5).[11] The MPT is an important event leading to both necrotic and apoptotic cell death.[17]

Confocal Imaging of Ca^{2+}

Nonratiometric Imaging

Green-fluorescing Fluo 3 and red-fluorescing Rhod 2 are useful visible-wavelength fluorophores for imaging Ca^{2+} by laser scanning confocal microscopy. Ca^{2+} binding produces a greater than 50-fold increase in fluorescence with a K_d of 300–600 n*M*.[18,19] No shift of the fluorescence spectrum of Fluo 3 and Rhod 2 occurs after Ca^{2+} binding, and calibration relies on measurements *in situ* of maximal and minimal fluorescence after respective additions of calcium ionophore and EGTA.[18,20] Even without calibration, Fluo 3 and Rhod 2 are useful to determine relative changes of free Ca^{2+}.

[15] G. Zahrebelski, A.-L. Nieminen, K. Al-Ghoul, T. Qian, B. Herman, and J. J. Lemasters, *Hepatology* **21,** 1361 (1995).

[16] M. Zoratti and I. Szabo, *Biochim. Biophys. Acta* **1241,** 139 (1995).

[17] J. J. Lemasters, A.-L. Nieminen, T. Qian, L. C. Trost, S. P. Elmore, Y. Nishimura, R. A. Crowe, W. E. Cascio, C. A. Bradham, D. A. Brenner, and B. Herman, *Biochim. Biophys. Acta* **1366,** 177 (1998).

[18] A. Minta, J. P. Y. Kao, and R. Y. Tsien, *J. Biol. Chem.* **264,** 8171 (1989).

[19] R. P. Haugland, "Handbook of Fluorescent Probes and Research Chemicals," 6th Ed. Molecular Probes, Eugene, Oregon, 1996.

[20] I. S. Harper, J. M. Bond, E. Chacon, J. M. Reece, B. Herman, and J. J. Lemasters, *Basic Res. Cardiol.* **88,** 430 (1993).

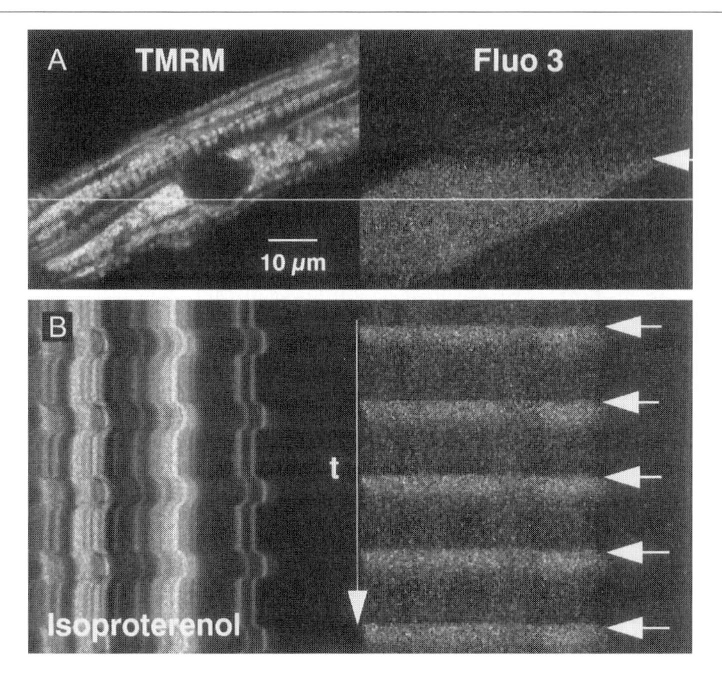

FIG. 6. Line-scanning confocal microscopy of mitochondrial and cytosolic Ca^{2+} transients during electrical stimulation. An adult rabbit cardiac myocyte was loaded with 10 μM Fluo 3-AM for 2 hr at 4° followed by 600 nM TMRM, and simultaneous red (TMRM) and green (Fluo 3) fluorescence confocal images were collected. (A) x–y images of TMRM (*left*) and Fluo 3 (*right*), and the effect of a single-field stimulation (arrow) as the 1.5-sec scan was collected. In (B), 1 μM isoproterenol was added, and the myocyte was stimulated at 0.5 Hz (arrows) as a line-scan image at 25 msec per line was collected. The region scanned is indicated by the white line in (A). [After Ohata *et al.*[12]]

Measurement of Ca^{2+} Transients Using Line Scanning

To increase temporal resolution of rapid Ca^{2+} transients, confocal images can be collected in the line-scanning mode whereby the x axis is scanned every 25 msec at the same y-axis position to create x-versus-time images.[12,21] Figure 6 illustrates simultaneous red and green line-scan images for a cardiac myocyte coloaded with TMRM and Fluo 3 at 4°.[12] An x-axis position is selected that crosses the cytosol, interfibrillar mitochondria, perinuclear mitochondria and the nucleus, as shown in Fig. 6A. In the line-scan images, red and green fluorescence is collected simultaneously (Fig. 6B). Red TMRM fluorescence appears as wavy vertical stripes. Each stripe is a mitochondrion, and each wave in the stripes is cell movement from an

[21] H. Cheng, W. J. Lederer, and M. B. Cannell, *Science* **262,** 740 (1993).

electrically stimulated contraction. By contrast, green Fluo 3 fluorescence does not show vertical striations but only horizontal banding corresponding to Ca^{2+} transients after each field stimulation. Importantly, the transients of Fluo 3 fluorescence occur both in pixels corresponding to TMRM-labeled mitochondria and in those of the TMRM-unlabeled cytosol. These results demonstrate that both cytosolic and mitochondrial Ca^{2+} transients occur after field stimulation of cardiac myocytes.

Using image analysis software, regions corresponding to the cytosol, the nucleus, and interfibrillar and perinuclear mitochondria can be selected on the basis of the brightness of TMRM fluorescence and the position in the cell. Bright regions in the red TMRM image are mitochondria, whereas areas of low fluorescence intensity are cytosol and nucleus. Pixels of intermediate intensity are excluded from analysis because they likely represent the overlap of mitochondrial and cytosolic domains. Using morphological criteria, high-TMRM regions are subdivided into interfibrillar and perinuclear mitochondria, and low-fluorescence regions are assigned to the nucleus and cytosol. The corresponding regions in the green Fluo 3 fluorescence image are then averaged in each horizontal line. Relatively inexpensive software packages, such as NIH Image (National Institutes of Health, Bethesda, MD) for Macintosh computers and Image PC (Scion Corp., Frederick, MD) for Windows-based personal computers, can perform this analysis. In this way, average Fluo 3 fluorescence for the cytosol, nucleus, and interfibrillar and perinuclear mitochondria is calculated every 25 msec. When the repeating Ca^{2+} transients are averaged and normalized as a percentage of baseline fluorescence, the peak change in fluorescence after field stimulation is greatest in the cytosol and least in perinuclear mitochondria (Fig. 7). Isoproterenol increases the peak intensity and rate of decay of Fluo 3 fluorescence in the cytosol and nucleus with a lesser effect on interfibrillar mitochondria and virtually no effect on perinuclear mitochondria. Because TMRM and Fluo 3 fluorescence are collected simultaneously, these calculations automatically correct for movement of mitochondria laterally or into and out of the optical section during cell contraction.[12]

Quantification of Free Ca^{2+} by Ratio Imaging of Indo 1-Loaded Myocytes

Ca^{2+} binding causes an increase of Fluo 3 fluorescence but no wavelength shift, preventing calibration by ratio imaging. To estimate absolute Ca^{2+} concentrations, myocytes can be loaded with Indo 1 AM,[22] which distributes into both the cytosol and the mitochondria even after ester loading at 37°. Using 351-nm excitation from a UV argon laser, Indo 1 fluorescence

[22] G. Grynkiewicz, M. Poenie, and R. Y. Tsien, *J. Biol. Chem.* **260,** 3440 (1985).

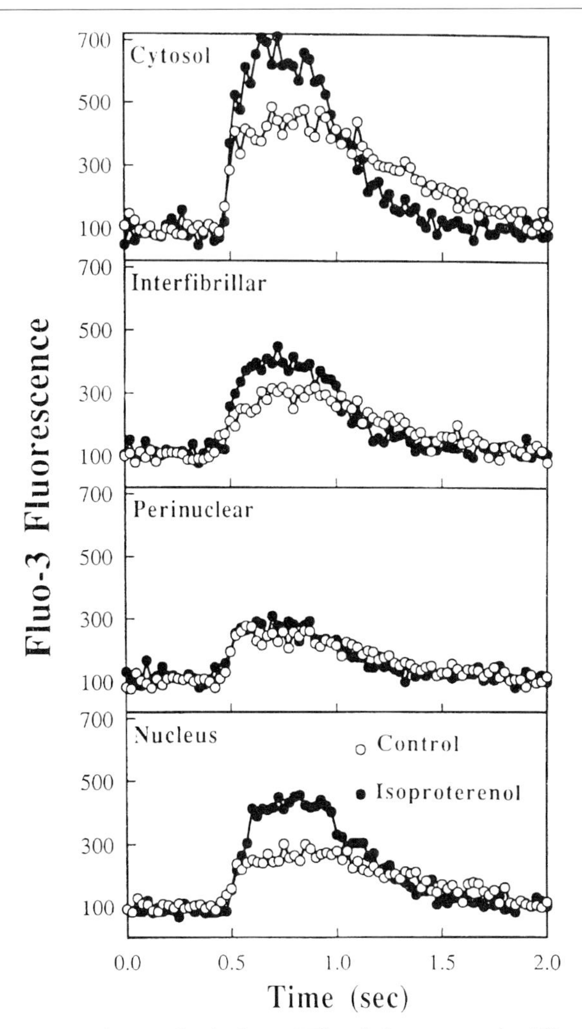

FIG. 7. Averaged and normalized plots of Fluo 3 fluorescence in different subcellular regions during electrical stimulation of a cardiac myocyte. From line-scan images in the presence and absence of 1 μM isoproterenol (see Fig. 6B), TMRM fluorescence was used to identify cytosol, interfibrillar mitochondria, perinuclear mitochondria, and the nucleus. Pixels corresponding to each region were superimposed on the corresponding Fluo 3 image, and Fluo 3 pixel intensity was calculated using NIH Image. Average pixel intensity minus background for each region in the line-scan image was determined. Fluorescence transients after electrical stimulation in each region were averaged, normalized as a percentage of diastolic intensity, and plotted versus time. [After Ohata *et al.*[12]]

increases at 405 nm and decreases at 480 nm as Ca^{2+} increases. Ratio images of Indo 1 fluorescence show a relatively uniform intracellular free Ca^{2+} concentration of about 200 nM in both the cytosol and mitochondria of resting myocytes (Fig. 8, see color insert). This value is uncorrected for possible increases in the binding constant of Indo 1 for Ca^{2+} that occurs when Indo 1 is in the intracellular environment.[23–25] Ca^{2+} sensitivity of Indo 1 and other BAPTA-derived probes is also affected by pH, but significant changes in K_d for Ca^{2+} occur only when the pH falls below 6.8.[26,27] Because mitochondrial pH is alkaline relative to the cytosol (Fig. 3), pH should not affect the K_d for Ca^{2+} binding inside mitochondria. In any case, ratioing of Indo 1 fluorescence corrects for the effect of variable loading of fluorophore into various cellular compartments and confirms that Ca^{2+} transients occur both in the cytosol and the mitochondria after electrical stimulation (Fig. 8).

Cold Loading of Rhod 2 followed by Warm Incubation

Another approach to measure mitochondrial Ca^{2+} uses cold loading of Rhod 2 AM followed by warm incubation. After loading in this way, Rhod 2 fluorescence in unstimulated myocytes shows a distinctly mitochondrial pattern, comparable to that observed when myocytes are loaded with TMRM or rhodamine 123 [Fig. 9, see color insert; compare with Fig. 6].[28] When the myocytes are electrically stimulated as 16-sec confocal scans are collected, mitochondrial Rhod 2 fluorescence increases after each electrical stimulation, producing horizontal bands in the image (Fig. 9). When the pacing frequency is doubled, banding frequency also doubles. The largest fluorescence transients occur in the bright regions corresponding to mitochondria. Fluorescence in areas between mitochondria either remains dark or increases moderately after electrical stimulation. Addition of the Ca^{2+} ionophore Br-A23187 verifies Rhod 2 localization (Fig. 9). After raising Ca^{2+} in all compartments to levels that saturate Rhod 2, the distribution of red fluorescence represents the intracellular distribution of the fluorophore. In this experiment, CCCP, a powerful protonophoric uncoupler, is also added to depolarize mitochondria and block cycling of free Ca^{2+} after

[23] M. W. Roe, J. J. Lemasters, and B. Herman, *Cell Calcium* **11,** 63 (1990).

[24] A. J. Baker, R. Brandes, J. H. Schreur, S. A. Camacho, and M. W. Weiner, *Biophys. J.* **67,** 1646 (1994).

[25] J. W. Bassani, R. A. Bassani, and D. M. Bers, *Biophys. J.* **68,** 1453 (1995).

[26] F. A. Lattanzio and D. K. Barschat, *Biochem. Biophys. Res. Commun.* **171,** 102 (1991).

[27] T. Kawanishi, A.-L. Nieminen, B. Herman, and J. J. Lemasters, *J. Biol. Chem.* **266,** 20062 (1991).

[28] D. R. Trollinger, W. E. Cascio, and J. J. Lemasters, *Biochem. Biophys. Res. Commun.* **236,** 738 (1997).

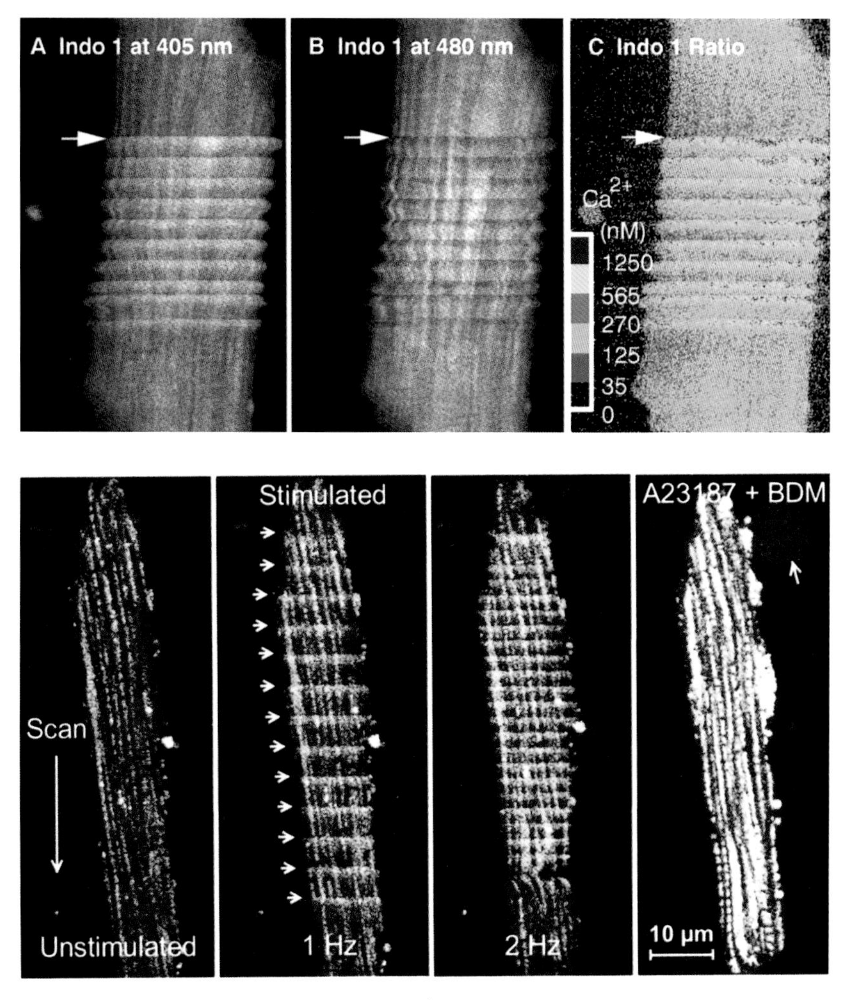

FIG. 8. Indo 1 ratio imaging of cytosolic free Ca^{2+} in a cardiac myocyte during electrical stimulation. Adult rabbit cardiac myocytes were loaded with 5 μM Indo 1 AM for 1 hr at 37°. Using 351-nm excitation from a UV-argon laser, fluorescence images of Indo 1 were collected at emission wavelengths of 395–415 nm (A) and 470–490 nm (B) as the cell was electrically stimulated at 0.5 Hz for 10 sec during a 40-sec scan. (C) The ratio image after background subtraction scaled to Ca^{2+} concentration. Electrically stimulated Ca^{2+} transients occurred in both mitochondrial and cytosolic regions, as identified by rhodamine 123 labeling (not shown). [After Ohata *et al.*[12]]

FIG. 9. Mitochondrial Ca^{2+} transients after cold Rhod 2 loading followed by warm incubation. An adult rabbit cardiac myocyte was cold loaded with Rhod 2 AM and then incubated at 37° for 3 hr prior to confocal fluorescence imaging. In the unstimulated myocyte, Rhod 2 fluorescence has a mitochondrial pattern. After stimulation at 1 Hz (arrows) and then 2 Hz, Rhod 2 fluorescence transiently increased in the bright mitochondrial regions and then returned to baseline. Oligomycin (10 μM), CCCP (10 μM), butanedione monoxime (20 mM), and then Br-A23187 (20 μM) were added to increase free Ca^{2+} in all compartments. Under these conditions, saturated Rhod 2 fluorescence was confined to mitochondria. Cytosolic areas, such as in blebs (arrow), were only faintly fluorescent. Each 16-sec confocal scan proceeds from top to bottom. The images are pseudocolored using the black body look-up table of Photoshop (Adobe Systems, San Jose, CA). [After Trollinger *et al.*[28]]

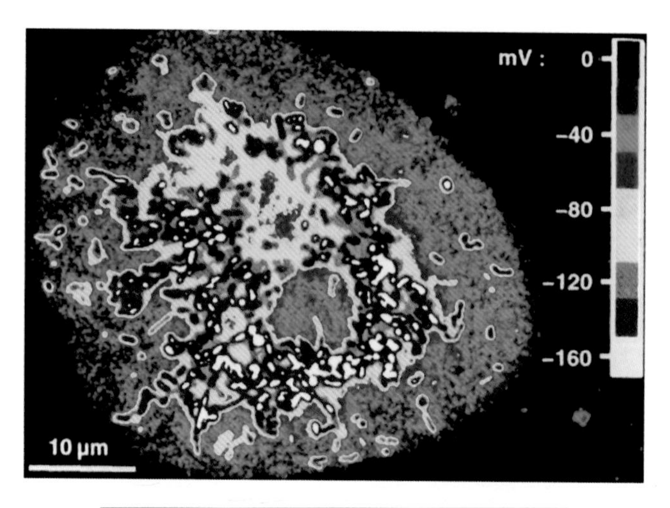

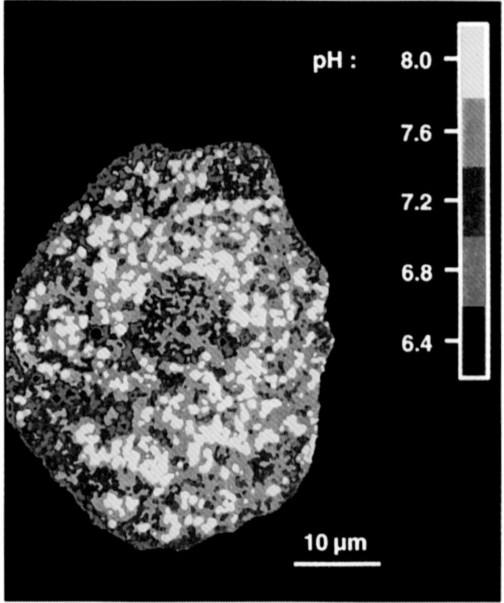

FIG. 2. Distribution of electrical potential in a cultured rat hepatocyte. A 1-day cultured rat hepatocyte was loaded with 500 n*M* TMRM for 15 min, and distribution of electrical potential was determined by laser scanning confocal microscopy using 568-nm excitation from an argon–krypton laser, as described in text. [After J. J. Lemasters, E. Chacon. J. M. Reece, A.-L. Nieminen, and G. Zahrebelski, unpublished.]

FIG. 3. Intracellular pH in a cultured rat hepatocyte. A 1-day cultured rat hepatocyte was loaded with 5 μM SNARF 1 AM for 45 min at 37°. Using 568-nm excitation light, a confocal fluorescence image at 584 nm (10-nm bandpass) was divided by the image at >620 nm after background subtraction. Using an *in situ* calibration, distribution of pH within the cell is represented in pseudocolor. [After J. J. Lemasters, E. Chacon, J. M. Reece, A.-L. Nieminen, and G. Zahrebelski, unpublished.]

Br-A23187 treatment. Glucose and oligomycin are also added. In the presence of a glycolytic substrate, oligomycin prevents ATP depletion by inhibiting the uncoupler-stimulated mitochondrial ATPase.[29–31] In addition, butanedione monoxime (BDM) is added prior to Br-A23187 to prevent myocyte contracture.[32] After Br-A23187 treatment in this way, Rhod 2 fluorescence shows a characteristic mitochondrial pattern of staining (Fig. 9). By contrast, cytosolic areas between mitochondria remain dark. Occasional toxic blebs form after Br-A23187 treatment (Fig. 9, arrow). The contents of these blebs are an extension of the cytosol.[33] These blebs in Br-A23187-treated myocytes remain dark, confirming the virtual absence of cytosolic Rhod 2 loading.

During electrical stimulation, Rhod 2 fluorescence increases sharply in the bright regions corresponding to mitochondria. However, areas between mitochondria sometimes appear to increase moderately in fluorescence, an apparent contradiction of the conclusion that Rhod 2 is confined virtually exclusively to mitochondria. This inconsistency is explained by the fact that the confocal slice thickness is finite, in the range of 0.8–0.9 μm. Cardiac mitochondria are approximately 1 μm in diameter and many mitochondria are only partially sectioned within the confocal slice. Parts of confocal images containing partially sectioned mitochondria will be darker than areas with fully sectioned mitochondria. However, fluorescence will nonetheless increase in darker regions containing partially sectioned mitochondria as Ca^{2+} increases. The observation that ruthenium red, an inhibitor of mitochondrial Ca^{2+} uptake, blocks mitochondrial but not cytosolic Ca^{2+} transients after electrical stimulation further confirms the mitochondrial localization of the fluorophore.[34]

Cold loading of Rhod 2 followed by warm incubation to measure mitochondrial Ca^{2+} has advantages over coloading with Fluo 3 and TMRM. With the latter approach, the contribution of partially sectioned mitochondria to the analysis must be excluded. Only Fluo 3 pixels corresponding to the brightest and darkest TMRM pixels can be relied on to represent, respectively, mitochondrial and cytosolic Ca^{2+}. Pixels corresponding to intermediate TMRM fluorescence must be discarded. With cold loading of Rhod 2

[29] A.-L. Nieminen, T. L. Dawson, G. J. Gores, T. Kawanishi, B. Herman, and J. J. Lemasters, *Biochem. Biophys. Res. Commun.* **167,** 600 (1990).

[30] D. M. Bers, J. W. Bassani, and R. A. Bassani, *Cardiovasc. Res.* **27,** 1772 (1993).

[31] A.-L. Nieminen, A. K. Saylor, B. Herman, and J. J. Lemasters, *Am. J. Physiol.* **267,** C67 (1994).

[32] S. C. Armstrong and C. E. Ganote, *J. Mol. Cell Cardiol.* **23,** 1001 (1991).

[33] J. J. Lemasters, S. Ji, and R. G. Thurman, *Science* **213,** 661 (1981).

[34] D. R. Trollinger, W. E. Cascio, and J. J. Lemasters, *Biophys. J.* **74,** A356 (1998).

followed by warm incubation, all pixels can be used for analysis, which increases signal strength and signal to noise.

Conclusion

Confocal microscopy has become an essential tool to study the physiology of single living cells. As more parameter-indicating fluorophores are discovered, the range of applications of confocal microscopy can only increase. Uniquely, confocal microscopy permits observation of the physiology and metabolism of single organelles inside cells, as illustrated here for measurements of Ca^{2+}, pH, $\Delta\Psi$, and membrane permeability of mitochondria. Overall, the impact of confocal microscopy on experimental cell physiology may someday compete with that of single-cell electrical recording.

Acknowledgments

This work was supported, in part, by Grants DK37034, HL27430, AG07218, and AA11605 from the National Institutes of Health and by Grant N00014-96-1-0283 from the Office of Naval Research.

[30] *In Vivo* Retroviral Transduction and Expression of Green Fluorescent Protein

By JOHN P. LEVY, REBECCA R. MULDOON, ILYA A. MAZO, STEVEN R. KAIN, and CHARLES J. LINK, JR.

Introduction

Murine retroviral vectors (RVs) containing optimized green fluorescent protein (GFP) genes have been used to study experimental retroviral gene transfer and expression in living cells. The expression of a single integrated humanized red-shifted GFP gene in eukaryotic cells provides a powerful tool with which to study gene transfer and expression, and to perform functional studies both *in vitro* and *in vivo*.[1,2] Such retroviral GFP expression in transduced cells is also stable *in vivo*.[3] GFP has provided a tracking

[1] R. R. Muldoon, J. P. Levy, S. R. Kain, P. A. Kitts, and C. J. Link, Jr., *BioTechniques* **22,** 162 (1997).

[2] J. P. Levy, R. R. Muldoon, and C. J. Link, *Nature Biotechnol.* **14,** 610 (1996).

[3] E. S. Kandel, B. D. Chang, B. Schott, A. A. Shtil, A. V. Gudkov, and I. B. Roninson, *Somatic Cell Mol. Genet.* **23,** 325 (1997).

0076-6879/99 $30.00

method by which to visualize metastatic cancer cells *in vivo*.[4] GFP has been used in a reporter system for detecting transduction by the human immunodeficiency virus type 1 (HIV-1).[5,6] HIV lentiviral vectors have been used to transduce terminally differentiated neurons of the brain *in vivo*.[7] Such lentiviral GFP vectors have been transduced and expressed in the liver for more than 22 weeks.[8] Hematopoietic stem cell retroviral transduction is an important aspect of gene therapy. An essential aim of gene therapy is to optimize retroviral vectors. GFP retroviral transduction of hematopoietic stem cells has allowed for rapid assessment of improved retroviral vectors.[9] Choice of 3′ long terminal repeat (LTR) and elimination of internal promoters dramatically improve EGFP transduction of stem cells.[10] Splenic hematopoietic colonies and peripheral blood cells from mice engrafted with GFP-transduced marrow maintain high levels of GFP fluorescence.[11] Highly purified $CD34^+$ cells transduced with a compact MFG-GFP retroviral vector generate fluorescent $CD4/8^+$ T, natural killer, and dendritic cells *in vitro*.[12] Inducible transgene expression within retroviral vectors has been rapidly optimized by selecting GFP as the marker transgene.[13] Attempts have been made to combine the GFP gene with a suicide gene within the context of the same retroviral vector in order to develop a rapidly selectable gene transfer system that also provides therapeutic effects for gene therapy of solid tumors.[14] *In vivo* expression of GFP appears to be nontoxic. Transgenic mouse lines that express EGFP in nearly all tissues appear brightly fluorescent and are viable throughout development and adulthood.[15] However,

[4] T. Chishima, *et al., Clin. Exp. Metastasis* **15,** 547 (1997).

[5] D. I. Dorsky, M. Wells, and R. D. Harrington, *J. Acquir. Immune Defic. Syndr. Hum. Retrovirol.* **13,** 308 (1996).

[6] K. A. Page, T. Liegler, and M. B. Feinberg, *AIDS Res. Hum. Retroviruses* **13,** 1077 (1997).

[7] H. Miyoshi, M. Takahashi, F. H. Gage, and I. M. Verma, *Proc. Natl. Acad. Sci. U.S.A.* **94,** 10319 (1997).

[8] T. Kafri, U. Blomer, D. A. Peterson, F. H. Gage, and I. M. Verma, *Nature Genet.* **17,** 314 (1997).

[9] L. Cheng, C. Du. D. Murray, X. Tong, Y. A. Zhang, B. P. Chen, and R. G. Hawley, *Gene Ther.* **4,** 1013 (1997).

[10] A. Limon, J. Briones, T. Puig, M. Carmona, O. Fornas, J. A. Canetas, M. Nadal, J. Garcia, F. Rueda, and J. Barqvinero, *Blood* **90,** 3316 (1997).

[11] D. A Persons, J. A. Allay, E. R. Allay, R. J. Smeyne, R. A. Ashmum, B. P. Sorrentino, and A. W. Nienhuis, *Blood* **90,** 1777 (1997).

[12] B. Verhasselt, M. De Smedt, R. Verhelst, E. Naessens, and J. Plum, *Blood* **91,** 431 (1998).

[13] T. Watsuji, Y. Okamoto, N. Emi, Y Katsuoka, and M. Hagiwara, *Biochem. Biophys. Res. Commun.* **234,** 769 (1997).

[14] U. Nestler, M. Heinkelein, M. Lucke, J. Meixensberger, W. Scheurlen, A. Kretschmer, and A. Rethwilm, *Gene Ther.* **4,** 1270 (1997).

[15] M. Okabe, M. Ikawa, K. Kominami, T. Nakanishi, and Y. Nishimune, *FEBS Lett.* **407,** 313 (1997).

in vivo adenoassociated viral transmission of GFP into the lungs of mice may elicit a mild immune response.[16] GFP has been expressed in live zebra fish embryos,[17] sea urchin,[18] *Drosophila,*[19] and as a lineage marker in *Xenopus.*[20]

Green Fluorescent Protein Retroviral Vector Design

To achieve the highest titer the retroviral vector should contain only the transgene GFP gene between the 5′ and 3′ LTRs. Such a simplified vector design has been used extensively in the past with the MFG retroviral vector.[21] The first report of GFP being used in a minimal retroviral vector, LEL, for *in vivo* gene transfer reported significantly improved transfer and expression of GFP compared with retroviral vectors containing internal promoters used to coexpress drug resistance genes such as neomycin phosphotransferase.[22] The LEL retroviral vector in this study is also discussed in [28] in this volume.[23] A similar vector has been used to efficiently transfer GFP into hematopoietic stem cells.[10] Because GFP provides a separable marker for transduction there is no need for traditional drug selection markers, provided that either a fluorescence-activated cell sorter (FACS) is on site or physical cell separation methods similar to limiting cell dilution can be used. Expression of a second transgene within the retroviral vector can be accomplished by constructing an inframe fusion of GFP to the second transgene or by cloning an internal ribosome entry sequence between GFP and the second gene. The choice of exactly which retroviral vector to use depends on the cell targeted for transduction. The Moloney retroviral vector is generally useful for transduction of most murine and human cell lines; however, retroviral vectors that are composed of LTRs derived from murine stem cell virus provide higher transgene expression when transduced in murine hematopoietic stem cells.[10] Therefore, it may be necessary to survey

[16] T. R. Flotte, S. E. Beck, K. Chestnut, M. Potter, A. Poirier, and S. Zolotukhin, *Gene Ther.* **5,** 166 (1998).

[17] A. Amsterdam, S. Lin, and N. Hopkins, *Dev. Biol.* **171,** 123 (1995).

[18] M. I. Arnone, L. D. Bogorad, A. Collqrozo, C. V. Kirchhamer, R. A. Cameron, J. P. Rast, A. Gregorians, and E. H. Davidson, *Development* **124,** 4649 (1997).

[19] A. Brand, *Trends Genet.* **11,** 324 (1995).

[20] M. Zernicka-Goetz, J. Pines, K. Ryan, K. R. Siemering, J. Haselof, M. J. Evans, and J. B. Gordon, *Development* **122,** 3719 (1996).

[21] D. S. Ory, B. A. Neugeboren, and R. C. Mulligan, *Proc. Natl. Acad. Sci. U.S.A.* **93,** 11400 (1996).

[22] R. R. Muldoon, J. P. Levy, and C. J. Link, *J. Virol.* Submitted (1999).

[23] I. A. Mazo, J. P. Levy, R. R. Muldoon, C. J. Link, and S. R. Kain, *Methods Enzymol.* **302,** [28], 1999 (this volume).

the known literature to best match the retroviral vector system to the target cell desired.

Construction of Enhanced Green Fluorescent Protein Retroviral Vectors

Plasmid pEGFP C-1 is obtained from Clontech (Palo Alto, CA). Cloned construct plasmid DNA is transformed into DH5α competent cells and colonies grown on plates containing L broth supplemented with ampicillin (50 mg/ml; LB/AMP). Positive colonies are grown in LB/AMP broth and plasmid DNA is isolated using the Qiagen plasmid kit (Qiagen, Chatsworth, CA). Plasmid pEGFP C-1 is used as the template to amplify by polymerase chain reaction (PCR) the CMV-EGFP fragment. The forward primer sequence is 5′-GGA TCT AGA GGA TCC GCG GCC GCC TAG TTA TTA ATA GTA ATC AAT TAC GGG GTC-3′. This primer adds the restriction endonuclease sites *Xba*I, *Bam*HI, and *Not*I to the 5′ end of the cytomegalovirus (CMV) promoter. The reverse primer is 5′-GGC AAG CTT GGC CGC TCA CTT GTA CAG CTC GTC CA-3′, which adds three in-frame stop codons to the 3′ end of EGFP followed by a *Hin*dIII restriction endonuclease site. The CMV-EGFP 1.2-kb PCR product is then ligated into the vector pPCR3 (Invitrogen, Carlsbad, CA) to produce a directional clone, pPCR3-EGG7. This vector is then restriction digested with *Bam*HI and *Hin*dIII enzymes to liberate the CMV-EGFP fragment. This fragment is then gel isolated (Jetsorb; Genomed, Raleigh, NC). Plasmid pLNCX is restriction digested with *Bam*HI and *Hin*dIII to remove the CMV fragment from this cloning vector. This vector fragment is then gel isolated (Jetsorb; Genomed). The CMV-EGFP fragment is then directionally cloned into the vector fragment to produce the vector pLNCE.

The vector pLNCE is then restriction digested with *Eco*47III and *Hin*dIII to liberate the EGFP open reading frame fragment. The Jetsorb-isolated fragment is treated to blunt the *Hin*dIII 3′ end of the gene. The vector pLXSN is then digested with the *Hpa*I enzyme, followed by ligation with this EGFP fragment to produce the vector pLESN. The vector pLESN is then digested with *Bam*HI and *Rsr*II to remove the simian virus 40 (SV40) and the majority of the *neo* gene from the plasmid. The 5.3-kb vector containing EGFP is then gel isolated (Jetsorb; Genomed), Klenow treated, and ligated together to produce pLEL (Fig. 1).

Production of LEL Retrovirus

A375 is a human melanoma target cell line [American Type Culture Collection (ATCC), Frederick, MD]. PA317 is a murine amphotrophic, RV packaging cell line (kindly provided by A. D. Miller, University of

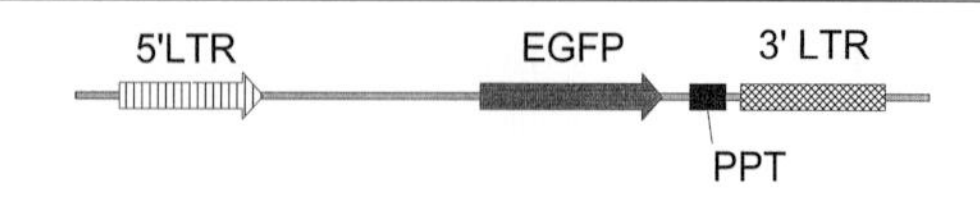

FIG. 1. LEL retroviral vector. The LEL vector contains 5′ and 3′ LTRs, a packaging sequence that extends into the *gag* open reading frame, and the EGFP gene. The LEL vector does not contain a drug-selectable marker or internal promoters. The splice donor and splice acceptor sites are unchanged from those of the wild-type Moloney retrovirus. PPT, polypurine tract.

Washington, Seattle, WA). GP+E86 is a murine ecotropic, RV packaging cell line (kindly provided by A. Banks, Columbia University, New York, NY). Cells are grown in RPMI or Dulbecco's medium supplemented with 10% fetal calf serum (FCS) (all obtained from GIBCO-BRL, Gaithersburg, MD) in monolayers at 37° and 5% CO_2. All cells are passaged and harvested by standard trypsin (GIBCO-BRL) digestion at 37°. Cells are routinely passaged at 80–90% confluence.

The EGFP plasmid constructs are then used to establish these EGFP retroviral vectors in the cell lines GPE86, PA317, and A375. GP+E86 cells are plated in a six-well dish (Falcon; Becton Dickinson Labware, Lincoln Park, NJ) 24 hr before transfection. Cells that are 30–50% confluent are transfected with 5 μg of plasmid DNA and 15 μl of DOTAP cationic liposome-mediated transfection reagent (Boehringer Mannheim, Indianapolis, IN) according to the manufacturer protocol. After 12 hr, the cells are rinsed and 1 ml of fresh medium is placed in each well. RV supernatant is collected 24 hr later, filtered (0.45-mm pore size; Costar, Cambridge, MA) and protamine sulfate solution (10 mg/ml) is added. This supernatant is transferred to a six-well dish containing PA317 cells at 30–50% confluency. After 24 hr the cell lines containing the vectors pLNCE and pLESN are selected in G418 (geneticin, 1 mg/ml; GIBCO, Grand Island, NY) for 10–14 days. The selected LNCE and LESN vector produces cells and the nonselected LEL vector producer cells are trypsin digested and plated by limiting dilution in 96-well plates to establish clonal cell lines; some of the cells are transferred to an 80-cm^2 flask (Fig. 2).

In Vivo LEL Transduction of Human Tumors in Mice

Transduction of A375 Human Tumor Cell Lines

PA317-EGFP retroviral vector cells are grown to 80–95% confluency in 80-cm^2 flasks. The medium is replaced with 10 ml of fresh medium and

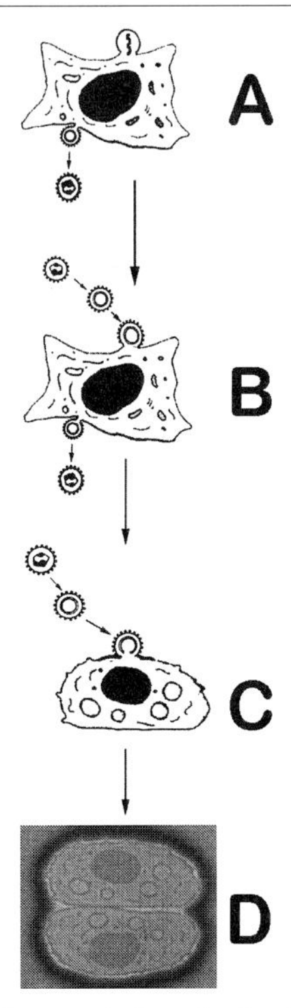

FIG. 2. Construction of the LEL retroviral packaging line. (A) Liposome transfection of ecotropic GP+E86 packaging cell; (B) retrovirus transduction of PA317 amphotropic packaging cell; (C) transduction of target cell lines; (D) expression of GFP in target cell line.

24 hr later the RV supernatant is collected, filtered, and supplemented with protamine sulfate (10 mg/ml). A375 cells are plated into a six-well dish (Falcon). Two milliliters of supernatant is added to each well of the six-well plates. The supernatant is removed 24 hr later and replaced with fresh medium containing G418 (1 mg/ml) for the LNCE and LESN group and selected for 7–14 days. The LEL supernatant is removed and replaced with fresh medium. After cell line selection is complete, the lines are cloned by limiting dilution in 96-well plates to establish individual cell clones.

Tumor Formation in Mice

Six- to 8-week-old athymic nude mice (*nu/nu*) are obtained from Harlan Sprague-Dawley (Indianapolis, IN). They are housed in cages with microisolator tops and all food, water, and bedding are autoclaved before use. A375, A375-LNCE, A375-LESN, or A375-LEL cells (1×10^5 to 20×10^6) are trypsin digested and washed with Hanks' balanced salt solution (HBSS).

Cells are resuspended in 1 ml of HBSS and injected (25-gauge needle) into the intraperitoneal cavity of each mouse. Animals are euthanized by using CO_2 and necropsies are performed. Tumors are excised from the peritoneal cavity, rinsed with HBSS, and processed for histopathology or tissue culture. Tumors are photographed with a Canon (Lake Success, NY) AE-1 camera using 400 Fujichrome film (Fuji, Tokyo, Japan).

In Vivo Transduction

Athymic nude mice are injected with 1×10^5, 1×10^6, or 2.5×10^6 A375 cells into the intraperitoneal cavity as described above. Beginning on day 6 the animals are injected (25-gauge needle) intraperitoneally with 1 ml of EGFP retroviral supernatant each day for 3–6 days. When tumor formation is evident, the mice are euthanized as described above and the tumors are processed. Individual tumors and tumor nodules are processed as a mixture or as individual samples.

Histology

Excised tumor samples are frozen at $-20°$ and placed in a mounting block with O.C.T. compound (Tissue-Tek; Miles, Elkhart, IN). Thin sections are sliced using a Tissue-Tek cryostat (Miles), and placed on slides for microscopic observations and photographs. Photographs are taken using 1600 Fujichrome slide film using a Nikon (Melville, NY) camera F-mounted to a Nikon Labophot-2 fluorescence microscope (Fig. 3A and E, see color insert for parts B–D).

Preparation of Tumor Cell Suspensions

The enzyme solutions used for tumor lysis contain 600 mg of collagenase, 60 mg of DNase, and 0.625 mg of hyaluronidase. These solutions are dissolved in 300 ml of HBSS, filter sterilized, and aliquoted into 15-ml portions (in 50-ml conical tubes) and stored in $-20°$. Fifteen milliliters of complete medium with 2× Pen/Strep (GIBCO, Gathersburg, MD) is added to each 15-ml tube of thawed enzyme solution. The tumor specimen is minced with scissors and placed in the tube with lysis medium. The tumors are allowed to dissociate on a rocker for 30–60 min at room temperature. Larger tissue pieces are filtered out using gauze, or are allowed to settle to the bottoms of the tubes and disposed of. Medium containing the cells is then centrifuged at 1500 rpm for 10 min at 15° to pellet the cells. Cells are resuspended in complete medium with 2× Pen/Strep in either six-well plates or 25-cm^2 flasks, or analyzed by FACS.

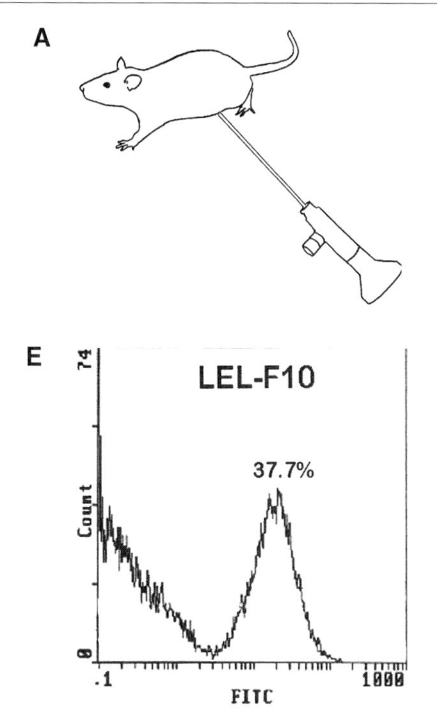

FIG. 3. Imaging of LEL-transduced tumors *in vivo*. (A) Schematic of endoscope probe used to view GFP tissue or tumor. (E) FACs analysis of collagenase-explanted LEL-transduced tumor cells.

Titer Assays

A375 cells are plated at 1×10^5 cells well in a six-well plate. Twenty-four to 48 hr later 1 ml of EGFP vector producer cell supernatant diluted to 1×10^4, 1×10^5, or 1×10^6 is added to the A375 cells. Twenty-four hours later the supernatant is removed and either 2 ml of complete medium containing G418 (1 mg/ml) or complete medium without G418 is added. Cells selected with G418 are grown for 10–14 days to obtain neomycin-resistant colonies. These colonies are stained with crystal violet and counted. Titers are expressed as colony-forming units per milliliter of supernatant.

To determine the titer of retroviruses containing GFP only, an assay[10] based on FACS analysis is used. Viral titer is determined as (cell number × fluorescent cells)/ml supernatant. A known number of NIH 3T3 cells is transduced with various volumes of retroviral supernatant. Data are collected 48 hr after transduction by FACS analysis. Increasing volumes of supernatant are plotted against the percentage of fluorescent target cells. The titer is calculated from volumes corresponding to the linear slope of the curve according to the preceding formula.

Fluorescence Detection of Green Fluorescent Protein-Expressing Cells

The EGFP-expressing cells are visualized with a Nikon Labophot-2 fluorescence microscope (Fryer Company, Huntley, IL). The filter cube used in our microscope is the fluorescein isothiocyanate (FITC) dichroic filter set (excitation at 450–490 nm and emission at 520 nm). Photographs are taken using the Nikon Microflex AFX-DX systems (Fryer Company).

Fluorescence-Activated Cell Sorter Analysis of Enhanced Green Fluorescent Protein-Transduced Mammalian Cells

FACS cytometry of transduced cells is performed with an Epics XL analyzer (Coulter, Hialeah, FL) with an excitation source of 488 nm. Cells are analyzed using a 525-nm bandpass filter set (Coulter). Cultures of cells that are 80–90% confluent are trypsin digested, washed with Hanks' balanced salt solution, and resuspended at a concentration of approximately 1×10^6 cells/ml. All FACS analyses use the FL1 emission channel to monitor green fluorescence (normally an FITC monitor).

Ex Vivo LEL Transduction of Hematopoietic Stem Cells

Isolation of Either Mouse Stem Cells or Human CD34$^+$ Stem Cells

Murine stem cells are harvested from marrow obtained from female C57/BL6 mice between the ages of 8 and 12 weeks. The marrow is harvested 2 days after injection with 5-fluorouracil (150 mg/kg). Human stem cells can be obtained from human cord blood and apheresis samples. Isolation of CD34$^+$ stem cells is achieved through a combination of immunomagnetic separation and FACS. The immunomagnetic isolation is best performed using an Isolex kit (Baxter Biotech Immunotherapy Division, McGaw Park, IL). These immunoisolated cells can be further purified by immunofluorescently labeling them with a CD34 sulforhodamine antibody and the FITC-labeled lineage (Lin) antibodies CD2, CD4, CD14, CD15, CD16, CD19, and glycophorin A. CD34$^+$Lin$^-$ cells are obtained by sorting for cells that are positive for the rhodamine label and negative for the FITC label.

Culturing Human or Mouse CD34$^+$ Hematopoietic Cells

Mouse stem cells or human CD34$^+$ cells are cultured in either Iscove's modified Eagle's or Dulbecco's medium (IMEM or IMDM) at a cell density between 1×10^4 and 5×10^4. Medium is supplemented with 12.5% heat-inactivated fetal calf serum, penicillin (100 IU/ml), and streptomycin (100 μg/ml). Culturing occurs at 37° in a humidified atmosphere containing 7.5% CO_2/air (v/v). Cell samples can be cryopreserved in liquid nitrogen and stored until needed for further experimentation.

Green Fluorescent Protein Retroviral Transduction of CD34$^+$ Cells

Cytokine Activation. Human CD34$^+$ cells cultured for 12–24 hr under the preceding culture conditions are supplemented with interleukins 3 and 6 (IL-3 and IL-6, 10 ng/ml each) and stem cell factor (100 ng/ml).

Transduction. Transduction is accomplished by mixing fresh or freshly thawed frozen supernatants 1 : 1 with target cells suspended in DMEM and Polybrene (8 μg/ml; Sigma, St. Louis, MO). This supernatant–target cell mixture is centrifuged at 1700*g* for 4 hr at 32°. Resuspend the stem cells in CD34$^+$ culture medium for 24 hr, then repeat the preceding transduction protocol once again. Resuspend and allow stem cells to grow in culture for 2–3 days following transduction. Alternatively, transductions can be repeated daily for six consecutive days, but with a decreased centrifugation time of 5 min each time. Another alternative: GFP retroviral vector producer cells can also be lethally irradiated (1200 cGy) and cocultured with cytokine-activated CD34$^+$ cells for 48 hr. Transduced, nonadherent CD34$^+$ cells can then be carefully decanted as a suspension culture away from the GFP-adherent vector producer cells. The transduced CD34$^+$ cells are then treated as described above.

In Vitro Clonogenic Progenitor Assays

Colony-forming unit cells (CFU-Cs), between 500 and 5000, are cultured in 3 ml of methylcellulose culture medium (Stem Cell Technologies, Vancouver, Canada). The medium is supplemented with the following cytokines: IL-3, IL-6, granulocyte-macrophage colony-stimulating factor (GM-CSF, 10 ng/ml), stem cell factor (100 ng/ml), and erythropoietin (2 units/ml). After mixing the cells are plated in 35-mm suspension culture dishes at 37° and 5% CO_2. Allow the cells to colonize for 10–14 days, then count total cells and fluorescent cells in each of three plates.

In Vivo Analysis of Engrafted LEL Stem Cells

Injection of CD34$^+$ Green Fluorescent Protein Cells

Transduced mouse stem cells are washed and resuspended in phosphate-buffered saline and 1–2% fetal calf serum. These cells are counted and

$1–6 \times 10^6$ cells are tail vein-injected into mice with radioablated bone marrow. Blood samples can be obtained from each animal 4–6 weeks postinjection of GFP stem cells. To determine spleen colony-forming units, obtain fresh spleen samples from animals and freeze the samples on coverslips with O.C.T. compound; section at 10-μm thickness. View uncoverslipped specimens by FITC microscopy using a 488-nm FITC filter. Images can be captured either with standard F-mounted slide or print cameras or with fluorescence sensitive digital cameras [Dage-MTI (Michigan City, IN), Hamamatsu (Bridgewater, NJ), Princeton Instruments (Trenton, NJ), etc.].

In Vivo LEL Transduction of Murine Embryos

lacZ has been used extensively in the past for transduction of murine and chicken embryos.[24] Clearly, GFP provides advantages over such a histochemical marker system. Essentially, the *in vivo* transduction techniques will be the same but the analysis of fluorescent tissue will be conducted similarly to that discussed in the section on GFP retroviral transduction of tumors *in vivo*. Imaging of GFP in the whole animal will be similar to that done for EGFP transgenic mice.[15]

In Vivo Imaging of LEL-Transduced Tumors and Tissue

To excite and detect the fluorescent signal of GFP expressed in the tissues or tumor of a living animal it is necessary to use a modified exploratory endoscope. Two main components of the standard endoscope must be retrofitted with new hardware. First, a 480-nm filter must be constructed to filter the light coming from the xenon light source and a 510-nm filter must be fashioned to fit directly in front of the charge-coupled device (CCD) camera that captures the GFP fluorescence. The appropriate filters can be obtained from Chroma (Brattleboro, VT). A system such as this has been used to visualize adenoassociated GFP viral gene transfer and expression within the bronchial epithelium of rabbits.[16] Second, the CCD camera must have high sensitivity to capture the weak fluorescent GFP signal. It is possible to use an inexpensive black-and-white CCD camera to capture green fluorescence if the only objective is to measure light intensity. There are a number of cameras available that are rated between 0.1 and 0.01 lux, which is sufficient to image green fluorescence through a filter-modified endoscope. If color images are necessary, much more costly

[24] C. L. Cepko, C. P. Austin, C. Waish, E. F. Ryder, A. Halliday, and S. Feilds-Berry, *Cold Spring Harbor Symp. Quant. Biol.* **55,** 265 (1990).

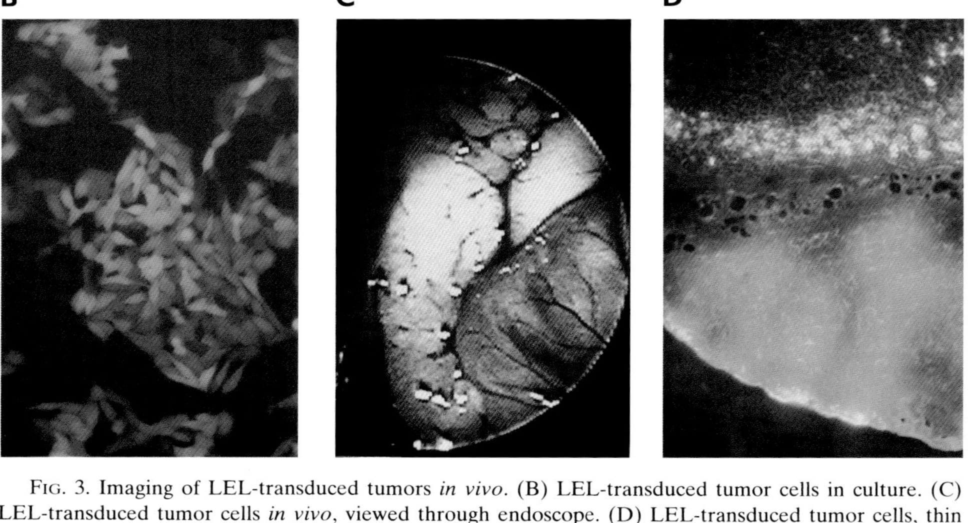

FIG. 3. Imaging of LEL-transduced tumors *in vivo*. (B) LEL-transduced tumor cells in culture. (C) LEL-transduced tumor cells *in vivo*, viewed through endoscope. (D) LEL-transduced tumor cells, thin sectioned and viewed with FITC microscope.

three-CCD chip slow-scan cooled cameras may be required to adequately detect *in vivo* green fluorescence through such a medical imaging device (Fig. 3).

Conclusion

The ability to transfer the powerful GFP gene into living cells and tissue via a retrovirus provides researchers with the distinct advantage of stable integration into the host genome, thus establishing a permanent vibrant marker. The fact that retrovirus tends to integrate only a single-gene copy, combined with the versatility and lack of toxicity of GFP, allow for stable and viable cells to be produced from GFP retroviral transductions. This marker system has distinct advantages that make it well suited for as many *in vitro* and *in vivo* experiments as the imagination will allow.

Acknowledgments

We thank Jeannine Malatesta, Ginger Dreifurst, and Julie Seiwert for technical assistance with flow cytometry, as well as Josh Dady and Feng Zhang for digital imaging GFP tumors.

Section IV

Mutants and Variants of Green Fluorescent Protein

[31] S147P Green Fluorescent Protein: A Less Thermosensitive Green Fluorescent Protein Variant

By YUKIO KIMATA, CHUN REN LIM, and KENJI KOHNO

Introduction

The efficiency with which newly synthesized green fluorescent protein (GFP) polypeptides mature into the fluorescent active form depends largely on the temperature at which the cells or organisms expressing the gene are cultured. As the culture temperature increases, the maturation rate is retarded. This temperature sensitivity exhibited by the GFP molecule causes some difficulty in the practical use of GFP. In this chapter we explain the basis of this problem and show how it may be addressed by use of a mutation, S147P, which alleviates the temperature sensitivity of the GFP molecule.

Maturation of Green Fluorescent Protein Molecule

The main modification responsible for GFP fluorescence is broadly believed to be the posttranslational cyclization and oxidation of three amino acid residues (–Ser-65–Tyr–Gly–) to form a chromophore within the primary structure of GFP (A. B. Cubitt *et al.,*[1] and references therein). The chromophore is surrounded by an 11-stranded β barrel and therefore completely shielded from bulk solvent.[2,3] In other words, newly synthesized GFP polypeptides need to mature properly into the chromophore structure before emitting fluorescence. It is notable that mature GFP is highly stable (A. B. Cubitt *et al.,* and references therein), retaining its fluorescence even when shifted to a temperature of 65°. On the other hand, when GFP is heterogeneously expressed as a reporter gene, one should consider that the intensity of its fluorescence depends not only on the stability of the mature fluorescent GFP molecule but also on the maturation efficiency of newly synthesized GFP polypeptides. Hence, the efficiency of maturation is an important consideration in understanding and using GFP.

[1] A. B. Cubitt, R. Heim, S. R. Adams, A. E. Boyd, L. A. Gross, and R. Y. Tsien, *Trends Biochem. Sci.* **20,** 448 (1995).
[2] M. Ormö, A. B. Cubitt, K. Kallio, L. A. Gross, R. Y. Tsien, and S. J. Remington, *Science* **273,** 1392 (1996).
[3] F. Yang, L. G. Moss, and G. N. Phillips, Jr., *Nature Biotechnol.* **14,** 1246 (1996).

METHODS IN ENZYMOLOGY, VOL. 302 0076-6879/99 $30.00

Temperature Sensitivity of Green Fluorescent Protein Maturation

The optimum temperature for growth of the yeast *Saccharomyces cerevisiae* spans a wide range, from 15 to 37°. Using a strain of *S. cerevisiae*, (SKEl; a derivative of S288C) we performed experiments to examine the relation between fluorescence intensity and culture temperature.[4] Yeast cells expressing wild-type GFP were cultured at various temperatures between 15 and 37°. Cell lysates were then prepared from these cultures, and their fluorescence intensity was measured with a spectrofluorometer [F4500 (Hitachi, Tokyo, Japan); photomultiplier, 900 V). We found that the fluorescence intensity of the 37° sample was only one-sixteenth that of the 15° sample. Concomitantly, immunoblot detection of GFP in the same lysates indicated that the 37° sample contained approximately one-fourth the amount of GFP as in the 15° sample. We surmise that this difference between the fluorescence intensity and the amount of GFP is due to the accumulation of the nonfluorescent form of the GFP molecule in the 37° sample. In other words, the ratio of fluorescent to nonfluorescent GFP decreases with increasing culture temperature. Together with our other data, this implies that at high temperatures such as 37°, the intracellular environment is unfavorable for the maturation of newly synthesized GFP polypeptides into the fluorescent form. Although we used a fusion protein consisting of GFP and nucleoplasmin in this experiment, similar results were obtained from cells expressing nonchimeric GFP, including S65T GFP, one of the most widely used GFP variants (described in [37] in this volume[5]). Furthermore, similar observations were consistently obtained in *Escherichia coli* and mammalian[6] cells.

The temperature-sensitive maturation rate of GFP molecules can lead to at least two problems. First, most mammalian cells are cultured at 37°, and therefore a high level of GFP expression is required to detect a significant fluorescent signal. Second, in experiments involving changes in culture temperature, such as when employing temperature-sensitive mutants, GFP fluorescence intensity is easily affected, and this makes comparison of results difficult and inaccurate.

Green Fluorescent Protein S147P Mutation

Low-fidelity polymerase chain reaction (PCR) amplification[7] of the GFP gene was used to generate a plasmid libary of GFP variants in *E. coli*. From

[4] C. R. Lim, Y. Kimata, M. Oka, K. Nomaguchi, and K. Kohno, *J. Biochem.* **118,** 13 (1995).
[5] S. Inouye, K. Umesono, and F. I. Tsuji, *Methods Enzymol.* **302,** [37], 1999 (this volume).
[6] H. Ogawa, S. Inouye, F. I. Tsuji, K. Yasuda, and K. Umesono, *Proc. Natl. Acad. Sci. U.S.A.* **92,** 11899 (1995).
[7] C. W. Dieffenbach and G. S. Dveksler, "PCR Primer: A Laboratory Manual," 1st Ed. Cold Spring Harbor Laboratory Press, Cold Spring Harbor, New York, 1995.

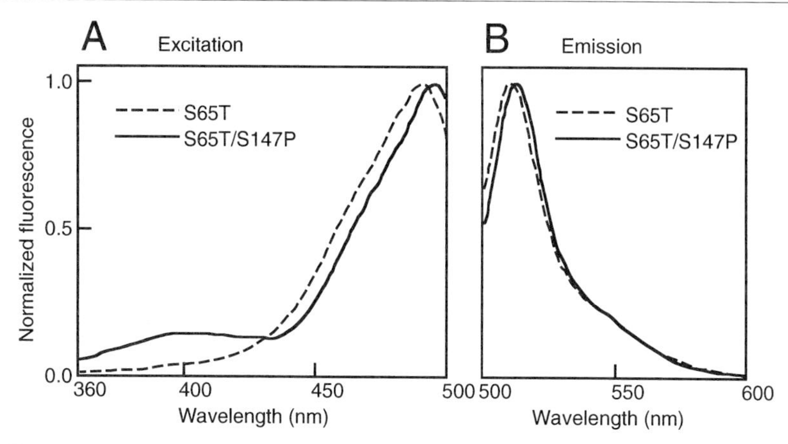

FIG. 1. Excitation and emission spectra of GFP variants. DNAs encoding S65T and S65T/S147P GFP were cloned into pGEX5X-2 (a bacterial expression vector for GST-tagged protein; Pharmacia) to obtain plasmids pGG1 and pGG2, respectively. GST-tagged GFP variants were individually expressed in *E. coli* cells containing these plasmids and purified by chromatography on immobilized glutathione (bulk GST purification module; Pharmacia). Shown here are the excitation (A) and emission (B) spectra of each sample when the emission is fixed at 510 nm (A) or the excitation is fixed at 480 nm. (B). All spectra were obtained with a spectrofluorometer [F4500 (Hitachi); photomultiplier, 900 V] and arbitrarily normalized to a maximum value of 1.0.

this library, the clone that emitted the strongest fluorescence at a culture temperature at 37° was selected. Sequencing of this GFP variant revealed a T-to-C transition at nucleotide position 439. This transition led to a change of serine at position 147 to proline (S147P) in the amino acid sequence. To evaluate the effects of the S147P mutation,[8] we combined it with S65T to create a novel GFP variant, S65T/S147P GFP. The excitation and emission spectra of the S65T/S147P and S65T GFP are shown in Fig. 1. While they share identical emission spectra with a peak at 512 nm, the S147P mutation has altered the excitation peak of the S65T GFP from 490 nm to the slightly longer wavelength of 496 nm. Nevertheless, no specific adjustment is required to detect the fluorescence of the S147P/S65T GFP, i.e., excitation and emission are similar to those of S65T GFP, and thus an authentic fluorescein isothiocyanate (FITC) filter set [e.g., Carl Zeiss (Thornwood, NY) filter set 09] can be used for detection of this variant under a fluorescence microscope. In addition, excitation by a 488-nm argon ion laser beam, typically used for fluorescence-activated cell sorting (FACS) and confocal microscopy, is practical. Furthermore, as long as the protein is abundant,

[8] Y. Kimata, M. Iwaki, C. R. Lim, and K. Kohno, *Biochim. Biophys. Res. Commun.* **232,** 69 (1997).

TABLE I
FLUORESCENCE INTENSITY ASSAY OF GREEN FLUORESCENT PROTEIN VARIANTS EXPRESSED IN *Escherichia coli*[a]

Expression plasmid	Mutations in GFP	Culturing temperature (°C)	Cell lysates: Relative fluorescence intensity[b]	Cell lysates: Relative GFP conc.[c]	Specific fluorescence activity (SFA)[d]
pGG1[e]	S65T	37	164	344	0.48
pGG2[e]	S65T/S147P	37	751	436	1.72
pGG1	S65T	23	2744	556	4.94
pGG2	S65T/S147P	23	2239	572	3.91
pGEX5X-2[e]	No GFP	37	13	<3	

[a] *Escherichia coli* XL1-blue cells containing expression plasmids for the GFP variants were cultured for 4 hr in the presence of isopropyl β-D-thiogalactopyranoside (IPTG) at the indicated temperatures.

[b] Measured with excitation at 488 nm and emission at 510 nm with a spectrofluorometer [F4500 (Hitachi); photomultiplier, 900 V].

[c] Estimated from quantitative immunoblot detection of GFP.

[d] Obtained by dividing the relative fluorescence intensity by the relative GFP concentration in each cell lysate.

[e] See Fig. 1.

excitation in the UV range is possible because the S65T/S147P GFP has a comparatively stronger excitation in the UV range than does S65T GFP.

Specific Fluorescence Activity

To examine the increased maturation efficiency of the GFP variant carrying the S147P mutation, we cultured *E. coli* clones expressing either S65T or S65T/S147P GFP at 23 or 37°. Relative fluorescence intensity of the lysates was measured with excitation at 488 nm and emission at 510 nm. To quantitate the intrinsic fluorescence of these variants, we introduced a new unit, specific fluorescence activity (SFA), which is determined by dividing the relative fluorescence intensity by the relative GFP concentration in each cell lysate (Table I). When these clones were cultured at 37°, the SFA of cells expressing the S65T/S147P GFP variant was 3.6-fold higher than that of cells expressing the S65T GFP variant. However, at a culture temperature of 23°, the SFA of cells expressing the S65T GFP variant was slightly higher than that of cells expressing the S65T/S147P GFP variant. Comparison of the same GFP variant at different temperatures revealed that the SFA of cells expressing S65T GFP at 37° was only one-tenth of

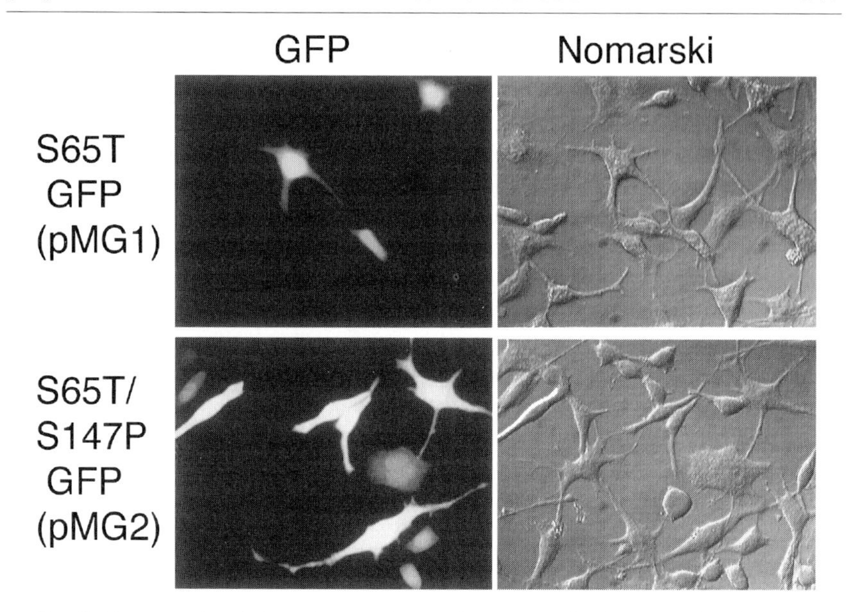

FIG. 2. Fluorescent emission of mammalian cells transfected with expression plasmids for GFP variants. DNAs encoding S65T and S65T/S147P GFPs were cloned into mammalian expression vector pCAGGS[9] to obtain plasmids pMG1 and pMG2, respectively. After transfecting 10^5 mouse L cells with 2 μg of DNA (pMG1 and pMG2), cells were further incubated at 37° for 48 hr. These cells were then illuminated by blue light (450–490 nm) and observed through a low-fluorescence filter (passing wavelengths >520 nm). [A fluorescence microscope, Axiophoto (Carl Zeiss), with filter set 09 was used.] To obtain fluorescent images, a CCD video camera (ZVS-47DE; Carl Zeiss) was used (exposure time, 1 sec).

that of cells at 23°. In contrast, cells expressing S65T/S147P GFP at 37° maintained an SFA level of approximately 50% relative to those cultured at 23°. These results clearly show that S65T/S147P GFP is stably expressed with strong fluorescence throughout a wide range of culture temperatures.

To investigate if the preceding observations could be reproduced in a mammalian cell system, we transiently transfected mouse L cells with the plasmid pMG1, which expresses the S65T GFP variant, or with pMG2, which expresses the S65T/S147P GFP variant. After 2 days of incubation at 37°, cells transfected with pMG2 exhibited a brighter fluorescent signal than those transfected with pMG1 (Fig. 2).[9]

[9] H. Niwa, K. I. Yamamura, and J. Miyazaki, *Gene* **108,** 193 (1991).

Conclusion

Our results show that the S147P mutation alters some properties of GFP, including increased maturation efficiency at 37° and a 5-nm shift in the peak of excitation. Three-dimensional structural analysis of GFP[2,3] shows that the Tyr-145 and His-148 amino acid residues near Ser-147 are in close proximity to the chromophore, and it is likely that the environment around the chromophore is significantly altered by the S147P mutation.

This chapter not only presents a new GFP variant, but also describes how its novel mutation may be useful to enhance the fluorescence properties of other GFP variants at 37°. It is hoped that the weak GFP signals encountered when GFP proteins are expressed in cell line systems incubated at comparatively high temperatures can be alleviated by use of this S147P mutation.

Acknowledgments

We are grateful to M. Iwaki and R. Ando for their experimental contributions. This research has been supported by CREST of Japan Science and Technology Corporation (JST) and Grants-in-Aid from the Ministry of Education, Science, Sports, and Culture of Japan, and has been partly supported by the Sasakawa Scientific Research Grant from The Japan Science Society.

[32] Spectral Variants of Green Fluorescent Protein

By GOTTFRIED J. PALM and ALEXANDER WLODAWER

Introduction

The use of green fluorescent protein (GFP)[1] for many applications in biological sciences strongly depends on its spectral properties. GFP originates from the jellyfish *Aequorea victoria,* in which its fluorescence serves to convert the blue light produced by aequorin to green light. Purified

[1] Abbreviations: GFP, green fluorescent protein; wt, wild type; Ser-65, etc., serine in position 65; Y66H, etc., Tyr-66 to His-66 mutation or protein with this mutation; Ala1b, insertion of an alanine after residue 1; CA, etc., atom names used by the Protein Data Bank (PDB); rms, root mean square.

0076-6879/99 $30.00

protein, which retains fluorescence, has been the object of intense investigations since the 1970s. Denaturation by harsh treatment, however, can abolish fluorescence and change the absorption spectrum of GFP. The spectrum of the native protein is also affected by many other parameters, such as mutations, protein concentration, ionic strength, temperature, and pH. However, the stability of GFP in a wide pH range (pH 5–12), at high temperature (T_m = 78°),[2] in chaotropic reagents (8 *M* urea), and against proteolysis makes it attractive for practical use.

The cloning and sequencing of GFP in 1992 opened new possibilities for applications and analyses.[3] Only 2 years later, Chalfie *et al.*[4] proved that expression in prokaryotes and eurkaryotes could lead to the formation of mature protein. This finding implicated autocatalytic chromophore formation, which allows reconstitution of GFP in virtually all organisms. The chemical nature of the chromophore was determined to be 5-(*p*-hydroxybenzylidene)-3,5-dihydro-4*H*-imidazol-4-one, within the peptide chain.[5,6] The chromophore formation is now believed to proceed in three steps: Cyclization and dehydration of the main chain atoms of Ser-65–Tyr-66–Gly-67 to an imidazolone ring after the protein is folded, and oxidation of the CA–CB bond to residue 66 to acquire the extended conjugated system necessary for long-wavelength absorption (Fig. 1).[7,8] Although the first applications used wild-type GFP (wt GFP),[9] the limitations of such constructs soon became obvious: Expression levels of mature protein are limited by poor folding at higher temperatures, extinction coefficients at commonly used excitation wavelengths (especially 488 nm from an argon laser) are low, and different constructs cannot be distinguished optically.

Considerable progress has been made by mutational approaches (see Refs. 10 and 11 for reviews). Expression has been optimized by mutations

[2] W. W. Ward, H. J. Prentice, A. F. Roth, C. W. Cody, and S. C. Reeves, *Photochem. Photobiol.* **35,** 803 (1982).

[3] D. C. Prasher, V. K. Eckenrode, W. W. Ward, F. G. Prendergast, and M. J. Cormier, *Gene* **111,** 229 (1992).

[4] M. Chalfie, Y. Tu, G. Euskirchen, W. W. Ward, and D. C. Prasher, *Science* **263,** 802 (1994).

[5] O. Shimomura, *FEBS Lett.* **104,** 220 (1979).

[6] C. W. Cody, D. C. Prasher, W. M. Westler, F. G. Prendergast, and W. W. Ward, *Biochemistry* **32,** 1212 (1993).

[7] R. Heim, D. C. Prasher, and R. Y. Tsien, *Proc. Natl. Acad. Sci. U.S.A.* **91,** 12501 (1994).

[8] B. G. Reid and G. C. Flynn, *Biochemistry* **36,** 6786 (1997).

[9] A. B. Cubitt, R. Heim, S. R. Adams, A. E. Boyd, L. A. Gross, and R. Y. Tsien, *Trends Biochem. Sci.* **20,** 448 (1995).

[10] R. Tsien and D. Prasher, *in* "Green Fluorescent Protein" (M. Chalfie and S. R. Kain, eds.), pp. 97–118. In press (1999).

[11] R. Y. Tsien, *Annu. Rev. Biochem.* **67,** 509 (1998).

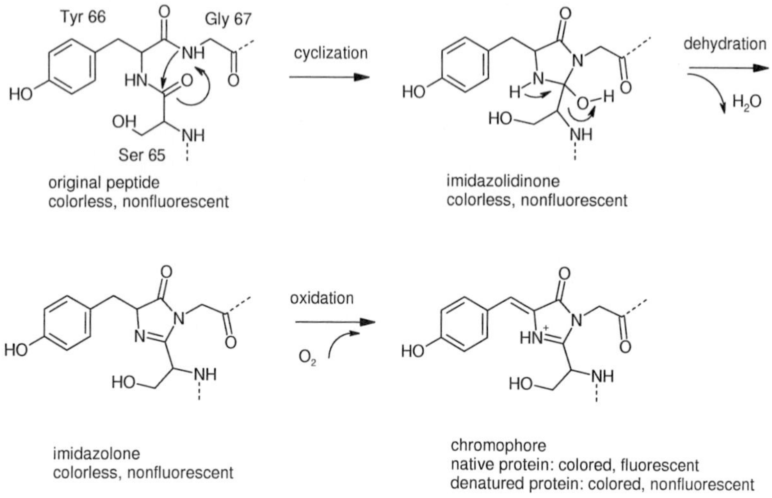

FIG. 1. Steps in the formation of the chromophore.

that change the promoter, ribosome binding site,[12] or codon usage[13–15]; eliminate splicing sites[16]; or enhance folding. Even though such procedures can drastically increase the overall fluorescence by GFP per cell, they are not discussed here, as the spectra of purified GFP remain the same in all these cases. Numerous mutations in and around the chromophore, however, do change the spectra. Taken together, these advances have enabled the use of GFP for detection of promoter activity and for localization of fusion proteins. Furthermore, new applications such as simultaneous localization of two fusion proteins[17,18] and fluorescence resonance energy transfer (FRET) for detection of Ca^{2+} concentration[19,20] or proteolytic

[12] M. Kozak, *J. Cell Biol.* **108,** 229 (1989).

[13] J. P. Levy, R. R. Muldoon, S. Zolothukin, and C. J. J. Link, *Nature Biotechnol.* **14,** 610 (1996).

[14] A. Crameri, E. A. Whitehorn, E. Tate, and W. P. C. Stemmer, *Nature Biotechnol.* **14,** 315 (1996).

[15] S. Zolothukin, M. Potter, W. Hauswirth, J. Guy, and N. Muzyczka, *J. Virol.* **70,** 4646 (1996).

[16] J. Haseloff and B. Amos, *Trends Genet.* **11,** 328 (1995).

[17] R. Heim and R. Y. Tsien, *Curr. Biol.* **6,** 178 (1996).

[18] T. T. Yang, S. R. Kain, P. Kitts, A. Kondepudi, M. M. Yang, and D. C. Youvan, *Gene* **173,** 19 (1996).

[19] V. A. Romoser, P. H. Hinkle, and A. Persechini, *J. Biol. Chem.* **272,** 13270 (1997).

[20] A. Miyawaki, J. Llopis, R. Heim, J. M. McCaffery, J. A. Adams, M. Ikura, and R. Y. Tsien, *Nature (London)* **388,** 882 (1997).

activity[17,21] have become possible through the use of GFP mutants of different colors.

Spectrum of Wild-Type Green Fluorescent Protein

Isolated *Aequorea* GFP is excited at ~395 nm (peak A) and 475 nm (peak B, minor peak) and emits light with a maximum at 507 nm.[22] Recombinant GFP has identical properties.[23] The isolated proteolytic fragment containing the chromophore (chromopeptide), as well as the denatured protein, absorbs at 382 nm at low pH and at 445 nm at high pH.[22] The pK_a for this transition is 8.1 (Fig. 2 and Ref. 23) and can be assigned to the phenol group of the chromophore originating from Tyr-66. Accordingly, in the native protein, peak A is attributed to GFP with a neutral phenol group and peak B to GFP with a phenolate group. To determine the protonation state of the other ring system of the chromophore, the imidazolone, it is more straightforward to analyze Y66F mutants, which do not have the complication of an ionizable side chain. For these mutants, the denatured protein exhibits spectral changes at extreme pH values, implicating pK_a values of about 0 and 13.5 (Fig. 2). Because the chromophore of Y66H mutants has only two protonatable groups, the carbonyl oxygen and the nitrogen (from residue 66) of the imidazolone, it is singly protonated at neutral pH. The neutral net charge of the Tyr-66 chromophore at basic pH, compared with the positive net charge of the Phe-66 chromophore, shifts the upper pK_a out of the range that is measurable in aqueous solution (ca. 14–15). Assuming there is no major change in pK_a by the protein environment, we predict that native GFP contains protonated imidazolone (Fig. 3). Theoretical calculations of the absorption and fluorescence spectra strongly suggest that, in all analyzed mutants, the imidazolone ring is indeed protonated on the secondary nitrogen.[24] Even though wt GFP is excited at two different wavelengths, only one emission peak is observed under normal conditions. Chattoraj *et al.*[25] and Lossau *et al.*[26] showed convincingly by spectroscopic methods that on excitation of the A form (i.e., the form absorbing at peak A, 395 nm), GFP undergoes proton transfer and a confor-

[21] R. D. Mitra, C. M. Silva, and D. C. Youvan, *Gene* **173,** 13 (1996).

[22] W. W. Ward, C. W. Cody, R. C. Hard, and M. J. Cormier, *Photochem. Photobiol.* **31,** 611 (1980).

[23] S. Inouye and F. I. Tsuji, *FEBS Lett.* **341,** 277 (1994).

[24] A. Voityuk, M. E. Michel-Beyerle, and N. Roesch, *Chem. Phys. Lett.* **272,** 162 (1997).

[25] M. Chattoraj, B. A. King, G. U. Bublitz, and S. G. Boxer, *Proc. Natl. Acad. Sci. U.S.A.* **93,** 8362 (1996).

[26] H. Lossau, A. Kummer, R. Heinecke, F. Pollingerdammer, C. Compa, G. Bieser, T. Jonsson, C. M. Silva, M. M. Yang, D. C. Yuovan, and M. E. Michel-Beyerle, *Chem. Phys.* **213,** 1 (1996).

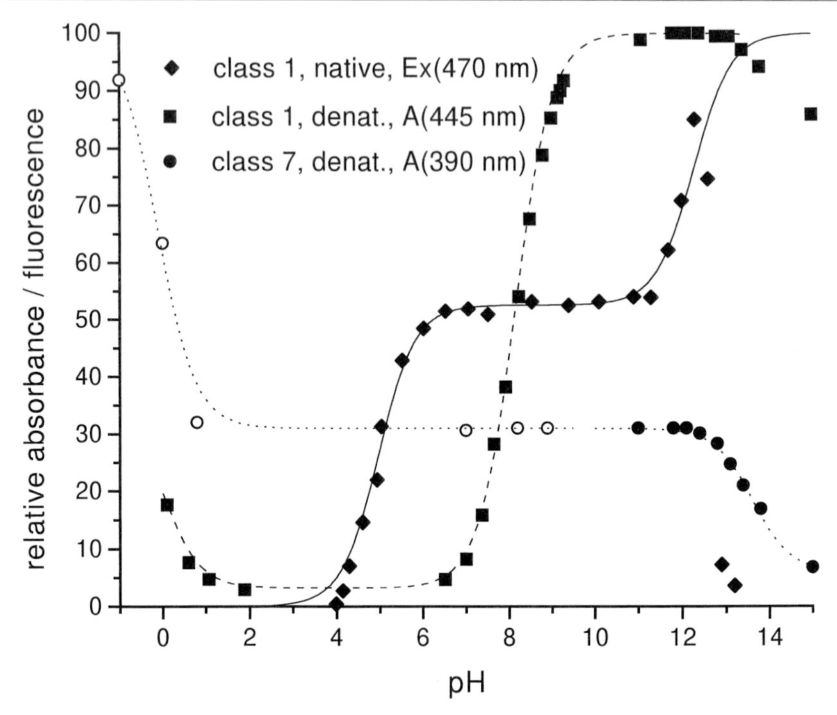

FIG. 2. pH dependence of spectra of GFP. Fluorescence (excitation) of native GFP for a class 1 mutant (—) and absorbance of denatured GFP for class 1 (– – –) and class 7 (· · ·) mutants. Extreme pH values (>12, <2) were calculated from the NaOH and HCl concentrations, respectively. The mutants used were Ala1b/F64L/Q80R/I167T/K238N (class 1) for native protein, and Ala1b/F64L/V68L/Q69L/Q80R/I167T/K238N (class 1) and Ala1b/F64L/Y66F/Q80R/I167T/K238N (class 7) for denatured protein. The spectral change at low pH (<1) for the class 7 mutant was corrected for chromophore degradation (○).

mational change and then emits at 507 nm. Thus, the excited A form becomes acidic enough to lose a proton and achieve the same conformation as the excited B form.

Structure of Wild-Type Green Fluorescent Protein

A number of wt and mutant forms of GFP have been crystallized and their structures have been determined.[27–32] The quaternary structure is

[27] M. Ormö, A. B. Cubitt, K. Kallio, L. A. Gross, R. Y. Tsien, and S. J. Remington, *Science* **273,** 1392 (1996).

[28] F. Yang, L. G. Moss, and G. N. Phillips, Jr., *Nature Biotechnol.* **14,** 1246 (1996).

[29] K. Brejc, T. K. Sixma, P. A. Kitts, S. R. Kain, R. Y. Tsien, M. Ormö, and S. J. Remington, *Proc. Natl. Acad. Sci. U.S.A.* **94,** 2306 (1997).

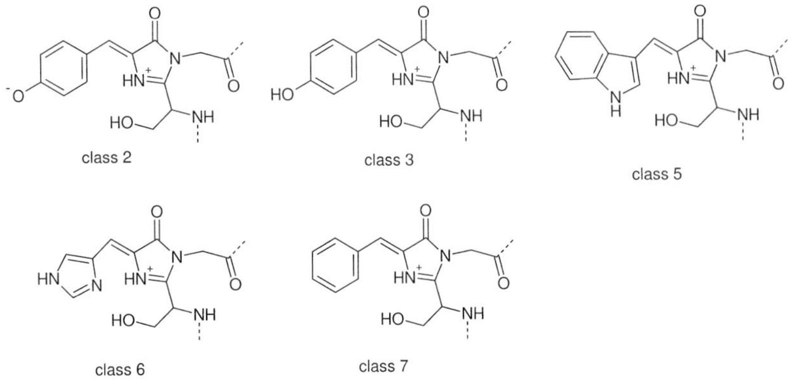

FIG. 3. Chemical formulas and assumed protonation state of the chormophore in various classes of mutants. Class 1 mutants are a mixture of class 2 and 3 mutants; in class 4 mutants, an additional aromatic group (side chain of residue 203) π-stacks with the side chain ring (J. Remington, personal communication, 1998); structures of class 8 and 9 mutants have not been solved yet.

either a monomer or a dimer (the effect of ionic strength on dimerization is discussed in Ionic Strength, below). The tertiary structure is similar in all crystal structures [pairwise root mean square (rms) deviations of CA < 0.4 Å]. The fold of GFP is a very regular 11-stranded β barrel (named β-can[28]), with a central helix passing through the center of the barrel and containing the chromophore. Several loops and two short α helices cap the barrel. Unexpectedly, many polar groups, including several water molecules, are located around the buried chromophore. Nevertheless, the chromophore environment is rigid and solvent inaccessible, and thus prevents both the quenching of fluorescence by small molecules and the dissipation of vibrational energy. The conformational freedom in the core of the protein, necessary for conformational changes during chromophore formation, might be provided by the presence of these water molecules in the vicinity of the chromophore.

The structure of the chromophore, as determined by analysis of the chromopeptide,[5] was confirmed by the X-ray structure (Fig. 4). The aromatic side chain of residue 66 and the imidazolone ring form a planar,

[30] G. J. Palm, A. Zdanov, G. A. Gaitanaris, R. Stauber, G. N. Pavlakis, and A. Wlodawer, *Nature Struct. Biol.* **4,** 361 (1997).

[31] R. M. Wachter, B. A. King, R. Heim, K. Kallio, R. Y. Tsien, S. G. Boxer, and S. J. Remington, *Biochemistry* **36,** 9759 (1997).

[32] C. K. Wu, Z. J. Liu, J. P. Rose, S. Inouye, F. Tsuji, R. Y. Tsien, S. J. Remington, and B. C. Wang, *in* "Bioluminescence and Chemoluminescence" (J. W. Hastings, L. J. Kricka, and P. E. Stanley, eds.), pp. 399–402. Wiley, Chichester, 1997.

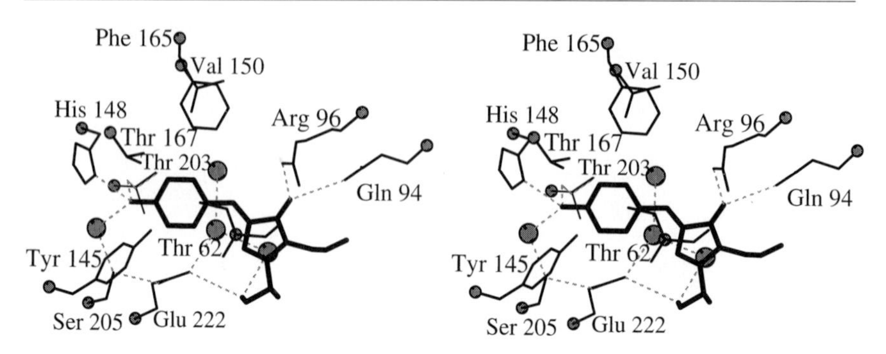

FIG. 4. Stereoview of the chromophore region taken from PDB entry 1EMC. Thr-203 has two conformations in the mutant Ala1b/F64L/Q80R/I167T/K238N (class 1), with the black conformation occurring in the A form and the gray conformation occurring in the B form. The side chain of Ile-167 in wt GFP superimposes well on the atoms of Thr-167 but has an additional atom (CD) pointing toward the chromophore.

conjugated system. The imidazolone oxygen is hydrogen bonded to Arg-96 and Gln-94. In the A form, the hydroxyl of the phenol group is coordinated by His-148 ND1 and by a water molecule. This water molecule is part of a larger chain, consisting of a phenol group, a water molecule, Ser-205, Glu-222 (deprotonated, accepting three hydrogen bonds from Ser-65, Ser-205, and a water molecule), and Ser-65. In the minor conformation (B form), the anionic phenolate is coordinated by a third ligand, Thr-203, which has rotated its side chain compared with the major conformation (gray in Fig. 4).[29,30] Spectroscopic studies predicted a conformational change,[25] which might correspond to the rotation of the threonine side chain. In the B form, Glu-222 makes only two hydrogen bonds (Ser-65 and a water molecule) and is neutral. The two states (A and B forms), related by proton transfer in the interior of the protein, explain why the phenol group is not titratable without denaturing GFP. The involvement of a conformational change (via the Thr-203 side chain) slows the interconversion.

Mutants of Green Fluorescent Protein

Analyses of wt GFP spectra and structures have identified the molecular species responsible for absorption at peaks A and B, as well as for emission at 507 nm. Moreover, the proton transfer and conformational change predicted from spectroscopic results have been explained structurally (see above). In addition to wt GFP, a large number of mutants have been made, many of them leading to new spectral types (Table I, Fig 5). Tsien[11] has

TABLE I
SPECTRAL PROPERTIES OF GREEN FLUORESCENT PROTEIN MUTANTS

Mutation	In mutant (common name)	Class	λ_{Ex} (ε)[a]	λ_{Em} (q)[b]	Ref.
wt	(wt, BioGreen)	1	395 (21.0) 475 (7.1)	508 (0.77)	17
	(Photoactivated wt)	9	525	600	39
Ala1b	Ala1b/F64L/Q80R (sg12)	1	396 (18.3) 473 (4.0)	508 (0.85)	30
W57L	Ala1b/W57L/F64L/Y66H/Q80R (sg84)	—	Not folding	Not folding	*c*
F64L	Ala1b/F64L/Q80R (sg12)	1	396 (18.3) 473 (4.0)	508 (0.85)	30
F64M	F64M/Y66H/V68I (BFP5)	6	385	450	21
S65T	S65T (GFP-S65T)	2	489 (39.2–55.0)	509 (0.64–0.66)	*d*, 17
	F64L/S65T (GFP Mut1)	2	488	507	36
	F64L/S65T (EGFP)	2	489 (53.0)	509 (0.60)	*d*
S65G	F64M/S65G/Q69L (RSGFP4, BioYellow)	2	490	505	*e*
S65C	Ala1b/F64L/S65C/Q80R/I167T/K238N (sg25)	2	474 (26.2)	506 (0.76)	30
	S65C	2	479	507	35
S65L	S65L	2	484		35
S65A	S65A	2	471	503	35
S65R	S65R	—	No fluorescence detected	No fluorescence detected	35
S65N	S65N	—	No fluorescence detected	No fluorescence detected	35
S65D	S65D	—	No fluorescence detected	No fluorescence detected	35
S65F	S65F	—	No fluorescence detected	No fluorescence detected	35
S65W	S65W	—	No fluorescence detected	No fluorescence detected	35
Y66W	Y66W/N146I/M153T/V163A/N212K (W7)	5	433 (18.0) 453 (17.1)	475 (0.67) 501	17
	Ala1b/F64L/Y66W/Q80R/I167T/K238N (sg122)	5	444 (18.4)	487 (0.05)	*c*
Y66H	Y66H (BFP, P4, BioBlue)	6	384 (21.0)	448 (0.24)	*f*
	Y66H/Y145F	6	382 (22.3)	446 (0.30)	*f*
	Ala1b/F64L/Y66H/Q80R (sg42)	6	384 (15.6)	450 (0.24)	30
	Ala1b/F64L/Y66H/Q80R/I167T/K238N (sg124)	6	381 (19.8)	449 (0.11)	*c*
	F64L/S65T/Y66H/Y145F (EBFP)	6	380–383 (26.3–31.0)	440–447 (0.17–0.26)	*d*, *f*
Y66F	Ala1b/F64L/Y66F/Q80R/I167T/K238N (sg123)	7	356 (9.8)	428 (0.013)	*c*

(continued)

TABLE I (*continued*)

Mutation	In mutant (common name)	Class	λ_{Ex} (ε)[a]	λ_{Em} (q)[b]	Ref.
Y66L	Ala1b/F64L/Y66L/Q80R/I167T/K238N (sg86)	8	413 (0.6)	461 (0.015)	*c*
V68I	F64M/Y66H/V68I (BFP5)	6	385	450	21
V68L	F64L/V68L/S72A (GFP Mut2)	2	481	507	36
Q69K	S65G/V68L/Q69K/S72A/T203Y (10C Q69K)	4	516 (62.0)	529 (0.71)	*f*
Q69L	F64M/S65G/Q69L (RSGFP4, BioYellow)	2	490	505	*e*
S72A	F64L/S72A (GFP Mut3)	2	501	511	36
	F64L/V68L/S72A (GFP Mut2)	2	481	507	36
K79R	V1b/S65G/S72Q/K79R/N149K/T203Y/H231L (Topaz)	4	514 (94.5)	527 (0.60)	*f*
Q80R	Ala1b/F64L/Q80R (sg12)	1	396 (18.3) 473 (4.0)	508 (0.85)	30
R96C	S65T/R96C	2	472	503	*f*
F99S	F99S/M153T/V163A (α GFP, cycle 3 GFP, GFPuv)	1	397 (30.0) 475 (6.5)	506 (0.79)	*d*
Y145F	Y66H/Y145F (P4-3)	6	382 (22.3)	446 (0.30)	*f*
Y145H	Y66W/I123V/Y145H/H148R/M153T/V163A/N212K (W2)	5	432 (10.0) 453 (9.6)	480 (0.72)	17
N146I	F64L/S65T/Y66W/N146I/M153T/V163A (W1B, ECFP)	5	434 (32.5) 452	476 (0.40) 505	*f*
S147P	S147P	1	Increased temperature tolerance	Increased temperature tolerance	33
H148R	Y66W/I123V/Y145H/H148R/M153T/V163A/N212K (W2)	5	432 (10.0) 453 (9.6)	480 (0.72)	17
N149K	V1b/F64L/S65T/S72A/N149K/M153T/I167T/H231L (Emerald)	2	487 (57.5)	509 (0.68)	*f*
Y151L	Ala1b/F64L/S65C/Q80R/Y151L/I167T/K238N (sg100)	2	474	*g*	*c*
M153A	S65T/M153A/K238E (P4-1)	1	396 (8.6) 504 (14.5)	514 (0.54)	17
M153T	F99S/M153T/V163A (α GFP, cycle 3 GFP, GFPuv)	1	397 (30.0) 475 (6.5)	506 (0.79)	*d*
V163A	V163A	1	Increased temperature tolerance	Increased temperature tolerance	34
	F64L/S65T/V163A	2	488 (42.0)	511 (0.58)	*f*

TABLE I (*continued*)

Mutation	In mutant (common name)	Class	λ_{Ex} (ε)[a]	λ_{Em} (q)[b]	Ref.
	Ala1b/F64L/Y66H/Q80R/ V163A (sg50)	6	384 (16.5)	450 (0.25)	30
I167T	I167T	1	396 (minor)	507	7
			471 (major)	502	
I167V	I167V	1	396 (minor)	507	7
			471 (major)	502	
Y200L	Ala1b/F64L/Q80R/I167T/ Y200L/K238N (sg43)	1	396 (9.7)	507 (0.96)	c
			469 (12.4)		
S202F	S202F/T203I (H9)	3	399 (20.0)	511 (0.60)	f
T203I	T203I	3	400	512	h
	Ala1b/F64L/S65C/Q80R/ I167T/T203I/K238N (sg115)	1	396 (14.8)	513 (0.56)	c
			506 (16.7)		
T203H	S65T/T203H (5B, 9B)	4	512 (19.4)	524	27
T203Y	S65T/T203Y (6C)	4	513 (14.5)	525	27
	F64L/S65G/S72A/T203Y (10B)	4	513 (30.8)	525	27
	S65G/V68L/S72A/T203Y (10C, EYFP)	4	513 (36.5)	527 (0.63)	27, 41
			514 (83.4)	527 (0.61)	
T203F	S65G/S72A/T203F	4	405	455	37, 41
			512 (66.5)	522 (0.70)	
T203W	S65G/S72A/T203W (11)	4	502 (33.0)	512	27
E222G	E222G	2	481	506	h
	Ala1b/F64L/S65C/Q80R/ I167T/E222G/K238N (sg104)	2	484 (14.5)	g	c
H231L	V1b/S72A/Y145F/T203I/ H231L (Sapphire, H9-40)	3	399 (29.0)	511 (0.64)	f
K238N	Ala1b/F64L/Q80R/I167T/ K238N (sg11)	1	395 (11.5)		c
			468 (12.7)	508 (0.85)	
K238E	S65T/M153A/K238E (P4-1)	2	504 (14.5)	514 (0.54)	17

[a] Wavelength (nm) of excitation peak(s) and extinction coefficient (M^{-1} cm^{-1}).
[b] Wavelength (nm) of emission peak and quantum yield. Excitation at the wavelength shown in the λ_{Ex} column.
[c] G.J.P., unpublished results (1998).
[d] G. H. Patterson, S. M. Knobel, W. D. Sharif, S. R. Kain, and D. W. Piston, *Biophys. J.* **73,** 2782 (1997).
[e] S. Delagrave, R. E. Hawtin, C. M. Silva, M. M. Yang, and D. C. Youvan, *Bio/Technology* **13,** 151 (1995).
[f] A. B. Cubitt, R. Heim, and L. A. Woollenweber, *Methods Cell Biol.* in press (1999).
[g] Not determined.
[h] T. Ehrig, D. J. O'Kane, and F. G. Prendergast, *FEBS Lett.* **367,** 163 (1995).

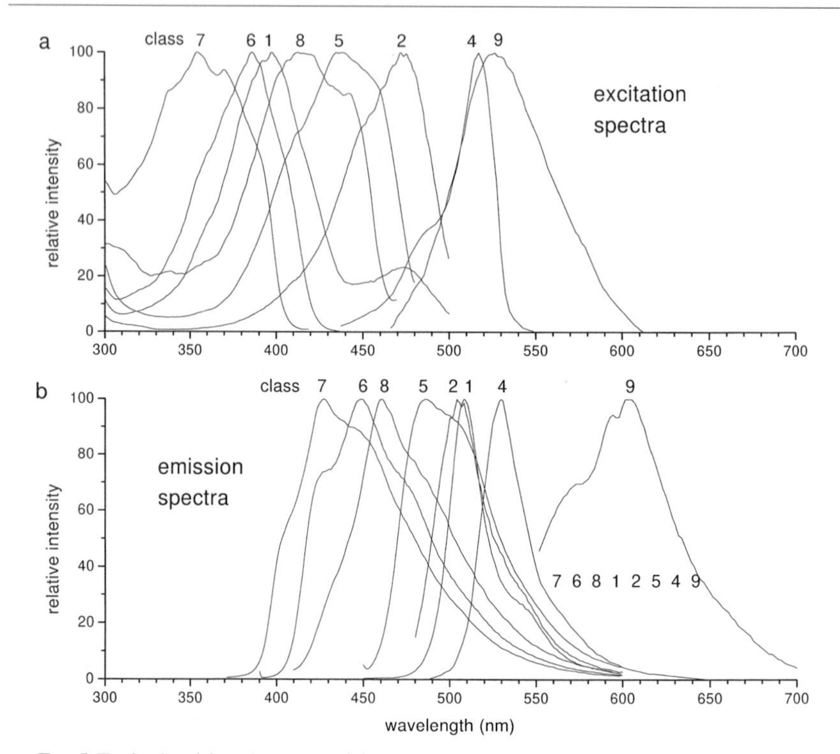

FIG. 5. Excitation (a) and emission (b) spectra of various mutants representing most classes. For class 3 mutants (not shown) the spectra resemble those of class 1 mutants, except for the missing peak B in the excitation spectrum. The mutants used as examples for the different classes are as follows: class 1, Ala1b/F64L/Q80R; class 2, Ala1b/F64L/S65C/Q80R/I167T/K238N; class 4, S65T/T203Y; class 5, Ala1b/F64L/Y66W/Q80R/I167T/K238N; class 6, Ala1b/F64L/Y66H/Q80R/I167/K238N; class 7, Ala1b/F64L/Y66F/Q80R/I167T/K238N; class 8, Ala1b/F64L/Y66L/Q80R/I167T/K238N; class 9, photoactivated purified wt GFP.

grouped the mutants into seven classes with similar excitation and emission spectra. The desired effects of new spectra—high extinction coefficients, good quantum yields, and efficient folding—could often be achieved only by multiple mutations. In these complex cases, it is not always obvious which mutation is directly responsible for specific effects.

The single mutations F64L,[30] S147P,[33] and V163A[34] increased the amount of expressed mature protein. F64L and V163A were shown not to induce any structural changes[30] and displayed wt spectra. This is a good

[33] Y. Kimata, M. Iwaki, C. R. Lim, and K. Kohno, *Biochem. Biophys. Res. Commun.* **232,** 69 (1997).

[34] K. R. Siemering, R. Golbig, R. Sever, and J. Haseloff, *Curr. Biol.* **6,** 1653 (1996).

indication that their folding is improved compared with that of wt GFP. In the context of multiple mutations, F64C/M/G/V, S72A/P, F99S, M153A/T, and S175G also appear to be responsible for better folding. These "folding" mutations are not further discussed here. The spectral classes of GFP mutants are described in more detail, together with single mutations, if they can be associated with a class

Class 1: Tyr-66, Neutral and Anionic Phenol Group

Wild-type GFP falls into this class and has been discussed above as a reference for all of the classes. Some mutations do favor either the anionic or the neutral phenol group (see classes 2 and 3), but do not tip the balance completely toward one form in a specific mutation background. Combinations can also synergistically change peak B. Ala1b/F64L/S65T/Q80R/V163A/T203I/K238N, for instance, has class 2 (S65T) and class 3 (T203I) mutations, giving rise to two absorption peaks at 396 and 507 nm. Peak B is further redshifted than what would be expected to arise from any of the single mutations.

Class 2: Tyr-66, Anionic Phenol Group

The A form with a deprotonated phenol group is responsible for the 475-nm absorption peak. This peak intrinsically has a higher extinction coefficient, making this the most successful class for practical applications. Mutations that stabilize the phenolate give enhanced or exclusive long-wavelength absorption. The hydrophobic part of Ile-167 (atom CD1) makes an unfavorable short contact to the phenol hydroxyl group [3.5 Å in PDB entry 1EMM (Ala1b/F64L/Q80R)]. With a negative charge on the phenolate, this distance increases [4.1 Å in PDB entry 1EMA (S65T)]. A larger separation can also be achieved by mutation to a smaller residue [4.9 Å from the Thr-167 (atom CG2) to the phenolate oxygen in PDB entry 1EME (Ala1b/F64L/Q80R/I167T/K238N)]. Because the protonation state of the phenol group is linked to that of Glu-222, stabilization of the phenolate can also be achieved by destabilizing the glutamate. E222G is a rational mutation to this end. Because no negative charge can reside on a glycine, E222G absorbs only at peak B. Ser-65 also has been shown to be critical for stabilizing the negative charge on Glu-222.[35] Only serine can make a hydrogen bond to the glutamate, such that two more hydrogen bonds can be formed (to Ser-205 and a water molecule). Any other residue fails to, either by not being able to form a hydrogen bond (Gly, Ala, Leu, and Cys), by forcing the carboxylate into a new conformation (no hydrogen bond to

[35] R. Heim, A. B. Cubitt, and R. Y. Tsien, *Nature* (*London*) **373,** 663 (1995).

Ser-205 in S65T), or by inhibiting chromophore formation (or folding) completely (Arg, Asn, Asp, Phe, and Trp). Even though the mutations S72A[36] and M153A[17] are well separated from the chromophore, each also leads to class 2 mutants. Most of the class 2 mutants also exhibit a significant redshift of peak B from 475 to ~490 nm, and in extreme cases up to 507 nm (see class 1). Peak B is thus not only more redshifted in the folded protein compared with the denatured protein, but also much more susceptible to further variations by such parameters as protein concentration, ionic strength, and mutations.

Class 3: Tyr-66, Neutral Phenol Group

A chromophore with a neutral phenol group gives rise to 395-nm absorption (peak A). Mutations that do not stabilize the anionic phenolate favor this A form. In T203I, one of the important ligands of the phenolate is missing and, thus, the phenol does not deprotonate easily. Compared with all other classes, these mutants are special in that their fluorescence process involves a change of the chromophore charge by proton transfer. This transfer implies energy loss and thereby results in the largest Stokes shift for all classes (the difference between the excitation and emission maxima is 110 nm).

Class 4: Tyr-66, Aromatic Residue 203

On the basis of structural considerations, Ormö *et al.*[27] have engineered a new class of mutants. An aromatic side chain in position 203 provides a significant redshift of peak B. Combining this mutation with those responsible for class 2 gives a single excitation peak at a long wavelength (the largest published redshift has been excitation 516 nm, emission 529 nm for mutant 10C Q69K[27]). A yellow color ("yellow fluorescent protein") is caused by the green emission peak tailing into the yellow range. Why Thr-203 mutations cause this redshift is not clear. Aromaticity might be involved, as reasoned by Ormö *et al.*,[27] whereas steric restraints could be another factor. Both the side chain size and the redshift increase in the series S65T/Thr-203 (488 nm), S65T/T203I (507 nm), and S65T/T203H/F/Y (512–513 nm).

Several new classes of mutations have been found by mutating the principal residue that creates the chromophore, Tyr-66. A fluorescent chromophore is created by placement of any aromatic amino acid in this position. Comparing the excitation and emission wavelengths clearly shows the similarity of all classes (except classes 3 and 9). As expected, the excitation

[36] B. P. Cormack, R. H. Valdivia, and S. Falkow, *Gene* **173,** 33 (1996).

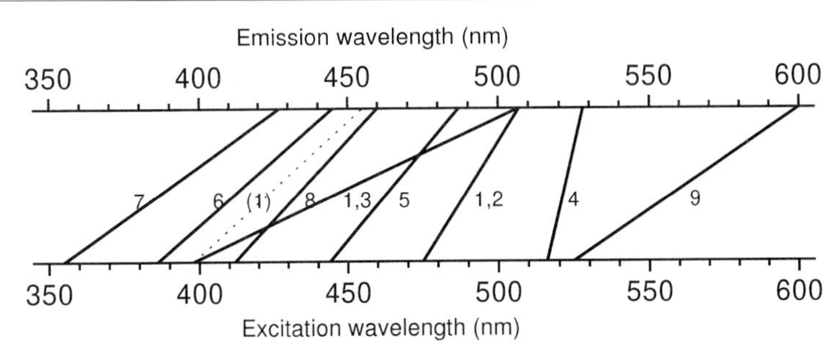

FIG. 6. Excitation and emission and wavelengths for all classes. The minor emission for peak A excitation of wt GFP is shown as a dashed line (see text).

wavelength generally increases with the size of the delocalized system: Phe-66 (355 nm) < His-66 (386 nm) < Tyr-66 (395 nm) < Trp-66 (436 nm) < Tyr-66 (488 nm) (Fig. 6). Class 3 mutants deviate from this pattern and emit at longer wavelengths because of the proton transfer in the excited state. If the transfer is hindered—for instance, at low temperatures[25] or in T203F mutants[37]—400 nm excitation/455 nm emission is in fact observed (dotted line in Fig. 6), a wavelength pair that fits in the series as expected.

Class 5: Trp-66

The mutation Y66W gives rise to excitation and emission wavelengths of ~436 and 476 nm, respectively. The blue–green emission led to the name "cyan fluorescent protein." The indole ring, which is the largest of all aromatic side chains in proteins, extends toward Val-150 and Phe-165. As for all available mutant structures, the imidazolone ring superimposes perfectly with that in the wt structure, but the indole group is slightly out of plane (G.J.P., unpublished results, 1998).

Class 6: His-66

The mutation Y66H gives rise to excitation at 385 nm and emission at 445 nm (blue), hence the name "blue fluorescent protein" for mutants in this class, even though the same name had been used earlier for apoaequorin. The imidazolyl group of these mutants is coplanar with the phenol group of the Tyr-66 mutants, but at physiological pH the smaller ring and the lack of a hydroxyl group leave room for an additional water molecule.[30] This molecule topologically replaces the phenol oxygen and hydrogen bonds to the NE2 atom of the chromophore, His-148, and a water molecule, as

[37] R. M. Dickson, A. B. Cubitt, R. Y. Tsien, and W. E. Moerner, *Nature (London)* **388,** 355 (1997).

does the chromophore phenol oxygen in class 3 mutants. The water molecule could be responsible for the failure to find Y66H mutants with high quantum yield. Denatured GFP with His-66 shows pK_a values of 4.9 and 12.0 for the imidazolyl side chain.[31] Native protein is thus expected to have a neutral imidazole (Fig. 3).

Class 7: Phe-66

The Y66F mutants provide the shortest wavelength spectra, with excitation at 355 nm and emission at 428 nm (UV). Because the excitation is far in the UV, where autofluorescence and damage to cells are more severe, this class of mutants has not been analyzed extensively. The missing hydroxyl group at the tip of the chromophore (compared with wt GFP) leaves a cavity close to the chromophore, as seen in the X-ray structure (G.J.P., unpublished results, 1998). This cavity could increase the vibrational freedom of the phenol ring and explain the very low quantum yields.

Class 8: Nonaromatic Residue 66

Mutation of Tyr-66 to nonaromatic residues does not allow formation of a chromophore with an extended conjugated system. Nevertheless, Y66L was found to have weak absorption at 410 nm (G.J.P., unpublished results, 1998). Fluorescence was also unambiguously detected at 460 nm. No structure of this mutant has been reported so far, so it is not known whether other residues substitute for the loss of the aromatic side chain.

Class 9: Photoactivated Green Fluorescent Protein

A new class of redshifted species has been described.[38,39] Mutants with Tyr-66 undergo a significant redshift in spectra when irradiated under anaerobic conditions (photoactivation). The resulting species absorb at 525 nm and emit at 600 nm. This form of GFP develops with some delay ($\tau = 0.7$ sec) after irradiation[39] but does not revert to its original spectrum unless brought in contact with O_2. A chemical reaction is probably involved in the formation of a modified chromophore. Irradiation could start this chemical reaction by creating a phenol radical. It is not known how mutation of Tyr-66 or Tyr-145 influences photoactivation of GFP.

[38] K. E. Sawin and P. Nurse, *Curr. Biol.* **7,** R606 (1997).
[39] M. B. Elowitz, M. G. Surette, P. E. Wolf, J. Stock, and S. Leibler, *Curr. Biol.* **7,** 809 (1997).

Effect of Various Parameters on Spectra

A number of physicochemical parameters affect the spectra of GFP. The most important parameters are summarized below.

Protein Concentration

Increasing the protein concentration favors the A form over the B form.[2] In class 1 mutants, this effect can already be seen at moderate concentrations, whereas for a class 2 mutant (Ala1b/F64L/S65C/Q80R/I167T/K238N), only a short-wavelength shoulder of peak B appears at high concentration (17 mg/ml).[30] It is not clear whether this effect is of physiological significance, nor is the structural reason for it obvious, but, interestingly, the proton transfer occurring between the A and B forms moves a charge exactly along the twofold axis relating the two monomers in the dimer. This is the direction with the maximum effect on the dipole moment of the dimer.

Ionic Strength

High salt concentration leads to a decrease of the peak B/peak A ratio and a redshift of peak B.[2] In crystals, high-salt crystallization conditions produce dimers, whereas monomer were always crystallized from organic solvents.[40] It thus seems that the spectral effect of high salt is mostly due to dimerization, although salt and protein concentration do not have exactly the same effect on peak ratio and redshift.[30]

Temperature

Higher temperatures increase the peak B/peak A ratio in wt GFP.[2]

pH

In contrast to denatured GFP, in which the chromophore has a well-defined pK_a of 8.1, folded protein shows no large change in the peak B/peak A ratio between pH 6 and 11 (see excitation at peak B in Fig. 2). In mutants with only one absorption peak (classes 2 and 3), the other peak does not appear in a wide pH range (G.J.P., unpublished results, 1998). The phenol group is not titratable by the solvent, and, most likely, neither are other buried groups, such as the imidazolone ring, Glu-222, and Arg-96. At low and high pH, however, wt GFP shows strong spectral changes.[2] At high pH, class 1 mutants show a decrease of peak A and an increase

[40] G. N. Phillips, Jr., *Curr. Opin. Struct. Biol.* **7,** 821 (1997).

of peak B (pK_a 12.3 for Ala1b/F64L/Q80R/I167T/K238N) (Fig. 2). This effect, too, is probably not a simple deprotonation of the chromophore, because Ala1b/F64L/S65C/Q80R/I167T/K238N (class 2 with deprotonated chromophore in the neutral range) and Ala1b/F64L/Y66W/Q80R/I167T/K238N (class 5) also show fluorescence changes: a decrease with pK_a 12.6 and an increase with pK_a >12, respectively. Deprotonation of a surface residue (a good candidate is Arg-168) is more likely to be primarily responsible for these fluorescence changes. Another fluorescence change occurs at low pH independently of the class. The pK_a for this transition is between 4.9 and 6.8 (G.J.P., unpublished results, 1998). His-148 could be responsible for this pK_a. In structures measured at low pH, this residue adopts a different conformation than at high pH.[30,31]

Acknowledgments

We thank G. A. Gaitanaris, R. Stauber, and G. N. Pavlakis for GFP constructs (sg series). We also thank A. Cubitt, J. Remington, R. Tsien, and their colleagues for material made available before publication. We are grateful to A. Arthur for editorial comments. The research conducted in our laboratory was sponsored by the National Cancer Institute, DHHS, under contract with ABL. The contents of this publication do not necessarily reflect the views or policies of the Department of Health and Human Services, nor does mention of trade names, commercial products, or organizations imply endorsement by the U.S. Government.

[33] Expression of Green Fluorescent Protein Using Baculovirus Vectors

By LINDA A. KING, CAROLE J. THOMAS, NICOLA WILKINSON, and ROBERT D. POSSEE

Introduction

The expression of heterologous genes in insect cells using baculovirus vectors is commonly used to achieve a high level of recombinant protein production.[1–3] The expression system exploits the promoters from a number of hyperexpressed virus genes that are activated in the very late stages of the *Autographa californica* nucleopolyhedrovirus (AcMNPV) replication cycle. The most widely used of these are the polyhedrin gene (*polh*) and

[1] L. A. King, R. D. Possee, D. S. Hughes, S. A. Marlow, C. P. Palmer, D. P. Miller, A. E. Atkinson, A. M. Lawrie, J. Pickering, K. A. Joyce, and D. J. Beadle, *Adv. Insect Physiol.* **25,** 1 (1994).

[2] L. K. Miller (ed.), "The Baculoviruses." Plenum Press, New York, 1997.

[3] R. D. Possee, *Curr. Opin. Biotechnol.* **8,** 569 (1997).

0076-6879/99 $30.00

p10 (*p10*) gene promoters.[4] Replacement of either the *polh* or *p10* coding region with that of a foreign gene results in the formation of a recombinant virus with the foreign gene under control of the hyperexpressed promoter. Recombinant viruses may be propagated in insect cells cultured *in vitro,* for example, *Spodoptera frugiperda* or *Trichoplusia ni* cells, or may be used to infect insect larvae.[2,4] In the latter case, recombinant viruses are usually prepared in which the *polh* and *p10* remain intact, so that larvae may be infected by the natural route of feeding upon a polyhedra-containing diet. In this case the foreign gene is expressed under control of a copy of the *polh* or *p10* promoter, located in a nonessential region of the virus genome, for example, just upstream of the *polh* promoter.[4]

Expression of Native Green Fluorescent Proteins in Insect Cells

A number of reports have demonstrated the efficient synthesis of green fluorescent protein (GFP) in insect cells or larvae using a variety of different baculovirus vectors.[5-14] The GFP has been synthesized in cultured insect cells[5-10,12,14] and in larvae[11,13] using standard *polh*-negative viruses. In most of these reports the GFP was synthesized and accumulated to large quantities (up to an estimated 30% total cell protein) in the cytoplasm of virus-infected cells. However, Laukkanen *et al.*[9] reported that GFP could be secreted from *T. ni* Hi Five cells, but not Sf9 cells, by the addition of signal peptides derived from the AcMNPV ecdysteroid UDP-glucosyltransferase and the rat brain glutamate receptor. As in other expression systems, GFP can be detected easily in recombinant baculovirus-infected cells because of its intrinsic fluorescence after excitation with an appropriate blue light source. Reports have shown that the fluorescence emission spectrum of

[4] L. A. King and R. D. Possee, "The Baculovirus Expression System: A Laboratory Guide." Chapman & Hall, London, 1992.

[5] L. E. Wilson, N. Wilkinson, S. A. Marlow, R. D. Possee, and L. A. King, *BioTechniques* **22,** 674 (1997).

[6] H. Reilander, W. Hasse, and G. Maul, *Biochem. Biophys. Res. Commun.* **219,** 14 (1996).

[7] S. Eriksson, E. Raivio, J. P. Kukkonen, K. Eriksson, and C. Lindqvist, *J. Virol. Methods* **59,** 127 (1996).

[8] C. Okerblom, A. Orellana, and K. Keinanen, *FEBS Lett.* **389,** 238 (1996).

[9] M. L. Laukkanen, C. Okerblom, and K. Keinanen, *Biochem. Biophys. Res. Commun.* **226,** 755 (1996).

[10] C. Wu, H. Z. Liu, R. Crossen, S. Gruenwald, and S. Singh, *Gene* **190,** 157 (1997).

[11] H. J. Cha, M. Q. Pham, G. Rao, and W. E. Bently, *Biotechnol. Bioeng.* **56,** 239 (1997).

[12] D. Mottershead, I. van der Linden, C. H. von Bonsdorff, K. Keinanen, and C. Okerblom, *Biochem. Biophys. Res. Commun.* **238,** 717 (1997).

[13] J. W. Barrett, A. J. Brownwright, M. J. Primavera, and S. R. Palli, *J. Virol.* **72,** 3377 (1998).

[14] H. J. Cha, T. Gotoh, and W. E. Bently, *BioTechniques* **23,** 782 (1997).

recombinant GFP expressed in insect cells is virtually identical to that of the native protein.[6–9] Alternative forms of GFP have also been expressed in insect cells including a bright GFP mutant,[8] and mutants that fluoresce blue under ultraviolet (UV) light and yellow in normal daylight.[10] The autofluorescent nature of GFP means that it can also be easily detected in virus-infected cells using laser-scanning confocal microscopy (Fig. 1).

Expression of Green Fluorescent Protein-Fusion Proteins in Insect Cells

The use of GFP as a reporter to detect the cellular targeting of recombinant proteins has also been described. In these cases, the *gfp* or *gfp* mutant has been cloned in frame with the gene of interest to allow visualization of the GFP-fusion protein using UV light.[10] Wu *et al.*[10] also inserted an in-frame 6× His tag between *gfp* and the foreign gene to facilitate purification of the fusion protein on nickel–nitrilotriacetic acid agarose matrices. Mottershead *et al.*[12] have produced recombinant baculoviruses displaying GFP on the surface of baculovirus virions and virus-infected cells by producing a GFP fusion with the AcMNPV major envelope glycoprotein, gp64. The GFP–gp64 fusion was detected by Western blot analysis of purified budded virus. The ability to tag recombinant proteins with GFP should allow the precise localization of these proteins without the requirement for the production of protein-specific antibodies.

Assessing Virus Titers Using Green Fluorescence

An advantage of recombinant baculoviruses expressing *gfp* is that virus titers can be quickly determined by exposing plaque assays to long-wave UV light, which visualizes plaques as discrete green fluorescent patches (see Fig. 2).[5,7,14]

Isolation of Recombinant Baculoviruses Using Green Fluorescent Protein as a Selectable Marker

We have shown that GFP can also be used as an efficient selective marker in the production of recombinant baculoviruses expressing other heterologous genes.[5] One of the most widely quoted methods for preparing recombinant baculoviruses utilizes a modified AcMNPV in which the *polh* coding region has been replaced by the *Escherichia coli lacZ* coding region, thus the parental virus gives blue plaques on staining cells with the chromogenic substrate 5-bromo-4-chloro-3-indolyl-β-D-galactopyranoside (X-Gal). After homologous recombination with the appropriate transfer vec-

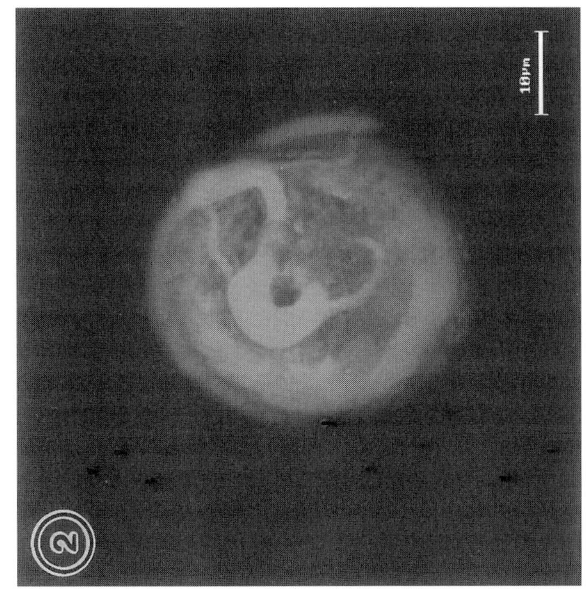

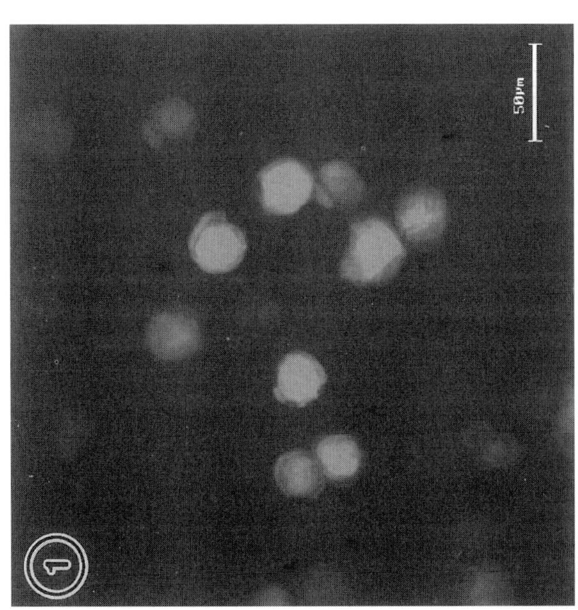

FIG. 1. Sf9 cells infected with AcMNPV.GFP[5] shown at 72 hr postinfection. GFP fluorescence was visualized by confocal laser scanning microscopy (Zeiss LSM410) using fluorescein-optimized filter sets. Panel (1) shows a low-power view of AcMNPV.GFP[5]-infected cells (bar = 50 μm); panel (2) shows a high-power view of an infected cell (bar = 10 μm).

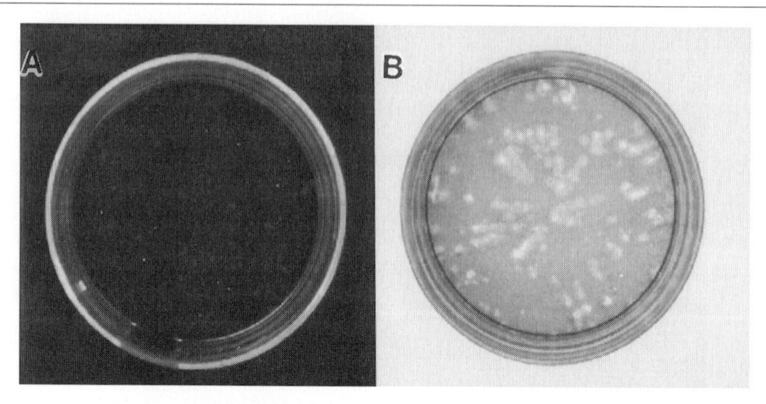

FIG. 2. A plaque assay of AcMNPV.GFP[5] in Sf9 cells is shown at 96 hr postinfection. In (A) the plaques were visualized by detection of green fluorescence after illumination with long-wave UV light, and in (B) the same plaques are shown after staining with neutral red.

tor, recombinant viruses produce colorless or clear plaques in the presence of X-Gal. The fortuitous discovery that *lacZ* contained a unique restriction site not found in the AcMNPV genome (*Bsu*36I) allowed Kitts *et al.*[15] to describe a method for the enhanced recovery of recombinant baculoviruses by linearization of the parental virus DNA prior to homologous recombination with the transfer plasmid. Kitts and Possee[16] further demonstrated that recovery of recombinant virus could be enhanced to almost 100% by the addition of two additional *Bsu*36I sites into essential virus genes flanking the *polh* locus. After cotransfection of insect cells with the linearized virus DNA (BacPAK6) and the transfer vector, parental and recombinant viruses were separated by plaque assay and staining with X-Gal. It is this final stage of the selection process that can be made more efficient by replacing *lacZ* with *gfp*. Wilson *et al.*[5] demonstrated that the recovery of recombinant viruses remains high (80–90%) when *gfp* replaces *lacZ* in the parental virus (BacGFP1). The virus DNA is linearized using the two *Bsu*36I sites in the genes flanking the *polh* locus and after cotransfection and plaque assay, recombinant viruses are identified as nonfluorescing plaques in a background of green-fluorescing parental virus plaques. Plaques identified in this way can be observed and picked about 1 day earlier than those visualized by staining with X-Gal. In the methodologies that follow we describe techniques for the infection of insect cells with recombinant baculoviruses expressing *gfp* together with protocols for the detection of green fluorescence in plaque assays and using confocal microscopy.

[15] P. A. Kitts, M. D. Ayres, and R. D. Possee, *Nucleic Acids Res.* **18,** 5667 (1990).
[16] P. A. Kitts and R. D. Possee, *BioTechniques* **14,** 810 (1993).

Procedures

Detailed methods for the production of the modified BacGFP1 virus can be found in Wilson *et al.*[5] and many of the basic procedures for the propagation of baculoviruses and insect cell culture can be found in King and Possee[4] and King *et al.*[17,18] The protocols that follow are those directly related to the infection of insect cells with baculoviruses and the subsequent detection of GFP.

Infection of Insect Cells with a Recombinant Baculovirus Expressing Green Fluorescent Protein

Required

Recombinant baculovirus expressing *gfp* (e.g., BacGFP1, AcMNPV.GFP as described in Wilson *et al.*[5]) of known titer
Healthy culture of *Spodoptera frugiperda* (Sf9 or Sf21) cells
Growth medium [e.g., TC100 containing 10% fetal calf serum (FCS) or Sf900II serum-free medium; both from GIBCO-Life Technologies, Gaithersburg, MD]
Sterile cell culture dishes (35 mm)
Long-wave UV light source (e.g., 365-nm hand-held UV light from UV Products, Cambridge, U.K.)

Method

1. Healthy Sf9 or Sf21 cells should be counted using a standard hemocytometer and 1×10^6 viable cells should be placed into a 35-mm cell culture dish in a final volume of 1.5 to 2 ml of fresh growth medium. We usually culture Sf21 cells in TC100 with 5–10% FCS and Sf9 cells in Sf900II serum-free medium at 27–28°. The cells can be maintained as suspension (Sf21), shake (Sf9), or monolayer cultures (see King and Possee[4] for further details).

2. After allowing the cells to adhere for 1–2 hr at ambient temperature, the culture medium is removed and the virus inoculum is added. To obtain a good infection of the cells, the virus inoculum should be added to a multiplicity of infection of 5–10 plaque-forming units per cell (PFU/cell). The total volume of inoculum added should not be less than 0.1 ml and should not be more than 0.25 ml. The virus is allowed to adsorb to the cells for between 30 and 60 min. The inoculum should be evenly distributed

[17] L. A. King, S. A. Marlow, L. A. Obosi, and R. D. Possee, *in* "Cell Biology: A Laboratory Handbook" (J. Celis, ed.), 2nd Ed., p. 211. Academic Press, San Diego, California, 1998.

[18] L. A. King, S. A. Marlow, L. E. Wilson, and R. D. Possee, *in* "Cell Biology: A Laboratory Handbook" (J. Celis, ed.), 2nd Ed., p. 204. Academic Press, San Diego, California, 1998.

over the surface of the monolayer, but continuous rocking is not required during the infection period. At the end of the inoculating period, remove the inoculum and discard into appropriate disinfectant (e.g., hypochlorite solution or Virkon from Antec International LTD, Sudbury, UK).

Note: When analyzing recombinant protein production always prepare control dishes of mock-infected cells (use growth medium instead of inoculum) and a wild-type virus control (use wild-type AcMNPV or BacPAK6).

3. After removing the inoculum, add 1.5 to 2.0 ml of fresh growth medium and incubate the cells at 27–28° for the required length of time. It is useful to place the dishes of cells in a sandwich box prior to placement in the incubator or warm room. Note that insect cells should not be placed in a CO_2 incubator.

4. If the foreign gene, in this case *gfp,* is under control of a very late *polh* or *p10* promoter, protein synthesis will increase from 24 hr postinfection until cell lysis at about 96 hr postinfection. It may be desirable to harvest dishes at different times (24, 48, 72, 96 hr postinfection) to assess the optimum level of recombinant protein production.

5. The GFP is synthesized in the cytoplasm of insect cells; thus, to analyze GFP synthesis by sodium dodecyl sulfate–polyacrylamide gel electrophoresis (SDS–PAGE) Western blotting, the cells will need to be scraped from the surface of the dish (although well-infected cells tend to lift off by themselves). A rubber policeman or blue pipettor tip works well for this. Pellet the cells using a low speed (about 3000–5000 rpm) and discard the culture medium into disinfectant. Wash the cell pellet with phosphate-buffered saline or TE [10 m*M* Tris-HCl (pH 8.0), 1 m*M* EDTA] and resuspend the cell pellet in protein solubilization buffer. Further details of the analysis of proteins by SDS–PAGE and/or Western blotting can be found in many other manuals, for example, King and Possee.[4] A rabbit anti-rGFP antiserum is available from Clontech (Palo Alto, CA).

6. The GFP may also be visualized in virus-infected cells simply by placing the dish of cells on a transilluminator fitted with a long-wave UV light source or with a hand-held UV light. Infected cells emit a typical bright green fluorescence. Alternatively, cells may be prepared for light or confocal microscopy as detailed in the following section.

Preparation of Recombinant Baculovirus Using Linearized BacGFP1 DNA

Required

Purified BacGFP1 DNA (Wilson *et al.*[5]; see King and Possee[4] for detailed protocols)

Transfer vector containing foreign gene of interest under control of *polh* promoter (see King and Possee[4] for further details)

*Bsu*36I restriction enzyme
*Bsu*36I enzyme buffer, 10× (as recommended by manufacturer)
Sterile water
Lipofectin (GIBCO-Life Technologies) or similar liposome reagent for cotransfection of insect cells
Healthy Sf9 cells seeded into 35-mm dishes (see Infection of Insect Cells, above)
Sf900II serum-free medium
Sterile disposable polystyrene container for cotransfection (e.g., bijoux or universal containers; do not use sterile microcentrifuge tubes)

Method

1. Using aseptic technique, digest 1–2 μg of BacGFP1 DNA with an excess of *Bsu*36I (10 units) in a final volume of 100–200 μl, for 5–6 hr at 37°. Check that the virus DNA has been fully digested by analyzing 100 ng in a mini-agarose gel. If necessary, add more enzyme and continue the digest for 2–3 hr. After digestion store the DNA at 4°. The DNA is stable for at least 1 year. Maintain aseptic technique because the DNA will be introduced into insect cells.

2. Using aseptic technique, add 100 ng of linearized virus DNA to 500 μl of serum-free medium. In a separate tube, add 5 μl of lipofectin to 500 μl of serum-free medium. Mix the two together and incubate at ambient temperature for 30 min.

Note: If using Sf21 cells, TC100 medium without FCS must be used at this stage.

3. Remove the medium from a dish of subconfluent Sf9 cells (1×10^6 cells per dish). Add the transfection mix and incubate for 5 hr (or overnight) at 27–28°. After 5 hr (or overnight) add a further 1 ml of medium and continue the incubation until 48 hr postinfection.

Note: If using Sf21 cells in TC100 plus FCS growth medium, the cell monolayer must be washed twice with serum-free TC100 prior to adding the transfection medium. After the 5-hr (or overnight) incubation, the additional 1 ml of medium should contain 10% FCS.

4. Harvest the culture medium at 48 hr postinfection and store at 4°. The medium will contain a mixture of parental (less than 10%) and recombinant virus particles. These should be separated by plaque assay (as described in the next section).

Plaque Assay to Separate Recombinant and Parental (GFP-Containing) Baculoviruses following a Cotranfection

Required

Sf9 or Sf21 cells (1×10^6 cells per 35-mm dish)

TC100 medium with 10% FCS (whether using Sf9 or Sf21 cells)
Low gelling temperature agarose (e.g., SeaPlaque; FMC, Philadelphia, PA), 2% (w/v) in water (autoclave in 10-ml aliquots; if solidified, melt in a microwave or boiling water bath and hold at 37–45°)
Harvested cotransfection medium (see previous section)
Phosphate-buffered saline (PBS)
Neutral red (0.5%, w/v; filter sterilized)
UV light source

Note. In our experience Sf21 cells produce more well-defined plaques than Sf9 cells, thus making the screening process for recombinant plaques much easier. We find that if Sf9 cells are used, the use of TC100–FCS medium (rather than serum-free medium) in the agarose overlay helps to improve the plaque assay.

Method

1. Prepare the required number of 35-mm dishes with 1×10^6 Sf21 or Sf9 cells per dish. Allow the cells to adhere for 1–2 hr before using. About 16 or 17 dishes will be needed to screen a transfection using BacGFP1 as the parental virus DNA.

2. Prepare dilutions of the harvested cotransfection medium. Make the dilutions in TC100 medium containing 10% FCS. Dilutions from 10^{-1} to 10^{-4} are normally sufficient, and 0.5 ml of each dilution will be required for the plaque assay.

3. Remove the medium from the dishes of cells and add 0.1 ml of the appropriate dilution to each dish. We would normally prepare three or four dishes for each dilution, three or four dishes of undiluted cotransfection medium, and one dish as a control (use 0.1 ml of culture medium). Leave the virus to adsorb at ambient temperature for 30–60 min.

4. Remove the inoculum completely (this is important) and discard safely. Overlay the cell monolayers gently with 2 ml of agarose overlay. The agarose overlay is prepared by mixing TC100–10% FCS and 2% (w/v) agarose in water at a ratio of 1:1. The agarose will solidify fairly rapidly after adding the growth medium, so do not prepare too much at one time. The overlay can be kept from solidifying by standing the stock bottle in a water bath at 37–42°. However, unless the water bath is clean this practice may lead to bacterial or fungal contamination of the plaque assay.

5. When the agarose overlay has set (5–10 min) add a further 1 ml of liquid overlay (TC100–10% FCS). This medium will permeate the agarose overlay and feed the monolayer of cells. The plaque assay will not work without this additional medium.

6. Incubate the cells for 3–4 days at 27–28°. If the seeding density of the cells was correct, large round, discrete plaques will form in 3–4 days (see Fig. 2). If the cells were too sparse or quite large, ragged looking

plaques will form with ill-defined edges. If the monolayer seeding density was too high, tiny plaques, or no plaques at all, will form.

7. Plaques containing GFP may be visualized simply by the use of a hand-held UV light or by placing the dishes on a transilluminator with a long-wave UV light source. To distinguish GFP-containing plaques from those that contain no GFP, it may be easier to stain the cells with neutral red. This will produce colorless plaques in a background of pink-stained cells. On the application of UV light, the parental, GFP-expressing plaques will be easily distinguished from those containing the recombinant virus that is to be isolated. To stain the cells with neutral red, simply dilute the stock stain 1:20 with PBS and add 1 ml to each dish of cells. After 2–3 hr, decant the liquid overlay containing stain, invert the dishes, and allow the plaques to clear for 1–24 hr. During this period it is best if the cells are kept in the dark (e.g., in a cupboard or wrapped in foil).

8. Using a sterile pasteur pipette, take up a plug of agarose from putative recombinant plaques (i.e, plaques that do not fluoresce green) and transfer to a sterile tube containing 0.5 ml of growth medium (TC100 with serum or Sf900II). Vortex to release virions from the agarose into the culture medium. Store at 4°. We routinely pick 5–10 plaques from each cotransfection and check that these are true recombinants by dot-blot or polymerase chain reaction (PCR) before proceeding to examine protein synthesis.

Note. If well-isolated plaques are picked from dishes that have no parental virus plaques (i.e., green plaques), then there is no need for further plaque purification before proceeding to amplify the recombinant virus. If the range of dilutions stated above is plated out, it should be possible to achieve this. If not, it will be necessary to replaque the plaque picks (plate out 10^{-1} to 10^{-5} dilutions of each plaque pick) until a virus is obtained that is free from parental virus.

Amplification of Recombinant Virus Stocks

Required

- Shake cultures of Sf9 cells grown in serum-free medium or stirred cultures of Sf21 cells grown in TC100–10% FCS (see King and Possee[4] for more details), or monolayer cultures of Sf9/Sf21 in 25- or 75-cm^3 flasks
- Serum-free medium (e.g., Sf900II) or TC100–10% FCS
- Facilities to shake cultures or stir cultures at about 130 rpm at 27–28°
- Disposable polycarbonate shake flasks (e.g., from Corning Costar, High Wycombe, UK) of varying sizes, or flat-bottomed flasks containing magnetic stirring bars for stirring cultures of Sf21 cells (commercially available flasks are also available, e.g., the Techne biologi-

cal stirrer system). Further details can be found in King and Possee (Techne, Cambridge, UK).[4]

Method

1. To amplify virus from a plaque pick, it is best to inoculate cells seeded in a 25-cm^3 flask. Place 1×10^6 cells in a flask and after 1–2 hr remove the culture medium and replace with 0.25 ml of plaque pick (see previous protocol). Leave the virus to adsorb at ambient temperature for 60 min, remove the inoculum, and replace with 5 ml of fresh culture medium. Incubate the infected cells for 5–7 days, until the cells are well infected. Harvest the culture medium, remove any cells by centrifugation, and store the amplified virus (P1 stock) at 4°. The titer of the virus can be determined by plaque assay (as described in the next protocol). At this stage the virus titer will not be high (about $1–5 \times 10^7$ PFU/ml). Before performing any experimental work it will be necessary to amplify a high-titer P2 stock.

Note: The cells discarded from this P1 amplification are a good source of virus DNA or intracellular proteins for preliminary analyses to confirm the identity of the recombinant virus (by PCR or dot-blot, for example) or protein expression (by Western blot, for example). Details of extracting crude virus DNA from virus-infected cells can be found in King and Possee.[4]

2. To produce a high-titer stock of virus, prepare 50–100 ml of cells in shake (Sf9) or stirred (Sf21) culture. The Sf9 cells grown in serum-free medium will normally give higher virus titers. Seed shake cultures of Sf9 cells at 2×10^5 cells/ml and allow the cells to grow until they reach a density of 2×10^6 cells/ml. Infect the cells with P1 virus stock (as described above) or with any other virus that requires amplification at a multiplicity of infection (MOI) of 0.1 PFU/cell (except plaque picks, which are best amplified to the P1 stage in monolayer culture first). Simply add the inoculum to the cells. Allow the cells to shake quite vigorously at 130–140 rpm until they are well infected (4–5 days). Using a low MOI allows multiple rounds of virus multiplication, which produces higher titers of virus and reduces the chances of obtaining defective virus particles. Harvest the amplified virus, remove cells by low-speed centrifugation, and store the virus in the dark at 4°. Determine the titer of the virus by plaque assay before undertaking experimental work (described in the next protocol).

Note: If using Sf21 cells, infect the cells when they reach a density of 5×10^5 cells/ml.

Plaque Assay to Determine Titer of Baculoviruses Expressing Either Native Green Fluorescent Protein or Green Fluorescent Protein as a Fusion Protein

Before performing experimental work with recombinant baculoviruses it is essential that the titer of the virus stock be determined. If the titer is assumed to be high but is in fact 10 times lower than expected, recombinant protein may not be detected. In our experience, problems encountered in failing to detect foreign proteins in recombinant baculovirus-infected cells are nearly always due to the use of virus stocks that have not been titrated. To examine optimal synthesis of recombinant proteins, cells should be infected with an MOI of 5–10 PFU/cell. This is much more easily achieved if the titer of the virus stock is high (1×10^8 PFU/ml or higher).

Required

Sf9 or Sf21 cells (1×10^6 cells per 35-mm dish)
TC100 medium with 10% FCS (whether using Sf9 or Sf21 cells)
Low gelling temperature agarose (e.g., SeaPlaque), 2% (w/v) in water (autoclave in 10-ml aliquots; if solidified, melt in a microwave or boiling water bath and hold at 37–45°)
Virus(es) to be titrated
Phosphate-buffered saline (PBS)
Neutral red (0.5%, w/v; filter sterilized)
Long-wave UV light source

Method

1. Prepare the required number of 35-mm dishes with 1×10^6 Sf21 or Sf9 cells per dish. Allow the cells to adhere for 1–2 hr before using. About 12 dishes will be needed to determine the titer of one recombinant virus.
2. Prepare dilutions of the virus to be titrated in TC100 medium containing 10% FCS. Dilutions from 10^{-1} to 10^{-8} should be prepared, and 0.3 ml of the 10^{-4} to 10^{-8} dilutions will be required for the plaque assay.
3. Remove the medium from the dishes of cells and add 0.1 ml of the appropriate dilution to duplicate dishes (10^{-4} to 10^{-8}). As a control, use 0.1 ml of culture medium on two dishes. Leave the virus to adsorb at ambient temperature for 30–60 min.
4. Remove the inoculum completely (this is important) and discard safely. Overlay the cell monolayers gently with 2 ml of agarose overlay. The agarose overlay is prepared by mixing TC100–10% FCS and 2% (w/v) agarose in water at a ratio of 1:1. The agarose will solidify fairly rapidly after adding the growth medium, so do not prepare too much at one time. The overlay can be kept from solidifying by standing in a water bath at 37–42°, but unless the water bath is clean this practice may lead to contamination of the plaque assay.
5. When the agarose overlay has set (5–10 min) add a further 1 ml of liquid overlay (TC100–10% FCS). This medium will permeate the agarose overlay and feed the monolayer of cells.
6. Incubate the cells for 3–4 days at 27–28°. If the seeding density of the cells was correct, large round, discrete plaques will form in 3–4 days.

If the cells were too sparse or quite large, ragged looking plaques will form with ill-defined edges. If the monolayer seeding density was too high, tiny plaques, or no plaques at all, will form.

7. Plaques containing GFP may be visualized simply by the use of a hand-held UV light or by placing the dishes on a transilluminator with a long-wave UV light source (see Fig. 2A). The titer of the virus is determined by finding the dishes that have a countable number of plaques (5–20) and calculating the average number of plaques at a particular dilution from the duplicate dishes. This number is then multiplied by 10 (as only 0.1 ml was added to the dish) and by the reciprocal of the appropriate dilution factor. Thus 10 plaques on the 10^{-6} dish would give a titer of $10^6 \times 10 \times 10$, which equals 1×10^8 PFU/ml.

8. Virus stocks may be stored in the dark at 4° for at least 1 year, but virus infectivity will decrease slowly, so stocks that have been stored for more than 1 year should be retitrated prior to experimental use. It may be worth storing a sample of virus at −80°. In this way virus stocks may be kept for several years. If the virus was prepared using serum-free medium, serum should be added to 10% prior to freezing the virus at −80°. Do not store virus at −20°.

Confocal Microscopy of Recombinant Baculoviruses Expressing Either Native Green Fluorescent Protein or Green Fluorescent Protein as a Fusion Protein

Visualization of GFP-expressing recombinant baculoviruses can be achieved by fluorescence microscopy. Furthermore, microscopy enables the localization of GFP-tagged proteins in virus-infected cells, without the need for immunofluorescence staining. However, it should be remembered that the addition of GFP may potentially affect either the localization or the function of the protein of interest.

The protocol we follow for visualizing GFP expressed by recombinant baculoviruses is adapted from the method of Karlsson and Pines.[19] The presence of GFP can be observed by microscopy using conventional fluorescein-optimized filter sets.

Required

Recombinant baculovirus expressing *gfp* (e.g., BacGFP1, AcMNPV.GFP as described in Wilson *et al.*[5]) of known titer
Healthy culture of *S. frugiperda* (Sf9 or Sf21) cells
Growth medium (e.g., TC100 containing 10% FCS or Sf900II serum-free medium; both from GIBCO-Life Technologies)

[19] C. Karlsson and J. Pines, *in* "Cell Biology: A Laboratory Handbook" (J. Celis, ed.), 2nd Ed., p. 204. Academic Press, San Diego, California, 1998.

Sterile cell culture dishes (35 mm)
Sterile 13-mm-diameter glass coverslips [such as those supplied by BDH (Poole, England), recommended thickness No. 0]
Phosphate-buffered saline (PBS) or 0.1 *M* piperazine-*N,N'*-bis(2-ethanesulfonic acid) (PIPES) buffer, pH 6.9
Coverslip mountant (e.g., CitiFluor from Agar Scientific LTD, Stansted, UK)
Glass microscope slide (plain)

Method

1. Place sterile 13-mm-diameter glass coverslips into 35-mm tissue culture dishes. *Note:* We seed cells for confocal microscopy onto sterile glass coverslips in small tissue culture dishes prior to infection with recombinant baculovirus, as plastic tissue culture dishes will autofluoresce.
2. Seed the required number of dishes with 1×10^6 Sf21 or Sf9 cells (per dish) and allow the cells to adhere to the glass coverslips for approximately 1–2 hr.
3. Remove medium from the dishes and replace with virus inoculum at an MOI of 5–10 PFU/cell. Allow the virus to adsorb to the cells for 1 hr at ambient temperature.
4. Remove the virus inoculum and add 1.5 ml of growth medium to the dishes. Allow the cells to incubate at 27–28° until the appropriate time postinfection.
5. Carefully remove the glass coverslips from the tissue culture dish and rinse once in PBS or 0.1 *M* PIPES buffer (pH 6.9) by pipetting reagents onto and off of the coverslips. We place the glass coverslips onto fixed glass rod supports to enable ease of pipetting solutions onto the coverslips.
6. Remove any excess liquid using filter paper and briefly air dry. Mount the coverslips onto a glass microscopy slide, using a small drop of mountant.

Notes

- The coverslips should not be sealed with nail varnish, as the acetone will destroy GFP fluorescence. If necessary, seal with a small amount of petroleum jelly.
- At late times in baculovirus infections, insect cells tend to round up and may lift from the surface to which they previously adhered. Therefore, if observations are to be made late in the infection cycle it may be of benefit to fix the cells prior to visualization (as described in the next section).

Fixation of Green Fluorescent Protein-Expressing Virus-Infected Cells

The protocol below describes the procedure we use for the fixation and permeabilization of virus-infected cells. This will enable further immunofluorescence staining of the cultured cells with antibodies or organelle markers.

Required

Materials as listed in previous section
Paraformaldehyde, 4% in 0.1 *M* PIPES buffer, pH 6.9 (or in PBS)
Triton X-100, 0.1% in 0.1 *M* PIPES buffer, pH 6.9 (or in PBS)

Method

1. Infect Sf9 or Sf21 cells, seeded on coverslips, with recombinant virus as described in the previous section. Remove the glass coverslips from cell culture dishes and rinse the cells once in PBS.
2. Fix the cells in 4% paraformaldehyde in 0.1 *M* PIPES buffer, pH 6.9, or in PBS for 30 min at ambient temperature. *Note:* Organic solvents, such as acetone and methanol, should not be used for fixation as they do not preserve GFP fluorescence.
3. Wash the cells twice with PBS or 0.1 *M* PIPES, pH 6.9, for 5 min.
4. Permeabilize the cells for 10 min at ambient temperature using 0.1% Triton X-100 (in 0.1 *M* PIPES, pH 6.9, or PBS).
5. Wash several times with PIPES buffer or PBS (three times, 5 min each time), and mount or stain with an appropriate antibody.

Acknowledgments

The authors' work described here was supported in part by the U.K. Biotechnology and Biological Sciences Research Council (BBSRC) and Natural Environmental Research Council. Nicola Wilkinson was supported by a BBSRC Studentship.

[34] Green Fluorescent Protein Forms for Energy Transfer

By ROGER HEIM

Introduction

Fluorescence resonance energy transfer (FRET) is a process that can occur between two different chromophores, usually denoted donor and acceptor, if they are in close proximity with suitable orientation. The donor is a fluorescent group whose excited state energy can be transferred to an acceptor if the energy corresponds to the energy required for excitation of

0076-6879/99 $30.00

the acceptor. Such transfer is manifested by a reduction in the fluorescence intensity of the donor and also results in a shorter lifetime of the excited state. If the acceptor is an absorbing group, then it is called a quencher. Alternatively, if the acceptor is a fluorophore, longer wavelength light, characteristic of the acceptor, is reemitted. With increasing distance between the partners, usually beyond a few nanometers, FRET is dramatically reduced. Therefore, the fluorescence spectrum is indicative of the distance between donor and acceptor. FRET can occur intra- and intermolecularly. For intramolecular FRET, donor and acceptor groups are attached to the same molecule. Detection of conformational changes and cleavage of the molecule between the donor and acceptor groups are typical applications. For intermolecular FRET, molecules of interest are usually labeled individually with donor or acceptor in order to follow their association and dissociation.

Prior to the advent of fluorescent protein technologies, the coupling of a donor or acceptor group to a molecule of interest was often challenging and restrictive. Advances in automated DNA synthesis and the site-specific incorporation of fluorescent nucleotides have renewed interest in the application of FRET for structural and functional studies of DNA and RNA.[1] Peptides can be labeled by incorporating unnatural chromophoric amino acids, assembled by solid-phase peptide synthesis, which poses certain problems of effort and expense. The labeling of whole proteins with chromophores is even more difficult because this technique typically relies on the availability of exposed cysteine side chains and is limited to isolated, well-characterized proteins. Such chemically labeled biological probes can be used for *in vitro* experiments. For cellular experiments, both small (peptides) and large probes (proteins) frequently must be microinjected into individual cells. Such an approach was used for protein kinase A,[2] the subunits of which were labeled with fluorescein and rhodamine, then associated *in vitro* and microinjected into cultured cells. Because increased levels of the second messenger cyclic AMP promote dissociation of the protein kinase A subunits, the measured emission ratio between fluorescein and rhodamine indicates the concentration of cAMP through loss of FRET between labeled subunits.

The functional expression of the green fluorescent protein (GFP) in heterologous systems[3,4] offered the labeling of a protein of choice with

[1] M. Yang and D. P. Millar, *Methods Enzymol.* **278,** 417 (1997).

[2] S. R. Adams, A. T. Harootunian, Y. J. Buechler, S. S. Taylor, and R. Y. Tsien, *Nature (London)* **349,** 694 (1991).

[3] M. Chalfie, Y. Tu, G. Euskirchen, W. W. Ward, and D. C. Prasher, *Science* **263,** 802 (1994).

[4] S. Wang and T. Hazelrigg, *Nature (London)* **369,** 400 (1994).

GFP by fusing their cDNAs and expressing the resulting GFP fusion protein in cells, thus avoiding laborious organic synthesis and microinjection. Wild-type GFP has an excitation spectrum with two peaks whose relative amplitudes depend on several factors including protein concentration,[5] buffer composition[5] and illumination history of the sample.[6] These properties can make the use of wild-type GFP as an FRET partner complex. Furthermore, wild-type GFP is not very bright relative to synthetic labels and does not fold well at 37°. A second fluorescent label with a different color as a suitable FRET partner would still be required. Since the initial demonstration of successful expression of wild-type GFP in heterologous cell systems, many mutants have been found that provide simplified spectra, improved brightness and folding, as well as variety of spectral colors. From this collection, several combinations of GFP mutants as donor and acceptor are theoretically possible. To select suitable FRET pairs, however, many parameters must be considered, and they are discussed in this chapter. Practical aspects of using GFP in FRET are emphasized rather than the quantum mechanics of the system. The more difficult lifetime measurements of chromophores are not discussed here. More thorough discussions of GFP mutants, including molecular and structural aspects, can be found in other articles.[7,8] As new GFP mutants develop in future, an evaluation of other potential donor/acceptor combinations should be possible, using this chapter as a reference.

Selection of Fluorescence Resonance Energy Transfer Pairs

The suitability of an FRET pair depends primarily on how much FRET can be maximally obtained with a particular combination, and on how well FRET changes can actually be measured. In addition, other parameters such as absolute brightness, efficiency of folding, resistance to photobleaching, or sensitivity to pH can have an influence on the successful use of GFP mutants for FRET.

The efficiency of FRET E is defined in the Foerster equation[9]:

$$E = R_0^6/(R^6 + R_0^6) \tag{1}$$

where R is the distance between donor and acceptor, and R_0 is the distance at which energy transfer is 50%. From Eq. (1) with R rising above R_0, E

[5] W. W. Ward, H. J. Prentice, A. F. Roth, C. W. Cody, and S. C. Reeves, *Photochem. Photobiol.* **35,** 803 (1982).

[6] A. B. Cubitt, R. Heim, S. R. Adams, A. E. Boyd, L. A. Gross, and R. Y. Tsien, *Trends Biochem. Sci.* **20,** 448 (1995).

[7] A. B. Cubitt, L. A. Woollenweber, and R. Heim, *Methods Cell Biol.* **58,** 19 (1998).

[8] G. J. Palm and A. Wlodawer, *Methods Enzymol.* **302,** [32], 1999 (this volume).

[9] T. Foerster, *Ann. Phys.* **2,** 55 (1948).

rapidly becomes small. A high value for R_0 indicates high efficiency of energy transfer, and means that efficient FRET still occurs at distances up to R_0. With the spectral parameters of donor and acceptor known, R_0 can be calculated (in angstroms) in Eq. (2):

$$R_0 = 9.79 \times 10^3(\kappa^2 Q J n^{-4})^{1/6} \tag{2}$$

$$J = \int_{\infty}^{0} F_D(\lambda)\varepsilon_A(\lambda)\lambda^4 \, d\lambda \tag{3}$$

The relevant parameters are the quantum yield Q of the donor in the absence of acceptor, the refractive index n of the intervening medium, the orientation factor κ^2, and the overlap integral J. As the orientation between the chromophores is usually not known, the value of 2/3 for κ^2 for freely mobile components is often used in calculations. The overlap integral J expresses the degree of spectral overlap between the donor emission and the acceptor absorption, wherein $F_D(\lambda)$ is the normalized donor fluorescence at wavelength λ, and $\varepsilon_A(\lambda)$ is the extinction coefficient of the acceptor at the same wavelength, as summarized in Eq. (3).

From a library of interesting GFP mutants, quantum yield and extinction coefficients have been measured and are given in Table I. For certain combinations of mutants that are potentially appropriate for FRET, R_0 values have been calculated, with the assumption of free mobility of the chromophores. The results are summarized in Table II. The best value of R_0, 60 Å, was calculated for the combination of Sapphire and Topaz as donor and acceptor, respectively. This high value is mostly due to the almost perfect overlap between the emission spectrum of Sapphire and the excitation spectrum of Topaz (see Fig. 1A), the high extinction coefficient of Topaz and the favorable quantum yield of Sapphire. However, the donor–acceptor pair with the highest degree of energy transfer is not necessarily the best combination in practice, because other factors play a role that are illustrated below.

Tandem Green Fluorescent Proteins

The most directed method to experimentally measure FRET between a given pair of GFP mutants is to link their cDNAs together and express them as a fusion protein. This way, donor and acceptor are physically forced to be close to each other, and some energy transfer can be expected. The amount of energy transfer will also depend on the nature of the linker. Therefore, only constructs with the same linker can be directly compared. Because GFP is extremely resistant to most proteases,[5,10] linkers cleavable

[10] R. Heim and R. Y. Tsien, *Curr. Biol.* **6,** 178 (1996).

TABLE I

SPECTROSCOPIC AND BIOCHEMICAL PROPERTIES OF SELECTED GREEN FLUORESCENT PROTEIN MUTANTS

Mutations[a]	Common name	Molar extinction[b]	QY[c]	Excitation maximum (nm)[d]	Emission maximum (nm)[e]	Relative fluorescence at 37°[f]
S65T type (GFP)						
S65T, S72A, N149K, M153T, I167T	Emerald	57,000	0.68	487	509	100
F64L, S65T, V163A		42,000	0.58	488	511	54
F64L, S65T	EGFP	55,900	0.60	488	507	20
S65T		52,000	0.64	489	511	12
T203I type (GFP)						
T203I, S72A, Y145F	H9–40	29,000	0.64	399	511	100
S202F, T203I	H9	20,000	0.60	399	511	13
T203Y type (YFP)						
S65G, S72A, K79R, T203Y	Topaz	94,500	0.60	514	527	100
S65G, V68L, S72A, T203Y	10c	83,400	0.61	514	527	58
S65G, V68L, Q69K, S72A, T203Y	10cQ69K	62,000	0.71	516	529	50
S65G, S72A, T203H		48,500	0.78	508	518	12
S65G, S72A, T203F		65,500	0.70	512	522	6
Y66H type (BFP)						
F64L, Y66H, Y145F, V163A	P43E	22,000	0.27	384	448	100
F64L, Y66H, Y145F	EBFP	26,300	0.26	383	447	82
Y66H, Y145F	P43	22,300	0.30	382	446	51
Y66H	BFP, P4	21,000	0.24	384	448	15
Y66W type (CFP)						
S65A, Y66W, S72A, N146I, M153T, V163A	W1C	21,200	0.39	435	495	100
F64L, S65T, Y66W, N146I, M153T, V163A	W1B, ECFP	32,500	0.4	434 452	476 505	80
Y66W, N146I, M153T, V163A	W7	23,900	0.42	434 452	476 505	61
Y66W		ND	ND	436	485	ND

[a] Most data are from A. B. Cubitt, L. A. Woollenweber, and R. Heim, *Methods Cell Biol.* **58,** 19 (1998). Note that this is not a complete list.

[b] Molar extinction coefficients are in units of M^{-1} cm^{-1}. Some values are slightly higher than originally reported values. To provide accurate comparisons, absorption and protein concentration measurements were carried out under the same conditions. Deviations from the values reported here are most likely due to variations in the quality of the protein preparation or to variations in the protein determination. ND, Not determined.

[c] Quantum yields were calculated by comparison with the standard fluorescein or 9-amino acridine.

[d] Excitation maxima are in nanometers. Two numbers on separate lines indicate two distinct peaks in the excitation spectrum.

[e] Emission maxima are in nanometers. Two numbers on separate lines indicate two distinct peaks in the emission spectrum.

[f] Relative fluorescence intensities of GFP mutants expressed in *E. coli* at 37° from the same vector under similar conditions. These values represent not only the intrinsic brightness measured by extinction and quantum yield, but also the folding efficiency at 37°. They have been normalized to 100% for the most fluorescent mutant of each class.

TABLE II
COMPARISON OF DIFFERENT FLUORESCENCE RESONANCE ENERGY TRANSFER PARTNERS

Donor : acceptor pair[a]	Maximum ratio change[b]	R_0 (Å)	Emission wavelengths
P4-3//W7	1.48	33	444//476 (505)
Y66H//S65C	3.93	37	448//507
P4-3//S65C	4.6	40	446//507
P4-3//S65T	ND	43	446//511
P4-3//Topaz	5.2	49	446//527
W7//S65T	2.84	49	476 (505)//511
W7//10cQ69K	4.29	49	476 (505)//529
W7//Topaz	4.2	52	476 (505)//527
W1B//Topaz	3.76	52	476 (505)//527
W1C//Topaz	3.4	ND	495//527
Sapphire//Topaz	2.2	60	511//527
BFP5//RSGFP4[d]	(1.8)[d]	ND	450//505

[a] All GFP tandem proteins have an N-terminal histidine tag and a 25-amino acid linker, except for BFP//RSGFP4 (see footnote *c*), and were purified from bacteria by nickel chelate chromatography.

[b] The maximum ratio change was calculated from the acceptor/donor emission ratio before and after cleavage by trypsin to completion. ND, Not determined.

[c] Values in parentheses indicate secondary emission maxima.

[d] BFP5//RSGFP4 has a 20-amino acid linker that is cleaved by factor Xa [R. D. Mitra, C. M. Silva, and D. C. Youvan, *Gene* **173,** 13 (1996)]. Because this linker is different from the linker in the other tandem proteins, the maximal ratio change cannot be directly compared with the other values.

by trypsin[10] or factor X_a[11] have been inserted between donor and acceptor to demonstrate protease-induced FRET changes. After cleavage by the respective protease, energy transfer is lost, manifested in a changed ratio between donor and acceptor emission.

To compare different combinations of GFP mutants with respect to FRET, fusion proteins with a variety of donor and acceptor GFPs, connected by a linker cleavable by trypsin, were expressed in bacteria, and purified. Emission ratios were then measured before and after cleavage by trypsin, and the changes of these ratios calculated (Table II). The linker used in this study consists of 25 amino acids (sequence SSMTG GQQMG RDLYD DDDKD PPAEF) and is derived from the vector pRSETB (Invitrogen, Carlsbad, CA) as previously described.[10] From these data, it is apparent that the best R_0 values do not always determine the best measured ratio changes. This is the case because the ratio change obtained is not only dependent on the amount of FRET before cleavage, but also on

[11] R. D. Mitra, C. M. Silva, and D. C. Youvan, *Gene* **173,** 13 (1996).

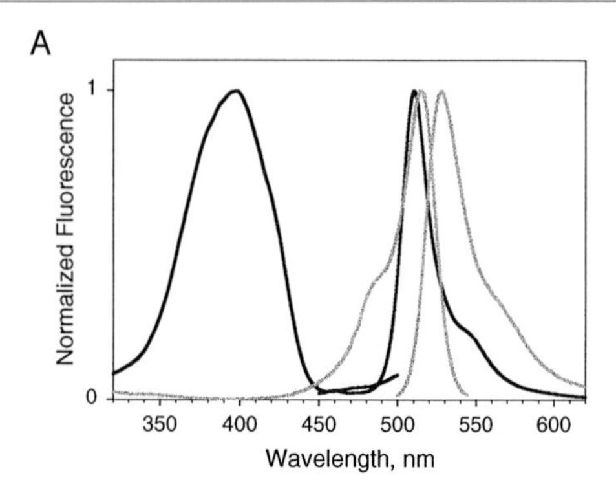

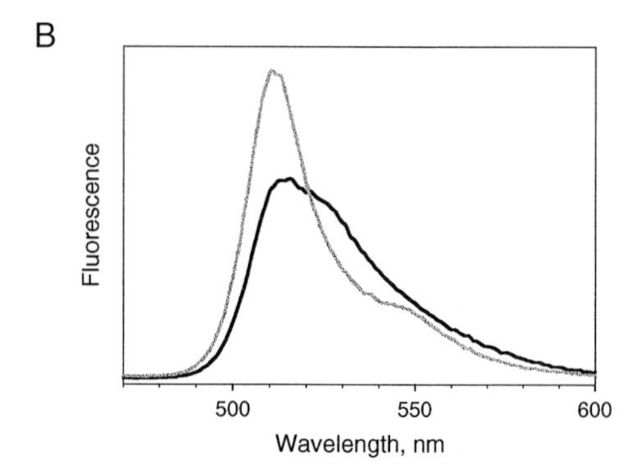

FIG. 1. Tandem GFP Sapphire–Topaz: Spectral overlap and cleavage by trypsin. This combination is an example of a good FRET-promoting overlap for donor emission and acceptor excitation, but it has a large interfering emission overlap. The result is a relatively poor ratio change after cleavage, making this combination useless in practice. (A) Normalized excitation and emission spectra; the donor Sapphire in black, the acceptor Topaz in gray. (B) Emission spectrum of the tandem Sapphire–Topaz before (black) and after (gray) cleavage of the 25-amino acid linker by trypsin. Excitation was at 400 nm.

how well it can be measured. This is illustrated by three examples of donor–acceptor combinations shown in Figs. 1–3. To facilitate comparisons, normalized excitation and emission spectra are shown in Figs. 1A, 2A, and 3A.

Three types of spectral overlap can play a role in FRET: (1) donor emission and acceptor excitation, (2) donor and acceptor emission, and (3) donor and acceptor excitation. Let us examine the combination of Sapphire–Topaz (Fig. 1A), which shows almost perfect spectral overlap

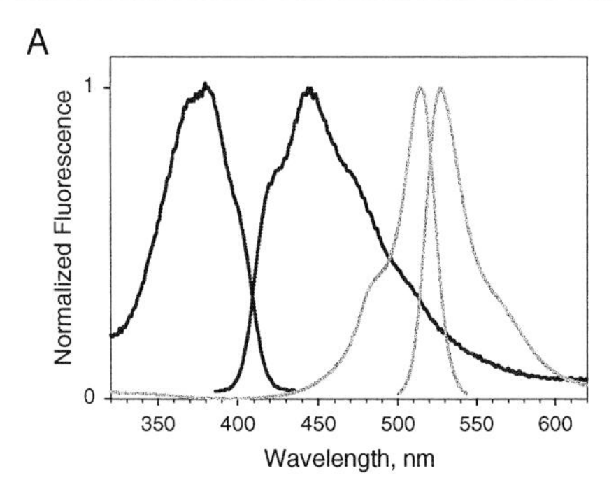

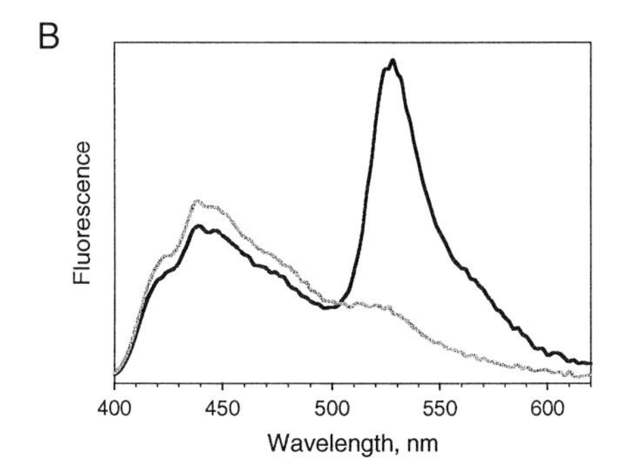

FIG. 2. Tandem GFP P43–Topaz: Spectral overlap and cleavage by trypsin. This combination is an example of a relatively poor FRET-promoting overlap of donor emission and acceptor excitation, but with a good separation of the emission spectra. The result is a good ratio change after cleavage, mostly owing to a relatively large change in acceptor emission. The absolute signals, however, are fairly low. (A) Normalized excitation and emission spectra; the donor P43 in black, the acceptor Topaz in gray. (B) Emission spectrum of the tandem P43–Topaz before (black) and after (gray) cleavage of the 25-amino acid linker by trypsin. Excitation was at 382 nm.

between donor emission and acceptor excitation. As described in Eqs. (1)–(3), this overlap is related to how much FRET can be obtained and thus is reflected in a high R_0 value. The high quantum yield of Sapphire and the high extinction coefficient of Topaz also contribute to this good value. However, the obtained ratio change in the cleavage experiment with trypsin is rather poor (Fig. 1B), because the emission spectra of Sapphire and Topaz overlap too much (Fig. 1A). As the amplitude of the donor emission peak increases during cleavage, the shoulder of the donor emission

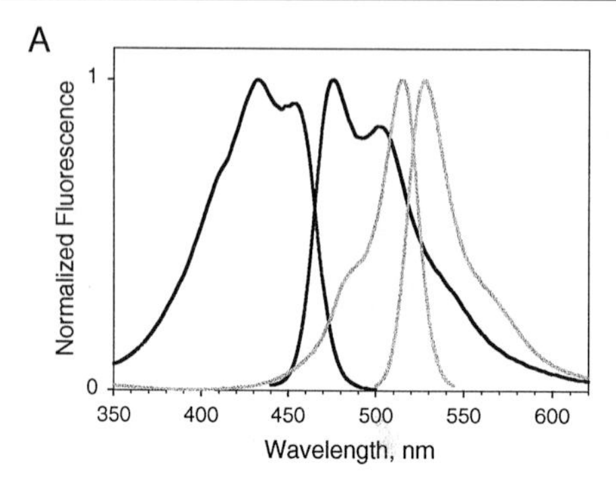

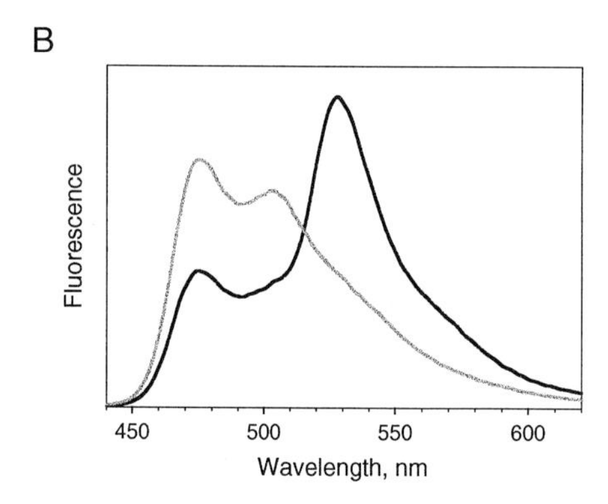

Fig. 3. Tandem GFP W7–10cQ69K: Spectral overlap and cleavage by trypsin. This combination compromises FRET-promoting overlap of donor emission and acceptor excitation with some interfering overlap of the two emission spectra. The result is a good ratio change after cleavage. The signal intensities are relatively high with these mutants. Therefore, this combination is preferred for *in vitro* experiments. For cellular applications, related spectral mutants that fold more efficiently are required. (A) Normalized excitation and emission spectra; the donor W7 in black, the acceptor 10cQ69K in gray. (B) Emission spectrum of the tandem W7–10cQ69K before (black) and after (gray) cleavage of the 25-amino acid linker by trypsin. Excitation was at 432 nm.

also increases and partially compensates for the decrease in the acceptor emission, and the observed net change is smaller than expected. Similarly, the change in the acceptor emission interferes with the change in donor emission, thus reducing the net change. Better separation between the emission spectra would make Sapphire–Topaz a better FRET pair combination. This would require a larger Stokes shift in Topaz. The overlap of donor and acceptor excitation is small enough not to cause any significant spectral cross-talk with Topaz because Sapphire can be selectively excited around 400 nm (Fig. 1A).

The second example illustrates an opposite case of spectral overlap. The blue–yellow combination P43–Topaz (Fig. 2A) has only a small overlap between donor emission and acceptor excitation. However, this pair exhibits a relatively high ratio change (Fig. 2B) measured for the combinations tested with the same linker. This most likely is due to the wide separation of donor and acceptor emission (Fig. 2A) that allows efficient resolution of the two. The high extinction coefficient of Topaz also has a positive influence on R_0 and FRET. Replacement of the acceptor Topaz by a shorter wavelength mutant such as S65T would be expected to increase FRET owing to a better overlap of acceptor excitation and P43 emission. However, the long emission tail of P43 would overlap more with the acceptor emission and therefore interfere with ratio change measurements. The excitation spectra are again sufficiently separated to minimize cross-talk (Fig. 2A). However, P43 is a relatively dim mutant, thus restricting the absolute brightness and detectability achieved with this combination, and it is also more susceptible to photobleaching than other mutants.[12]

The third example, the cyan–yellow combination W7–10cQ69K (Fig. 3A), is a compromise for the three types of spectral overlap discussed as well as for brightness. The donor emission overlaps the acceptor excitation reasonably well. The left side of the biphasic donor emission helps to obtain a useful resolution of the two emission spectra. There is a small overlap between the two excitation spectra, but cross-talk between GFP mutants can be kept to a minimum if care is taken to excite at relatively short wavelengths within the donor excitation spectrum. The obtained ratio change in the cleavage experiment (Fig. 3B) was slightly lower than with P43–Topaz. Overall, the cyan mutants are generally better for FRET-based applications than P43 and other blue mutants because they are brighter and have higher quantum yields and improved photostability. Good separation of emission spectra and of excitation spectra is particularly important if bandpass filters must be used as in a microscope or a microplate reader, rather than distinct wavelengths as is the case with a fluorometer. Combined curves of cleavage experiments as in Figs. 1B, 2B, and 3B are quite helpful in designing emission filters for FRET. They indicate in which spectral range on each side of the cross-over point the intensity changes of donor and acceptor can most efficiently be recorded.

On the basis of the results presented above, there is room for developing additional mutants for use in energy transfer experiments. For FRET,

[12] R. Rizzuto, M. Brini, F. DeGiorgi, R. Rossi, R. Heim, R. Y. Tsien, and T. Pozzan, *Curr. Biol.* **6,** 183 (1996).

fluorophores are generally preferred that have narrow peaks of excitation and emission with large Stokes shifts. All blue and cyan fluorescent proteins (BFPs and CFPs, respectively) have wide emission spectra with long tails. In addition, BFPs also have a low quantum yield. Sapphire could be a good donor with its long Stokes shift, spectral shapes, and quantum yield, but until now there has been no suitable GFP-based acceptor available for Sapphire. The acceptor candidate Topaz and other yellow fluorescent proteins (YFPs) have small Stokes shifts; thus further red-shifting may be an improvement. Mutagenesis may lead to additional mutants. How new GFP mutants fit as potential FRET partners can be determined by the considerations outlined above.

Influence of Linker; Dimerization

From the Foerster equation it is evident that the amount of FRET in a tandem GFP depends on the sequence and the length of the linker, because they determine distance and orientation of the FRET partners. Therefore, evaluations of GFP FRET pairs with regard to efficiency of energy transfer should always be done with the same linker to allow direct comparison. With floppy linkers, the GFP portions might weakly interact with each other, because native GFP has been found to self-associate in solution under a variety of conditions,[5] and recombinant wild-type GFP,[13] P43,[14] and 10c[15] crystallize in dimers. Short linkers would sterically disable this intramolecular association. Indeed, the ratio change progressively dropped with linker lengths shorter than 24 amino acids in a comparative study. Part of this effect could have been caused by poor folding and fluorescence development leading to reduced FRET. Rigid linkers with a distinct mode of three-dimensional structure do not allow this intramolecular association. However, certain linkers of this type can also increase energy transfer by bringing the FRET partners particularly close together with favorable orientation, as ratio change values of up to 9 have been measured with specific linkers using the GFP mutants W1B and Topaz.[16]

GFP association, which has been reported to be concentration depen-

[13] F. Yang, L. G. Moss, and G. N. Phillips, *Nature Biotechnol.* **14,** 1246 (1996).

[14] F. M. Wachter, B. A. King, R. Heim, K. Kallio, R. Y. Tsien, S. G. Boxer, and S. J. Remington, *Biochemistry* **36,** 9759 (1998).

[15] S. J. Remington, personal communication (1998).

[16] A. Miyawaki, personal communication (1998).

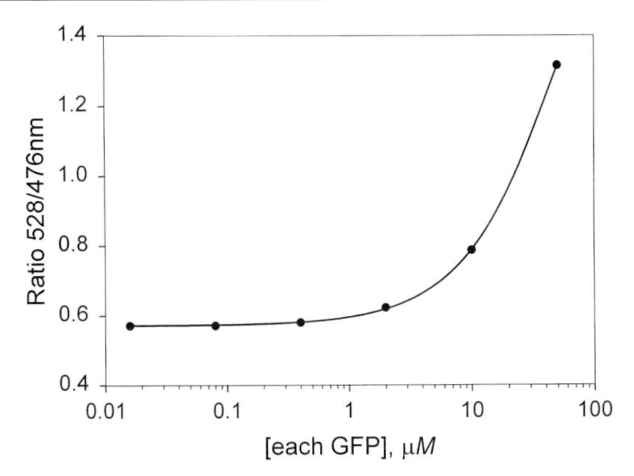

FIG. 4. Effect of high concentration of GFP on dimerization and FRET. To test whether high concentrations of GFP can cause dimerization of donor and acceptor that may result in FRET, emission spectra of mixtures of the donor W1B and the acceptor Topaz at various concentrations, but always in equimolar ratios, were measured while exciting the donor. The ratio between acceptor and donor emission as an indicator for FRET was calculated and plotted against the GFP concentration. There was significant energy transfer occurring at concentrations exceeding 2 μM for each GFP (4 μM total concentration); at concentrations up to 2 μM for each GFP, however, dimerization is negligible. For other GFP mutants, the curve could be slightly shifted in either direction, because small changes in the surface structure could affect dimerization. Excitation was at 432 nm.

dent,[5] raises the possibility that GFP-induced dimerization could interfere with FRET, especially within cells. To determine whether and at what concentrations dimerization-induced FRET might interfere with experiments, the cyan blue mutant W1B and the yellow version Topaz were mixed at various concentrations in a 1:1 molar ratio, and emission spectra were measured while exciting the donor W1B. As shown in Fig. 4, the acceptor/donor emission ratio increases with increasing concentration of GFP, indicating association of GFP at high concentrations. The effect, however, seems to be negligible at concentrations of up to about 2 μM; GFP concentrations below this amount usually give sufficient signal for use in *in vitro* experiments. Results from crystal structures indicate that GFP is not an obligate dimer. Whereas wild-type GFP,[13] P43,[14] and 10c[15] were found to crystallize in dimers, monomeric crystals were also formed by S65T[17] and wild-type GFP,[18] although under different crystal growth condi-

[17] M. Ormo, A. B. Cubitt, K. Kallio, L. A. Gross, R. Y. Tsien, and S. J. Remington, *Science* **273,** 1392 (1996).

[18] K. Brejc, T. K. Sixma, P. A. Kitts, S. R. Kain, R. Y. Tsien, M. Ormo, and S. J. Remington, *Proc. Natl. Acad. Sci. U.S.A.* **94,** 2306 (1997).

tions. Thus, it is unclear whether and how much diversity in dimer formation is dependent on the nature of the mutant. Unfortunately, there are no comparison studies of mutants in solution available. The crystal structure of wild-type GFP suggests which residues are likely to be important for dimerization.[13] The hydrophobic side chains from A206, L221, and F223, and several hydrophilic residues in the range from amino acids 142 to 151, among others, are located in the dimer interface. Targeted mutagenesis of some of these residues could be an approach to reduce the effect of dimerization.

Application of Fluorescence Resonance Energy Transfer

After careful consideration and planning of the aforementioned properties of GFP mutants, the use of GFP for different types of FRET—intra- and intermolecular FRET, *in vitro* and in cell cultures—employing several different mutants has been successful.

We have used a variety of tandem GFP combinations as discussed above for *in vitro* cleavage experiments and had best success with the cyan blue–yellow combination W7–10cQ69K, with regard to ratio change and signal intensity. The *in vitro* experiments were all done with proteins purified from bacteria that were expressing at room temperature. Some of these mutants are not particularly suitable for mammalian expression. In cell-based experiments, folding mutants such as W1B and Topaz, or closely related ones, are preferred over W7–10cQ69K because they give a brighter signal, both in transient and stable transfected mammalian cells, as found by fluorescence-activated cell sorting (FACS) analysis. These mutants also give a higher yield of soluble protein in bacteria expressing at room temperature. To a smaller extent, the introduction of mammalian codon usage in most folding mutants also helps to improve signals in mammalian cells and does not have negative effects on expression in bacteria.

Such a system of tandem GFPs with a cleavable linker can provide a facile method to assay for any protease that has a recognition sequence. Fluorometric protease assays that had been developed previously employed, typically, the fluorophore 5-(2-aminoethylamino)naphthalene-1-sulfonic acid (EDANS) and the quencher 4-(4-dimethylaminophenylazo)-benzoyl (DABCYL), coupled to a peptide, and relied on the absolute measurement of fluorescence from the donor.[19] A ratiometric, GFP-based system overcomes the problems of background absorbances of the sample, fluctuations in the excitation intensity, and variations in the absolute amount of substrate. Furthermore, it has been shown to work in cells without the

[19] C. G. Knight, *Methods Enzymol.* **248,** 18 (1995).

need for organic chemistry and microinjection. Indeed, one report describes a substrate for CPP32 (caspase-3) using GFP-based FRET.[20] The blue and green mutants EBFP and EGFP (Table I), respectively, have been fused to the termini of a cleavable linker consisting of 18 amino acids with the target sequence DEVD in the center. This substrate was transfected into mammalian 293 cells for FACS analysis, and the cells were compared with cells in which apoptosis had been induced by coexpression of the protein kinase Rip. Only nonapoptotic cells showed up in the acceptor channel on donor excitation, indicating that these cells were exhibiting FRET that was lost in apoptotic cells owing to cleavage of the substrate by activated CPP32. No data from the donor channel, or spectroscopic data, were shown. Thus, a comparison with other FRET studies is not possible, but the results still demonstrate the successful use of GFP-based FRET to measure cellular protease activity.

A FRET pair consisting of the mutant W1B (for CFP) and a mutant closely related to Topaz (for YFP) has been used in another case of intramolecular FRET to measure conformational changes. Miyawaki *et al.*[21] have developed calcium sensors—designated cameleons—consisting of calmodulin linked to the calmodulin-binding domain of myosin light chain kinase (called M13) labeled with CFP and YFP on opposite sides of the molecule. In the absence of calcium, the molecule is in a relaxed, extended conformation. Calcium induces a contraction of the sensor as the M13 peptide binds to calmodulin, thus bringing CFP and YFP closer together, resulting in increased FRET. A Ca^{2+}-dependent maximal ratio change of 1.5 was observed with this probe in the cuvette, and a comparison was made with the donor–acceptor combination EBFP and EGFP (Table I), which yielded a slightly higher ratio change of 1.8. Both combinations of GFP mutants proved to be useful in HeLa cells whose calcium levels were modulated by several agonists and ionophores as measured by a microscope imaging system. The CFP–YFP combination, however, was preferred in cellular experiments owing to higher and more consistent signals, and yielded a maximal ratio change of 1.6, demonstrating accordance with the *in vitro* data. A convenient feature of these GFP-based Ca^{2+} probes is that they can be targeted to different subcellular compartments to measure calcium selectively, which differs from organelle to organelle. With specific localization signals, the cameleons were targeted to the nucleus or to the endoplas-

[20] X. Xu, A. L. V. Gerard, B. C. B. Huang, D. C. Anderson, D. G. Payan, and Y. Luo, *Nucleic Acids Res.* **26,** 2034 (1998).

[21] A. Miyawaki, J. Llopis, R. Heim, J. M. McCaffery, J. A. Adams, M. Ikura, and R. Y. Tsien, *Nature* (*London*) **388,** 882 (1997).

mic reticulum. Without any localization signal, the probe resided in the cytoplasm because it was too large to diffuse through nuclear pores.

The same authors also successfully demonstrated intermolecular FRET using the same elements of the described Ca^{2+} GFP probe. They coexpressed CFP-tagged calmodulin and YFP-labeled M13 from individual plasmids. A Ca^{2+}-induced ratio change was similarly observed, both *in vitro* and in HeLa cells, using these noncovalently attached partners. The maximal change was slightly larger than for the fused version, probably because the fusion protein does not completely lose FRET in the extended conformation at low Ca^{2+}. If there was significant GFP-induced dimerization, either in the fused or the nonfused system, it apparently did not severely interfere with the functioning of the cameleon sensor. If at all, it may have slightly decreased the dynamic range.

The blue and green mutants P43 and S65T, respectively, were also used by Mahajan *et al.*[22] to demonstrate another protein–protein interaction in several mammalian cell lines. Bcl-2 and Bax are known to act antagonistically in regulation of apoptosis. When tagged with P43 and S65T, Bcl-2 and Bax were found, by using a microscope to measure FRET, to interact with each other. The signal in the acceptor channel was compared with signals obtained from noninteracting proteins, but no ratio measurements were reported. The GFP tagged, overexpressed proteins also proved to be functionally similar to the corresponding native proteins in apoptosis regulation.

Mutants and Future

The successful use of the GFP FRET technology is dependent on the mutants used and on the particular application. The available equipment, including optics, and the biological system used (e.g., intra- versus extracellular systems) may determine which mutants are preferable to use. For *in vitro* experiments, folding efficiency and expression levels are less important because sufficient amounts of protein can be expressed in large bacterial cultures with subsequent concentration. Experiments in cell cultures, however, often benefit from higher expression levels. Two types of mutants have been found to improve functional expression in mammalian cells. One type involves silent mutations changing the AT-rich codons of *Aequorea victoria* to codons more typical of mammalian genes. They lower the number of incompletely synthesized molecules and thus increase the expression level. Other organisms such as yeast or plants, however, have a different

[22] N. P. Mahajan, K. Linder, G. Berry, G. W. Gordon, R. Heim, and B. Herman, *Nature Biotechnol.* **16,** 547 (1998).

codon usage and require other changes for the same effect. The other type of mutation causes substitution of amino acids that influence spectral properties or the folding of the protein. Combinations of different folding mutations showed that they are only partially cooperative. There is definitely a demand for further GFP mutants for FRET. In particular, further red-shifted acceptors or acceptors with bigger Stokes shift could allow the combination with green donors or at least provide a better separation of donor and acceptor emission. The currently used donors of the BFP and CFP groups all have wide spectra. Sharper peaks and better quantum yields, especially in the case of BFP, would improve their use in FRET. The crystal structures give us some ideas about changes that could be promising for further testing. Interestingly, several mutations that affect spectral properties, with the exception of T203 mutants, were found before the crystal structure was solved, mostly by random mutagenesis. However, many mutations create species that are dim or nonfluorescent, probably by destabilizing the barrel structure surrounding the chromophore. Secondary mutations might be required to restore stability. Thus, a combination of targeted and random mutagenesis might be the key.

For some specific applications of FRET, when extreme conditions are expected, it is advisable to test GFP mutants for suitability. For instance, some mutants, including S65T, EGFP, EBFP, and Topaz, were found to be more sensitive to pH values below 7.0[23,24] Thus, for experiments in acidic compartments, suitable mutants should be selected also by the criterion of pH stability.

In summary, the results obtained to date with GFP-based FRET have established the application of this technique in the investigation of *in vitro* and intracellular biological phenomena and have pointed at areas of further improvement. Biochemical applications using purified GFPs have been useful in delineating issues regarding GFP mutants to use, linker composition, etc. Importantly, cellular applications seem to be particularly well suited for these genetically encoded dyes.

Acknowledgments

I thank Drs. Andrew Cubitt and Brian Pollok for helpful advice and discussion.

[23] G. H. Patterson, S. M. Knobel, W. D. Sharif, S. R. Kain, and D. W. Piston, *Biophys. J.* **73,** 2782.

[24] J. Llopis, J. M. McCaffery, A. Miyawaki, M. G. Farquhar, and R. Y. Tsien, *Proc. Natl. Acad. Sci. U.S.A.* **95,** 6803 (1998).

[35] Use of Codon-Modified, Red-Shifted Variants of Green Fluorescent Protein Genes to Study Virus-Mediated Gene Transfer

By CHARLES J. LINK, JR., SUMING WANG, REBECCA R. MULDOON, TATIANA SEREGINA, and JOHN P. LEVY

Introduction

One central requirement of gene therapy is the need to analyze gene transfer efficiency into target cells either *in vitro* or *in vivo*. The most common methods of gene transfer employ modified viruses as delivery vectors. Viral vectors have achieved widespread use in phase I and II clinical trials. These initial trials indicate that a great deal of further effort is needed to optimize viral vectors. Continued improvements in the available viral vector classes are actively pursued, as well as the development of new types of viral vectors. In this chapter we review examples of exploiting the green fluorescent protein (GFP) marking system to study viral vector gene transfer relevant to human gene therapy. Retroviral vectors can efficiently transduce rodent and human lymphocytes. Several groups now employ gene-modified lymphocytes as adoptive immunotherapy for patients who have failed allogeneic bone marrow transplantation. In addition to such *ex vivo* therapeutic approaches, a primary focus of more recent approaches for cancer has been the development of improved *in vivo* gene delivery vector systems. One such system is the herpes simplex vector amplicon system. Red-shifted, codon-modified GFP genes permit the rapid visualization of herpes vectors within a few hours after target cell transductions. The small gene size and remarkable ease of visualizing red-shifted GFP variants using standard fluorescein isothiocyanate (FITC) filter systems makes them ideal for studying virus-mediated gene transfer.

Green Fluorescent Protein

The green fluorescent protein has provided a myriad of applications for biological systems.[1,2] This chapter reviews the use of codon-optimized, red-shifted variants of the GFP gene and their application to the study of virus-mediated gene transfer. The red-shifted mutant was developed by the

[1] D. C. Prasher, V. K. Eckenrode, W. W. Ward, F. G. Prendergrast, and M. J. Cormier, *Gene* **111,** 229 (1992).
[2] D. C. Prasher, *Trends Genet.* **11,** 320 (1995).

0076-6879/99 $30.00

random bacterial mutagenesis of the natural GFP sequence.[3] The serine-to-threonine mutation changes the excitation wavelength so that the protein emits fluorescence at 511 nm.[3] Previous studies demonstrated the utility of condon modification of nonmammalian genes to optimize RNA translation (Fig. 1B).[4] These two approaches combined in concert provide rapid, effective GFP translation in mammalian cells that is easily visualized using standard FITC filter sets (Fig. 1C).[5]

Gene Transfer Efficiency Evaluated by Green Fluorescent Protein

A key part of evaluating and developing any gene therapy strategy is to measure gene transfer efficiency accurately. These data are essential for making predictions about the amount of vector transduction required to obtain a therapeutic level of the transgene *in vivo*. The two steps of this process are to obtain a standardized stock of vector supernatants and then to measure transgene activity. Gene transfer can be evaluated by using marker genes to visualize gene transfer and expression easily. Mammalian codon-optimized, red-shifted mutants of GFP make excellent marker genes because of their greatly enhanced fluorescence characteristics.[4–7] This novel marker protein does not require any special staining or fixation procedures. Single-copy gene expression can be visualized from viral vectors within 5 to 24 hr after transduction.

Fluorescence Detection of Green Fluorescent Protein-Expressing Cells or Tissues

Cell slide preparations or frozen tissue sections (e.g., tumor biopsies) can be visualized with a Nikon Labophot-2 fluorescence microscope (Fryer Company, Huntley, IL). The filter cube in the microscope is the FITC dichromic filter set (excitation at 450–490 nm and emission at 520 nm). Photographs are taken using a Nikon Microflex AFX-DX system (Fryer Company). Selected high-power fields can be photographed with and without background light to calculate the percentage of GFP-positive cells. Cytometry of transduced cells is performed in an Epics Profile II analyzer (Coulter, Hialeah, FL), with an excitation light of 488 nm. Cells are analyzed using a 525-nm bandpass filter set (Coulter). All fluorescence-activated cell

[3] R. Heim, A. B. Cubitt, and R. Y. Tsien, *Nature (London)* **373,** 663 (1995).

[4] J. Haas, E. C. Park, and B. Seed, *Curr. Biol.* **6,** 315 (1996).

[5] S. Zolotukhin, M. Potter, W. W. Hauswirht, J. Guy, and N. Muzyczka, *J. Virol.* **70,** 4646 (1996).

[6] J. P. Levy, R. R. Muldoon, S. Zolotukhin, and C. J. Link, *Nature Biotechnol.* **14,** 610 (1996).

[7] R. R. Muldoon, J. P. Levy, S. R. Kain, P. A. Kitts, and C. J. Link, *BioTechniques* **22,** 162 (1997).

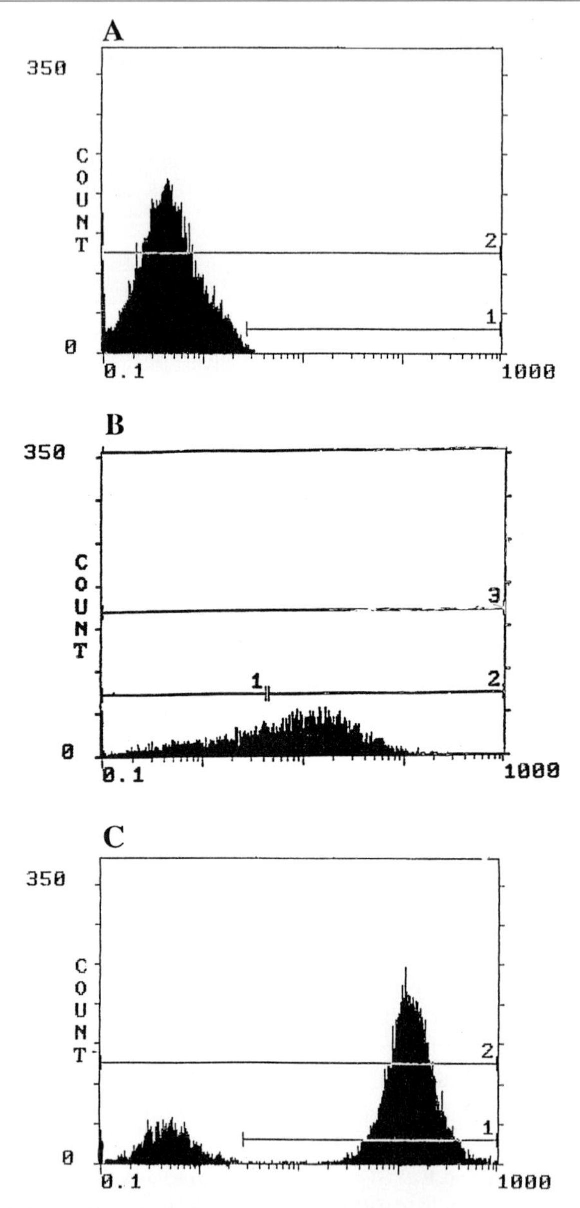

FIG. 1. FACS analysis of PA317 murine vector producer cells expressing codon-modified GFP genes. (A) PA317 packaging cells without GFP (negative control. No significant fluorescence is detected. (B) PA317 cells containing a condon-optimized "humanized" GFP gene.[4] Note the 2-log shift in detected fluorescence with relatively wide variation in the cell population. (C) PA317 cells expressing a humanized, red-shifted mutant of GFP.[5,6] Note the much greater shift in average detected fluorescence in these murine cells. The numerals 1, 2, and 3 refer to the gates measured to compare positive and negative cell populations.

sorting (FACS) analyses use the FL1 emission channel to monitor green fluorescence (FITC filter monitor).

Retroviral Gene Transfer

Murine retroviral vectors have been available for more than a decade.[8] These vectors employ packaging cell lines that are designed to provide viral structural proteins in trans and contain retroviral vectors that include a ψ sequence or a packaging signal to permit vector RNA incorporation into virions. Murine retroviral vectors designed to minimize the possibility of recombination resulting in regeneration of a replication-competent virus have proved to be extremely efficient for gene transfer into mammalian cells, with efficiencies as high as 95% in cultured murine fibroblast cell lines.[8-10] For more detailed information on such vectors, see the chapter by Mazo and colleagues ([28] in this volume[11]). Murine amphotropic retroviral vectors are the most commonly used vectors in human gene therapy clinical trials.[12] Most of these protocols employ *ex vivo* modification of hematopoietic cells in an attempt to mark the transferred cells or to provide a therapeutic gene. This includes the first demonstration of gene transfer and expression in human lymphocytes as part of adoptive immunotherapy strategies for solid tumors.[13]

Lymphocyte Suicide Gene Therapy Strategy in Adoptive Immunotherapy Using Gene-Modified Lymphocytes in Patients with Relapsed Leukemia

The increasing interest in gene transfer into human lymphocytes has been spurred by the data from gene therapy trials designed to reduce the lethal complications of graft-versus-host disease (GvHD) that are associated with bone marrow transplantation (BMT). Allogeneic BMT cures leukemia by means of myeloablation induced by the preparative regimen and by transfer in the bone marrow allograft of immunocompetent donor cells that exert an antileukemic effect called graft versus leukemia (GvL). The central goal of the research in this field is to augment the GvL effect and separate it from the GvHD. The removal of T cells from the marrow graft represents the best way to prevent the development of acute and

[8] A. D. Miller, *Curr. Top. Microbiol. Immunol.* **158,** 1 (1992).

[9] A. D. Miller and C. Buttimore, *Mol. Cell. Biol.* **6,** 2895 (1986).

[10] I. M. Verma, *Sci. Am.* **263,** 68 (1990).

[11] I. A. Mazo, J. P. Levy, R. R. Muldoon, C. J. Link, Jr., and S. R. Kain, *Methods Enzymol.* **302,** [28], 1999 (this volume).

[12] J. A. Roth and R. J. Cristiano, *J. Natl. Cancer Inst.* **89,** 21 (1997).

[13] S. A. Rosenberg, P. Aebersold, K. Cornetta, A. Kasid, R. Morgan, R. Moen, E. Karson, M. T. Lotze, J. C. Yang, S. Topalian, M. J. Merino, K. Culver, A. D. Miller, R. M. Blaese, and W. F. Anderson, *N. Engl. J. Med.* **323,** 570 (1990).

chronic GvHD. However, T cell depletion results in increased graft failure and leukemia relapse so that overall survival is not improved. Kolb *et al.* renewed interest in the use of donor leukocyte infusions to treat allogeneic BMT patients after relapse.[14] Many groups around the world are now intensively investigating this approach and the optimal dose and interval of donor lymphocyte infusions are unknown. Strategies, pioneered by Bordignon and colleagues, have suggested a way to combat severe GvHD disease, should it occur in these cellular therapy attempts, by using selective suicide ablation of transferred lymphocytes.[15] Suicide genes have been developed that permit the selective destruction of target cells after gene transfer. The herpes simplex virus thymidine kinase (HSV *tk*) gene is a selectable marker or "suicide" gene that sensitizes transduced cells to ganciclovir (GCV). GCV is a substrate for phosphorylation by HSV *tk* gene product, resulting in inhibition of DNA polymerase and cell death.[16,17] Because the human thymidine kinase enzyme, which is present in all human cells, has low affinity for GCV, little systemic toxicity is observed. Lymphocytes are logical targets for gene transfer and were used in the first gene transfer experiment in humans.[13] Several groups have already reported progress by demonstrating that T cells could be transduced with a retroviral vector containing the HSV *tk* gene.[15,18,19] Initial clinical results demonstrate the feasibility of this approach in arresting GvHD in patients after donor lymphocyte infusions.[20] GFP expression with the HSV *tk* gene will permit sorting of such gene-modified lymphocytes in future protocol designs.

Gene Transduction Efficiency in Rat Lymphocytes

Whole blood is extracted by cardiac puncture after complete anesthesia of adult rats (the animals are then euthanized in CO_2). Blood aliquots (10

[14] H. J. Kolb, J. Mittermuller, C. H. Clemm, E. Holler, G. Ledderose, M. Brehm, and W. Wilmanns, *Blood* **76,** 2462 (1990).

[15] C. Bordignon, C. Bonini, S. Verzeletti, N. Nobili, D. Maggioni, C. Traversari, R. Giavazzi, P. Servida, E. Zappone, E. Benazzai, B. Massimo, F. Porta, G. Ferrari, F. Mavilio, S. Rossini, R. M. Blaese, and F. Candotti, *Hum. Gene Ther.* **6,** 813 (1995).

[16] F. L. Moolten, *Cancer Res.* **46,** 5276 (1986).

[17] F. L. Moolten and J. M. Wells, *J. Natl. Cancer Inst.* **82,** 297 (1990).

[18] P. Tiberghien, C. W. Reynolds, J. Keller, S. Spence, M. Deschaseaux, J. Certoux, E. Contassot, W. J. Murphy, R. Lyons, Y. Chiang, P. Herve, D. Longo, and F. W. Ruscetti, *Blood* **84,** 1333 (1994).

[19] C. J. Link, Jr., R. K. Burt, A. E. Traynor, W. R. Drobyski, T. Seregina, J. P. Levy, L. Gordon, S. T. Rosen, W. H. Burns, B. Camitta, J. Casper, M. Horowitz, M. Juckett, C. Lawton, D. Margolis, D. Pietryga, P. Rowlings, C. Taylor, M. Furtado, J. Stefka, S. Gupta-Burt, H. Kaiser, and D. H. Vesole, *Hum. Gene Ther.* **9,** 115 (1998).

[20] C. Bonini, G. Ferrari, S. Verzeletti, P. Servida, E. Zappone, L. Ruggieri, M. Ponzoni, S. Rossini, F. Mavilio, C. Traversari, and C. Bordignon, *Science* **276,** 1719 (1997).

ml) are diluted in tissue culture medium (RPM 1640 or equivalent) (no serum), layered over 4 ml of lymphocyte separation medium (Lymphoprep; GIBCO-BRL, Gaithersburg, MD), and centrifuged (2000 rpm), and then the interface layer is extracted. Cells are then checked for viability and suspended in fresh medium at 10^6 cells/ml. Cells are then stimulated in flasks with phytohemagglutinin at 5 μg/ml and human recombinant interleukin 2 (rIL-2; Cetus, Emeryville, CA) at 100 units/ml for 48 to 96 hr. After stimulation, cells are maintained at a density of about 10^6 cells/ml.

Standard Single Infection Cycle for Vector-Mediated Gene Transfer. Stimulated leukocytes are suspended in a viral supernatant for 24 hr with protamine (10 μg/ml). Lymphocytes are incubated at a density of 1×10^6 cells/well in 24-well plates with 1 ml of viral supernatant. The titers of the viral supernatants used should be greater than 10^5 colony-forming units (CFU)/ml for gene transfer into rat lymphocytes. After infection cells are washed by centrifugation and suspended in RPMI 1640 supplemented with 10% fetal bovine serum (FBS) and rhIL-2 (100 mg/ml) in the presence or absence of G418 (0.4 mg/ml). This generally results in a transduction frequency of 1 to 3%.

Repeated Cycles of Infection and Cocultivation of Rat Lymphocytes with Retroviral Vector Producer Cells. To increase the gene transfer rate into lymphocytes a number of additional methods can be attempted.

Multiple infection cycles: The lymphocytes are repeatedly exposed to fresh viral supernatants. *Cocultivation with irradiated retroviral vector producer cells:* Lymphocytes can be cocultivated with irradiated vector producer cells (VPCs, 3000 rad) for 48 to 72 hr with protamine and then selected in G418.

Cocultivation in Transwell plates: Cocultivation in Transwell plates (Costar, Cambridge, MA) permits vector passage but prevents cell-to-cell contact by means of a separating barrier. VPCs are seeded in the cluster plate wells and incubated overnight. Stimulated lymphocytes (5×10^5 cells) are placed into Transwells and grown for 72 hr with protamine. To estimate infection frequencies 24 to 48 hr after infection, lymphocytes are analyzed by FACS using standard FITC filters to detect the humanized, red-shifted GFP-mediated fluorescence. Cell growth is evaluated on day 14 after plating. No surviving cells should be detected in wells containing nontransduced cells cultured in the presence of G418.

With these methods, the transduction frequency can be increased to between 8 and 14%. GFP-positive lymphocytes are easily observed by fluorescence microscopy or analyzed by FACS. Efficient gene transfer can be accomplished with GFP retroviral vectors into mammalian T cells and this method can be used to develop improved gene transfer procedures. FACS of these fluorescent cells may eliminate the time-consuming requirement for drug selection altogether.

Transduction of Human Peripheral Blood Lymphocytes

Human peripheral blood mononuclear cells are isolated by Ficoll-Hypaque density gradient centrifugation, washed in serum-free RPMI 1640 (GIBCO-BRL), and cultured in RPMI 1640 supplemented with 5% human serum, phytohemagglutinin (1:100, v/v; GIBCO-BRL), and a 100-IU/ml concentration of recombinant human interleukin-2 (rhIL-2; Chiron, Emeryville, CA) at 37° in 5% CO_2. Forty-eight hours later rhIL-2 is added to a final concentration of 500 IU/ml. On day 6, cells are transduced at 37° with the retroviral vector. Lymphocytes are suspended in vector supernatant daily for 3 days. Approximately 48 hr later the cells are suspended in medium containing G418 for 72 hr. Cells are then grown without G418 and 4 days later can be tested by quantitative polymerase chain reaction (PCR) or FACS for the presence of the GFP gene. Of note, gene-modified lymphocytes remain alloreactive after retroviral transduction. Transduced, selected lymphocytes, when analyzed for the percentage of cells with different phenotypes, usually show both CD4 and CD8 subpopulations being represented. Studies have employed these methods and others to optimize gene transfer into lymphocytes. For example, in one study targeting lymphocytes as a therapeutic approach for acquired immunodeficiency syndrome (AIDS) patients, investigators have combined phosphate depletion (to upregulate the phosphate transporter that serves as the viral receptor), lower-temperature tissue culture (to increase viral titer), and pseudotyped envelopes (to increase efficiency) to obtain transduction frequencies of up to 40%.[21] GFP will permit FACS of transduced, gene-modified human lymphocytes in future trials now being designed by our group and others. For solid tumor gene therapy applications, retroviral vector transduction is less than optimal, especially when *in vivo* gene delivery is required. These challenges have led investigators to explore alternative viral vector systems such as those based on the herpes simplex virus.

Herpes Simplex Viral Vectors

Herpes Simplex Viral Vectors for in Vivo Gene Delivery

Modified herpes simplex virus type 1 (HSV-1) vectors can infect mammalian cells efficiently. HSV-1 is a double-stranded DNA virus of approxi-

[21] B. A. Bunnell, L. M. Muul, R. E. Donahue, R. M. Blaese, and R. A. Morgan, *Proc. Natl. Acad. Sci. U.S.A.* **92,** 7739 (1995).

mately 152 kb, encoding 81 known genes. Its genome contains three replication origins and several proteins that regulate viral and cellular gene expression from three different transcription units termed late, early, or immediate early (IE). Several different helper viruses have been constructed with deletions in essential genes, such as the IE3 gene, that are required for virus replication. Such partially deleted viruses can replicate only in helper cells that complement the gene (e.g., ICP4 protein for IE3-deleted helper virus). The first HSV vectors to be developed carried transgenes in their genomes to transfer expression to target cells. These replication-defective vectors have high titer [10^8 to 10^9 plaque-forming units (PFU)/ml], but sometimes induce significant cell toxicity in the transduced target cells. This class of virus is distinct from retroviruses in that viral genome information is not integrated into the host genome.

Members of the second category of HSV-based vectors are amplicon based. The vectors are generated from plasmids that incorporate both the "a" HSV-1 packaging signal sequence and the viral lytic replication origin (*oriS*). These vectors can replicate in the presence of helper virus and are packaged into infectious HSV-1 virions. These plasmid-based shuttle vectors permit rapid cloning and transfer of genetic information between prokaryotic cells and eukaryotic cells. The amplicon viral stocks tend to have lower vector titers. The titers of amplicon vectors can be increased by serial passaging of the amplicon viral stocks on the complementing cell line. However, this process also increases the probability of generating wild-type virus within the amplicon stocks. In addition, the ratio of amplicon to helper virus changes with each passage. Both HSV vectors and HSV amplicon vectors have demonstrated efficient gene transfer *in vivo*.

Our laboratory and others have developed a second generation of the HSV amplicon system.[22,23] Plasmids containing the HSV-1 lytic replication origin (*oriS*) and the HSV-1 terminal packaging signal sequences can be amplified and packaged into infectious HSV-1 virions in the presence of transactivating helper virus.[24–27] This novel HSV vector, pHE700, also contains Epstein–Barr virus (EBV) sequences that maintain the plasmid as an episome in the transfected cell nucleus.[22] EBV contains a latent replication origin (*oriP*) that directs viral self-replication and maintenance in cells

[22] S. Wang, W.-B. Young, C. Jacobson, and C. J. Link, *Gene Ther.* **4,** 1132 (1997).

[23] S. Wang and C. J. Link, Jr., *Proc. Annu. Meet. Am. Assoc. Cancer Res.* **37,** A2373 (1996).

[24] R. R. Spaete and N. Frenkel, *Cell* **30,** 295 (1982).

[25] A. D. Kwong and N. Frenkel, *J. Virol.* **51,** 595 (1984).

[26] A. I. Geller and X. O. Breakefield, *Science* **241,** 1667 (1988).

[27] A. Geller, K. Keyomarsi, J. Bryan, and A. B. Pardee, *Proc. Natl. Acad. Sci. U.S.A.* **87,** 8950 (1990).

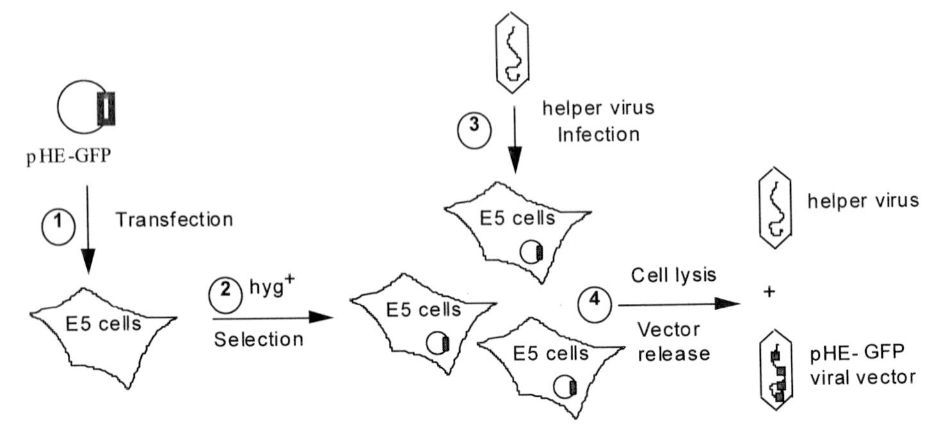

FIG. 2. HSV amplicon packaging sytem. (1) pHE-GFP plasmid is transfected into E5 helper cells that express the IE3 gene in trans. (2) Transfected E5 cells are subjected to hygromycin selection to obtain population of cells containing the pHE vector. (3) Cells are infected with helper virus that provides replication and packaging functions. (4) pHE-GFP vector and helper virus replicate and induce lytic release.

without entering the lytic cycle.[28–30] The Epstein–Barr virus nuclear antigen 1 (EBNA-1) encodes a DNA-binding transactivator for *oriP*.[28–30] The EBV *oriP* and EBNA-1 gene provide more stability to eukaryotic expression vectors. The ΔEBNA-1 gene sequence is a modified version of the native EBNA-1 sequence that has less cytotoxicity compared with the original EBNA-1 gene [B. Sugden (University of Wisconsin, Madison, WI), personal communication, 1997]. The combination of the HSV amplicon with the EBV sequences improves the ease of use of the HSV amplicon system (Fig. 2). The hygromycin resistance gene (hyg^+) was driven by the HSV-1 thymidine kinase (HSV *tk*) promoter and terminated by the HSV *tk* poly(A) signal. The maintenance of the pHE vector as an episome occurs after transfection of pHE700 into E5 cells and selection with hygromycin (Fig. 2). Our replication-incompetent pHE700 vectors maintained wide tropism for delivering transgene(s) with high efficiency both *in vitro* and *in vivo* into both dividing and quiescent cells of the rat brain caudate nucleus or hippocampus.[22] Previously reported amplicon systems were unable to package and maintain large DNA fragments (>15 kb) efficiently.[25] Our improved vector could be produced at high titer and stably packages a vector with a 21-kb DNA insert.[22]

[28] D. Reisman, J. Yates, and B. Sugden, *Mol. Cell. Biol.* **5,** 1822 (1985).
[29] B. Sugden, K. Marsh, and J. Yates, *Mol. Cell. Biol.* **5,** 410 (1985).
[30] J. L. Yates, N. Warren, and B. Sugden, *Nature* (*London*) **313,** 812 (1985).

Construction of pHE Vectors

pHE700 is a 10,617-bp plasmid generated by the combination of components from several other plasmids.[22] Plasmid pTO11 [kindly provided by N. Stow, Medical Research Council (MRC), Glasgow, U.K.] was the source of the HSV "a" packaging signal and the origin of replication, *oriS*. The ΔEBNA-1 and *oriP* sequences were from the plasmid p205 (kindly provided by B. Sugden, University of Wisconsin). The resultant plasmid, pHE100, was restricted and ligated to a fragment from pREP10 (Invitrogen, San Diego, CA) containing the hygromycin resistance gene and a multicloning site. The plasmid generated (pHE600) was again restricted and ligated to the *Sal*I fragment of pCEP4 (Invitrogen) containing the cytomegalovirus (CMV) promoter expression cassette to generate the vector pHE700 (CMV cassette in 5′-to-3′ orientation). pHE700-GFP was constructed by inserting the red-shifted, humanized GFP gene (kindly provided by S. Zolotukhin, University of Florida, Gainsville, FL) into the *Hin*dIII site of the pHE700 vector.

Cells and Viruses

All cells are grown and maintained in Dulbecco's modified Eagle's medium (DMEM; GIBCO-BRL) containing 10% fetal bovine serum (FBS; GIBCO-BRL), glutamine, and penicillin–streptomycin, and incubated at 37° in a humidified, 5% CO_2 incubator. VA13 cells are simian virus 40 (SVR40)-transformed human WI38 fibroblast cells [American Type Culture Collection (ATCC), Rockville, MD]. IGROV is a human ovarian carcinoma cell line. The d120 ($IE3^-$) HSV helper virus and E5 helper cells were kindly provided by N. A. DeLuca (University of Pittsburgh, Pittsburgh, PA). CgalΔ3 is an $IE3^-$ helper virus containing a *lacZ* expression cassette (kindly provided by P. A. Johnson, University of California, San Diego, CA). E5 cells are resistant to G418 and are efficient hosts for ICP4 (IE3)-deficient viruses. All viral stocks are grown and titered in E5 cells.

Herpes Viral Vector Production

The procedure for the generation of viral stocks is shown in Fig. 2. The maintenance of pHE700 vector as an episome is demonstrated by transfection of pHE700-GFP into E5 cells with LipofectAMINE per the manufacturer protocol (GIBCO-BRL). Transfected cells are placed under selection with hygromycin B (ICN Biomedical, Aurora, OH) 1 day after plating. By day 16 of drug selection, almost all cells express the reporter gene.[22] Cells are trypsinized and plated at 6×10^5 cells per 100-mm dish 2 days after transfection. To generate viral stocks, these selected E5 cells containing pHE700-GFP plasmid are infected with d120 helper virus (kindly provided by N. A. DeLuca, University of Pittsburgh). The colonies are trypsinized and 3×10^6 cells are plated on a 100-mm dish. When the cells are nearly confluent, a multiplicity of infection (MOI) of 0.01 to 0.1 of d120 helper virus in 1 ml of Opti-MEM (GIBCO-BRL) is added. The viruses are allowed to adsorb to the cells for 2 hr at 37° in a humidified, 5% CO_2 incubator. The virus solutions are removed and 10 ml of DMEM with 10% FBS is added and incubated for an additional 24 to 36 hr. The resulting supernatants contain both the pHE700-GFP vector and helper virus (Fig. 2). The average titers obtained are about 10^6 green-forming units (GFU)/ml.

Herpes Amplicon Vector Concentration

Viral supernatant and cell debris are collected from infected cell monolayers after complete cell lysis. The lysate is frozen and thawed three times at 37° and then centrifuged at 3200 rpm at room temperature for 5 min to pellet the cell debris. The supernatant is filtered through a disposable Nalgene 0.8-mm pore size nylon filter bottle (Nalgene, Rochester, NY) to remove any remaining cell debris. Thirty-four milliliters of viral vector supernatant is added to a Beckman (Palo Alto, CA) polyallomer Quick-Seal centrifuge tube (40-ml volume) on top of a 6-ml cushion of 25% sucrose in 1× phosphate-buffered saline (PBS). The tubes are then ultracentrifuged at 26,100 rpm (75,000 *g*) at 4° in a Beckman 55.2 Ti rotor for 16 hr to pellet the virus particles. The pellet is suspended with 0.2–0.5 ml of sterile Hanks' buffered saline solution (HBSS) for further applications.

Herpes Vector Titration

The viral stocks are serially diluted in Opti-MEM and placed onto confluent E5 cell monolayers in six-well plates. The viruses are allowed to adsorb to the cells for 2 hr at 37°. The virus solutions are aspirated and the cells are washed with HBSS and overlaid with 2 ml of DMEM containing 5% FBS and 0.3% methylcellulose and then incubated for an additional 3 days. Cells are fixed and plaques visualized by staining with 0.5% crystal violet for 10 min. The titers are reported as plaque-forming units or green-forming units per milliliter. pHE700-GFP viral stocks are titered 24 hr after infection of VA13 cells in 24-well plates. Cells are rinsed with HBSS and are observed for fluorescence.

In Vitro Gene Transfer with pHE700-GFP

The pHE700-GFP-containing supernatants can be used to transduce human target cells. Cultured human cell lines are trypsinized, counted, and

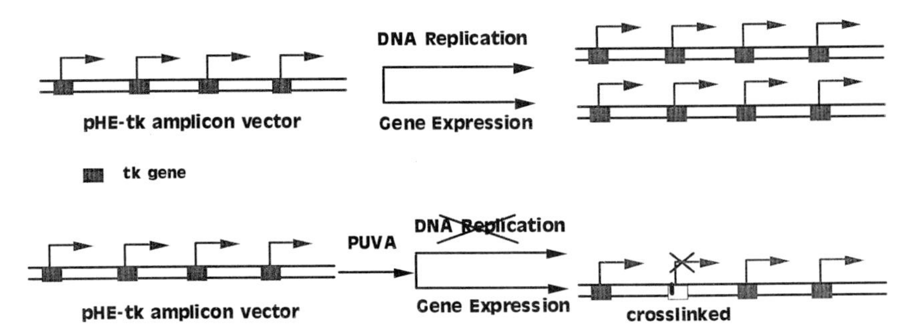

FIG. 3. Proposed mechanism of the effects of photochemical modification of pHEtk vector containing tandem repeats of the HSV *tk* gene expression unit. In permissive cells transduced with unmodified vector, both DNA replication and gene expression occur (*top*). After PUVA treatment, interstrand DNA cross-links inhibit viral replication, but permit transgene expression from unaffected transcription units.

seeded. When cells are nearly confluent, pHE700-GFP viral stock is added into the cells at 3–100 MOI of GFU for 2 hr. The GFP fluorescent activity is measured 5 to 24 hr later to document the transduction efficiency. At MOIs of 3–10, >95% of human tumor cells are readily observed to be transduced. This compares favorably with the higher MOIs often required for adenoviral vectors (sometimes as high as 3000 MOI *in vitro*).[31] Viral stocks used for these *in vitro* transducer experiments contain a helper virus-to-vector ratio 1 : 1. Only at higher ratios has significant toxicity been noted *in vitro*.

Herpes Simplex Virus Amplicon Vector Cytotoxicity

Helper virus d120, necessary to package amplicon vector, has deletions of both IE3 gene loci to prevent viral replication in normal cells, but permits replication in the E5 helper cell line expressing the IE3 gene. This helper virus causes substantial cytotoxicity to infected normal cells *in vitro* (data not shown). Unfortunately, there is no published method currently available to separate HSV amplicon vectors from helper virus. Because of this viral toxicity concern, we developed a novel method that uses psoralen and UV-A light (PUVA) to reduce the replication-competent outbreak of HSV vectors but retain high-level gene expression (Fig. 3). Psoralens are polycyclic planar molecules that form covalent, cyclobutane-type linkages. Previ-

[31] S. Chen, X. H. Li Chen, Y. Wang, K. Kosai, M. J. Finegold, S. S. Rich, and S. L. Woo, *Proc. Natl. Acad. U.S.A.* **92,** 2577 (1995).

ous studies applying cross-linking methods with psoralen and UV-A completely inactivated virus by blocking DNA replication and viral gene expression. In our experiments, the appropriate PUVA dose induces DNA cross-links in the vector that result in differential inactivation of viral replication and transgene expression. PUVA exposure inhibits replication in the E5 cells while retaining reporter gene expression. To evaluate the selective elimination of cytotoxicity of an HSV amplicon vector while maintaining the transgene activity, a cell proliferation assay that measured cellular DNA synthesis is used as previously described. The pHE-tk vector contains the HSV *tk* gene in the pHE700 vector.[22] Cells expressing HSV *tk* can phosphorylate GCV to a toxic form that induces cell death. The tests are performed with IGROV1 ovarian carcinoma cells transduced with pHE-tk vector packaged with herpes virus d120. Untreated pHE-tk vector stocks significantly reduced cellular proliferation, indicating the cytotoxicity of the helper virus and pHE-tk vector to the infected cells (Fig. 4). However, cells infected with the PUVA-treated vector substantially maintained their proliferation rate. Importantly, while the cytotoxicity from PUVA-treated pHE-tk vector is reduced to nearly that of cells not exposed to vector, we found that the vector retained functional transgene expression. GCV reduced cellular proliferation under conditions in which only small inhibitory effects from the modified vector are found. This antiproliferative effect of GCV on cells transduced by PUVA-treated vector indicates HSV *tk* enzyme function

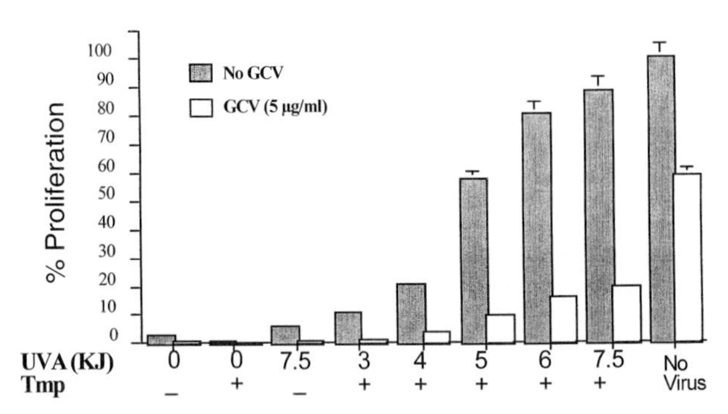

FIG. 4. HSV *tk*-encoded enzyme activity expressed from PUVA-modified vectors in IGROV1 ovarian carcinoma cells. IGROV1 cell proliferation after exposure to pHEtk vector, measured by [^{3}H]thymidine incorporation. Vectors treated with trimethoxypsoralen (Tmp, 1 μg/ml) and UV-A exposure (0–7.5 kJ/m^2). Tumor cells (10^4 cells/well) were evaluated in proliferation assays as previously described. Differences with GCV (0 or 5 μg/ml) illustrate the decrease in vector cytotoxicity while still expressing functional HSV *tk* activity. Incorporated [^{3}H]thymidine is measured as counts per minute (cpm) and expressed as a percentage of control (no GCV).

(Fig. 4). Thus, PUVA treatment prevented vector-mediated cytotoxicity, but permitted HSV *tk* gene expression. Our results demonstrate that the combination of trimethoxypsoralen (Tmp) and UV-A light results in the differential inactivation of HSV vectors. We postulate that interstrand DNA cross-links block viral DNA replication and randomly inactivate different regions of the vector genome (Fig. 3), genes located near the cross-links are inactivated, but gene expression can occur from other transcription units in amplicon vectors. Thus, the cross-linked DNA retains transgene expression while prohibiting vector replication. This photochemical modification should prevent any lytic viral replication that might occur owing to the rescue of recombinant wild-type virus both *in vitro* and *in vivo*. One alternative to this cross-linking approach is the generation of helper virus-free amplicons using plasmids, although the titer is about 10-fold lower than in our current system.[32]

Other Viral Vectors

The use of modified green fluorescent proteins is not limited to murine retroviral vectors and herpes amplicon vectors. As discussed previously, the first application of these GFP variant genes *in vivo* was in an adenoassociated viral vector transduced into the guinea pig eye.[5] GFP could be readily observed. Several groups have incorporated modified GFP genes into lentiviral vectors. An example is a study using a GFP gene under the control of the rhodopsin promoter and the demonstration of predominant expression in photoreceptor cells.[33] Adenoviral vectors have also been constructed to contain the GFP gene, and efficient vector transfer into human umbilical vein endothelial cells has been demonstrated.[34] Undoubtedly, almost every new gene delivery system will be explored and optimized using the versatile GFP gene.

[32] C. Fraefel, S. Song, F. Lim, P. Lang, L. Yu, Y. Wang, P. Wild, and A. I. Geller, *J. Virol.* **70,** 7190 (1996).

[33] H. Miyoshi, M. Takahashi, F. Gage, and I. Verma, *Proc. Natl. Acad. Sci. U.S.A.* **94,** 10319 (1997).

[34] R. de Martin, M. Raidl, E. Hofer, and B. Binder, *Gene Ther.* **4,** 493 (1997).

[36] Generation of a Destablilized Form of Enhanced Green Fluorescent Protein

By XIAONING ZHAO, XIN JIANG, CHIAO-CHIAN HUANG, STEVEN R. KAIN, and XIANQIANG LI

Introduction

The green fluorescent protein (GFP) from the jellyfish *Aequorea victoria* is a useful reporter, widely applied to the study of gene expression and protein localization *in vivo*. This wide application is due to the autofluorescent activity of the protein, which does not require any substrate or cofactor.[1-3] Variants with mutations affecting the chromophore structure and the expression of the protein in human cells have been engineered; these mutations have enhanced the expression level and modified the fluorescence properties of GFP in mammalian cells. Enhanced GFP (EGFP) is such a mutant with a 35-fold increase in fluorescence intensity.[4] Its enhanced fluorescence intensity makes EGFP a popular reporter in many applications. By fusion with other proteins, EGFP is widely used to monitor the expression and distribution of the target proteins in live cells.[5] By linking it with a promoter/enhancer element, expression of EGFP can be used to measure the promoter activity as well as to determine its regulation. The introduction of EGFP into transgenic animals permits the study of gene expression and regulation of stage-dependent and spatially distributed genes.

The limitation of EGFP as a reporter in certain cases is due to its stability. Crystallographic structures of GFP and its S65T mutant reveal a compacted β-barrel structure.[6] This structure presumably contributes to its unusual resistance to proteolysis and denaturation. The stability of EGFP means a slower turnover *in vivo*, which limits its applications in certain cases, such as being a reporter in transient transcription studies. The accurate measurement of changes in mRNA level requires that a receptor protein

[1] M. Chalfie, G. Euskirchen, W. W. Ward, and D. C. Prasher, *Science* **263,** 802 (1994).

[2] J. Marshall, R. Molloy, G. W. Moss, J. R. Howe, and T. E. Hughes, *Neuron* **14,** 211 (1995).

[3] A. B. Cubitt, R. Heim, S. R. Adams, A. E. Boyd, L. A. Gross, and R. Y. Tsien, *Trends Biochem.* **20,** 448 (1995).

[4] C. W. Cody, D. C. Prasher, M. W. Wester, F. G. Prendergast, and W. W. Ward, *Biochemistry* **32,** 1212 (1993).

[5] S. R. Kain, M. Adams, A. Kondepuki, T. T. Yang, W. W. Ward, and P. Kitts, *Biotechniques* **19,** 650 (1995).

[6] M. Ormo, A. B. Cubitt, K. Kallio, L. A. Gross, R. Y. Tsien, S. J. Remington, *Science* **273,** 1392 (1996).

0076-6879/99 $30.00

turn over rapidly. Therefore, creation of a destabilized form of EGFP would significantly expand the applications of this reporter. By fusing a degradation domain of mouse ornithine decarboxylase with EGFP, we generated destabilized EGFP (dEGFP), which has a much shorter half-life. The application of dEGFP as a transcriptional reporter is discussed in [4] in this volume.[7]

Principle of Method

To generate the destabilized EGFP, a "degradation domain" from mouse ornithine decarboxylase (ODC) was fused to the C terminus of EGFP. ODC is the key enzyme in the biosynthesis of polyamines. The protein is known to be one of the most short-lived proteins in mammalian cells.[8] Its C terminus contains a PEST sequence that has been proposed to be a structural motif for short-lived protein. Removal of the C terminus of mouse ODC prevents its fast degradation, whereas addition of this domain to another, more stable protein, *Trypanosoma brucei* ODC, results in a less stable fusion protein.[9] In this chapter, we demonstrate that the addition of the degradation domain of ODC to EGFP shortens the half-life of EGFP.

Protein stabilities of EGFP and its destabilized form (dEGFP) were determined *in vivo* by expressing them in CHO-tTA cells. After turning off the protein synthesis by addition of cycloheximide (CHX), the half-lives of EGFP and dEGFP were measured by Western blots and fluorescence-activated cell sorting analysis. The destabilized EGFP has an apparent half-life of 2 hr, which is much shorter than that of unmodified EGFP.

Materials and Methods

Cycloheximide is obtained from Sigma (St. Louis, MO), and is made as a 10-mg/ml stock in phosphate-buffered saline (PBS). The Western Exposure chemiluminescence detection kit and the CLONfectin transfection kit are from Clontech (Palo Alto, CA).

Construction

The cDNAs encoding EGFP and the C terminus of mouse ODC are amplified by polymerase chain reaction (PCR) with PfuTurbo (*Pfu*) DNA

[7] X. Zhao, T. Duong, C.-C. Huang, S. R. Kain, and X. Li, *Methods Enzymol.* **302,** [4], 1999 (this volume).

[8] Z. Bercovich, Y. Rosenberg-Hasson, A. Ciechanover, and C. Kahana, *J. Biol. Chem.* **264,** 15949 (1989).

[9] L. Ghoda, T. Van Daalen, Wetters, M. Macrase, D. Ascherman, and P. Coffino, *Science* **243,** 1493 (1989).

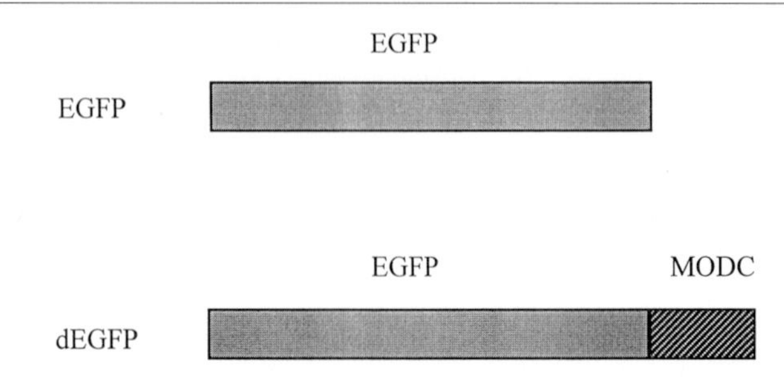

FIG. 1. Schematic map of EGFP fusion proteins with the degradation domain of mouse ODC (MODC).

polymerase (Stratagene, La Jolla, CA). EGFP is amplified with primers that incorporate a *Sac*II recognition sequence at the 5′ end and a *Hin*dIII site at the 3′ end. The stop codon of EGFP is deleted in order to make an open reading frame with the degradation domain of mouse ODC. The degradation domain of mouse ODC is amplified with primers that incorporate with a *Hin*dIII recognition sequence at the 5′ end and an *Eco*RI sequence at the 3′ end. The amplified PCR products are ligated at the *Hin*dIII site (Fig. 1) and the fusion is cloned into the pTRE expression vector for use in the tetracycline (Tc)-regulated expression system (Clontech).

Transfection and Fluorescence Analysis

1. CHO-K1 Tet-off cells (Clontech) are seeded sparsely in six-well plates 1 day before transfection.
2. Prepare the following two stock solutions:
   Medium A: Serum-free α-MEM (Sigma), 100 μl, to which 4 μg of construct DNA is added
   Medium B: Serum-free α-MEM, 100 μl, to which 1 μg of CLONfectin is added
3. Add the CLONfectin-containing medium B to the DNA-containing medium A dropwise, and mix well.
4. Let the DNA mixture stand at room temperature for 20 to 30 min.
5. Dilute the DNA mixture with 1.8 ml of serum-free medium. Replace the culture medium of the CHO-K1 Tet-off cells with 2 ml of the DNA-containing medium.
6. Keep the cells in a 37° incubator for 1 hr.
7. Wash the cells once with fresh medium and replace with fresh α-MEM containing 10% fetal calf serum.

8. One day after transfection, cycloheximide is added to a final concentration of 100 μg/ml. Cells are then collected for Western blot or fluorescence-activated cell sorting (FACS) analysis at different time intervals.

9. For fluorescence microscopy study, transfection is performed using cells grown on glass coverslips. Before examination, cells are fixed with 4% paraformaldehyde in PBS at room temperature for 30 min. After being rinsed twice with PBS, coverslips are mounted on a glass slide and sealed with rubber cement. The cells are then ready for examination under a fluorescence microscope.

10. For flow cytometry analysis, the transfected cells with or without CHX treatment are detached by 2 m*M* EDTA in PBS. Cells are collected and resuspended in 0.5 ml of PBS. The cell suspensions are then analyzed for fluorescence intensity using a FACS Calibur flow cytometer (Becton Dickson, San Jose, CA). EGFP is excited as 488 nm, and emission is detected using a 510/20 bandpass filter.

After transfection with either pTRE-EGFP or pTRE-dEGFP, CHO-K1 Tet-off cells are maintained in culture medium without tetracycline for 24 hr. To determine protein degradation, protein synthesis is first blocked by addition of cycloheximide. The rate of the fluorescence decay of EGFP and dEGFP is measured by flow cytometry (Fig. 2A). The analysis results for dEGFP-transfected cells indicate that its fluorescence activity decreases dramatically within a 4-hr range. However, EGFP-transfected cells show little change in fluorescence after 4 hr of CHX treatment. The time course of protein stability is plotted in Fig. 2B to estimate the half-life. It takes 2 hr for dEGFP activity to decrease by 50%, therefore its half-life is about 2 hr. In contrast, EGFP needs a much longer time, and its half-life is greater than 4 hr.

Under a fluorescence microscope, the decrease in fluorescence intensity of dEGFP-expressing cells after 4 hr of treatment is obvious in comparison with untreated cells, but such a decrease in fluorescence intensity is not observed in EGFP-expressing cells (Fig. 3). These results suggested that dEGFP has a faster degradation rate and, therefore, a shorter half-life.

Western Blot

1. Cells in six-well plates are collected by scraping and transferred into eppendorf tubes.

2. Cells are collected in a microcentrifuge tube by centrifuging them at maximum speed for 30 sec.

3. Supernatant is discarded and the cells are resuspended in 100 μl of PBS.

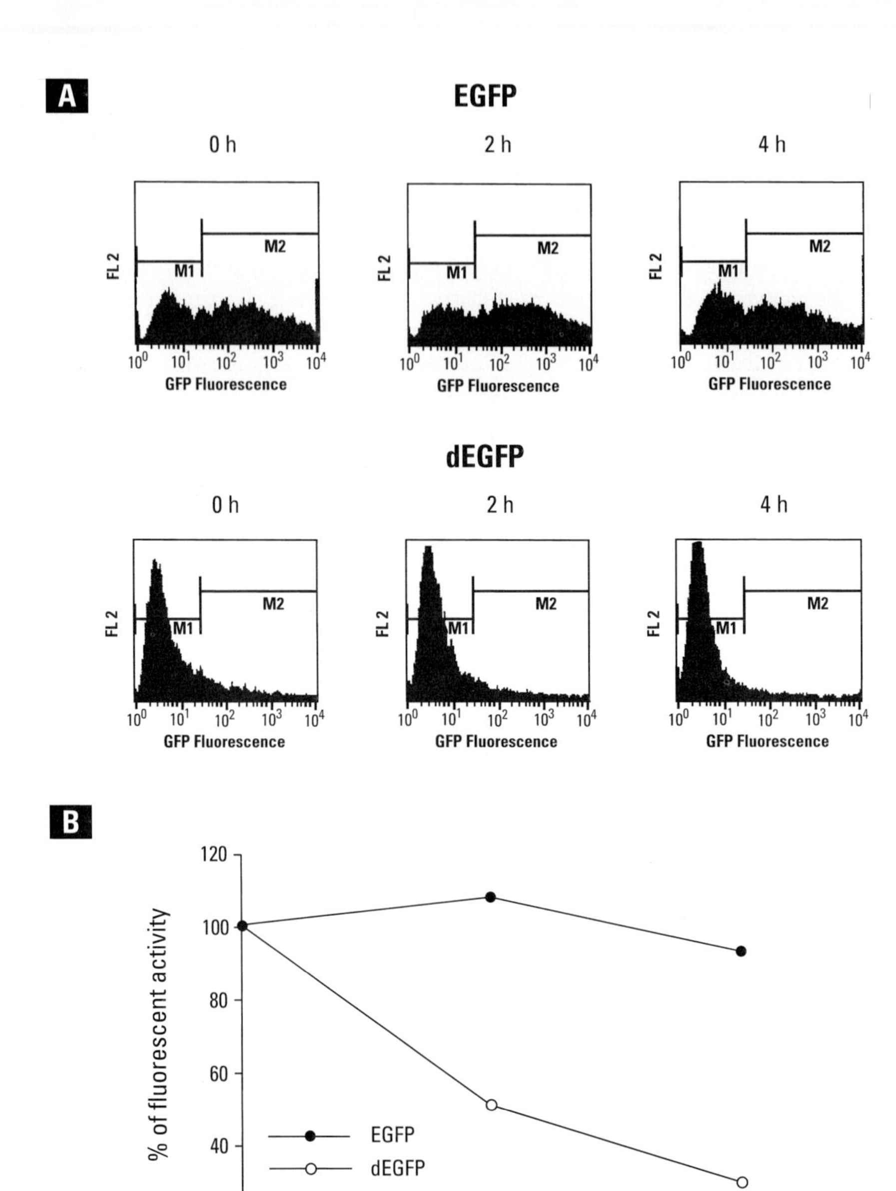

FIG. 2. Flow cytometric analysis of the fluorescence stabilities of EGFP and dEGFP. CHO-K1 Tet-off cells were transfected with pTRE-EGFP and pTRE-dEGFP. Twenty-four hours after transfection, cells were treated with CHX (100 μg/ml) for 0, 1, 2, and 3 hr. The treated cells were collected with EDTA and subjected to FACS analysis. (A) Result of FACS analysis of EGFP- and dEGFP-expressing cells at different time points. (B) Time course analysis of the percentage of fluorescent cells.

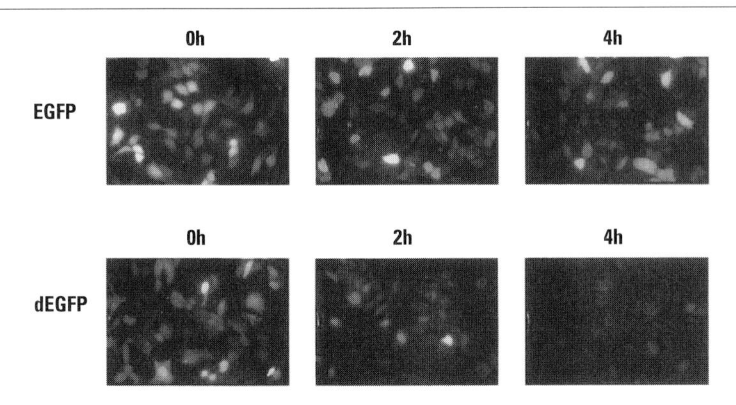

FIG. 3. The fluorescence stabilities of EGFP and dEGFP. Cells expressing EGFP or dEGFP were treated with CHX for 2 or 4 hr. The fluorescence activity was compared with that of untreated cells under a fluorescence microscope.

4. Cells are sonicated on ice for 45 sec.
5. Fifteen-microliter samples are mixed with equal volumes of 2× sodium dodecyl sulfate (SDS) sample buffer and loaded onto a 12% SDS–polyacrylamide gel.
6. After electrophoresis, proteins are transferred onto nitrocellulose paper, and blocked with 5% milk in Tris-buffered saline + 0.05% Tween-20 (TBST).
7. Anti-GFP monoclonal antibody (Clontech) is used as a primary antibody at 1:500 dilution. Alkaline phosphatase (AP)-conjugated goat anti-mouse IgG is used as the secondary antibody. The detection is visualized with the Western Exposure chemiluminescence detection kit (Clontech).

To determine whether the decrease in dEGFP fluorescence activity is correlated with the degradation of the protein, the amount of EGFP and dEGFP is detected using the Western blotting method. dEGFP is degraded over time in the presence of CHX (Fig. 4). After 4 hr, a low level of dEGFP is

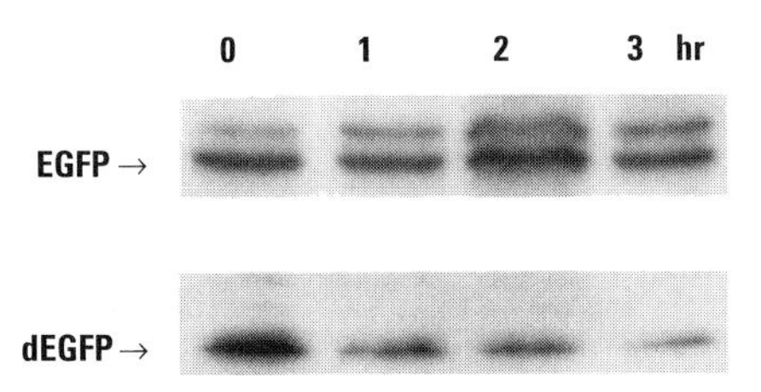

FIG. 4. Western blot analysis of protein stability. Cells expressing EGFP or dEGFP were treated with CHX for 0, 1, 2, and 3 hr. Cells were then collected and subjected to Western blotting. Monoclonal antibody against GFP was used.

detected, compared with that in untreated cells. In contrast, the decrease in EGFP expression is insignificant after 4 hr of treatment. The decrease in protein level is consistent with that of the fluorescence activity.

Conclusion and Remarks

By appending the degradation domain of mouse ODC, we engineered a destabilized EGFP (dEGFP) that yet retains its fluorescence intensity, making it a better reporter gene than EGFP when a short-lived reporter is desirable. We demonstrated that dEGFP has a faster degradation rate *in vivo* by FACS analysis; the half-life of dEGFP is about 2 hr, which is much shorter than that of EGFP. The faster turnover rate of dEGFP may result in less accumulation, therefore it may cause less toxicity to cells, which will facilitate the generation of stable-dEGFP-expressing cell lines. In addition, dEGFP would be a better reporter in the kinetic study of trans- or cis-acting regulatory elements. The dEGFP reporter can also be used in developmental biological studies, such as with transgenic animals. Application of dEGFP as a reporter gene is also discussed in [37] in this volume.[10]

[10] S. Inouye, K. Umesono, and F. I. Tsuji, *Methods Enzymol.* **302,** [37], 1999 (this volume).

[37] Special Properties of Green Fluorescent Protein-S65A

By SATOSHI INOUYE, KAZUHIKO UMESONO, and FREDERICK I. TSUJI

Introduction

The green fluorescent protein (GFP), isolated from the jellyfish *Aequorea victoria,* emits a greenish fluorescence on exposure to ultraviolet light, with fluorescence excitation maxima at 400 and 475 nm and an emission maximum at 508–509 nm.[1] The fluorescence is due to the presence of a chromophore in the protein, consisting of an imidazolone ring structure formed by a posttranslational cyclization reaction and an oxidation reaction

[1] H. Morise, O. Shimomura, F. H. Johnson, and J. Winant, *Biochemistry* **13,** 2656 (1974).

0076-6879/99 $30.00

involving the tripeptide -Ser^{65}-Tyr^{66}-Gly^{67}- in the primary structure.[2–4] The cDNA for GFP has been cloned and expressed, and the protein has been found to be useful as a marker in biological studies.[5–8] Cloning of the GFP has also made possible mutagenesis of the gene to obtain other mutant GFPs.[9] One such mutant, GFP-S65A, with alanine substituted for serine at position 65, has a red-shifted fluorescence excitation spectrum and a higher fluorescence intensity than wild-type GFP.[10,11]

GFP-S65A has been used to label neuronal cells by employing an adenoviral vector carrying a membrane-anchoring signal fused to GFP-S65A.[12] COS cells expressing the GFP-S65A gene had a brighter fluorescence than those expressing wild-type GFP and cells producing GFP-S65T were not significantly brighter. In another study, an alanine mutant with triple amino acid substitutions has been reported.[13] The mutant protein, GFP-S65A/V68L/S72A, had a red-shifted fluorescence excitation maximum (481 nm) and a fluorescence intensity 19-fold greater than that of wild-type GFP. The increased fluorescence was attributed, in part, to greater solubility and more efficient folding of the protein. A recombinant human immunodeficiency virus type 1 (HIV-1) expressing a GFP in which the serine at position 65 was substituted by alanine has also been reported.[14] H9 cells infected with the virus produced a 10- to 20-fold greater fluorescence intensity compared with wild-type GFP. GFP-S65A has been also incorporated into a bacterial cloning vector for use in selecting for inserted foreign genes.[15] Expression of the gene in *Escherichia coli* produced a much brighter fluorescence than did other GFP genes. It should be noted that the wild-type GFP (GFP-II)[8] used in the latter study[15] differs at seven amino acid positions

[2] O. Shimomura, *FEBS Lett.* **104,** 220 (1979).

[3] C. W. Cody, D. C. Prasher, W. M. Westler, F. G. Prendergast, and W. W. Ward, *Biochemistry* **32,** 1212 (1993).

[4] H. Niwa, S. Inouye, T. Hirano, T. Matsuno, S. Kojima, M. Kubota, M. Ohashi, and F. I. Tsuji, *Proc. Natl. Acad. Sci. U.S.A.* **93,** 13617 (1996).

[5] D. C. Prasher, V. K. Eckenrode, W. W. Ward, F. G. Prendergast, and M. J. Cormier, *Gene* **111,** 229 (1992).

[6] M. Chalfie, Y. Tu, G. Euskirchen, W. W. Ward, and D. C. Prasher, *Science* **263,** 802 (1994).

[7] S. Inouye and F. I. Tsuji, *FEBS Lett.* **341,** 277 (1994).

[8] S. Inouye and F. I. Tsuji, *FEBS Lett.* **351,** 211 (1994).

[9] R. Heim, D. C. Prasher, and R. Y. Tsien, *Proc. Natl. Acad. Sci. U.S.A.* **91,** 12501 (1994).

[10] S. Delagrave, R. E. Hawtin, C. M. Silva, M. M. Yang, and D. C. Youvan, *Bio/Technology* **13,** 151 (1995).

[11] R. Heim, A. B. Cubitt, and R. Y. Tsien, *Nature (London)* **373,** 663 (1995).

[12] K. Moriyoshi, L. J. Richards, C. Akazawa, D. D. M. O'Leary, and S. Nakanishi, *Neuron* **16,** 255 (1996).

[13] B. P. Cormack, R. H. Valdivia, and S. Falkow, *Gene* **173,** 33 (1996).

[14] K. A. Page, T. Liegler, and M. B. Feinberg, *AIDS Res. Hum. Retroviruses* **13,** 1077 (1997).

[15] S. Inouye, H. Ogawa, K. Yasuda, K. Umesono, and F. I. Tsuji, *Gene* **189,** 159 (1997).

from the wild-type GFP (GFP-I) used in another study with *Caenorhabditis elegans*.[6] However, the three-dimensional structures of all of the GFPs show a common 11-stranded β barrel with a central helix containing the chromophore.[16–18]

Expression and Purification of Recombinant Green Fluorescent Proteins

The expression system used for producing the histidine-tagged GFP (His-GFP) and the Ni(II)-chelate affinity column used for purifying the expressed His-GFP are essentially the same as previously described.[7,8] Briefly, the *Bam*HI–*Sal*I fragment obtained from various GFP cDNAs by the polymerase chain reaction, using GFP-A (5′ region) and GFP-B (3′ region) as primers,[7] is inserted into the *Bam*HI–*Xho*I site of pTrcHis-C (Invitrogen, Carlsbad, CA) to give the GFP expression vector. The expressed GFP has an extra 37-amino acid polypeptide possessing 6 histidine residues in tandem at the N terminus. Site-specific mutagenesis for producing the mutant GFP cDNAs is carried out by the polymerase chain reaction.

For overexpressing the protein, *E. coli* strain Top10 (Invitrogen), carrying the expression plasmid, is grown at 37° in 400 ml of Luria–Bertani broth containing anti-foam (Disfoam CE457; Nippon Oil and Fats, Tokyo, Japan) in a 3-liter flask with reciprocal shaking (180 rpm/min). After incubation for 2 hr (optical density at 600 nm = 0.3), isopropyl-β-D-thiogalactopyranoside (IPTG) is added to a final concentration of 0.2 m*M* and the incubation is continued for another 3 hr. The cells are harvested by centrifugation and suspended in 5 ml of 50 m*M* Tris-HCl (pH 7.5)–0.2 *M* NaCl. The suspension is placed in a tube immersed in an ice bath and the cells are disrupted by sonication using a Branson (Danbury, CT) model 250 sonifier (six times, 30 sec each time). After centrifugation at 12,000*g* for 20 min, the supernatant is applied directly to a Ni(II)-chelating Sepharose column (1.5 × 5 cm; Pharmacia, Piscataway, NJ) equilibrated with 50 m*M* Tris-HCl (pH 7.5)–0.2 *M* NaCl. After washing thoroughly with the buffer, the adsorbed proteins are eluted with 0.3 *M* imidazole, dialyzed against the same buffer, and reintroduced into a Ni(II)-chelating Sepharose column (0.8 × 4 cm). The His-GFP is eluted from the column using a linear gradient of imidazole (0–0.3 *M*). The fractions are analyzed by sodium dodecyl

[16] M. Ormo, A. B. Cubitt, K. Kallio, L. A. Gross, R. Y. Tsien, and S. J. Remington, *Science* **273,** 1392 (1996).

[17] F. Yang, L. G. Moss, and G. N. Phillips, Jr., *Nature Biotechnol.* **14,** 1246 (1996).

[18] C.-K. Wu, Z.-J. Liu, J. P. Rose, S. Inouye, F. Tsuji, R. Y. Tsien, S. J. Remington, and B.-C. Wang, *in* "Bioluminescence and Chemiluminescence" (J. W. Hastings, L. J. Kricka, and P. E. Stanley, eds.), p. 399. John Wiley & Sons, New York, 1997.

TABLE I

SPECTRAL PROPERTIES OF GREEN FLUORESCENT PROTEINS[a]

GFP	Absorbance (λ_{max}, nm)	Absorbance (peak ratio)	Extinction coefficient (M^{-1} cm^{-1})	Fluorescence emission (λ_{max}, nm)	WHMF (nm)	Fluorescence intensity (relative)	Fluorescence excitation (λ_{max}, nm)
Native GFP	277.2, 398.4, 472.4	0.79 (398/277)	20,500 (398)	510.0 (400)	24	1.00 (400)	400.0, 482.2 (510)
		0.40 (472/398)	9,100 (475)	505.8 (475)	36	0.62 (475)	
GFP-I	278.8, 397.2, 476.0	0.98 (397/279)	20,900 (397)	510.0 (400)	24	1.08 (400)	399.2, 481.2 (510)
		0.31 (476/397)	6,600 (475)	505.4 (475)	36	0.77 (475)	
GFP-II	278.0, 397.6, 475.2	0.77 (398/278)	19,400 (398)	509.8 (400)	24	1.03 (400)	399.2, 481.8 (510)
		0.37 (475/398)	6,900 (475)	504.4 (475)	36	0.72 (475)	
S65A	277.2, 475.6	1.52 (476/277)	39,000 (475)	504.6 (475)	38	3.20 (475)	474.6 (505)
S65T	278.0, 489.6	1.70 (490/278)	41,000 (489)	512.4 (490)	28	4.68 (490)	492.8 (510)
Y66H	278.0, 386.8	0.67 (387/278)	16,900 (387)	450.2 (385)	74	0.19 (385)	386.8 (450)
Y66F	281.2, 358.4	0.56 (358/281)	14,400 (358)	429.6 (360)	74	0.01 (360)	365.8 (430)
Y145F	277.2, 395.2, 471.5	0.78 (395/277)	19,400 (395)	508.4 (400)	26	1.02 (400)	398.4, 474.4 (510)
		0.45 (471/395)	8,800 (471)	503.6 (475)	36	0.73 (475)	
C48S	ND	ND	ND	509.8 (400)	24	ND	399.8, 474.8 (510)
C70S	ND	ND	ND	509.4 (400)	28	ND	400.4, 474.6 (510)

[a] The purified native GFP from *Aequorea* was kindly provided by J. Blinks (Friday Harbor Laboratories, University of Washington). GFP-I and GFP-II consist of two different wild-type GFPs with a common histidine tag. GFP-I is derived from one wild-type GFP [D. C. Prasher, V. K. Eckenrode, W. W. Ward, F. G. Prendergast, and M. J. Cormier, *Gene* **111,** 229 (1992)] and GFP-II is derived from another [S. Inouye and F. I. Tsuji, *FEBS Lett.* **351,** 211 (1994)]. GFP-I and GFP-II were prepared by expressing the cDNAs for GFP-I and GFP-II in *E. coli* and purifying the proteins by Ni(II)-chelate affinity chromatography as described in text. All protein spectra were measured in 100 m*M* ammonium bicarbonate, pH 8.0, at 25° using a Hitachi (Tokyo, Japan) U3210 spectrophotometer (bandpass, 2 nm; scan speed, 120 nm/min; response time, medium) and a Hitachi F4010 fluorescence spectrophotometer (excitation bandpass, 1.5 nm; emission bandpass, 3 nm; scan speed, 60 nm/min; response time, 2 sec). WHMF, Width at half-maximum of fluorescence band; ND, not determined.

sulfate–polyacrylamide gel electrophoresis and the His-GFP fractions showing >95% purity are combined and concentrated with an Amicon (Beverly, MA) Centricon 30 microconcentrator at 4°. The yield of His-GFP is 0.5–1.0 mg from 400 ml of cultured cells, except that the two cysteine mutant proteins (C48S and C70S) were produced primarily (>95%) as inclusion bodies under the preceding expression conditions.

Spectroscopic Measurements and Properties

Table I shows the spectral properties of GFP-S65A (mutant of GFP-II) and various other GFPs. The fluorescence excitation maxima of GFP-S65A and GFP-S65T (mutant of GFP-II) are seen to be red shifted, and the fluorescence intensity of GFP-S65A is found to be greater than the intensities of all other GFPs except that of GFP-S65T. Other values that have been reported for the peak absorbance ratio of native GFP are 0.7–0.9 (400/

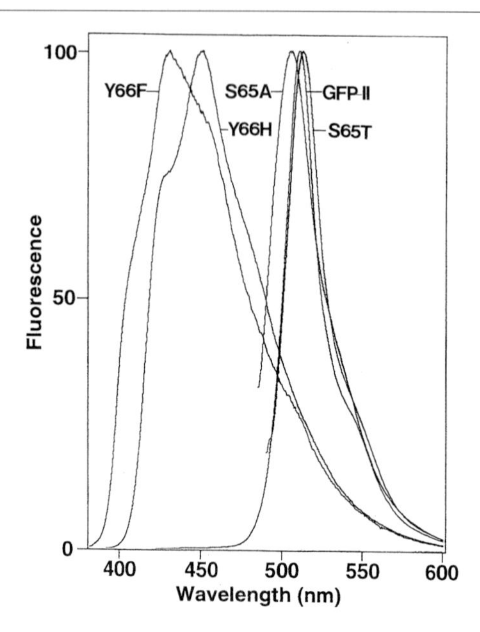

FIG. 1. Normalized fluorescence emission spectra of GFP-S65A and other GFPs. The spectra were measured as described in the footnote to Table I.

280),[1] 1.0 (400/280),[2] 1.10–1.25 (395/280),[3] and 1.0 (395/280).[19] Changes in absorption spectra of native GFP caused by varying the temperature, ionic strength, protein concentration, and pH have also been reported.[20] On the basis of fluorescence intensity, GFP-S65A and GFP-S65T may be considered good candidates for use as markers.

Figure 1 shows the normalized fluorescence emission spectrum of GFP-S65A, as well as those of other GFPs. The spectrum of GFP-S65A is found to be not very different from the spectra of GFP-II and GFP-S65T, with a maximum located above 500 nm. In contrast, GFP-Y66F and GFP-Y66H show blue-shifted spectra with large increases in the half-height fluorescence bandwidth.

Figure 2 shows the absorbance spectrum of GFP-S65A and other GFP proteins. The GFP-I and GFP-II proteins are observed to have very similar spectra, with each having a major and minor maxima. GFP-S65A and GFP-

[19] W. W. Ward and S. H. Bokman, *Biochemistry* **21,** 4535 (1982).

[20] W. W. Ward, H. J. Prentice, A. F. Roth, C. W. Cody, and S. C. Reeves, *Photochem. Photobiol.* **35,** 803 (1982).

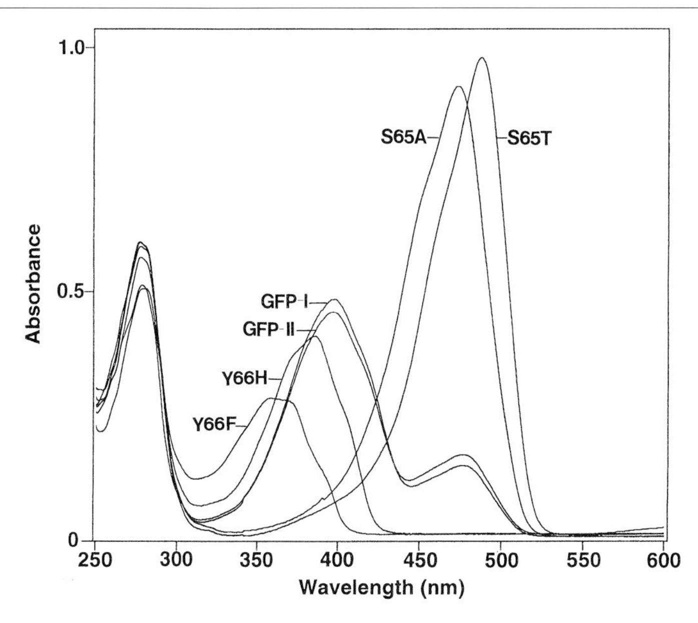

FIG. 2. Absorbance spectra of GFP-S65A and other GFPs. The spectra were measured as described in the footnote to Table I.

S65T are characterized by an absence of the major maximum at 400 nm and possessing a large maximum corresponding to the minor maximum of wild-type GFP-I and -II. Thus, substitution of serine at position 65 is seen to have a significant effect on the absorption spectrum.

Author Index

Numbers in parentheses are footnote reference numbers and indicate that an author's work is referred to although the name is not cited in the text.

A

B

C

D

E

F

G

H

I

J

K

L

M

N

O

P

Q

R

S

T

U

V

W

X

Y

Z

Subject Index

A

B

C

D

G

I

L

M

N

S

T

ISBN 0-12-182203-6